PHYSICS LIBRARY

D1616875

THE
INTERNATIONAL SERIES
OF
MONOGRAPHS ON PHYSICS

GENERAL EDITORS

J. BIRMAN S. F. EDWARDS R. H. FRIEND
C. H. LLEWELLYN SMITH M. REES D. SHERRINGTON
G. VENEZIANO

Stellarator and Heliotron Devices

MASAHIRO WAKATANI

New York Oxford
OXFORD UNIVERSITY PRESS
1998

Oxford University Press

Oxford New York
Athens Auckland Bangkok Bogota Bombay Buenos Aires
Calcutta Cape Town Dar es Salaam Delhi Florence Hong Kong
Istanbul Karachi Kuala Lumpur Madras Madrid Melbourne
Mexico City Nairobi Paris Singapore Taipei Tokyo Toronto Warsaw

and associated companies in
Berlin Ibadan

Copyright © 1998 by Oxford University Press, Inc.

Published by Oxford University Press, Inc.
198 Madison Avenue, New York, New York 10016

Oxford is a registered trademark of Oxford University Press

All rights reserved. No part of this publication may be reproduced, stored in a retrieval system, or transmitted, in any form or means, electronic, mechanical, photocopying, recording, or otherwise, without the prior permission of Oxford University Press.

Library of Congress Cataloging-in-Publication Data
Wakatani, Masahiro.
Stellarator and heliotron devices / by Masahiro Wakatani.
p. cm. – (International series of monographs on physics; 95)
ISBN 0-19-507831-4
1. Stellarators. 2. Plasma (Ionized gases) 3. Plasma confinement – Instruments.. I. Title. II. Series: International series of monographs on physics ; v. 95.
QC791.77.S7W35 1998
621.48′4 – dc21 97-6186

9 7 5 3 1 2 4 6 8
Printed in the United States of America
on acid-free paper

QC
791
.77
S7
W35
1998
PHYS

Dedicated to
Yasuko Wakatani

PREFACE

The "Stellarator" is a generic name for the nonaxisymmetric, toroidal, magnetic confinement devices used to produce high-temperature plasmas and to stably confine them by means of an externally generated magnetic field with specific properties. The purpose of such a confinement scheme is to yield a substantial amount of nuclear fusion between fuel ions such as deuterons and tritons inside the plasma, to generate electricity continuously. Investigations of high-temperature plasmas in stellarators have revealed numerous interesting properties pertinant to nonaxisymmetric, currentless toroidal plasmas. The "Heliotron" is one of the high-temperature plasma confinement devices similar to stellarators, which has been intensively studied since the 1970s at Kyoto University. The magnetic field of the heliotron has unique properties for both plasma stability and high-energy particle confinement compared to classic stellarators. The heliotron is named after *helios*, which means "sun" in Greek.

The basic theory of plasma dynamics was established at an early stage of plasma physics research, and several excellent introductory textbooks have been published on these subjects. More recently, some useful books devoted to specific applications of plasma physics, such as tokamak plasmas, laser-produced plasmas, and space plasmas, have been published. To my knowledge, however, no introductory and standard books are available which start from the very basic theory of high-temperature plasmas and subsequently deal with its applications to present-day stellarator research in a self-contained manner. As a result, many students and researchers who want to study stellarator plasmas encounter difficulties in understanding the relation between the basic plasma physics and the stellarator and heliotron devices. This book is written with the aim of filling this gap, by explaining plasma physics for stellarators from the fundamental level of plasma physics. It is known that there are many varieties of stellarators, since the three-dimensional magnetic configuration has large degrees of freedom. When the heliotron is adequate as an example of a stellarator, I have concentrated on it to explain the theory and experiments related to stellarator plasma.

It must be pointed out that both the tokamak and the stellarator are toroidal confinement systems and that they share many common aspects of plasma physics. Thus this book will also be useful in the study of tokamak plasmas, and both tokamaks and stellarators are discussed in some chapters.

Since plasma physics for stellarators is making progress every year, it is not an easy task to describe the most recent advances in this field. I thus decided to present only those subjects or results which are more or less established, and to avoid those which are still controversial. I intended the present book to be an introduction which will enable the reader to understand the literature on the more advanced topics related to stellarator and heliotron plasmas.

The present book aims to be a comprehensive textbook for a graduate-level plasma physics course relevant to thermonuclear fusion research, and particularly to the stellarator approach to the fusion reactor. The prerequisites are a knowledge of classical mechanics, electromagnetism, fluid dynamics, applied mathematics, and numerical analysis, which are normally taught at the undergraduate level for majors in physics and engineering. This work is also intended to be an introduction to magnetic confinement theory, which should allow the reader to understand review articles on specific subjects in both stellarator and tokamak research.

It is a pleasure to record my gratitude to former graduate students with whom I have enjoyed discussions on magnetic confinement theory and stellarator and heliotron plasmas, who also gave me critiques of the individual sections; and in particular to Yuji Nakamura, Hideo Sugama, Katsuji Ichiguchi, Masatoshi Yagi, Kiyomasa Watanabe, Yasutomo Ishii and Taro Matsumoto. Also, the careful reading of the manuscript by the graduate student Masayuki Yokoyama is gratefully acknowledged.

CONTENTS

Stellarator and
Heliotron Devices

1

INTRODUCTION

The three states of matter, solid, liquid, and gas, are well known to us. As the temperature is elevated, solid is liquefied and liquid is evaporated to form a gaseous state. If we further increase the temperature, the molecules constituting the gas are decomposed into atoms and the atoms are then decomposed into electrons and positively charged ions. The degree of ionization increases as the temperature rises. For the case of hydrogen gas at normal pressure, the ionization becomes almost complete at a temperature of about $2\text{–}3 \times 10^4$ K or 2–3 eV. The fully ionized gas formed in this way is called high-temperature plasma. It consists of a large number of negatively charged electrons and positively charged protons, both electron and protons moving at high speed randomly. The neg negative charge of the electrons cancels the net positive charge of the protons in the plasma. This is called the overall charge neutrality of the plasma.

There are other types of plasmas, such as the electron gas in metals and the plasma inside stars, in which the quantum effect is important due to extremely high densities. Also, electron rich nonneutral plasmas, positron plasmas, and quark-gruon plasmas are other examples. In this book we shall not deal with such plasmas, but restrict our attention to fully ionized high-temperature plasmas, which will be simply referred to as plasmas: these are relevant to the present mainstream research on controlled thermonuclear fusion, in which the hydrogen is replaced with a deuterium (D) and tritium (T) mixture or a deuterium (D) and helium (He^3) mixture.

Historically, plasma research was initiated by studies of gas discharges. Tonks and Langmuir (1929) observed an electric oscillation in a rarefied gas discharge. They referred to this oscillation as "plasma oscillation." Since then, the word *plasma* has been used to represent a conducting gas. Because it has properties that are quite different from those of ordinary neutral gas, the plasma is often called the fourth state of matter, distinct from the previously known three states of matter, solid, liquid and gas.

Experimental research on plasmas has made little progress over more than a quarter of the present century because of difficulties in controlling plasmas. On the other hand, the plasma has received considerable attention from theoreticians. A collisionless many-particle system consisting of a large number of charged particles cannot be treated by the standard theory of gases, based on the expansion in powers of the density. The difficulties are due to the long-range character of the Coulomb interactions, which intrinsically require many-body correlation effects. A number of theoretical efforts have been devoted to a formulation which properly takes into account the many-body effects. A theoretical model, treating plasmas in a strong magnetic field as a continuous electromagnetic fluid (magnetohydrodynamics, or MHD), has also been established by Alfvén (1951) and others.

Progress in plasma physics emerged in the second half of this century, motivated by two very important applications: space exploration and thermonuclear fusion. In the 1940s, radio-astronomers discovered that more than 99% of the universe is in a plasma state. Since the first Sputnik spacecraft was launched in 1954 into an orbity circling the Earth, a number of space probes have been used to collect data on the Earth's magnetosphere. This information has disclosed that the space beyond the Earth is not just a vacuum but is filled with an active dilute plasma composed of energetic charged particles. This plasma exhibits many interesting physical phenomena, such as the interaction of the solar wind plasma with the Earth's magnetic field, particle acceleration and trapping, wave excitation and propagation, and so on. Since diagnostics for solar plasma have made progress recently, the solar system is now considered to be the primary laboratory in which a rich variety of plasma processes can be studied.

Controlled thermonuclear fusion research was started in the early 1950s. (For an introduction to the thermonuclear fusion, it is recommended to read the first part of section 9.2.) In the beginning, it was conducted as classified research. Declassification took place in 1958, at the second Genera Conference on the Peaceful Uses of Atomic Energy. The stellarator had already been originated by L. Spitzer, and results were presented at this conference. The stellarator is a concept for confining high-temperature plasmas using an externally produced helical magnetic field, by means of figure of eight shaped solenoidal coils or by helical windings (helically deformed coils covering a toroidal plasma). It contrasts with the tokamak, which was also presented at the same conference, and requires a plasma current in the toroidal direction to produce the helical magnetic field inside the plasma. The tokamak is a typical axisymmetric torus. When "torus" or "toroidal plasma" appears, it is better to imagine the donut shape. Various ideas for confining high-temperature plasmas by magnetic fields were proposed in the 1960s, but experiments revealed unexpected difficulties in controlling plasmas, until the results of the former Soviet T-3 tokamak were reported in 1968. They showed a dramatic improvement in plasma confinement and promising scaling properties. Since then, a large number of tokamaks have been constructed all over the world and a substantial amount of experimental data has been accumulated.

In the Princeton Plasma Physics Laboratory, Princeton University, the C-Stellarator device was designed to prove L. Spitzer's stellarator principle. However, the confinement properties were governed by Bohm diffusion and were less attractive compared to the T-3 tokamak. In 1970 the C-Stellarator was changed to ST tokamak and a difficult period began for stellarator research in the U.S.A. However, stellarator research was continued in Japan, the former U.S.S.R., the former West Germany and the United Kingdom. In the 1970s, the most significant characteristic of stellarator plasma, in other words, confinement of currentless toroidal plasma, could not be demonstrated for high-density and high-temperature plasmas, because of the lack of a plasma heating technology. In around 1982, both the Heliotron E (Kyoto University) and the Wenderstein VII-A Stellarator (Max-Planck Institute) showed that the confinement characteristics of currentless plasma are superior to or comparable with those of tokamak plasmas of a similar size and with a similar magnetic field. In particular, currentless plasma production by high-power ECR heating, with gyrotrons radiating high-frequency electromagnetic waves, has become a standard technique in stellarator research. Due to the successful experimental results in both the stellarator and heliotron, stellarator research was restarted at the Oak Ridge National Laboratory in the U.S.A.

Also, new stellarator concepts, called the heliac, the modular stellarator, and the helias, were developed in the 1980s. From the point of view of nuclear fusion reactors, the stellarator became the most promising alternative to the tokamak. Stellarators have several intrinsic advantages compared to tokamaks. Since no plasma current is flowing in the toroidal direction in stellarators, current disruption is avoided in principle and the continuous containment of toroidal plasma is easier, if external coils are made of superconducting material.

Heliotrons have $\ell = 2$ helical coils and axisymmetric poloidal field coils for the vertical, quadrupole, and hexapole fields. The helical coils produce both the helical and the toroidal components simultaneously. From this point of view, the heliotron is the same as the torsatron developed by Gourdon, Marty, Maschke, and Touche (1971). Usually, magnetic surfaces have a roughly elliptical shape with a weak triangular deformation that depends on the poloidal field coil system. The main characteristics of the heliotron magnetic configuration are a large rotational transform and high shear. Heliotron E was built under the direction of Uo to prove the heliotron principle.

In the modular stellarator, the helical magnetic field is produced by a combination of a modular coil and a three-dimensionally deformed toroidal coil. The IMS Stellarator at Wisconsin University was designed to produce magnetic fields corresponding to a $\ell = 3$ classical stellarator, with separatrix for a divertor. The W7-AS Stellarator at the Max-Planck Institute was designed to produce magnetic fields reducing the Pfirsch–Schlüter current, which corresponds to an improvement of plasma confinement based on the neoclassical transport theory.

Heliacs have toroidal coils placed along an helical line surrounding a ring coil. When appropriate parameters are chosen for the coil currents in the toroidal coils and the ring coil, crescent-shaped magnetic surfaces are generated for plasma confinement. In heliacs a large rotational transform and a magnetic well are usually produced; however, magnetic shear is weak. Recently, the H-1 Heliac has been completed at the Australian National University and TJ-II Heliac is under construction at CIEMAT, Spain.

The helias concept emerged from extensive study of the W7-AS Stellarator, and a quasi-helically symmetric configuration is proposed to improve the neoclassical transport by reducing the toroidal effect in particle orbits significantly. Even in the toroidal configuration, the spectrum of the magnetic field strength in Boozer coordinates is very similar to that in a straight helically symmetric configuration. In both the heliac and the helias, the magnetic axis is nonplanar due to the $\ell = 1$ helical component, which is different from the heliotron and classical stellarator.

As a new stellarator, Drakon is being investigated theoretically at the Kurchatov Institute, Moscow. It aims to confine two straight mirror plasmas by connecting two ends of each mirror field with an appropriately designed helical magnetic field.

At present, two large stellarators are expected to produce high-density and high-temperature plasmas close to a fusion-graded plasma; one is the Large Helical Device (LHD) in Japan, which is under construction, and the other is the W7-X Stellarator in Germany. The LHD is a heliotron and the W7-X Stellarator is a helias, respectively. When the LHD is completed, the minor and major radii of the toroidal plasma will be about 0.5 m and 3.9 m, respectively, and the maximum magnetic field will be 3 Tesla. The plasma temperature is expected to be several KeV at a plasma density of $5\text{–}10 \times 10^{19}\,\mathrm{m}^{-3}$.

In stellarator and heliotron devices, the plasma is contained inside a family of toroidal magnetic surfaces produced by magnetic field lines. A radial outward shift of the toroidal plasma is compensated for by the rotational transform due to the helical windings. In tokamaks the rotational transform is generated by the net toroidal plasma current. In the stellarator and heliotron, three-dimensional effects become important. Issues of major interest for plasma confinement are magnetohydrodynamic equilibrium and stability, and transport. In analytic studies they have been investigated by means of expansions in small parameters, and analysis in two spatial dimensions after averaging over the rapid variation of the magnetic field strength is significantly useful for gaining an understanding of the physics. More recently, large computer codes have been developed, with acceptance as a tool to solve more difficult problems in three dimensions. The latter approach is also included in this book.

In chapter 2, we explain the fundamental properties of the stellarator and heliotron magnetic fields without plasma effects. The point that the vacuum magnetic field has characteristics capable of confining toroidal plasmas is unique and different from tokamaks. The most important finding in recent stellarator research is that, in practice, toroidal magnetic surfaces able to confine high-temperature plasmas exist in three dimensions. Rigorous mathematical theory suggests destruction of magnetic surfaces in many cases; however, its effect seems negligible when the scale of the destruction is less than the electron Larmor radius. The problem of destruction of magnetic surfaces is closely related to the appearance of stochasticity in nondissipative Hamiltonian systems. In chapter 3, we explain the theoretical description of magnetically confined plasmas. First, we discuss the basic properties of high-temperature plasmas, and then we set out the fundamental equations describing plasma dynamics in phase space. However, in order to study the physics in three-dimensional toroidal plasmas confined by magnetic fields, the fluid approximation is appropriate as a first step. We derive magnetohydrodynamic equations as moment equations of the Vlasov equation. To study transport phenomena in the collisionless regime, the drift-kinetic equation is appropriate. Finally, we derive averaged reduced MHD equations describing plasma dynamics in heliotrons.

In chapter 4, we describe the MHD equilibrium of stellarator and heliotron plasmas. First, the Grad–Shafranov equation for MHD equilibrium is shown to be extended to three-dimensional configurations such as stellarators and heliotrons. However, the averaged MHD equilibrium equation, similar to the usual Grad–Shafranov equation, is more useful in practice. Recent three-dimensional MHD equilibrium and stability codes are based on a variational formulation of magnetohydrodynamics. We explain the theoretical background to the variational principle briefly. The recent direction of stellarator research is toward helical axis stellarators, which include the heliac and the helias. In order to study the MHD equilibrium of the helical axis stellarator analytically, we give the Solov'ev–Shafranov equation here. Finally, we explain the Pfirsch–Schlüter current, which is essential in maintaining toroidal equilibrium and is related to neoclassical transport in chapter 7. The bootstrap current driven by density and temperature gradients, also discussed in chapter 7, has a connection with the Pfirsch–Schlüter current. When a rational magnetic surface with a rotational transform equal to a rational number exists inside the plasma column, the equilibrium properties change, particularly for finite resistivity. In chapter 5, we intensively discuss the MHD stability of stellarator and heliotron plasmas for the ideal and resistive modes. Here, magnetic surface breaking in finite-beta plasmas is pointed out, which may play a role in experiments. We also explain several methods to reduce the chaotic behavior of the magnetic field line. In chapter 6, we explain particle orbits in the stellarator and heliotron. There are two types of trapped particles in heliotrons, and collisionless transitions between these traping states may occur. Also, the radial electric field is effective in suppressing the loss of trapped particles. In chapter 7, we discuss the neoclassical transport theory for stellarator and heliotron plasmas. The appearance of the $1/\nu$ dependence of the transport coefficient is different from tokamaks, where ν is a collision frequency. However, this unfavorable neoclassical transport is mitigated by a careful choice of magnetic configuration. Here we also discuss the bootstrap current in heliotrons, which has a tendency to degrade the MHD stability. In chapter 8, we briefly discuss the heating of stellarator and heliotron plasmas. Then recent experimental results of heliotron and stellarator devices are reviewed. In chapter 9, we discuss the prospects of heliotron and stellarator devices for a future fusion reactor. Here, we also explain recent topics of alpha particle physics in toroidal plasmas.

BIBLIOGRAPHY

Alejaldre, C., Javier, J., Gozalo, A., Perez, J. B., Mogana, F. C., et al. (1990). TJ-II project (a flexible heliac stellarator). *Fusion Technol.*, **17**, 131.

Alfvén, H. (1951). *Cosmical electrodynamics*. Oxford University Press, Oxford.

Anderson, D. T., Derr, J. A., and Shohet, J. L. (1981). The interchangeable modular stellarator. *IEEE Trans. Plasma Sci.*, **PS-9**, 212.

Artimovich, L. A. (1972). Tokamak devices. *Nucl. Fusion*, **12**, 215.

Beidler, C., Grieger, G., Herrnegger, G., Harmeyer, E., Kisslinger, J., et al. (1990). Physics and engineering design for Wendelstein VII-X. *Fusion Technol.*, **17**, 148.

Carreras, B. A., Grieger, G., Harris, J. H., Johnson, J. L., Lyon, J. F., et al. (1988). Progress in stellarator/heliotron research (1981–1986). *Nucl. Fusion*, **28**, 1613.

Cowling, T. G. (1976). *Magnetohydrodynamics*. Adam Hilger, Bristol.

Furth, H. P. (1975). Tokamak research. *Nucl. Fusion*, **15**, 487.

Gourdon, C., Marty, D., Maschke, E. K., and Touche, J. (1971). The torsatron without toroidal field coils as a solution of the divertor problem. *Nucl. Fusion*, **11**, 161.

Hamberger, S. M., Blackwell, B. D., Sharp, L. E., and Shenton, D. B. (1990). H-1 design and construction. *Fusion Technol.*, **17**, 123.

Iiyoshi, A., Fujiwara, M., Motojima, O., Ohyabu, N., and Yamzaki, K. (1990). Design study for the Large Helical Device. *Fusion Technol.*, **17**, 169.

Kennel, C. F., Lanzerotti, L. J., and Parker, E. N., (eds.) (1979). Solar system plasma physics, Vols. I, II, and III. North-Holland, Amsterdam.

Melrose, D. B. (1980). *Plasma astrophysics*. Gordon and Breach, New York.

Miyamoto, K. (1978). Recent stellarator research. *Nucl. Fusion*, **18**, 243.

Proceedings of the Second United Nations International Conference on the Peaceful Uses of Atomic Energy in Geneva (1958). *Theoretical and experimental aspects of controlled nuclear fusion*, Vol. 31; *Controlled fusion devices*, Vol. 32. United Nations Publication, Geneva.

Rosenbluth, M. N., Hazeltine, R. D., and Hinton, F. L. (1972). Plasma transport in toroidal confinement systems. *Phys. Fluids*, **15**, 116.

Sapper, J., and Renner, H. (1990). Stellarator Wendelstein VII-AS (physics and engineering design). *Fusion Technol.*, **17**, 62.

Spitzer, L. (1958). The stellarator concept. *Phys. Fluids*, **1**, 253.

Tonks, L., and Langmuir, I. (1929). Oscillations in ionized gases. *Phys. Rev.*, **33**, 195.

Trubnikov, B. A., and Glagolev, V. M. (1984). The Drakon confinement system with a magnetic well and circular magnetic surface. *Sov. J. Plasma Phys.*, **10**, 167.

Uo, K. (1971). The helical heliotron field for plasma confinement. *Plasma Phys.*, **13**, 243.

Young, K. (1974). The C-Stellarator—A review of containment. *Plasma Phys.*, **16**, 119.

2

DESIGN PRINCIPLES OF COIL SYSTEMS IN THE STELLARATOR AND HELIOTRON

2.1 INTRODUCTION

The possibility that a magnetic bottle of the toroidal configuration (or a geometry topologically identical to the donut) may confine a high-temperature plasma makes a detailed investigation of the toroidal magnetic configuration important. The practical utilization of magnetic fields for plasma confinement depends on the solution of a number of problems. In the first place, it is necessary to determine precisely which properties of the magnetic field are responsible for stable confinement of individual charged particles and plasmas. Investigations of the confinement of individual charged particles (in chapter 6) and of the conditions necessary for the existence and stability of megnetohydrodynamic equilibrium (in chapters 4 and 5) have shown that the confinement properties of the magnetic field depend on the geometry of the magnetic lines of force and the spatial variation of the field intensity $B = |\mathbf{B}|$. These considerations lead to a detailed investigation of the structures of coils capable of producing favorable magnetic lines of force. Although the field intensity B is a local quantity, the lines of force are integral characteristics of the magnetic field **B** and the analysis becomes more complicated. Nonetheless, a large number of magnetic fields have been studied and the general features of the relation between the coil system and the structure of magnetic field are fairly well known now. It is considered that the condition $\nabla \cdot \mathbf{B} = 0$ means that magnetic lines of force must be closed or must go to infinity. However, it has been found that there are systems in which the magnetic lines of force do not close and in which they do not go to infinity. The next advance was made by Spitzer, who showed that there are irrotational toroidal fields in which the lines of force wind continuously around a circular toroidal axis, or the magnetic axis, and that to a high degree of accuracy these can be shown to lie on toroidal surfaces, or the magnetic surfaces.

The investigation of MHD equilibrium and stability in toroidal systems reveals that the important quantities related to the magnetic surface are the rotational transform, ι, on the magnetic surface, and the dependence of the rotational transform on the distance from the magnetic axis. When ι depends on the distance from the magnetic axis, this property is characterized by the magnetic shear, which is related to the radial derivative of the rotational transform. The investigation of plasma stability in toroidal systems leads to the concept of the magnetic well. These three concepts—the rotational transform, the magnetic shear, and the magnetic well—are the basis of stellarator and heliotron theory.

The notion of the separatrix was introduced to define the confinement region positively. This problem is also related to the distortion or destruction of magnetic surfaces by small perturbations. It should be noted that the mathematical problems that arise in connection with the structure of magnetic fields are particular examples of the general theory for dynamical systems described by the Hamiltonian. The basic qualitative features of a perturbed toroidal magnetic configuration are understandable from the theory of dynamical systems. When the small perturbations have a resonant component, magnetic islands appear in the vicinity of a particular resonant magnetic surface. If the magnetic islands appear on several resonant magnetic surfaces and they are sufficiently wide, the behavior of lines of force between the resonant surfaces becomes stochastic and the magnetic surfaces in this region are destroyed. This is also a mechanism that gives the last closed magnetic surface.

In the design of stellarators and heliotrons, coil systems to produce the magnetic surfaces with the desired characteristics are chosen and the magnetic fields generated by these coils are calculated with using the Biot–Savart law. When $\mathbf{B}(\mathbf{r})$ are given, a line-tracing calculation is applied to follow the lines of force and to show the existence of the magnetic surfaces, where $\mathbf{r}$ is a position vector. When they exist within the accuracy of numerical calculations, the rotational transform, the magnetic shear, and the magnetic well are also calculated. Recently, the Fourier spectrum of $B = |\mathbf{B}|$ in an appropriate coordinate system has been calculated, since it is directly related to charged particle orbits and neoclassical transport (see chapter 6). This stream of numerical calculations corresponds to the traditional approach to stellarator design. A new approach is first to give the Fourier components of the magnetic field with suitable properties for stable plasma confinement. After the last closed magnetic surface is defined, an attempt is made to find the coil system to produce the given Fourier spectrum of the magnetic field. By imposing several criteria for the shape of magnetic surfaces, the rotational transform, the magnetic shear and magnetic well, and so on, an obtimized coil system for stable plasma confinement will be found.

2.2 THE MAGNETIC SURFACE AND THE ROTATIONAL TRANSFORM

There is a tendency for the plasma pressure to be constant along a line of force, since charged particles can move freely in this direction. This result holds if the magnetic field confines the plasma for a time appreciably longer than a collision time. For this reason, toroidal magnetic systems, in which most of the lines of force do not intersect the chamber walls, are of special interest from the point of view of plasma confinement. Speaking more rigorously, the magnetic field produces a toroidal region, V_i, inside of which the lines of force do not move outside this region. The volume V_i is called the confinement region of toroidal plasma, while the remainder of the three-dimensional space, V_e, is called the exterior region. The surface that separates these regions is called the separatrix in helically symmetric systems, or is called the last closed magnetic surface in nonaxisymmetric tori. It is obvious that we are primarily interested in the interior region of the toroidal plasma V_i. Recently, a divertor to control particle and heat fluxes has been introduced in the region V_e, and is necessary for a continuously operated fusion reactor.

An understanding of the behavior of a line of force in the interior region of a toroidal configuration V_i can be obtained by examining the family of intersection points formed by the line of force as it passes through an arbitrary fixed plane P_ℓ perpendicular to the ζ-direction (see Fig. 2.1). We shall call these intersection points projection points. If a line of force of a given magnetic field closes on itself after 1, 2, or n circuits around the torus, the family of projection points consists of 1, 2, or n points, respectively. When the line of force does not close on itself, the family of projection points becomes infinite.

In each circuit of the torus, the family of projection points is transformed into itself. Hence, it is natural to introduce the concept of circuit transformation, T; this means the correspondence between the initial positions and the positions after one circuit. If a coordinate system (r, θ) is introduced in the P_ℓ plane, this transformation can be represented by $(r_2, \theta_2) = T(r_1, \theta_1)$. We may note one property of this transformation. If Σ_1 denotes some region in the P_ℓ plane and Σ_2 denotes the region into which Σ_1 goes after the circuit transformation, by virtue of $\nabla \cdot \mathbf{B} = 0$, it is evident that

$$\int_{\Sigma_1} \mathbf{B} \cdot d\mathbf{S} = \int_{\Sigma_2} \mathbf{B} \cdot d\mathbf{S} \tag{2.1}$$

or, in a differential form,

$$B_1(r_1, \theta_1) = B_2(r_2, \theta_2) \frac{\partial(r_2, \theta_2)}{\partial(r_1, \theta_1)}, \tag{2.2}$$

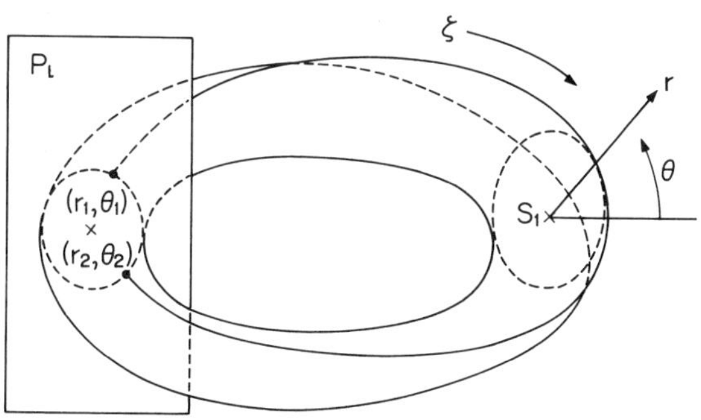

Fig. 2.1 The behavior of a line of force on the magnetic surface. The line of force starts at (r_1, θ_1) on the plane $P\ell$ perpendicular to the ζ-direction and arrives at (r_2, θ_2) after one circuit around the torus. S_1 is a cross-section of the magnetic surface.

where $|d\mathbf{S}|$ is an area element on the P_ℓ plane and $\partial(r_2, \theta_2)/\partial(r_1, \theta_1)$ corresponds to the Jacobian. The relation (2.1) defines the conservation of magnetic flux through the region Σ_1.

It is important to study the families of projection points that are produced by nonclosing lines of force in the toroidal geometry. In an axisymmetric magnetic field with respect to the ζ-direction, it can be shown that the infinite projection points lie exactly on a closed curve. This curve corresponds to the cross-section of the magnetic surface that is ergodically covered by the nonclosing line of force. In other words, a given line of force passes arbitrarily close to any arbitrary point on the magnetic surface. In a nonaxisymmetric magnetic field, if the deviation from the helical symmetry or axisymmetry is sufficiently small, magnetic surfaces seem to remain. The line of force that is surrounded by a magnetic surface, and transforms into itself in each circuit of the torus, is called the magnetic axis. There can be more than one magnetic axis in a given magnetic field configuration. Toroidal configurations with only one magnetic axis and consisting of a system of magnetic surfaces that nest in each other (see Fig. 2.2) are of great interest in connection with the problem of plasma confinement. This interest stems from the fact that the plasma pressure, P, is constant on such a magnetic surface, since the magnetic surface is everywhere covered densely by one line of force on which $P =$ constant. Thus in MHD equilibrium (see chapter 4), $P = P(\Psi)$, where Ψ is a quantity characterizing the magnetic surface.

A magnetic surface is characterized by the following quantities: the rotational transform, the longitudinal and azimuthal fluxes, and the specific volume for a magnetic surface.

In each circuit of the torus, the projection point of a line of force is displaced in the P_ℓ plane along a closed curve associated with the magnetic surface. This curves encircles the projection point corresponding to the magnetic axis. This displacement, or the rotation around the magnetic axis of the projection point on the P_ℓ plane, can be described in terms of the mean value of the angle of rotation of the projection point after $N \to \infty$ circuits of the torus,

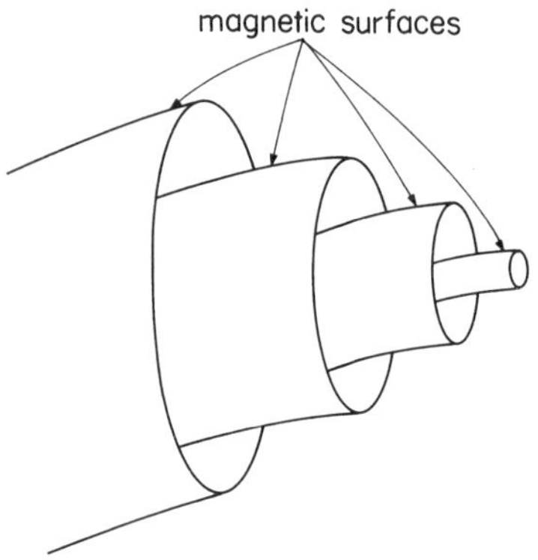

Fig. 2.2 Nested magnetic surface.

$$\iota = \lim_{N\to\infty} \frac{\sum_{k=1}^{N} \iota_k}{N}, \tag{2.3}$$

where ι_k is the rotation angle between the kth and $(k+1)$st circuits. When ι is divided by 2π, it is called a rotational transform. We are interested in nested toroidal magnetic surfaces with $\iota \neq 0$ produced by closed current-carrying conductors under the conditions of $\nabla \times \mathbf{B} = 0$ and $\nabla \cdot \mathbf{B} = 0$, or the vacuum magnetic field. Let us introduce the angle coordinates θ and ζ on the magnetic surface (see Fig. 2.1) and assume that ζ is a toroidal angle and θ is a poloidal angle. The longitudinal or toroidal magnetic flux Φ inside a specific magnetic surface is the flux through S_1; that is, through the cross-section $\zeta =$ constant (see Fig. 2.1), $\Phi = \int(\mathbf{B}\cdot\nabla\zeta)d\tau/2\pi$, where $d\tau$ is a volume element. The azimuthal or poloidal flux χ is the magnetic flux that threads the surface S_2 which extends on the $\theta =$ constant surface (see Fig. 2.3), $\chi = \int(\mathbf{B}\cdot\nabla\theta)d\tau/2\pi$. Using the quantities χ and Φ, we can give a different definition of the rotational transform which is equivalent to that given in (2.3),

$$\iota = d\chi/d\Phi. \tag{2.4}$$

We will now demonstrate the equivalence of these definitions of ι using a simple cylindrical plasma model with a uniform current density $J_z =$ constant in the presence of a uniform longitudinal field $\mathbf{B}_0 = B_0\mathbf{e}_z$. this configuration is considered as a cylindrical model of a tokamak with a circular cross-section, and is frequently used for MHD stability studies of the tokamak (see Fig. 2.4). The azimuthal or poloidal magnetic field produced by the current is $B_\theta = B_\theta(a)r/a$, and the equation for the line of force in the cylindrical coordinates (r, θ, z) is given by

$$\begin{cases} \dfrac{dr}{dz} = \dfrac{B_r}{B_0} = 0 \\ \dfrac{d\theta}{dz} = \dfrac{B_\theta}{rB_0} = \dfrac{B_\theta(a)}{aB_0}. \end{cases} \tag{2.5}$$

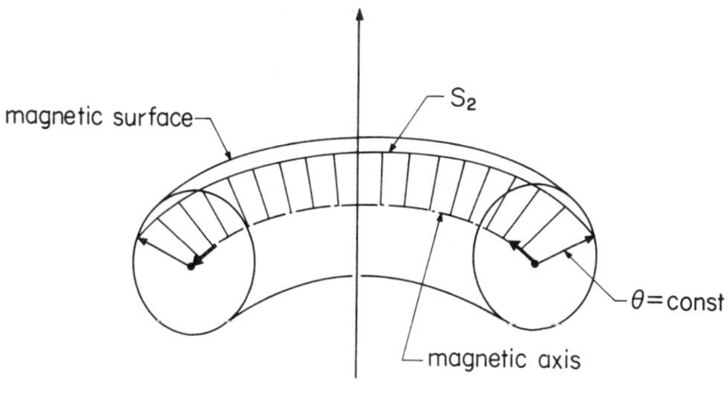

Fig. 2.3 The area S_2 for calculating the poloidal flux in a torus.

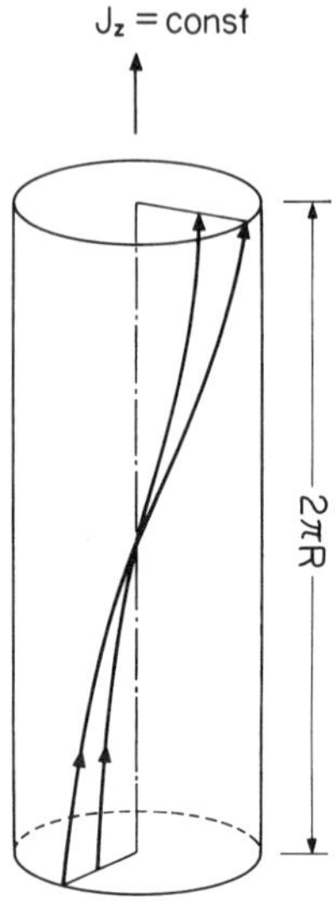

Fig. 2.4 A cylindrical tokamak model and the schematic behavior of the magnetic lines of force.

If L is the length of the corresponding torus $L = 2\pi R$, then

$$\iota = \frac{RB_\theta(a)}{aB_0}. \tag{2.6}$$

On the other hand,

$$\Phi = \pi r^2 B_0, \quad \chi = \pi B_\theta(a) r^2 \frac{R}{a}, \tag{2.7}$$

and, consequently, the definition (2.4) leads to the expression $\iota = d\chi/d\Phi = B_\theta(a)R/aB_0$, which coincides with (2.6).

If the geometric volume of a magnetic surface V is known, we can introduce the notion of a specific volume for this magnetic surface by defining the quantity as

$$U = dV/d\Phi. \tag{2.8}$$

The specific volume of a magnetic tube U is interpreted by the ratio of the geometric volume of the tube along the magnetic line of force, dV, to the magnetic flux enclosed by the magnetic tube, $d\Phi$. By requiring the conservation of magnetic flux inside the magnetic tube, we have

$$U = \frac{\int dS_n d\ell}{dS_n B} = \int \frac{dS_n d\ell}{dS_n B} = \int \frac{d\ell}{B}. \tag{2.9}$$

Here dS_n is the area of the normal cross-section of the magnetic tube. The integral in (2.9) is taken along the entire line of force if it is closed, or between certain appropriately specified points corresponding to the beginning and end of the magnetic tube. The role of the specific volume of the magnetic surface in the MHD stability of a toroidal plasma is discussed in chapter 5.

2.3 THE MAGNETIC WELL AND MAGNETIC SHEAR

The scalar function is introduced to describe the magnetic surface as

$$\Psi(\mathbf{r}) = \text{constant}, \tag{2.10}$$

and the magnetic surfaces are assumed to form a system of nested tori (see Fig. 2.2). Here $\mathbf{r}$ is a position vector. A function that depends only on Ψ is called a surface function. The basic characteristics of a toroidal configuration, which relate to the stability of the magnetic surface structure as well as the stability of the plasma confined in the toroidal configuration, are

$$V = \int d\tau, \tag{2.11}$$

the volume bounded by the surface $\Psi(\mathbf{r}) = \text{constant}$, the toroidal flux Φ and the poloidal flux χ. All of these quantities are independent and are surface functions.

There are important surface functions to characterize the stability of the toroidal configuration with respect to the magnetohydrodyamic perturbations; the rotational transform ι and the derivative of the volume with respect to the toroidal flux or the specific volume of the magnetic surface U.

Here the problem to give $\Psi(\mathbf{r})$ is considered. It is natural to take V, Φ, or χ as Ψ, when nested magnetic surfaces exist. In this section we will mostly assume that $\Psi = \Phi$.

Consider a rational (or resonant) magnetic surface, in which the magnetic lines of force are closed in n revolutions around the circuit of the torus in the toroidal direction and m revolutions around the circuit of the torus in the poloidal direction. In this case the rotational transform is given by the ratio

$$\iota = n/m. \tag{2.12}$$

On an irrational (or nonresonant) magnetic surface, on which the line of force covers the surface ergodically without closing, the rotational transform ι is given by (2.3).

The plasma has a tendency to expand and occupy the largest possible volume. If the plasma is collisionless or dissipationless, only the displacement is allowed for which the magnetic flux in a magnetic tube remains unchanged. Thus the plasma tends to be displaced into a region with maximum specific volume $U = dV/d\Phi$. For stable confinement, this region must be located in the central region of the magnetic confinement system. If the average value of the magnetic field $\bar{B}$ is defined by

$$\bar{B} = \bar{L}\frac{d\Phi}{dV} = \bar{L}/U, \tag{2.13}$$

where $\bar{L}$ is the length of the toroidal magnetic axis, the stability condition implies the existence of a minimum in $\bar{B}$ within the configuration or a *magnetic well* (more precisely, an average magnetic well). It will be evident that a toroidal magnetic configuration cannot satisfy the more stringent requirement; namely, that the magnetic field $B = |\mathbf{B}|$ increases in all directions going away from the magnetic axis. In the vicinity of magnetic axis the toroidal magnetic field is described by $B \simeq B_\zeta = B_0/R$, where $R = R_0 + r\cos\theta$ in the coordinates (r, θ, ζ). In the $\theta = 0$ direction or the outward direction from the magnetic axis, B decreases, while B increases in the $\theta = \pi$ direction.

The vacuum magnetic well is defined as

$$\hat{W} = 2\frac{V}{\langle B^2\rangle}\frac{d}{dV}\left\langle\frac{B^2}{2}\right\rangle, \tag{2.14}$$

where the average is defined as

$$\langle Q\rangle \equiv \int_0^L \frac{Q}{B}\,d\ell \Big/ \int_0^L \frac{d\ell}{B} \tag{2.15}$$

and $d\ell$ is the incremental arc length along the line of force. On the irrational magnetic surface $\langle B^2\rangle$ is evaluated in the limit as $L \to \infty$ in the integral of (2.15). For the rational magnetic surface, L corresponds to the complete revolution of the closed magnetic line of force.

The *magnetic shear* is a quantity that measures the change in pitch angle of a magnetic field line or in the rotational transform from one magnetic surface to the next. It plays an important role in stabilizing MHD instabilities, particularly those driven by the pressure gradient (see chapter 5) and drift instabilities. The magnetic shear is also a surface quantity defined on each flux surface

$$s(V) = 2\frac{V}{\iota}\frac{d\iota}{dV}. \tag{2.16}$$

It should be noted that the sign of the shear is defined to be positive in the case of $d\iota/dV > 0$, which means that $s(V) < 0$ in ordinary tokamaks, since $q = 1/\iota$ is an increasing function of radius. This q is called a safety factor in tokamak research. In stellarator research ι is more popular than q in discussing the MHD equilibrium and stability. As a general rule, large shear is favorable for stability. Heliotrons are characterized by their large shear, although the magnetic well is lost in the outer edge region with significantly large shear.

The magnetic well defined by (2.14) is related to the average curvature of the magnetic line of force. The arc length ℓ is defined by

$$\frac{d\mathbf{r}}{d\ell} = \mathbf{b} = \frac{\mathbf{B}}{B}. \tag{2.17}$$

This implies that

$$\mathbf{b} \cdot \nabla = \frac{d\mathbf{r}}{d\ell} \cdot \nabla = \frac{\partial}{\partial \ell}. \tag{2.18}$$

Consider a vacuum magnetic field in closed-line systems described by

$$\mathbf{B} = \nabla\Psi \times \nabla\beta = \nabla\phi, \tag{2.19}$$

where Ψ, β, and ϕ are scalar functions of position. The first expression of (2.19) automatically guarantees that $\nabla \cdot \mathbf{B} = 0$. The magnetic line of force lies on the surface of $\Psi =$ constant and $\beta =$ constant, and is defined as their cross-line, since $\mathbf{B} \cdot \nabla\Psi = 0$ and $\mathbf{B} \cdot \nabla\beta = 0$. The magnetic flux Ψ represents a radial-like variable and the function β a poloidal angle-like variable (see Fig. 2.5). The third coordinate that needs to be defined is a length-like variable measuring the distance in the toroidal direction. If it is shown by ζ, the coordinates (Ψ, β, ζ) correspond to one of the magnetic coordinates, since the magnetic surface is referred to in constructing the coordinates. The gradient operator is given by

$$\nabla = \nabla\Psi \frac{\partial}{\partial\Psi} + \nabla\beta \frac{\partial}{\partial\beta} + \nabla\zeta \frac{\partial}{\partial\zeta} \tag{2.20}$$

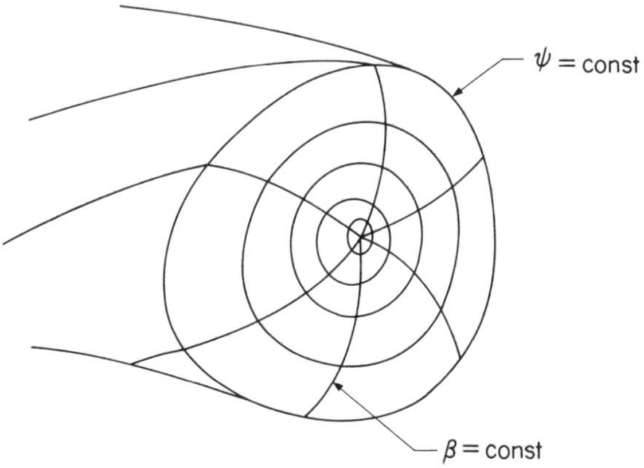

Fig. 2.5 $\Psi =$ constant corresponds to a magnetic surface and $\beta =$ constant denotes a poloidal angle line.

and volume element is given by

$$d\tau = Jd\Psi d\beta d\zeta, \tag{2.21}$$

where J is the Jacobian of the transformation

$$J = \frac{1}{\nabla\Psi \times \nabla\beta \cdot \nabla\zeta} = \frac{1}{\mathbf{B} \cdot \nabla\zeta}. \tag{2.22}$$

We note that the last expression of (2.19) suggests a choice of $\zeta = \phi$ as the third coordinate. Then the Jacobian becomes

$$J = \frac{1}{\mathbf{B} \cdot \nabla\phi} = \frac{1}{B^2}, \tag{2.23}$$

and the coordinates (Ψ, β, ϕ) are similar to the Boozer coordinates for the vacuum magnetic field. In the vacuum magnetic field the curvature vector of the magnetic field line is defined as

$$\boldsymbol{\kappa} = (\mathbf{b} \cdot \nabla)\mathbf{b}. \tag{2.24}$$

This may be expressed in a different way as

$$\boldsymbol{\kappa} = \frac{1}{B^2}\left[\nabla \frac{B^2}{2} - \mathbf{b}(\mathbf{B} \cdot \nabla B)\right]. \tag{2.25}$$

In the coordinates (Ψ, β, ϕ), the curvature vector is described by

$$\boldsymbol{\kappa} = \kappa_\psi \nabla\Psi + \kappa_\beta \nabla\beta, \tag{2.26}$$

where $\kappa_\psi = -(\mathbf{b} \times \nabla\beta \cdot \boldsymbol{\kappa})/B$ and $\kappa_\beta = (\mathbf{b} \times \nabla\Psi \cdot \boldsymbol{\kappa})/B$, since $\mathbf{b} \cdot \boldsymbol{\kappa} = 0$. From (2.25) κ_ψ also becomes

$$\kappa_\psi = \frac{1}{B}\frac{\partial B}{\partial \Psi}. \tag{2.27}$$

In order to obtain stability against interchange modes, the rational magnetic surface is considered, where the magnetic line of force is closed. The stability requires that the average curvature must be positive (see chapter 5),

$$\oint \kappa_\psi \frac{d\ell}{B} > 0, \tag{2.28}$$

which is obtained from (2.15). In the coordinates (Ψ, β, ϕ), the stability condition (2.28) yields

$$\oint \frac{1}{B^3} \frac{\partial B}{\partial \Psi} d\phi = -\frac{1}{2} \frac{\partial}{\partial \Psi} \oint \frac{d\phi}{B^2} > 0, \tag{2.29}$$

by using the relation

$$d\phi / B^2 = d\ell / B, \tag{2.30}$$

and exchanging the order between derivative and integral. The inequality (2.29) may be written as

$$\frac{\partial}{\partial \Psi} \oint \frac{d\ell}{B} < 0. \tag{2.31}$$

The inequality (2.31) corresponds to the condition of a magnetic well.

Here the integrations in the ϕ and ℓ coordinates are compared in a closed-line system. For any quantity Q, it follows that

$$\int_0^{\phi_0} \frac{Q d\phi}{B^2} = \int_0^L \frac{Q d\ell}{B}, \tag{2.32}$$

where ϕ_0 and L correspond to values for one complete circuit of the closed line of force. For the vacuum magnetic field,

$$\oint B d\ell = \int_0^L \mathbf{B} \cdot d\ell = \int_0^L \nabla\phi \cdot d\ell = \int_0^{\phi_0} d\phi = \phi_0 \tag{2.33}$$

along an arbitrary magnetic field line defined by $\Psi =$ constant and $\beta =$ constant. Here we consider the situation in which ϕ_0 is independent of Ψ or $\mathbf{B}$ is the vacuum magnetic field. Under the condition $\partial\phi_0/\partial\Psi = 0$, we divide (2.31) by ϕ_0. Then the stability condition has the form

$$\frac{\partial}{\partial \Psi} \langle B^2 \rangle > 0, \tag{2.34}$$

where, for any quantity Q,

$$\langle Q \rangle \equiv \oint \frac{Q}{B} d\ell \Big/ \oint \frac{d\ell}{B}. \tag{2.35}$$

The concept of a magnetic well is apparent for the inequality (2.34), which implies that $\langle B^2 \rangle$ increases away from the magnetic axis.

We will calculate the integral $\int d\ell/B$ along a line of force given by $\beta = \beta_0 + \iota(\Psi)\zeta$ on the magnetic surface $\Psi =$ constant, where β_0 is a constant.

It is noted that the magnetic line of force is straight since $d\beta/d\zeta = \iota(\Psi)$, where $\iota(\Psi)$ is a rotational transform. Since $d\ell/B = d\zeta/B^\zeta$,

$$\int \frac{d\ell}{B} = \int \frac{d\zeta}{B^\zeta} = 2\pi \int J d\zeta \tag{2.36}$$

where the contravariant component of the magnetic field in the ζ-direction is $B^\zeta = 1/(2\pi J)$, which is obtained from (2.22). From the expressions (2.11) and (2.21), the following relation is given:

$$dV/d\Psi = \int_0^{2\pi} \int_0^{2\pi} J d\beta d\zeta. \tag{2.37}$$

We now expand J in a doubler Fourier series in the coordinates where $0 \le \beta \le 2\pi$ and $0 \le \zeta \le 2\pi$,

$$J = Re \sum_{m,n} J_{mn} e^{i(m\beta - n\zeta)}. \tag{2.38}$$

It is evident that

$$dV/d\Psi = (2\pi)^2 J_{00}. \tag{2.39}$$

Thus the area integral in (2.37) now becomes

$$\int J d\beta d\zeta = \frac{dV}{d\Psi} + Re \sum_{m,n}{}' \int J_{mn} e^{i(m\beta - n\zeta)} d\beta d\zeta, \tag{2.40}$$

where the prime on the summation symbol indicates that the term with $m = 0$ and $n = 0$ does not appear in the summation. It is a simple matter to write the expression for the integral (2.36) taken along the line of force, $\beta = \beta_0 + \iota(\Psi)\zeta$, that passes through the point $\beta = \beta_0$ and $\zeta = 0$,

$$\begin{aligned} U(\Psi, \beta_0) &\equiv \frac{1}{2\pi N} \int_0^{2\pi N} \frac{d\zeta}{B^\zeta} \\ &= \frac{dV}{d\Psi} \left\{ 1 + \frac{1}{2\pi N J_{00}} Re \sum_{m,n}{}' J_{mn} \frac{e^{2\pi N i(m\iota - n)} - 1}{i(m\iota - n)} e^{im\beta_0} \right\}, \end{aligned} \tag{2.41}$$

for N turns around the torus. Assume that the line of force is closed after circulating the torus n_0 times around the minor circuit and m_0 times along the major circuit. It is evident from the expression for the line of force that the closure can occur on a rational surface for which $\iota(\Psi_r) = n_0/m_0$. Under these conditions, using (2.41) we find that

$$U(\Psi_r, \beta_0) \equiv \frac{1}{n_0}\int_0^{2\pi n_0} \frac{d\zeta}{B^\zeta} \tag{2.42}$$
$$= \frac{dV}{d\Psi}\left\{1 + \frac{1}{J_{00}}\, Re(J_{m_0 n_0} e^{im_0\beta_0} + J_{-m_0-n_0} e^{-im_0\beta_0})\right\}$$

for n_0. It is evident from this expression that the integral (2.42) taken along different lines of force on the same magnetic surface will, in general, give different values for $J_{m_0n_0} \neq 0$ or $J_{-m_0-n_0} \neq 0$. The fact that the lines of force on a rational magnetic surface do not have the same length for this situation shows some important consequences. It will be discussed below that, when a plasma is confined in such a configuration, the magnetic surfaces exhibit splitting, with the formation of magnetic island structures.

It should be noted that, for axisymmetric configurations such as a tokamak, the second term in (2.42) vanishes on the rational surfaces with $\iota = n/m$, since $J_{mn} = 0$ for $n \neq 0$. On an irrational surface, for a large number of circuits $N \to \infty$, the integration along the line of force leads to an averaging of the function $U(\Psi, \beta_0)$. It is evident from (2.41) that the summation over (m, n) remains bounded even for $N \to \infty$. In this limit we find that

$$U(\Psi) = \lim_{N\to\infty} \frac{1}{N}\int \frac{d\ell}{B} = \frac{dV}{d\Psi}. \tag{2.43}$$

This formula is useful for a numerical calulation of the specific volume U in the general toroidal configuration.

2.4 THE AVERAGE MAGNETIC SURFACE

There exists a wide class of magnetic fields for which approximate magnetic surfaces can be found by the method of averaging. This class includes magnetic fields that are periodic along the ζ-direction or the toroidal direction with a large longitudinal component of the magnetic field. For this type of magnetic configuration, the method of averaging makes it possible to obtain equations for averaged magnetic surfaces to an accuracy of the order of $(B_\perp/B_\parallel)^2$, where $B_\perp$ and $B_\parallel$ characterize the transverse and longitudinal magnetic fields, respectively. By improving the averaging method, the accuracy may be increased up to the higher order of $B_\perp/B_\parallel$. Here we consider the derivation of averaged magnetic surfaces for a quasi-straight field.

Suppose that the magnetic field has a large constant component B_0 directed along the z-coordinate of a cylindrical coordinate system (r, θ, z) and a small field $\mathbf{b}(r, \theta, z)$ that is periodic in a short scale in the z-coordinate, and $|\mathbf{b}|/|\mathbf{B}_0| \ll 1$. However, it changes slowly in a longer scale. In order to apply the averaging method, the field period length L must be much smaller than the characteristic scale of the slow variation. To terms of order $(b/B_0)^2$, the equations for the line of force (2.5) can be written in the form

$$\begin{cases} \dfrac{dr}{dz} = \dfrac{b_r}{B_0 + b_z} \simeq \dfrac{b_r}{B_0} - \dfrac{b_r b_z}{B_0^2} \\ \dfrac{d\theta}{dz} = \dfrac{1}{r}\dfrac{b_\theta}{B_0 + b_z} \simeq \dfrac{b_\theta}{rB_0} - \dfrac{b_\theta b_z}{rB_0^2}, \end{cases} \tag{2.44}$$

where B_0 is chosen in such a way that the mean value of b_z vanishes.

The canonical form for the method of averaging is the system of first-order equations

$$dx_k/dt = f_k(x_i, t, \alpha), \tag{2.45}$$

where $\alpha = t/\epsilon$. Here ϵ is a small parameter, while f_k is a periodic function of argument α, with period 2π, which depends on (x_1, x_2, x_3) and t. Here, the small parameter ϵ characterizes the high-frequency oscillation of the RHS of (2.45) and a solution is sought over time intervals t of the order of unity or larger. Equation (2.45) can be written in another form which is frequently used; specifically, if the argument is taken to be $\bar{t} = t/\epsilon$, then it becomes

$$dx_k/d\bar{t} = \epsilon f_k(x_i, \epsilon\bar{t}, \bar{t}), \tag{2.46}$$

in which the small parameter ϵ now characterizes the smallness of the RHS of (2.46). The existence of the second argument in f_k indicates that, in addition to the periodic change of the RHS, there can also be a slow variation of arbitrary nature. The solution of (2.45) which is valid over an interval $t \sim 1$ corresponds to a solution of (2.46) which is valid over an interval $\bar{t} \sim 1/\epsilon$.

In order to solve (2.45) we seek to replace the x_k with new averaged variables ξ_k, which will differ slightly from the x_k, but which satisfy equations that do not contain the rapidly varying phase α:

$$x_k = \xi_k + \epsilon g_{1k}(\xi_i, t, \alpha) + \epsilon^2 g_{2k}(\xi_i, t, \alpha) + \cdots, \tag{2.47}$$

$$d\xi_k/dt = h_{0k}(\xi_i, t) + \epsilon h_{1k}(\xi_i, t) + \epsilon^2 h_{2k}(\xi_i, t) + \cdots. \tag{2.48}$$

The equations (2.47) and (2.48) define new unknown functions g_{ik} and h_{ik}, and the problem is now reduced to the determination of these functions. Substituting the relations (2.47) and (2.48) into (2.45), and equating terms of same powers of ϵ, yields the following system of equations for the determination of g_{ik} and h_{ik}:

$$\begin{cases} h_{0k} + \dfrac{\partial g_{1k}}{\partial \alpha} = f_k, \\ h_{1k} + \dfrac{\partial g_{1k}}{\partial \xi_i} h_{0i} + \dfrac{\partial g_{1k}}{\partial t} + \dfrac{\partial g_{2k}}{\partial \alpha} = \dfrac{\partial f_k}{\partial \xi_i} g_{1i} \\ h_{2k} + \dfrac{\partial g_{1k}}{\partial \xi_i} h_{1i} + \dfrac{\partial g_{2k}}{\partial \xi_i} h_{0i} + \dfrac{\partial g_{2k}}{\partial t} + \dfrac{\partial g_{3k}}{\partial \alpha} = \dfrac{\partial f_k}{\partial \xi_i} g_{2i} + \dfrac{1}{2}\dfrac{\partial^2 f_k}{\partial \xi_i \partial \xi_j} g_{1i} g_{1j}, \text{ etc.} \end{cases} \tag{2.49}$$

Here a subscript that appears twice denotes summation, while $f_k = f_k(\xi_i, t, \alpha)$. We require that the function g_{ik} be a periodic function of the argument α, and we now introduce a new notation. For any periodic function $f(\xi_i, t, \alpha)$, we can write

$$f = \bar{f} + \tilde{f}, \tag{2.50}$$

where

$$\bar{f} = \frac{1}{2\pi}\int_0^{2\pi} f d\alpha.$$

Here and below, the integration is carried out for fixed arguments ξ_i and t; $\bar{f}$ is the constant part of f, and $\tilde{f}$ is the variable part of f. The following symbol $\hat{f}$ denotes the variable part of the integral of f:

$$\hat{f} = \int_0^{\alpha} \tilde{f} d\alpha. \tag{2.51}$$

We note that $\bar{\hat{f}} = 0$. The number of functions g_{ik} and h_{ik} introduced in (2.49) is such that we must make additional assumptions in order to determine these functions uniquely. The most reasonable assumption is to require that the constant part of g_{ik} must vanish. With this choice, ξ_k becomes a mean value of x_k (with respect to the variable α), $\bar{x}_k = \xi_k$, about which the true values of the solutions oscillate. However, it is evident that this choice is not the only possible one. In particular, if (2.45) are written in a Hamiltonian form, it is possible to choose g_{ik} in such a way that the averaged equation (2.48) is also written in the Hamiltonian form.

With $\bar{g}_{ik} = 0$ and requiring that (2.49) are satisfied identically, we equate separately the constant and variable parts in (2.49) in order to fix ξ_i and t. Equations for the constant parts give expressions for the h_{ik}, and integrating the remained relations with respect to α yields g_{ik}. Using the notations introduced above, we find systematically

$$h_{0k} = \bar{f}_k, \tag{2.52}$$

$$\begin{cases} g_{1k} = \hat{f}_k, \\ h_{1k} = \overline{\dfrac{\partial f_k}{\partial \xi_i} g_{1i}}, \end{cases} \tag{2.53}$$

$$\begin{cases} g_{2k} = \widehat{\dfrac{\partial f_k}{\partial \xi_i} g_{1i}} - \dfrac{\partial \hat{g}_{1k}}{\partial \xi_i} h_{0i} - \dfrac{\partial \hat{g}_{1k}}{\partial t}, \\ h_{2k} = \overline{\dfrac{\partial f_k}{\partial \xi_i} g_{2i} + \dfrac{1}{2}\dfrac{\partial^2 f_k}{\partial \xi_i \partial \xi_j} g_{1i} g_{1j}}. \end{cases} \tag{2.54}$$

In transforming these expressions, it is convenient to make use of the identity

$$\overline{\hat{a}b} = -\overline{a\hat{b}}, \tag{2.55}$$

which follows from the fact that the mean value of the following derivative must vanish:

$$\overline{\frac{\partial}{\partial\alpha}(\hat{a}\hat{b})} = 0. \tag{2.56}$$

In (2.55) we have considered $\overline{\hat{a}\bar{b}} = \overline{\bar{a}\hat{b}} = 0$. Finally, from (2.48) and (2.52)–(2.55), we obtain the averaged equations for the ξ_k,

$$\frac{d\xi_k}{dt} = \bar{f}_k + \epsilon\overline{\frac{\partial f_k}{\partial\xi_i}\hat{f}_i} + \epsilon^2\left\{-\overline{\hat{f}_i\frac{\partial\bar{f}_j}{\partial\xi_i}\frac{\partial\hat{f}_k}{\partial\xi_j}} + \overline{\bar{f}_i\frac{\partial\hat{f}_j}{\partial\xi_i}\frac{\partial\hat{f}_k}{\partial\xi_j}} + \overline{\frac{\partial\hat{f}_i}{\partial t}\frac{\partial\hat{f}_k}{\partial\xi_i}} + \frac{1}{2}\overline{\hat{f}_i\hat{f}_j\frac{\partial^2 f_k}{\partial\xi_i\partial\xi_j}}\right\} \tag{2.57}$$

and periodic corrections to the solution

$$x_k = \xi_k + \epsilon\hat{f}_k + \epsilon^2\left(\widehat{\frac{\partial f_k}{\partial\xi_i}\hat{f}_i} - \frac{\partial\hat{\hat{f}}_k}{\partial\xi_i}\bar{f}_i - \frac{\partial\hat{\hat{f}}_k}{\partial t}\right). \tag{2.58}$$

Now we apply (2.57) to the equations of line of force (2.44), neglecting $O(\epsilon^2)$ terms. Here ξ_k and t are considered to be $\bar{r}$, $\bar{\theta}$ and z, respectively. We can write (2.44) with reasonable accuracy in the following way:

$$\frac{d\bar{r}}{dz} = \frac{\overline{b_r}}{B_0} - \frac{\overline{b_r b_z}}{B_0^2} + \overline{\frac{\partial}{\partial\bar{r}}\left(\frac{b_r}{B_0}\right)\frac{\hat{b}_r}{B_0}} + \overline{\frac{\partial}{\partial\bar{\theta}}\left(\frac{b_r}{B_0}\right)\frac{\hat{b}_\theta}{\bar{r}B_0}}, \tag{2.59}$$

and

$$\frac{d\bar{\theta}}{dz} = \frac{\overline{b_\theta}}{\bar{r}B_0} - \frac{\overline{b_\theta b_z}}{\bar{r}B_0^2} + \overline{\frac{\partial}{\partial\bar{r}}\left(\frac{b_\theta}{\bar{r}B_0}\right)\frac{\hat{b}_r}{B_0}} + \overline{\frac{\partial}{\partial\bar{\theta}}\left(\frac{b_\theta}{\bar{r}B_0}\right)\frac{\hat{b}_\theta}{\bar{r}B_0}}. \tag{2.60}$$

In (2.59) and (2.60) we can replace the derivatives $\partial b_r/\partial r$ and $\partial b_\theta/\partial\theta$ by their values obtained from the condition

$$\nabla\cdot\mathbf{b} = \frac{1}{r}\frac{\partial}{\partial r}rb_r + \frac{1}{r}\frac{\partial b_\theta}{\partial\theta} + \frac{\partial b_z}{\partial z} = 0. \tag{2.61}$$

Then, taking account of $\partial\hat{a}/\partial\alpha = \tilde{a}$ and the relation (2.55) yields

$$\frac{d\bar{r}}{dz} = \frac{\bar{b}_r}{B_0} - \frac{1}{\bar{r}B_0}\frac{\partial}{\partial\bar{\theta}}\overline{\left(\frac{\hat{b}_r b_\theta}{B_0}\right)} \tag{2.62}$$

and

$$\frac{d\bar{\theta}}{dz} = \frac{\bar{b}_\theta}{\bar{r}B_0} + \frac{1}{\bar{r}B_0}\frac{\partial}{\partial\bar{r}}\overline{\left(\frac{\hat{b}_r b_\theta}{B_0}\right)}. \tag{2.63}$$

Introducing the vector potential $\bar{\mathbf{A}} = (0, 0, \bar{A}_z)$ and replacing $\bar{b}_r$ and $\bar{b}_\theta$ by the expressions

$$\begin{cases} \bar{b}_r = \dfrac{1}{r}\dfrac{\partial\bar{A}_z}{\partial\theta} \\ \bar{b}_\theta = -\dfrac{\partial\bar{A}_z}{\partial r}, \end{cases} \tag{2.64}$$

we write (2.62) and (2.63) in the form

$$\frac{d\bar{r}}{dz} = \frac{1}{\bar{r}B_0}\frac{\partial\Psi}{\partial\bar{\theta}} \tag{2.65}$$

and

$$\frac{d\bar{\theta}}{dz} = -\frac{1}{\bar{r}B_0}\frac{\partial\Psi}{\partial\bar{r}}, \tag{2.66}$$

where

$$\Psi = \bar{A}_z - \frac{\overline{\hat{b}_r b_\theta}}{B_0}. \tag{2.67}$$

Evidently, the integral of (2.65) and (2.66) becomes the equation for the *averaged magnetic surface*,

$$\Psi(\bar{r}, \bar{\theta}) = \text{constant}. \tag{2.68}$$

The behavior of the line of force, including the first-order periodic correction with $O(b/B_0)$, is given by

$$\begin{cases} r = \bar{r} + \dfrac{\hat{b}_r}{B_0} \\ \theta = \omega z + \dfrac{\hat{b}_\theta}{rB_0}, \end{cases} \tag{2.69}$$

when the function Ψ depends only on $\bar{r}$. It also follows from (2.65) and (2.66) that Ψ can be expressed in terms of the mean rate of rotation of the line of force around the z-axis,

$$\begin{cases} \Psi = -\displaystyle\int_0^r \bar{r} B_0 \frac{d\bar{\theta}}{dz}\, d\bar{r} \\ \dfrac{d\bar{\theta}}{dz} = \omega, \end{cases} \tag{2.70}$$

where ω is the mean rotation angle of the line of force.

2.5 THE HELICALLY SYMMETRIC MAGNETIC FIELD

We consider a magnetic field produced by a helical winding with a finite pitch length, which exhibits helical symmetry. In cylindrical coordinates (r, θ, z) the helical symmetry requires that the magnetic field depends only on two variables, r and $\zeta = \theta - \alpha z$, where α is a constant. Since the vacuum magnetic field is curl-free, $\nabla \times \mathbf{B} = 0$. By using $\mathbf{B} = \nabla\varphi$ for $\nabla \cdot \mathbf{B} = 0$, $\nabla^2\varphi = 0$ is obtained. The helical field satisfying Laplace's equation is conveniently described by means of a scalar potential expanded in an harmonic series with respect to ζ,

$$\varphi = B_0 z + \frac{1}{\alpha}\sum_{n=1}^{\infty} b_n I_n(n\alpha r)\sin n\zeta, \tag{2.71}$$

where I_n is a modified Bessel function. The magnetic field is also described by the vector potential $\mathbf{B} = \nabla \times \mathbf{A}$. Here we set $A_z = 0$. This is possible because we can add the gradient of any arbitrary function satisfying Laplace's equation. From $\nabla\varphi = \nabla \times \mathbf{A}$, we obtain

$$\begin{cases} \dfrac{\partial\varphi}{\partial r} = -\dfrac{\partial A_\theta}{\partial z}, \\ \dfrac{1}{r}\dfrac{\partial\varphi}{\partial\theta} = \dfrac{\partial A_r}{\partial z}. \end{cases} \tag{2.72}$$

Then the vector potential is also described as

$$\begin{cases} A_r = -\dfrac{1}{\alpha^2 r}\displaystyle\sum_{n=1}^{\infty} b_n I_n I(n\alpha r)\sin n\zeta, \\ A_\theta = \dfrac{B_0}{2} r - \dfrac{1}{\alpha}\displaystyle\sum_{n=1}^{\infty} b_n I_n'(n\alpha r)\cos n\zeta, \\ A_z = 0. \end{cases} \tag{2.73}$$

The components of the magnetic field are expressed in terms of the vector potential

$$\begin{cases} B_r = \dfrac{1}{r}\dfrac{\partial}{\partial\theta} A_z - \dfrac{\partial}{\partial z} A_\theta, \\ B_\theta = \dfrac{\partial}{\partial z} A_r - \dfrac{\partial}{\partial r} A_z, \\ B_z = \dfrac{1}{r}\dfrac{\partial}{\partial r}(rA_\theta) - \dfrac{1}{r}\dfrac{\partial}{\partial\theta} A_r. \end{cases} \tag{2.74}$$

By introducing

$$\Psi = \alpha r A_\theta + A_z, \tag{2.75}$$

we find the relations

$$B_r = \frac{1}{r}\frac{\partial\Psi}{\partial\zeta} \tag{2.76}$$

and

$$\alpha r B_z - B_\theta = \alpha\frac{\partial}{\partial r}(rA_\theta) + \frac{\partial A_z}{\partial r} = \frac{\partial\Psi}{\partial r}. \tag{2.77}$$

Equations (2.76) and (2.77) mean that Ψ = constant is a magnetic surface that is explicitly shown as

$$\Psi(r, \zeta) = \frac{\alpha r^2}{2} B_0 - r\sum_{n=1}^{\infty} b_n I_n'(n\alpha r)\cos n\zeta = \text{constant}. \tag{2.78}$$

We now consider the magnetic field due to the nth harmonic and uniform magnetic field $\mathbf{B}_0$ directed along the z-axis. We call this magnetic field an $\ell = n$ field, because it requires ℓ pairs of helical conductors with current flow in opposite directions in adjacent windings (see fig. 2.6). It is noted that, rigorously, the magnetic field due to ℓ conductors includes an infinite number of harmonics; however, the magnetic field may be approximated by a single harmonic in the region of $r \simeq 0$, since the contributions from the higher harmonics become negligible.

In this case, the scalar potential and the equation for the magnetic surfaces are

$$\varphi = B_0 z + \frac{b}{\alpha} I_n(n\alpha r)\sin n(\theta - \alpha z) \tag{2.79}$$

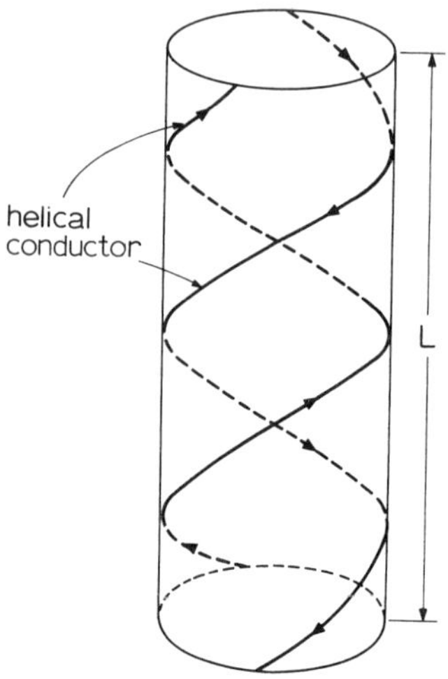

Fig. 2.6 Helical conductors with coil currents in opposite directions on a straight cylindrical column.

and

$$\begin{aligned}\Psi &= \frac{B_0}{2\alpha}\left\{(\alpha r)^2 - \frac{2b}{B_0}\,\alpha r I_n'(n\alpha r)\cos n(\theta - \alpha z)\right\} \\ &= \frac{B_0}{2\alpha}(\alpha\bar{r})^2.\end{aligned} \tag{2.80}$$

Here $\bar{r}$ is the average radius of the magnetic surface. In order to obtain the cross-sections of the magnetic surfaces, we consider the plane of $z = 0$. Here we use the fact that

$$I_n(n\alpha r) \simeq \frac{1}{n!}\left(\frac{n\alpha r}{2}\right)^n \tag{2.81}$$

for $r \to 0$. With this approximation, (2.80) becomes

$$(\alpha r)^2 - \frac{b}{B_0}\frac{1}{(n-1)!}\left(\frac{n}{2}\right)^{n-1}(\alpha r)^n \cos n\theta = \text{constant}. \tag{2.82}$$

Three important cases, $n = 1$, $n = 2$, and $n \geq 3$, can be distinguished. For $n = 1$, (2.82) becomes

$$r^2 - \frac{C}{\alpha}\,r\cos\theta = \left(x - \frac{C}{2\alpha}\right)^2 + y^2 - \frac{C^2}{4\alpha^2} = \text{constant}, \tag{2.83}$$

where $C = b/B_0$, and $x = r\cos\theta$, $y = r\sin\theta$, and $z = 0$. The equation (2.83) shows that the cross-sections of the magnetic surfaces comprise a family of off-center circles in the vicinity of the origin (see Fig. 2.7(a)). This property is usually seen in the $\ell = 1$ stellarator. It is noted that this result holds only for $b/B_0 \ll 1$.

For $n = 2$, (2.82) reduces to

$$r^2\left(1 - \frac{b}{B_0}\cos 2\theta\right) = \text{constant}. \tag{2.84}$$

It is evident that when $b/B_0 < 1$ the origin becomes an elliptic singular point, as shown in Fig. 2.7(b). For $b/B_0 > 1$, the origin becomes a hyperbolic singular point (Fig. 2.7(c)). The elliptic magnetic surfaces are usually seen in the $\ell = 2$ stellarator. For the last case of $n \geq 3$, (2.82) can be written in the form

$$r^2(1 - D(\alpha r)^{n-2}\cos n\theta) = \text{constant}, \tag{2.85}$$

where

$$D = \frac{b}{B_0}\frac{1}{(n-1)!}\left(\frac{n}{2}\right)^{n-1}.$$

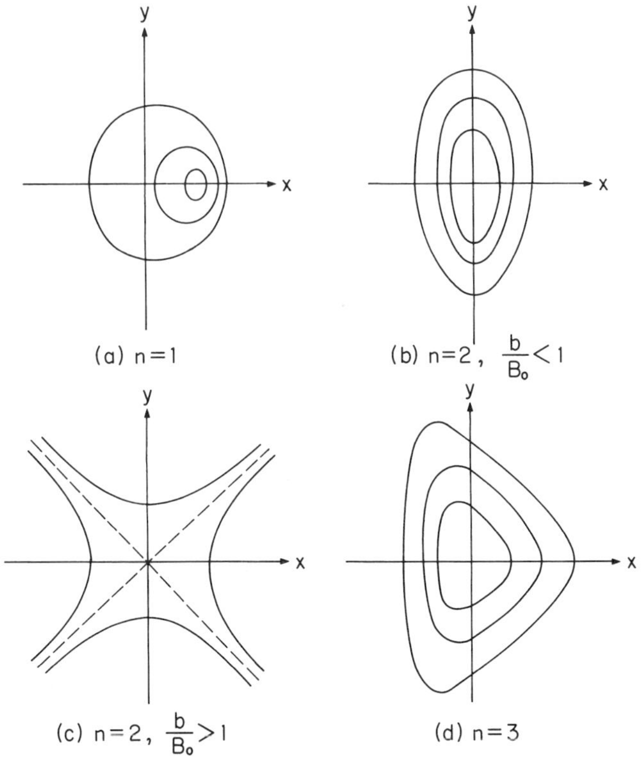

Fig. 2.7 Magnetic surfaces of a stellarator.

For any value of D and for sufficiently small r, the cross-sections of the magnetic surfaces encircle the origin. For the case $n = 3$, the cross-sections show a triangular shape (see Fig. 2.7(d)), which is a property of the $\ell = 3$ stellarator.

In the other limiting case of $r \to \infty$, we note that $I_n'(n\alpha r) \to \infty$ and the first term in (2.80) is negligible:

$$(\alpha r) I_n'(n\alpha r) \cos n(\theta - \alpha z) = \text{constant}. \tag{2.86}$$

It is evident that the curves $r = r(\theta)$ have asymptotes in the form of straight lines diverging at an angle θ_A satisfying $\cos(n\theta_A) = 0$ for $z = 0$. The change of the magnetic surfaces, which are closed in the vicinity of the origin in the case of $b/B_0 \ll 1$, to a magnetic structure divided into $2n$ groups by the straight lines at infinity occurs at the separatrix.

The singular points of the equation describing the magnetic surface (2.80) are determined from the equations

$$\begin{cases} \dfrac{\partial \Psi}{\partial r} = 0 \\ \dfrac{\partial \Psi}{\partial \zeta} = 0. \end{cases} \tag{2.87}$$

By using the differential equation for the modified Bessel function, $I_n(X)$,

$$\frac{d^2 I_n}{dX^2} + \frac{1}{X}\frac{dI_n}{dX} - \left(1 + \frac{n^2}{X^2}\right) I_n = 0, \tag{2.88}$$

(2.87) give

$$(\alpha r)\left[1 - \frac{nb}{B_0}\left(1 + \frac{1}{(\alpha r)^2}\right) I_n(n\alpha r) \cos n\zeta\right] = 0 \tag{2.89}$$

and

$$\sin(n\zeta) = 0. \tag{2.90}$$

It then follows from (2.89) and (2.90) that:

(i) For $n > 1$, the origin of coordinates is a singular point of the elliptic type and the z-axis corresponds to a magnetic axis.
(ii) For $n > 1$, the number of singular points other than the origin is n.
(iii) The distance from the origin of coordinates to the singular point r_s is related to b/B_0 by

$$\frac{b}{B_0} = \frac{(\alpha r_s)^2}{n(1 + (\alpha r_s)^2 I_n(n\alpha r_s))}. \tag{2.91}$$

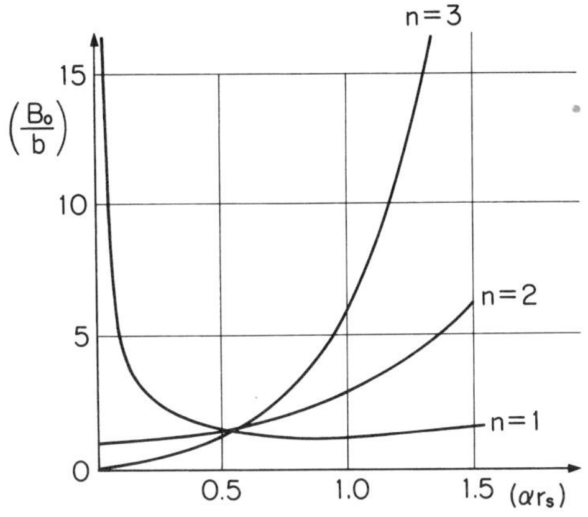

Fig. 2.8 B_0/b versus αr_s satisfying (2.91).

Curves satisfying (2.91) are described in the plane showing B_0/b versus αr_s for different values of n in Fig. 2.8. When $n > 2$, a separatrix exists for any b/B_0; however, for $n = 2$, a separatrix exists only when $b/B_0 < 1$. The behavior of the $n = 1$ case indicates that the separatrix vanishes if $b/B_0 > 0.9$; however, for $b/B_0 < 0.9$, there are two singular points. The first singular point is elliptic, while the second is hyperbolic and is the trace of the ridge on the separatrix. Here, we show schematic cross-sections for magnetic surfaces for several n and b/B_0 in Fig. 2.9.

Here, we apply the method of averaging explained in section 2.4 to the helically symmetric magnetic field, which is valid for $r \ll r_s$, where r_s is the characteristic radius of the separatrix given by (2.91). The equation for the

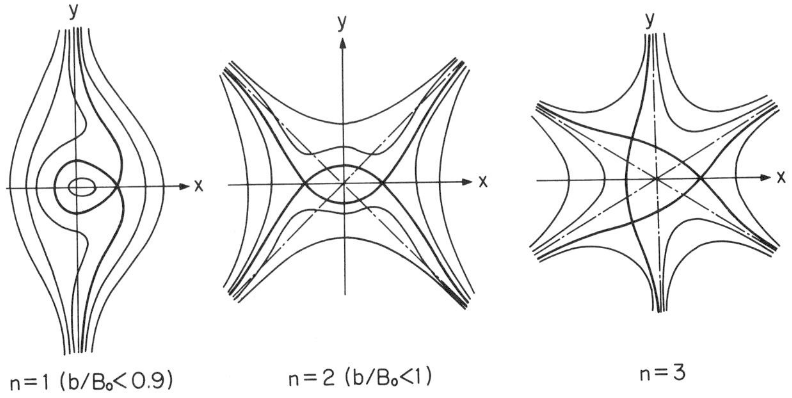

Fig. 2.9 Schematic cross-sections of magnetic surfaces and the separatrix for $n = 1$, $n = 2$, and $n = 3$.

average magnetic surface is given by

$$\Psi = -\frac{1}{B_0}\overline{\hat{B}_r B_\theta} \tag{2.92}$$

for the gauge condition $A_z = 0$ (see (2.67)). Substituting the components B_r and B_θ obtained from (2.74) for the vacuum helical magnetic field, we have

$$\Psi = -\frac{1}{4B_0\alpha}\sum_n b_n^2 \frac{1}{\rho}\frac{d}{d\rho} I_n^2(n\rho), \tag{2.93}$$

where $\rho = \alpha r$. Using the definitions (see (2.66) and (2.70))

$$\begin{cases} \dfrac{\partial \Psi}{\partial r} = -B_0 r \dfrac{d\theta}{dz} \\ \dfrac{d\theta}{dz} = \alpha\omega_0, \end{cases} \tag{2.94}$$

we find an expression for the rotational transform per field period ω_0,

$$\omega_0 = \sum_n \frac{b_n^2}{4B_0^2}\left(\frac{1}{\rho}\frac{d}{d\rho}\right)^2 I_n^2(n\rho). \tag{2.95}$$

In the vicinity of the magnetic axis we can write ω_0 for the nth harmonic in the following form by expanding $I_n(n\rho)$,

$$\omega_0 = \left(\frac{b_n}{B_0}\right)^2 \left(\frac{2}{n!}\left(\frac{n}{2}\right)^{n+1}\right)^2 \left\{\left(1 - \frac{1}{n}\right)\rho^{2n-4} + \frac{n}{2}\rho^{2n-2} + \cdots\right\}. \tag{2.96}$$

Thus, for $n = 1$ and $n = 2$, the rotational transform is nonzero at $r = 0$; however, for $n > 2$, the rotational transform vanishes at $r = 0$.

To an accuracy of $O(b_n)$, the equations for the lines of force are given by (2.69). For the magnetic field given by the scalar potential (2.71), they give

$$\rho = \rho_0 + \frac{1}{B_0}\sum_n b_n I_n'(n\rho)\cos(n\psi) \tag{2.97}$$

and

$$\theta = 2\pi\omega_0 \frac{z}{L} - \frac{1}{B_0\rho^2}\sum_n b_n I_n(n\rho)\sin(n\psi), \tag{2.98}$$

where $\psi = (\omega_0 - 1)\alpha z$ and $\alpha = 2\pi/L$. In (2.97) the prime denotes the derivative with respect to the argument. Using these approximate equations for the lines of force, we now determine the basic characteristics of the surface quantities of a straight helical magnetic field.

(i) The volume of the magnetic surface within one pitch length is

$$V = L\int_0^{2\pi} d\psi \,\frac{1}{2}\,(r(\psi))^2 = \frac{L}{2}\int_0^{2\pi} d\psi\left(r_0 + \frac{1}{\alpha B_0}\sum_n b_n I_n'(n\alpha r_0)\cos(n\psi)\right)^2$$
$$= \pi L\left(r_0^2 + \frac{1}{2\alpha^2 B_0^2}\sum_n b_n^2 I_n'^2\right). \tag{2.99}$$

(ii) The longitudinal magnetic flux inside the magnetic surface is

$$\Phi = \int_0^{2\pi} d\psi \int_0^{r(\psi)} B_z r dr = \int_0^{2\pi} d\psi \int_0^{r(\psi)} r dr\left(B_0 - \sum_n n b_n I_n(n\alpha r)\cos(n\psi)\right)$$
$$= B_0\,\frac{V}{L} - \int_0^{2\pi} d\psi \sum_n Y_n(r(\psi)) n b_n \cos(n\psi), \tag{2.100}$$

where

$$Y_n(r) = \int_0^r I_n(n\alpha r) r dr. \tag{2.101}$$

Substituting (2.97) into (2.101) and expanding r in the neighborhood of r_0 yields

$$Y_n(r) = Y_n(r_0) + r_0 I_n(n\alpha r_0)\sum_m \frac{b_m}{\alpha B_0}\, I_m'(n\alpha r_0)\cos(m\psi), \tag{2.102}$$

and consequently (2.100) becomes

$$\Phi = B_0\,\frac{V}{L} - \pi\sum_n \frac{n b_n^2}{\alpha B_0}\, r_0 I_n(n\alpha r_0) I_n'(n\alpha r_0). \tag{2.103}$$

(iii) If $n > 1$, the magnetic axis is straight and the azimuthal magnetic flux is given by

$$\chi = \int_0^L dz \int_0^{r(z)} B_\theta dr\Big|_{\theta=\text{constant}} \tag{2.104}$$
$$= \int_0^L dz \int_0^{r(z)} dr \sum_n n\,\frac{b_n}{\alpha r}\, I_n(n\alpha r)\cos(n\psi)$$
$$= \frac{L}{2}\sum_n \frac{n b_n^2}{\alpha^2 B_0 r_0}\, I_n(n\alpha r_0) I_n'(n\alpha r_0).$$

(iv) Here we find the specific volume by

$$U = \frac{dV}{d\Phi} = \frac{L}{B_0\left(1 - \sum_n \frac{nb_n^2}{\alpha B_0^2} \frac{d(r_0 I_n I_n')}{2r_0 dr_0}\right)}. \tag{2.105}$$

It is evident that the specific volume of the magnetic surface increases continuously with increasing distance from the magnetic axis.

By differentiating χ with respect to Φ in agreement with $\iota = d\chi/d\Phi$ given by (2.4), an expression for ω_0 coincides with (2.95), which was obtained with a different method.

We now discuss the behavior of the lines of force near a ridge of the separatrix. First, it is shown that the ridge of the separatrix is a line of force, i.e., the ridge of the separatrix given by

$$\begin{cases} r_c = \text{constant} \\ \theta - \alpha z = 0, \end{cases} \tag{2.106}$$

satisfies the equation for the lines of force

$$\begin{cases} \dfrac{dr}{dz} = \dfrac{B_r}{B_z}, \\ \dfrac{rd\theta}{dz} = \dfrac{B_\theta}{B_z}. \end{cases} \tag{2.107}$$

It follows from (2.106) and (2.107) that

$$\begin{cases} B_r(r_c, 0) = 0 \\ \alpha r_c = \dfrac{B_\theta(r_c, 0)}{B_z(r_c, 0)}. \end{cases} \tag{2.108}$$

It can be shown that these relations are equivalent to (2.87) by using (2.76) and (2.77).

From (2.107), we have

$$\frac{rd\theta}{dr} = \frac{B_\theta}{B_r}. \tag{2.109}$$

Substituting B_r and B_θ with keeping $\ell = n$ component only into this equation, we have

$$\frac{rd\theta}{dr} = \frac{I_n(n\alpha r)\cos n(\theta - \alpha z)}{\alpha r I_n'(n\alpha r)\sin n(\theta - \alpha z)}. \tag{2.110}$$

In order to examine the behavior of the lines of force near the ridge given by r_c and $\zeta \equiv \theta - \alpha z = 0$, we expand (2.110) in powers of $x = \alpha(r - r_c)$ and ζ, and retain the first nonvanishing terms only,

$$\frac{d\theta}{dx} = \frac{I_n(n\alpha r_c)}{(\alpha r_c)^2 I_n'(n\alpha r_c)n\zeta} \equiv \frac{F(\rho_c)}{\zeta}, \tag{2.111}$$

where $\rho_c = \alpha r_c$. On the other hand, expanding the equation for the magnetic surface in powers of x and ζ, and taking account of (2.87), yields

$$\Psi(r_c, 0) + \tfrac{1}{2}\frac{\partial^2\Psi(r_c, 0)}{\partial\zeta^2}\zeta^2 + \tfrac{1}{2}\frac{\partial^2\Psi(r_c, 0)}{\partial x^2}x^2 = \text{constant}. \tag{2.112}$$

From (2.80) we obtain

$$\tfrac{1}{2}\frac{\partial^2\Psi(r_c, 0)}{\partial\zeta^2} = \tfrac{1}{2}\frac{B_0}{2\alpha}\left(\frac{2b}{B_0}\rho_c I_n'(n\rho_c)n^2\right) \equiv \frac{1}{a^2} \tag{2.113}$$

and

$$\tfrac{1}{2}\frac{\partial^2\Psi(r_c, 0)}{\partial x^2} = \tfrac{1}{2}\frac{B_0}{2\alpha}\left(2 - \frac{2b}{B_0}\frac{\partial^2}{\partial\rho_c^2}\rho_c I_n'(n\rho_c)\right) \equiv -\frac{1}{b^2}. \tag{2.114}$$

We note that the second derivative of Ψ with respect to x is negative at the ridge of the separatrix by replacing b/B_0 with (2.91).

From (2.112) with (2.113) and (2.114), we have

$$\frac{x^2}{b^2} - \frac{\zeta^2}{a^2} = \text{constant}. \tag{2.115}$$

If the line of force lies on the separatrix, the RHS of (2.115) vanishes, and

$$\zeta = \pm\frac{a}{b}x. \tag{2.116}$$

Substituting (2.116) into (2.111), we have

$$x \propto \exp\left(\pm\frac{a\theta}{bF}\right) = \exp\left(\pm\frac{a}{bF}\alpha z\right). \tag{2.117}$$

It is then obvious that the quantity $x = \rho - \rho_c$ approaches zero or infinity as $z \to \infty$. Thus, if a line of force approaches the ridge as $z \to \infty$, it moves away from the ridge when $z \to -\infty$, as shown in Fig. 2.10.

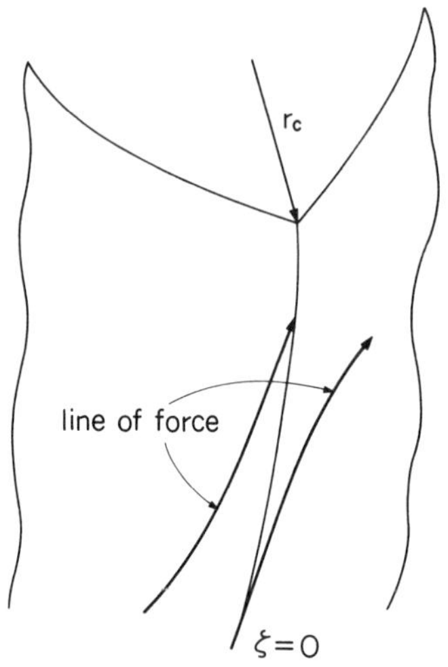

Fig. 2.10 The behavior of a line of force near the ridge given by $r = r_c$ and $\zeta = 0$.

2.6 THE MAGNETIC ISLAND AND DESTRUCTION OF THE MAGNETIC SURFACE

Let us consider an irrotational or vacuum magnetic field in the cylindrical coordinates (r, θ, z). It is a straight helical magnetic field with $\ell = n$ that is a combination of a homogeneous field B_0 directed along the z-axis and a magnetic field produced by conductors in the form of n helical windings with constant pitch $L = 2\pi/\alpha$. Such a magnetic field possesses helical symmetry and its scalar potential has the form (2.79) when higher harmonics are neglected in the expression (2.71). In (2.79), b is the amplitude of the field produced by the helical windings. From the equations of the magnetic field line (2.107) for the magnetic field given by $\mathbf{B} = \nabla\varphi$, we obtain

$$\begin{cases} B_z \dfrac{dr}{dz} = bnI_n'(n\alpha r)\sin n\zeta \\ B_z \dfrac{d\zeta}{dz} = -\alpha + bn\left(\alpha + \dfrac{1}{\alpha r^2}\right)I_n(n\alpha r)\cos n\zeta, \end{cases} \tag{2.118}$$

where $\zeta \equiv \theta - \alpha z$ and $B_z = \partial\varphi/\partial z$ is expressed as a function of r and ζ. Equations (2.118) have the integral

$$H = \tfrac{1}{2}\,\alpha^2 r^2 - b\alpha r I_n'(n\alpha r)\cos n\zeta, \tag{2.119}$$

since

$$\frac{dH}{dz} = \frac{\partial H}{\partial r}\frac{dr}{dz} + \frac{\partial H}{\partial \zeta}\frac{d\zeta}{dz} = 0.$$

Therefore, we may introduce canonical variables and interpret the equations of magnetic field line (2.118) in the form of Hamiltonian equations. First, we replace z by a time-like variable t, using

$$\frac{dz}{dt} = B_z = B_0 + \frac{bn}{\alpha} I_n(n\alpha r)\cos n\zeta. \tag{2.120}$$

By introducing a new variable $\rho = \frac{1}{2}\alpha r^2$, it is easily seen that (2.118) are equivalent to the following expressions:

$$\begin{cases} \dfrac{d\rho}{dt} = \dfrac{\partial H}{\partial \zeta} \\ \dfrac{d\zeta}{dt} = -\dfrac{\partial H}{\partial \rho}. \end{cases} \tag{3.121}$$

Thus the integral of "motion" (2.119) can be considered as the "Hamiltonian" for the field lines, while the variables (ρ, ζ) are canonical.

Let us now change the variables (ρ, ζ) to the action-angle variables (J, Θ). Using standard procedures,

$$\begin{cases} J = J(H) = \dfrac{1}{2\pi}\oint \rho d\zeta \\ \alpha\omega = \alpha\omega(J) = \dfrac{dH}{dJ}. \end{cases} \tag{2.122}$$

The equations of motion are written in these variables,

$$\begin{cases} \dfrac{dJ}{dt} = 0 \\ \dfrac{d\Theta}{dt} = \alpha\omega, \end{cases} \tag{2.123}$$

and the relationship between the old and the new phase variables is given by

$$\begin{cases} \Theta(J, \zeta) = \dfrac{\partial S(J, \zeta)}{\partial J} \\ S(J, \zeta) = \displaystyle\int \rho d\zeta, \end{cases} \tag{2.124}$$

where the second integral is an indefinite integral. Let us consider the relationship between the "frequency" ω and the angle of field line rotation $\alpha\omega_0 = \delta\theta/\delta z$ (see (2.94)). According to (2.123), we obtain

$$\alpha\omega = \frac{d\Theta}{dt} = \frac{\delta\Theta}{\delta\zeta}\frac{\delta\zeta}{\delta z}\frac{dz}{dt}. \tag{2.125}$$

Taking into account that $\zeta \equiv \theta - \alpha z$ and that $\delta\Theta = 2\pi$ when $\delta\zeta = 2\pi$, we obtain

$$\alpha\omega = \alpha(\omega_0 - 1)\,\frac{dz}{dt}. \tag{2.126}$$

Near the singular points of the separatrix, $\omega_0 \to 1$ and consequently $\omega \to 0$. In (2.125), by noting

$$\alpha\omega = \frac{\delta\Theta}{\delta\zeta}\,\frac{d\zeta}{dt}, \tag{2.127}$$

$d\zeta/dt \to 0$ near the singular points, since $\omega \to 0$, which is consistent with (2.106) for the ridge.

The equations (2.123) describe the behavior with time of the action J and the phase Θ of a conservative system with the Hamiltonian H. Let us now see how these equations change in the case of a nonstationary perturbation which is assumed to be small. We write the magnetic potential as

$$\varphi = \varphi_0 + \epsilon\varphi_1, \tag{2.128}$$

where the unperturbed potential φ_0 is the same as (2.79) and ϵ is the small parameter. Then the equations of magnetic field line "motion" become

$$\begin{cases} \dfrac{dr}{dt} = \dfrac{\partial\varphi_0}{\partial r} + \epsilon\,\dfrac{\partial\varphi_1}{\partial r} = \dfrac{dr_0}{dt} + \epsilon\,\dfrac{dr_1}{dt} \\ \dfrac{d\zeta}{dt} = -\alpha B_z^{(0)} - \epsilon\alpha\,\dfrac{\partial\varphi_1}{\partial z} + \dfrac{1}{r^2}\left(\dfrac{\partial\varphi_0}{\partial\theta} + \epsilon\,\dfrac{\partial\varphi_1}{\partial\theta}\right) = \dfrac{d\zeta_0}{dt} + \epsilon\,\dfrac{d\zeta_1}{dt}. \end{cases} \tag{2.129}$$

Let us now determine the action-angle variables (J, Θ) with respect to the unperturbed part of the equations of "motion" (2.129). For the change in (J, Θ) with time, (2.123) become

$$\begin{cases} \dfrac{dJ}{dt} = \dfrac{\partial J}{dr}\,\dfrac{dr}{dt} + \dfrac{\partial J}{d\zeta}\,\dfrac{d\zeta}{dt} = \dfrac{dJ}{dH}\left(\dfrac{\partial H}{\partial r_0}\,\dfrac{dr}{dt} + \dfrac{\partial H}{\partial\zeta_0}\,\dfrac{d\zeta}{dt}\right) \\ \dfrac{d\Theta}{dt} = \dfrac{d\Theta}{d\zeta}\,\dfrac{d\zeta}{dt} = \dfrac{d\Theta/dt}{d\zeta/dt}\,\dfrac{d\zeta}{dt}. \end{cases} \tag{2.130}$$

By noting the Hamiltonian equation,

$$\begin{cases} \dfrac{dr_0}{dt} = \dfrac{1}{r_0}\,\dfrac{\partial H}{\partial\zeta_0} \\ r_0\,\dfrac{d\zeta_0}{dt} = -\dfrac{\partial H}{\partial r_0}, \end{cases} \tag{2.131}$$

(2.130) are written in the form

$$\begin{cases} \dfrac{dJ}{dt} = \epsilon \dfrac{r_0}{\alpha\omega}\left(-\dfrac{d\zeta_0}{dt}\dfrac{dr_1}{dt} + \dfrac{dr_0}{dt}\dfrac{d\zeta_1}{dt}\right) \\ \dfrac{d\Theta}{dt} = \alpha\omega\left(1 + \epsilon\dfrac{d\zeta_1/dt}{d\zeta_0/dt}\right). \end{cases} \tag{2.132}$$

Substituting dr_1/dt and $d\zeta_1/dt$ obtained from (2.129) into (2.132) yields

$$\begin{cases} \dfrac{dJ}{dt} = \epsilon \dfrac{r_0}{\alpha\omega}\left[-\dfrac{d\zeta_0}{dt}\dfrac{\partial\varphi_1}{\partial r_0} + \dfrac{dr_0}{dt}\left(-\alpha\dfrac{\partial\varphi_1}{\partial z} + \dfrac{1}{r_0^2}\dfrac{\partial\varphi_1}{\partial\theta}\right)\right] \\ \dfrac{d\Theta}{dt} = \alpha\omega + \dfrac{\epsilon\alpha\omega}{(d\zeta_0/dt)}\left(-\alpha\dfrac{\partial\varphi_1}{\partial z} + \dfrac{1}{r_0^2}\dfrac{\partial\varphi_1}{\partial\theta}\right). \end{cases} \tag{2.133}$$

Here we consider the structure of magnetic surfaces in toroidal systems when the principal perturbation is a correction associated with bending of the straight helical field into a torus. Then the field line equations contain non-stationary perturbations depending on t due to the toroidicity in the toroidal helical system, and the method discussed above can be applied to them. We will show that there is always magnetic surface destruction due to the stochastic behavior of field lines near the separatrix.

We will assume

$$\begin{cases} \dfrac{b}{B_0} \ll 1 \\ \dfrac{r}{R} \ll 1 \end{cases} \tag{2.134}$$

for simplicity, where r is the minor radius of the torus and R is the major radius of the torus. When the scalar potential of a straight helical magnetic field is bent into the toroidal geometry, it is written as

$$\varphi = \varphi_0\left[1 + \sum_N (-1)^N \left(\frac{r}{R}\right)^N \cos^N(\zeta + \alpha z)\right]. \tag{2.135}$$

The index $N \geq 1$ corresponds to the order of the toroidal corrections. The potential perturbation in (2.128) is thus

$$\epsilon\varphi_1 = \sum_N (-1)^N \left(\frac{r}{R}\right)^N \frac{b}{\alpha} I_n(n\alpha r)\sin(n\zeta)\cos^N(\zeta + \alpha z). \tag{2.136}$$

Although an actual perturbation may be more complex, this approximation is adequate to describe the fundamental property for destruction of the magnetic surface.

We consider the effect of the perturbation (2.136) on the field lines near the magnetic axis or for small r. Substituting (2.136) into (2.133) and retaining leading terms in r, we obtain

$$\begin{cases} \dfrac{dJ}{dt} \simeq -\epsilon r_0 \dfrac{\partial \varphi_1}{\partial r_0} \\ \dfrac{d\Theta}{dt} \simeq \alpha\omega + \dfrac{\epsilon}{r_0^2} \dfrac{\partial \varphi_1}{\partial \theta}, \end{cases} \tag{2.137}$$

with $dr_0/dt \simeq 0$, $d\zeta_0/dt \simeq \alpha\omega$ and $\alpha \ll 1$. We change the variables on RHS of (2.137) to the (J, Θ) variables. With the help of the relation $J(H) = r_0^2/2$,

$$\begin{cases} \dfrac{dJ}{dt} = -\dfrac{1}{2} \epsilon J \dfrac{\partial \varphi_1}{\partial J} \\ \dfrac{d\Theta}{dt} = \alpha\omega + \dfrac{1}{2} \epsilon \dfrac{1}{J} \dfrac{\partial \varphi_1}{\partial \Theta} \end{cases} \tag{2.138}$$

are obtained from (2.137). We expand $\varphi_1(J, \Theta, t)$ into a Fourier series,

$$\varphi_1(J, \Theta, t) = \sum_{m,\ell} [\varphi_{1m\ell}(J) e^{i(m\Theta - \ell t)} + \varphi_{1m\ell}^*(J) e^{-i(m\theta - \ell t)}] \tag{2.139}$$

and study the behavior of the field lines in the vicinity of one of the resonances. In this case, only one resonance harmonic with $m = m_0$ and $\ell = \ell_0$ can be left in (2.139), and

$$\begin{cases} \dfrac{dJ}{dt} = -\epsilon J \left(\dfrac{\partial}{\partial J} \varphi_{1m_0\ell_0} \right) \cos(m_0\Theta - \ell_0 t + \Theta_0) \\ \dfrac{d\Theta}{dt} = \alpha\omega(J) + m_0 \epsilon J^{-1} \varphi_{1m_0\ell_0} \sin(m_0\Theta - \ell_0 t + \Theta_0), \end{cases} \tag{2.140}$$

are obtained, where Θ_0 is the phase of $\varphi_{1m_0\ell_0}$. With an accuracy up to $O(\epsilon)$, (2.140) have the integral of motion

$$I(J, \Theta, t) \equiv \epsilon \varphi_{1m_0\ell_0} \sin(m_0\Theta - \ell_0 t + \Theta_0) + \int \frac{\varphi_{1m_0\ell_0}}{J(\partial \varphi_{1m_0\ell_0}/\partial J)} (m_0 \alpha\omega - \ell_0) dJ = \text{constant}. \tag{2.141}$$

It is noted that $dI/dt = (\partial I/\partial J)(dJ/dt) + (\partial I/\partial \Theta)(d\Theta/dt) + \partial I/\partial t = 0$. In the $t =$ constant plane, the magnetic field line trajectories near the resonance with $J = J_r$, produce magnetic island structures, as shown in Fig. 2.11. Near the resonant surface, where $m_0 \alpha\omega(J_r) - \ell_0 = 0$ is satisfied, (2.141) has the following form:

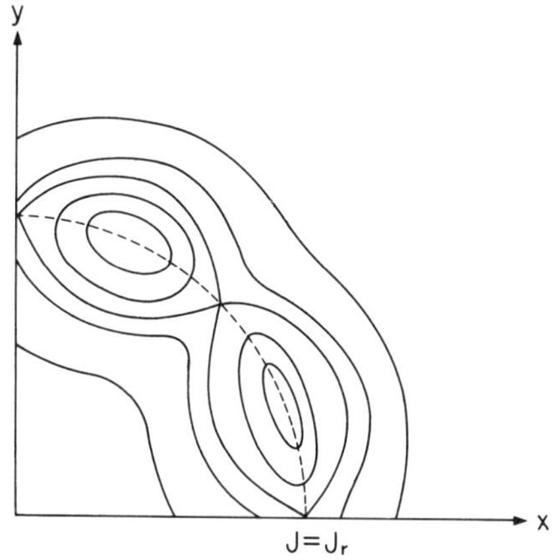

Fig. 2.11 Magnetic islands for a single resonance. $J = J_r$ denotes the resonant surface.

$$\epsilon\varphi_{1m_0\ell_0}(J_r)\sin(m_0\Theta - \ell_0 t + \Theta_0) + \frac{\varphi_{1m_0\ell_0}(J_r)}{J_r(\partial\varphi_{1\ell_0 m_0}/\partial J_r)} m_0\alpha \frac{d\omega}{dJ_r}\frac{1}{2}(\delta J)^2 = 0, \quad (2.142)$$

where the suffix r denotes the resonance. This shows that the island width $|\delta J|$ is proportional to $(\epsilon\varphi_{1m_0\ell_0})^{1/2}$.

Let us come back to the resonance for the perturbation (2.136). The variables (ζ, z) are transformed in (Θ, t) by

$$\begin{cases} \zeta = \Theta + g_1(J, n\Theta) \\ z = t + g_2(J, n\Theta), \end{cases} \quad (2.143)$$

where g_1 and g_2 are periodic functions with respect to $n\Theta$, which is related to the winding law of a helical coil. By expanding $\sin(sn\zeta)\cos^N(\zeta + \alpha z)$ in the expression (2.136) (the case of $s = 1$ is shown in (2.136)) into a Fourier series with respect to Θ and t, the resonance condition for which the perturbation becomes independent of t is obtained from the term with $((sn + N)\Theta - N\alpha t)$,

$$\omega(J_r) = \frac{N}{sn + N}. \quad (2.144)$$

It is noted that s denotes the harmonics of the $\ell = n$ helical magnetic field. Here $\Theta = \alpha\omega t + \Theta_0$ is used and s denotes the index corresponding to the harmonics of the helical magnetic field. This means that, when $s = 1$ and $n = 2$, resonances appear for $\frac{1}{3}, \frac{2}{4}, \frac{3}{5}, \frac{4}{6}, \ldots$; or when $s = 1$ and $n = 3$, resonances appear for $\frac{1}{4}, \frac{2}{5}, \frac{3}{6}, \frac{4}{7}, \ldots$. An example of resonance concentration in the vicinity of the separatrix is shown in Fig. 2.12. It is obvious that stochastic destruction of the magnetic surfaces may be expected near the separatrix, where $\omega \to 0$ or $s \gg 1$.

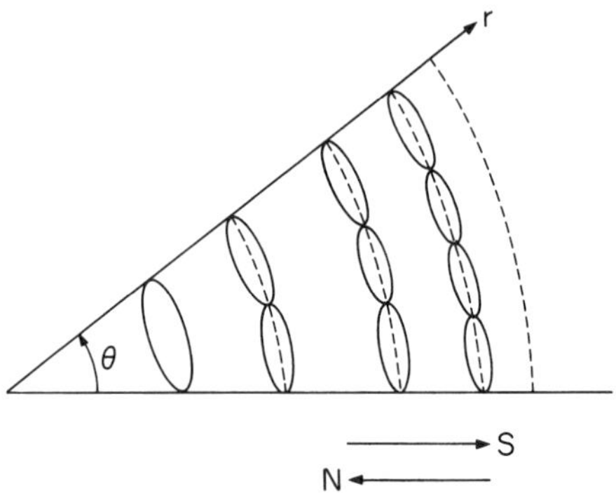

Fig. 2.12 Magnetic islands on resonant surfaces concentrated in the vicinity of the separatrix. When $N(s)$ is large, the magnetic island exists away from (close to) the separatrix.

Here $s \gg 1$ corresponds to the higher harmonics of the $\ell = n$ helical magnetic field. In this case $\omega(J_r^{(s)}) \sim 1/s$, and the nearest resonances satisfy

$$|\omega(J_r^{(s+1)}) - \omega(J_r^{(s)})| \simeq \frac{1}{s^2} \ll \omega(J_r^{(s)}). \tag{2.145}$$

The magnetic surfaces in $J_r^{(s)} < J < J_r^{(s+1)}$ may be destroyed when the sum of the magnetic island widths at $J_r^{(s)}$ and $J_r^{(s+1)}$ exceeds the distance between the nearest resonances. On the contrary, for the opposite case, part of the magnetic surfaces in $J_r^{(s)} < J < J_r^{(s+1)}$ may survive (see the condition (2.152)).

It is instructive to consider the magnetic island in a simpler situation. In the cylindrical tokamak model, the equation of the magnetic field line is given by (2.5), where the cylindrically symmetric magnetic field vector is assumed to be $\mathbf{B}_0 = (0, B_\theta^0(r), B_0)$ in cylindrical coordinates (r, θ, z), where B_0 is a constant. Here we consider the helical magnetic perturbation with the mode numbers (m, n) in the azimuthal and longitudinal directions, respectively, which satisfy $q = rB_0/(RB_\theta^0) = m/n$ at $r = r_s$. A perturbation with this property is called a resonant perturbation.

Let us consider the magnetic structure in the vicinity of r_s, when the resonant perturbation $\mathbf{B}_1$ is finite. In the cylindrical geometry, we consider the following magnetic field:

$$\mathbf{B}^*(r, \theta, z) = \mathbf{B}_0(r) + \mathbf{B}_1(r, \theta, z) - \frac{r}{r_s} B_\theta^0(r_s)\mathbf{e}_\theta - B_0\mathbf{e}_z, \tag{2.146}$$

where the third term corresponds to the poloidal magnetic field produced by a uniform longitudinal plasma current. First, the narrow region including the surface of $r = r_s$ in Fig. 2.13 is stretched in the form of a slab geometry by cutting the cylinder at the constant azimuthal angle shown by the lines *ab* or *cd*. If the resonant perturbation is neglected, the azimuthal magnetic field B_θ^{0*} is described in Fig. 2.14(a). It changes direction at $r = r_s$. Next, we assume that

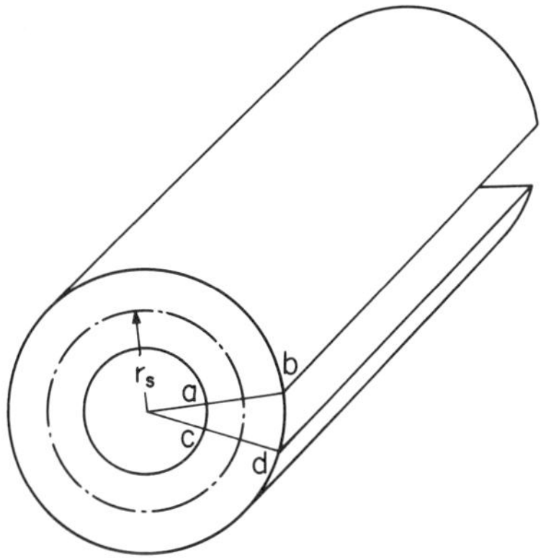

Fig. 2.13 An annular plasma obtained from a cylindrical tokamak model with the resonant surface $r = r_s$.

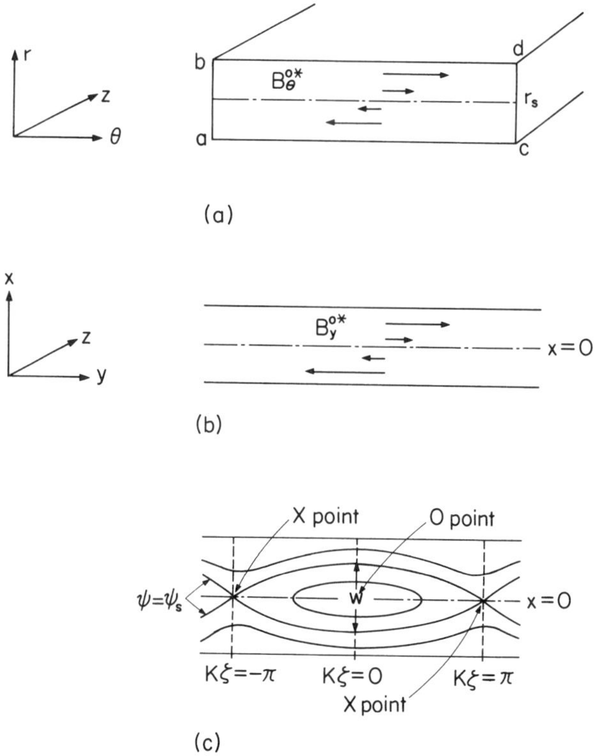

Fig. 2.14 (a) A slab model obtained from the cylindrical tokamak plasma. (b) A slab model infinitely extending in the y-direction. (c) A magnetic island in the slab model. Here O- and X-points are shown.

the θ-direction corresponds to the y-direction by imposing the periodic condition. Also, we set $x = r - r_s$ in Cartesian coordinates. In the case of Fig. 2.14(b) the magnetic field $B_y^{0*}(x)$ can be expanded as

$$B_y^{0*}(x) = \left.\frac{dB_y^{0*}}{dx}\right|_{x=0} x + \ldots . \tag{2.147}$$

Here we assume that $dB_y^{0*}/dx|_{x=0} < 0$. The resonant perturbation is given by

$$B_x^1 = b \sin\left(ky - \frac{n}{R} z\right) \tag{2.148}$$

and the other components B_y^1 and B_z^1 do not play a role here. We note that $B_x = -\partial\Psi/\partial\zeta$, $B_y = \partial\Psi/\partial x$, and $B_z = B_0$ satisfy $\nabla \cdot \mathbf{B} = 0$, where $\zeta = y - nz/(kR)$. The flux function Ψ, including the pertubation (2.148), is given by

$$\Psi = \Psi_0 + \frac{1}{2}\left.\frac{dB_y^{0*}}{dx}\right|_{x=0} x^2 + \ldots\ldots + \frac{b}{k}\cos k\zeta, \tag{2.149}$$

where Ψ_0 is a constant. For $dB_y^{0*}/dx|_{x=0} < 0$, $\Psi =$ constant contours in the (x, y) plane are shown in Fig. 2.14(c). The $\Psi =$ constant surface has saddle points at $x = 0$ and $k\zeta = \pm\pi$, and $\Psi_s = \Psi_0 - b/k$ on them. The surface $\Psi = \Psi_s$ corresponds to the separatrix, and the inside region surrounded by $\Psi = \Psi_s$ is the magnetic island, which is the same as in Fig. 2.11. By introducing w as the maximum width of the magnetic island, the separatrix is given by

$$\Psi_s = \Psi_0 + \frac{1}{2}\left.\frac{dB_y^{0*}}{dx}\right|_{x=0}\left(\frac{w}{2}\right)^2 + \ldots\ldots + \frac{b}{k} \tag{2.150}$$

for $x = \pm w/2$ and $k\zeta = 0$. By noting $\Psi_s - \Psi_0 = -b/k$, the magnetic island width w is given by

$$w = 4\left(b/k\left|\frac{dB_y^{0*}}{dx}\right|\right)^{1/2}. \tag{2.151}$$

In the cylindrical tokamak model, the plasma current in the z-direction produces $B_\theta^0(r)$ and the magnetic field is not irrotational. Here we consider a mechanism to produce a resonant helical perturbation. These are two possibilities: one is the current-driven tearing mode destabilized in a resistive tokamak plasma, which will be discussed in chapter 5. The other is a static externally imposed error field. In the former case, the magnetic islands are generated by reconnection of magnetic field lines driven by the tearing mode. In the latter case, the situation is very similar to the magnetic islands in stellarators and heliotrons.

Even for the cylindrical tokamak model, stochastic behavior of the magnetic field lines appears when many Fourier components are resonant at different radial positions; $q(r_s^1) = m_1/n_1$, $q(r_s^2) = m_2/n_2, \ldots, q(r_s^i) = m_i/n_i$, etc. When the sum of the island widths w^i at r_s^i and w^{i+1} at r_s^{i+1} satisfies

$$\frac{1}{2}\,(w^i + w^{i+1}) > |r_s^i - r_s^{i+1}|, \tag{2.152}$$

the magnetic surfaces in $r_s^i < r < r_s^{i+1}$ disappear. When this situation is created for several low-order resonant surfaces by the large-amplitude multi-helicity tearing modes, the magnetic surfaces are almost completely destroyed and the plasma confinement is suddenly lost. This is considered to be a mechanism of *disruption* and is occasionally observed in tokamaks (see chapter 5).

2.7 THE MAGNETIC SURFACE BY LINE TRACING CALCULATION

The numerical calculation to follow magnetic field lines for the given magnetic field vector is the foundation of the study of magnetic surfaces of stellarator and heliotron configurations. Here we describe a line tracing calculation for heliotron configurations. In heliotrons, the $\ell = 2$ helical winding, as shown in Fig. 2.15, produces a helical magnetic field $\mathbf{B}_h$. Since the coil current in the helical winding flows in the same direction, the toroidal component of the magnetic field $\mathbf{B}_t$ is also produced. The magnitude of the toroidal magnetic field due to the helical winding at the center of the vacuum chamber is denoted by $B_{h\phi 0}$. The magnetic field $\mathbf{B}_h(\mathbf{r})$ at the position $\mathbf{r}(r, \theta, \phi)$ produced by the N helical filamentary currents is given by the Biot–Savart law:

$$\mathbf{B}_h(\mathbf{r}) = \sum_{j=1}^{N} \frac{\mu_0 I_j}{4\pi} \int \frac{d\mathbf{r}_j' \times (\mathbf{r} - \mathbf{r}_j')}{|\mathbf{r} - \mathbf{r}_j'|^3}, \tag{2.153}$$

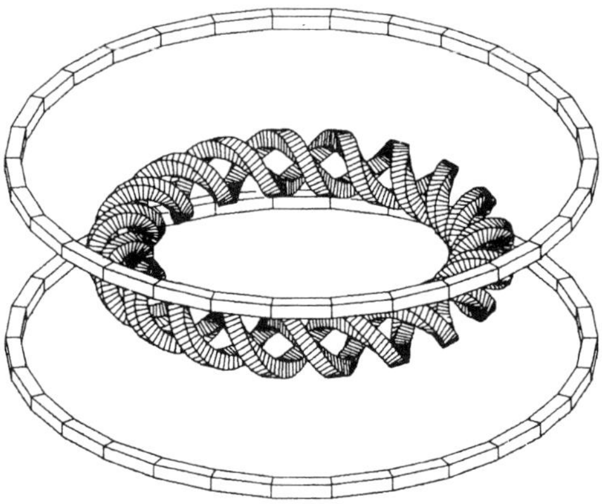

Fig. 2.15 The helical winding and vertical field coils of Heliotron E.

where $\mathbf{r}_j'$ is the position of the jth helical current, I_j is the current density, N is the total number of the current filaments, and μ_0 is the vacuum magnetic permeability. In the numerical model for calculating $\mathbf{B}_h(\mathbf{r})$ with the expression (2.153), the volume current of the finite size coil in experimental devices is approximated by several filamentary currents. The simplest, and somewhat crude, model is that one filamentary current represents the helical winding with the volume current. In order to obtain a more accurate result, 27 filamentary currents are used to simulate the finite-size coils shown in Fig. 2.16 for the Heliotron E.

In a line tracing calculation each helical filamentary current is divided into L equal segments in the ϕ direction and the integral in (2.153) is replaced with the summation

$$\mathbf{B}_h(r) = \sum_{j=1}^{N} \frac{\mu_0 I_j}{4\pi} \sum_{i=1}^{L} \frac{\dfrac{d\mathbf{r}_{ji}}{d\phi'} \times (\mathbf{r} - \mathbf{r}_{ji}')}{|\mathbf{r} - \mathbf{r}_{ji}'|^3} \Delta\phi', \tag{2.154}$$

where $\Delta\phi' = 2\pi/L$ and $\mathbf{r}_{ji}'$ refers the position vector of the ith element of the jth filamentary current. L is typically several hundreds and $\Delta\phi'$ is about one degree.

In order to obtain closed magnetic surfaces in heliotrons, the vertical magnetic field component is required to reduce the vertical magnetic field that appears when the helical symmetry is lost by bending a straight column into a torus. The vertical field coils are already shown in Fig. 2.15. It is noted

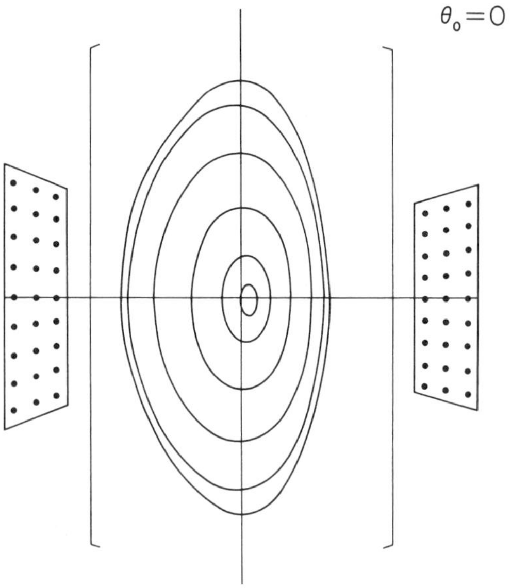

Fig. 2.16 A cross-section of the vacuum magnetic surface in the Heliotron E with 27 filamentary helical currents.

that the cross-section of the chamber with a racetrack shape in Fig. 2.16 rotates helically. We introduce a nondimensional parameter

$$\beta^* \equiv B_{v0}/B_{h\varphi 0}, \tag{2.155}$$

where B_{v0} is the magnetic field by the vertical field coil at the center of the toroidal chamber. The toroidal magnetic field can be added in the Heliotron E with the additional toroidal coils. We also introduce a nondimensional parameter

$$\alpha^* = B_{t0}/B_{h\varphi 0}, \tag{2.156}$$

where B_{t0} is the magnetic field at the center of the toroidal chamber produced by the toroidal coils. By changing α^* and β^* independently, the characteristics of the vacuum magnetic surfaces can be controlled. The magnetic field of the heliotron configuration is composed of the magnetic fields produced by three different coils described above,

$$\mathbf{B} = \mathbf{B}_h + \mathbf{B}_v + \mathbf{B}_t, \tag{2.157}$$

and the equation of the magnetic line of force is written as

$$\frac{d\mathbf{r}}{d\ell} = \frac{\mathbf{B}}{|\mathbf{B}|}, \tag{2.158}$$

where ℓ denotes the length along the mangetic line of force. In order to study the magnetic surfaces by solving (2.158) numerically, we must be careful about the accuracy of the numerical scheme. In the case of the Runge–Kutta method, the fourth-order or sixth-order accuracy is considerable.

The Runge–Kutta method for the vector differential equation

$$d\mathbf{x}/dt = \mathbf{f}(\mathbf{x}, t) \tag{2.159}$$

is obtained by the approximate expression of integration from $\mathbf{x}_0$ at time $t = t_0$ to $\mathbf{x}$ at time $t = t_0 + \delta t$,

$$\mathbf{x} = \mathbf{x}_0 + \mathbf{f}\left(\mathbf{x}_0 + \frac{\delta t}{2}\mathbf{f}(\mathbf{x}_0, t_0), t_0 + \frac{\delta t}{2}\right)\delta t, \tag{2.160}$$

which is the second-order scheme, and

$$\begin{cases} \mathbf{x} = \mathbf{x}_0 + (\mathbf{k}_1 + 2\mathbf{k}_2 + 2\mathbf{k}_3 + \mathbf{k}_4)/6 \\ \mathbf{k}_1 = \mathbf{f}(\mathbf{x}_0, t_0)\delta t \\ \mathbf{k}_2 = \mathbf{f}\left(\mathbf{x}_0 + \dfrac{\mathbf{k}_1}{2}, t_0 + \dfrac{\delta t}{2}\right)\delta t \\ \mathbf{k}_3 = \mathbf{f}\left(\mathbf{x}_0 + \dfrac{\mathbf{k}_2}{2}, t_0 + \dfrac{\delta t}{2}\right)\delta t \\ \mathbf{k}_4 = \mathbf{f}(\mathbf{x}_0 + \mathbf{k}_3, t_0 + \delta t)\delta t, \end{cases} \tag{2.161}$$

which is the fourth-order scheme.

In the stellarator and heliotron configuration that is periodic in the toroidal direction or ϕ-direction, the cross-points of the magnetic line of force on the poloidal plane $\phi =$ constant are plotted by exploiting the periodicity in order to save computational time. This means that the poloidal planes at ϕ_0 and $\phi_0 + m(2\pi/M)$ for $m = 1, \cdots, M-1$ are identical, where ϕ_0 is an arbitrary toroidal angle. When the cross-points draw a smooth closed line with increasing revolutions around the torus, it is considered that the magnetic surface exists. It should be noted that this criterion for the existence of the magnetic surface is practical. According to the KAM theorem, the magnetic surface may exist when a deviation from helically symmetric straight stellarators is sufficiently small. This suggests that magnetic surfaces exist for large aspect ratio nonaxisymmetric toroidal systems such as Heliotron E with $R/\bar{a} \simeq 10$–11, where $\bar{a}$ is an average minor radius and R is a major radius. Usually, the average plasma radius is used, since the cross-section of the magnetic surface is not circular in the Heliotron E, as shown in Fig. 2.16. When the initial point to calculate the magnetic field line is moved outward, the nested magnetic surface disappears at a particular radius. For this situation, the last closed magnetic surface is defined at the radius just before the disappearance of the magnetic surface. Outside the last closed magnetic surface, magnetic islands usually exist due to the perturbation of toroidal bending. The last closed magnetic surface position is easily determined to an accuracy of the order of 1 mm with the line tracing code.

The magnetic configuration of heliotrons depends on the pitch parameter γ as well as on α^* and β^*. It is defined as

$$\gamma = \frac{a_c}{R} \frac{M}{\ell}, \tag{2.162}$$

where M is the pitch number of the helical magnetic field, ℓ is the pole number, and a_c is the minor radius of the toroidal chamber in which the helical windings are mounted. For example, $\ell = 2$ and $M = 19$ for Heliotron E at Kyoto University; $\ell = 2$ and $M = 12$ for ATF (Advanced Toroidal Facility) at Oak Ridge National Laboratory; $\ell = 2$ and $M = 10$ for the LHD (Large Helical Device); and $\ell = 2$ and $M = 8$ for the CHS (Compact Helical System) at the National Institute for Fusion Science.

It is noted that the region surrounded by the magnetic surface disappears for $\gamma \lesssim 1$ at $\alpha^* = 0$ or in the case of no additional toroidal field. This is understable by the fact that the smaller γ corresponds to the smaller M. When M is small, $B_{h\varphi 0}$ decreases, which makes it difficult to obtain the magnetic surface. The extreme case is $\alpha^* = -1$. In this case $B_{h\varphi 0}$ is canceled by B_{t0}, and the magnetic surface does not exist for any γ. The above-mentioned heliotron devices belong to $1.20 \lesssim \gamma \lesssim 1.40$. Usually, for the smaller γ case, there is a tendency for the rotational transform to increase and for the average plasma radius to decrease. On the contrary, for the larger γ case, there is the tendency for the rotational transform to decrease and for the average plasma radius to increase.

In heliotrons the geometric center in the poloidal cross-section of the helical winding rotates helically in the form described as $r = a =$ constant and $\theta = (M/\ell)\phi$. In the case of Heliotron E, the poloidal cross-section of the vacuum chamber has a ractrack shape and the poloidal cross-section of the helical conductor has the shape shown in Fig. 2.16. The 27 coils, insulated from each other, are divided into three layers, and there are nine coils in each layer. Usually, a current filament is assumed at the center of the poloidal cross-section of each coil for the line tracing calculation. Figure 2.17 shows the relationship between the magnetic surface shape and the helical winding position, where R_c denotes the major radius and a_c the minor radius of the helical winding. An average radius $\bar{r}$ is estimated as $\pi\bar{r}^2 = \{S(0°) + 2S(45°) + S(90°)\}$, where $S(0°)$, $S(45°)$, and $S(90°)$ are the areas of the cross-sections of the magnetic surface at $\theta_0 = 0°$, $45°$, and $90°$, respectively. The angle θ_0 is given in Fig. 2.17.

The shear parameter Θ is given by

$$\Theta = \frac{\bar{r}^2}{R}\frac{d\iota}{d\bar{r}}, \tag{2.163}$$

where ι is the rotational transform. Figure 2.18 shows the rotational transform and the shear of the standard Heliotron E configuration with $\alpha^* = 0$ and $\beta^* = 0.185$. The high shear and the large rotational transform above 1 at the last closed magnetic surface are the main characteristics of Heliotron E.

We note that the vertical magnetic field is axisymmetric in the toroidal direction. In addition to the vertical magnetic field, it is possible to impose the axisymmetric quadrupole field. Here we consider the scalar potential.

$$\varphi = \frac{\mu_0 I}{\pi}\left(\frac{r}{a}\right)^2 \sin 2\theta, \tag{2.164}$$

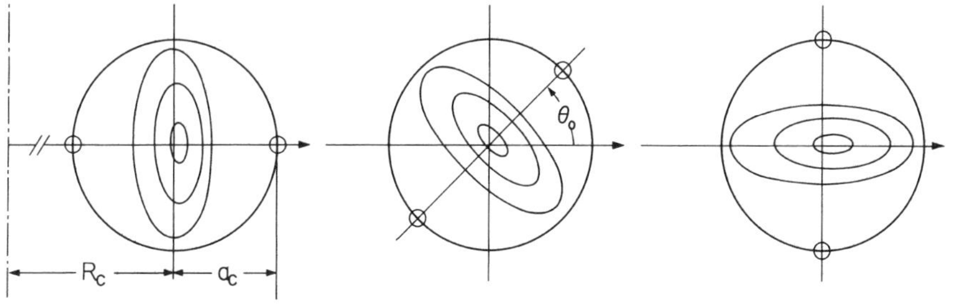

Fig. 2.17 Three different cross-sections of the magnetic surface are shown, where R_c denotes the major radius and a_c the minor radius, respectively, of the helical winding.

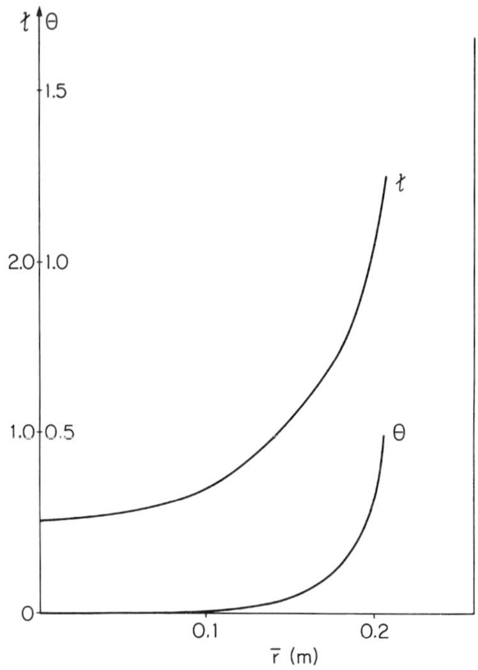

Fig. 2.18 The rotational transform and shear parameter of the Heliotron E shown in Fig. 2.16.

for $r \ll a$ in straight cylindrical coordinates (r, θ, z). This potential is produced by the straight coils with the positive current I located at $(r = a, \theta = 0)$ and $(r = a, \theta = \pi)$, and those with the negative current $-I$ at $(r = a, \theta = \pi/2)$ and $(r = a, \theta = 3\pi/2)$, as shown in Fig. 2.19. The quadrupole magnetic field $B_r = (2\mu_0 I/\pi a)(r/a) \sin 2\theta$ and $B_\theta = (2\mu_0 I/\pi a)(r/a) \cos 2\theta$ make an elliptic deformation for the circular magnetic surface through the change in the poloidal magnetic field (see Fig. 2.19). There are two types of elliptic deformation; one is vertically elongated and the other is horizontally elongated. By using the axisymmetric hexapole field, a triangular deformation of the magnetic surface is also possible. These additional axisymmetric field components can be used to optimize the shape of the magnetic surfaces in order to give attractive properties for high-temperature plasma confinement.

The other way in which to control the magnetic surfaces in heliotrons is to give a pitch modulation to the helical winding, which follows

$$\theta = (M/\ell)\phi + \sum_n \alpha_n \sin(nM\phi/\ell). \tag{2.165}$$

Usually only α_1 is considered, and $\alpha_1 \simeq \bar{a}/R_0$ has been chosen to restore the property of helical symmetry approximately even in the toroidal geometry. It can be shown that $\alpha_1 < 0$ is favorable to magnetic well generation. The recent devices designed to use pitch modulation are the CHS (Compact Helical

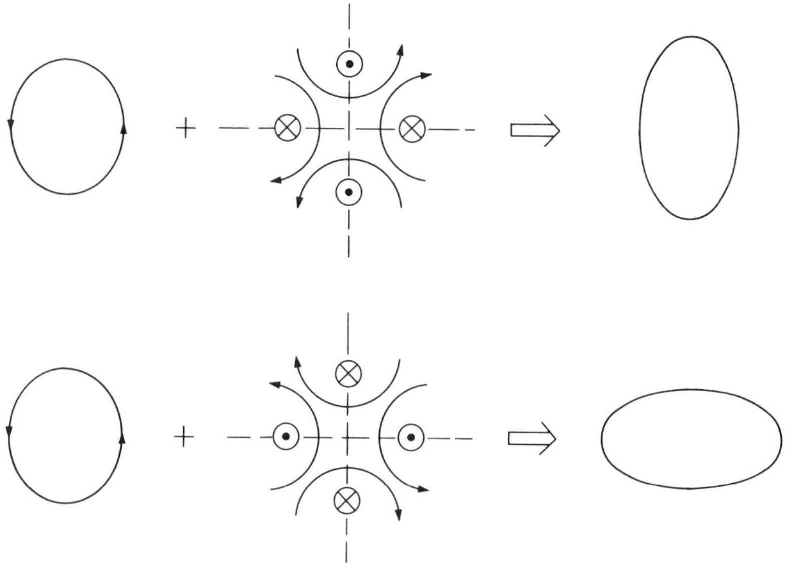

Fig. 2.19 Coil positions to produce the quadrupole field, and schematic views of an elliptic deformation by the quadrupole field.

System) at the National Institute for Fusion Science and the LHD. The $\ell = 2$ helical coils with $M = 10$ for LHD in Fig. 2.20 show the effect of $\alpha_1 > 0$, which is a larger poloidal rotation of the helical coil in the outer region than that in the inner region. Figure 2.20 also shows the rotational transform and the magnetic well depth given by $(dV(0)/d\Phi - dV(\bar{r})/d\Phi)/(dV(0)/d\Phi)$, where $\bar{r}$ is the average radius of the magnetic surface. When the magnetic well depth has a positive (negative) gradient, the region corresponds to magnetic well (hill).

In heliotrons, the magnetic axis is circular from the top view and is located on a flat plane. An interesting result is that the rotational transform appears when the magnetic axis becomes a helix, as shown in Fig. 2.21. Here the magnetic field **B** is produced by toroidal coils set along the helix. In Fig. 2.21 the helix corresponding to the helical magnetic axis is located on the cylinder with a radius a. Here the pitch length of the helix is λ. The helically deformed plasma column is also shown. Both curvature k and torsion κ are finite for such a helical magnetic axis and are given by

$$k = \frac{a}{a^2 + (\lambda/2\pi)^2} \tag{2.166}$$

$$\kappa = \frac{\lambda/2\pi}{a^2 + (\lambda/2\pi)^2}. \tag{2.167}$$

It is known that the normal direction of the helix rotates with the magnitude of κL, when the helix is followed from the bottom plane to the top plane of the

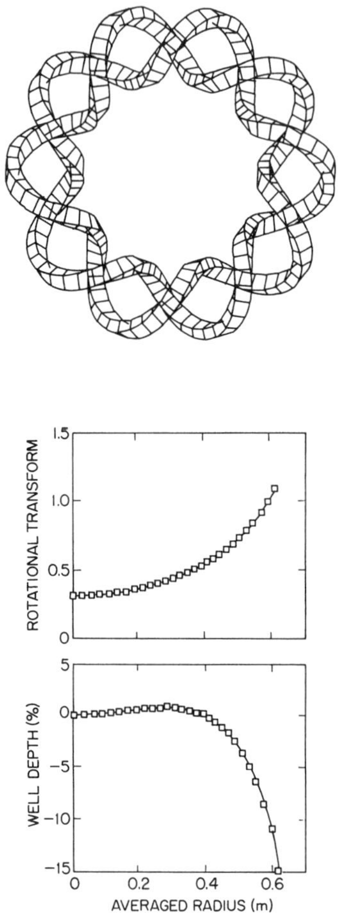

Fig. 2.20 The helical coils and characteristics of the vacuum magnetic configuration corresponding to the LHD. The rotational transform and magnetic well depth given by $(dV(0)/d\Phi - dV(r)/d\Phi)/(dV(0)/d\Phi)$ are shown, where $\bar{r}$ is an average radius.

cylinder in Fig. 2.21. Here L is the length along the helix for one pitch length λ, $L = (4\pi^2 a^2 + \lambda^2)^{1/2}$. It is thought that the magnetic field line in the vicinity of the helical magnetic axis also rotates with an angle almost equal to κL. From the definition of the rotational transform per pitch length,

$$\iota = \kappa L/(2\pi) = \frac{\lambda}{\sqrt{4\pi^2 a^2 + \lambda^2}} \tag{2.168}$$

is given. For a full torus of a helical axis stellarator with M pitches, the total rotational transform becomes ιM. With an increase of the pitch number M, the total rotational transform becomes large; however, the aspect ratio $R/\bar{a}$ also

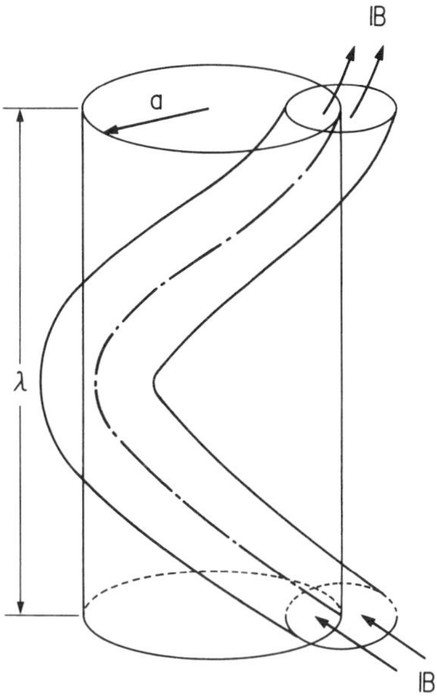

Fig. 2.21 A helical magnetic axis stellarator in the straight approximation.

becomes large. It is noted that the characteristics of this type of helical axis stellarator are similar to those of the $\ell = 1$ stellarator. Although a stellarator with a large rotational transform is possible, it is not easy to make a magnetic well in a $\ell = 1$ stellarator with almost circular magnetic surfaces (see Fig. 2.7). In order to obtain the magnetic well, triangular deformation of the magnetic surfaces by the axisymmetric quadrupole and hexapole fields is necessary.

Recently, several new stellarators with helical magnetic axes have been invented. One is the heliac. In the straight helically symmetric limit, the heliac consists of toroidal coils located along a helix around a straight center conductor, as shown in Fig. 2.22. It is noted that the center conductor exists inside the toroidal coils at any cross-section. The plasma has a bean-shaped cross-section. The indentation of the bean shape increases as the current in the center conductor is increased at a constant solenoidal field produced by toroidal coils. The degree of indentation is related to the depth of the magnetic well. In a real system, the above straight heliac is bent into a torus and vertical field coils are added as in heliotrons. The H-1 Heliac at the Australian National University has three pitches and the TJ-II Heliac at CEAMAT, Madrid, has four pitches. In these devices a spiral coil is mounted on the center conductor to increase the flexibility of the magnetic surface characteristics.

The other new device is the helias, which has a different magnetic surface cross-section at different toroidal positions. An example of the cross-section of a helias is shown in Fig. 2.23. The cross-sections are optimized to satisfy the criteria for high-beta plasma confinement with the minimum orbit loss of high-energy particles and the minimum bootstrap current. When the toroidal effect

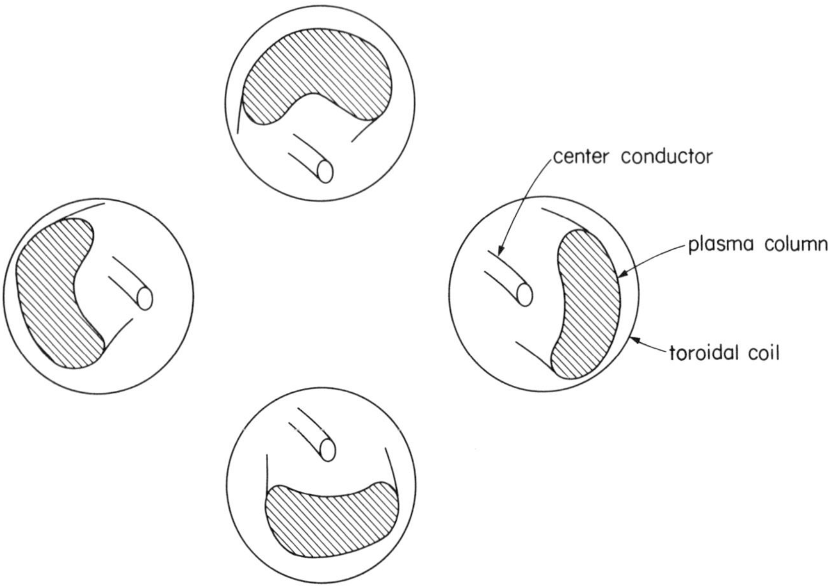

Fig. 2.22 Schematic figures showing the center conductor, the plasma column, and the toroidal coil in the heliac.

is minimized in a toroidal geometry, the obtained configuration is called a quasi-helically symmetrical stellarator. The W7-X Stellarator has been designed based on the helias at the Max-Planck Institute.

A helical magnetic axis configuration is also possible in heliotrons with an even pitch number M by arranging for there to be different coil currents in the two helical coils; however, the helical axis effects on MHD equilbrium and stability are not optimum.

The line tracing calculation is also useful to construct magnetic coordinates or Boozer coordinates for heliotrons. The magnetic field in the Boozer coordi-

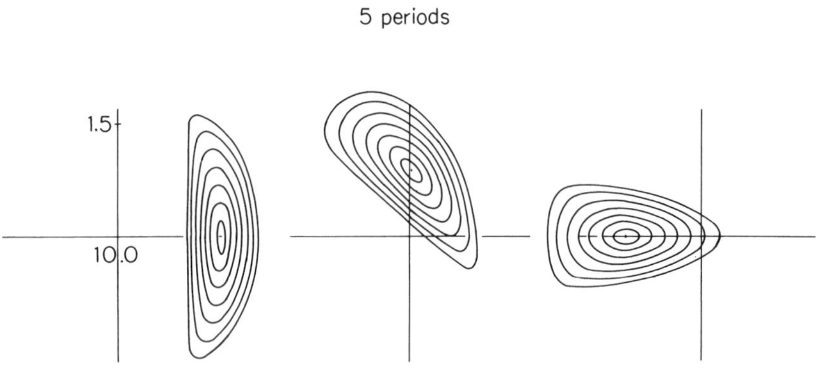

Fig. 2.23 An example of the magnetic surfaces of a helias with five periods.

nates (Ψ, θ_B, ϕ_B) will be given by (6.22) in chapter 6, where $\theta_B(\phi_B)$ is a poloidal (toroidal) angle variable. By using the line tracing calculation, the magnetic surface $\Psi = \text{constant}$ and the rotational transform $\iota(\Psi)$ are obtained. Transformation between (x, y, z) and (Ψ, θ_B, ϕ_B) by means of line tracing and Fourier decomposition has been established by Boozer (1982). Because of the double periodicity, the three Cartesian components, x, y, and z, of the position vector $\mathbf{R}$ can be written as a double Fourier series:

$$\mathbf{R}(\Psi, \theta_B, \phi_B) = \sum_{m,n} \mathbf{R}_{mm}(\Psi) e^{i(m\theta_B - n\phi_B)}. \tag{2.169}$$

The Boozer coordinates use a function χ as a parameter to measure the length along a field line,

$$\chi = \int_0^\ell B d\ell', \tag{2.170}$$

and χ reduces to the magnetic scalar potential if the plasma is negligible, $\mathbf{B} = \nabla\chi$. There is a relation between χ, θ_B, and ϕ_B,

$$\chi = \frac{\mu_0}{2\pi}(I^i_{tor}\theta_B + I^e_{pol}\phi_B), \tag{2.171}$$

where I^i_{tor} is a toroidal current inside the magnetic surface and I^e_{pol} is a poloidal current outside the flux surface. For the currentless plasma, $I^i_{tor} = 0$ and $I^e_{pol} \neq 0$. In (2.171), μ_0 is the vacuum magnetic permeability. Since the equation of a straight field line in Boozer coordinates is given by

$$\theta_B = \iota\phi_B + \theta_{B0}, \tag{2.172}$$

we can write the exponent of the Fourier series in terms of θ_{B0} and χ,

$$\mathbf{R}(\Psi, \theta_B, \phi_B) = \sum_{m,n} \mathbf{R}_{mm}(\Psi) \exp\left[i \frac{(mI^e_{pol} + nI^i_{tor})\theta_{B0} + (m\iota - n)2\pi\chi/\mu_0}{I^i_{tor}\iota + I^e_{pol}} \right]. \tag{2.173}$$

To evaluate the Cartesian coefficients $x_{mn}(\Psi)$, $y_{mn}(\Psi)$, and $z_{mn}(\Psi)$ of $\mathbf{R}_{mn}(\Psi)$, a field line integration is performed which starts from an arbitrary point. It is chosen that χ and θ_{B0} are zero at the starting point. In this case χ is determined by the integration (2.170). Thus, the field line integration determines $\mathbf{R}(\Psi, \theta_{B0} = 0, \chi)$. In (2.173) the LHS is now known numerically from the field line integration, and

$$\mathbf{R}(\Psi, \theta_{B0} = 0, \chi) = \sum_{m,n} \mathbf{R}_{mn}(\Psi) \exp\left\{ i \frac{2\pi(m\iota - n)\chi}{\mu_0(I^i_{tor}\iota + I^e_{pol})} \right\}. \tag{2.174}$$

This Fourier decomposition of $\mathbf{R}(\Psi, \theta_{B0} = 0, \chi)$ with respect to χ may have distinct peaks. The amplitudes of the peaks determine the $\mathbf{R}_{mn}(\Psi)$, and their frequencies determine m and n. The numerical decomposition is carried out with the help of a suitable window function. In practice, about five poloidal and toroidal circuits of a field line must be followed to carry out the Fourier decomposition to a reasonable accuracy. Thus $\mathbf{R}(\Psi, \theta_B, \phi_B)$ has been obtained. This method outlined above is often applicable for construction of the Boozer coordinates where (2.171) holds. We may generate other magnetic coordinates, such as Hamada coordinates, by transformation from Boozer coordinates (see chapter 7). The above explanation is in principle only valid for irrational (or nonresonant) surfaces; i.e., surfaces that are traced out by a single field line. This is not the case for rational (or resonant) surfaces on which ι is the ratio of two integers. However, as long as the number of poloidal and toroidal transits is sufficiently large, the rational surface is adequately covered by a magnetic field line for all practice purposes. For ι equal to the ratio of two low integers, we note that the real number system is dense, and any rational number can be approximated by a very close irrational number.

BIBLIOGRAPHY

Boozer, A.H. (1981). Plasma equilibrium with rational magnetic surfaces. *Phys. Fluids*, **24**, 1999.

Boozer, A. H. (1982). Establishment of magnetic coordinates for a given magnetic field. *Phys. Fluids*, **25**, 520.

Freidberg, J. P. (1987). *Ideal magnetohydrodynamics*. MIT Press, Cambridge.

Mackay, R. S., and Meiss, J. D. (1987). *Hamiltonian dynamical systems*. Adam Hilger, Bristol.

Miyamoto, K. (1976). *Plasma physics for nuclear fusion*. MIT Press, Cambridge.

Morozov, A. I., and Solovév, L. S. (1966). The structure of magnetic field, in *Reviews of plasma physics*, Consultants Bureau, New York, vol. 2, p. 1.

Rosenbluth, M. N., Sagdeew, R. Z., Taylor, J. B., and Zaslavski, G. M. (1966). Destruction of magnetic surfaces by magnetic field irregularities. *Nucl. Fusion*, **6**, 297.

Solovév, L. S., and Shafranov, V. D. (1970). Plasma confinement in closed magnetic systems, in *Reviews of plasma physics*, Consultants Bureau, New York, vol. 5, p. 1.

Spitzer, L. (1958). The stellarator concept. *Phys. Fluids*, **1**, 253.

Wesson, J. (1987). *Tokamaks*. Clarendon Press, Oxford.

3

A DESCRIPTION OF MAGNETICALLY CONFINED PLASMAS

3.1 INTRODUCTION

In this chapter we introduce fundamental equations for applications in the rest of this book. First, we describe the properties of high-temperature plasmas in section 3.2. For the study of collective phenomena, the Vlasov equation is introduced by using a phase space description for a many-particle system in section 3.3. Then magnetohydrodynamic (MHD) approximation is described. For the investigation of plasma dynamics in an inhomogeneous magnetic field, or particularly in a three-dimensional magnetic field such as the stellarator and the heliotron, one-fluid magnetohydrodynamic equations are the most appropriate and reasonable theoretical tool, and are explained in section 3.4. An extension to include the anisotropic pressure in MHD approximation is also given. The anisotropy appears between the parallel and perpendicular pressures, where the direction of the magnetic field vector is the reference for these pressures. When the MHD equations are linearized for the study of the dynamics of small-amplitude perturbations, three types of MHD waves appear, as shown in section 3.5 these are also important ingredients in the study of MHD instabilities in chapter 5. As an approximation to the Vlasov equation, the drift-kinetic equation is shown in section 3.6, which is useful in the study of collisional transport of magnetically confined toroidal plasmas. In order to study the transport phenomena of toroidal plasmas, the collision term in the phase space description is important. In the transport equations based on the moments of the Boltzmann equation (or the Vlasov equation with the collision term) collisional effects are included through resistivity, viscosity, and transport coefficients for density and temperature (see section 3.7). Finally, for an approximate description of heliotron plasma, we introduce averaged MHD equations in section 3.8: these are widely used to discuss the macroscopic behavior of heliotron plasma in this book.

3.2 THE BASIC PROPERTIES OF HIGH-TEMPERATURE PLASMAS

Plasma is generated under the condition that the average kinetic energy of an electron substantially exceeds the average Coulomb energy needed for an ion to bind an electron. The average kinetic energy of an electron is represented by the electron temperature, T_e. Here, and throughout this book we shall use temperature as an energy unit in which the Boltzmann constant is included, for example, $T_e = 1\,\text{keV}$. The average Coulomb energy needed for an ion to bind an electron is represented by $e^2/(4\pi\varepsilon_0 d)$, where d is the average distance between the nearest electron and ion, e is the elementary charge, and ε_0 is the permittivity of a vacuum. In this book we assume that the ions are usually singly ionized, such as protons (H^+), deuterons (D^+), and tritons (T^+), since the nuclear fusion reaction appropriate for the first-generation fusion reactor is primarily $D^+ + T^+ \rightarrow {}^4He^{++} + n$. The important doubly ionized ions are ${}^4He^{++}$ and ${}^3He^{++}$. The alpha particles ${}^4He^{++}$ are produced by the D–T reaction, and ${}^3He^{++}$ is a fuel for a D–^{3}He fusion reaction; $D^+ + {}^3He^{++} \rightarrow {}^4He^{++} + H^+$. This reaction is particularly useful to reduce the radioactivity generated by the neutrons (n) in the *D–T* fusion ractor. Since the plasma parameters required for the D–He^3 fusion reactor are more severely constrained than those for the D–T fusion reactor (see chapter 9), fusion reactors based on the D–He^3 reaction are considered in an advanced reactor.

Let n_0 be the average number of electrons or ions per unit volume: then we can estimate d by the relation $(4\pi d^3/3)n_0 = 1$. The condition that plasma is formed can then be written as

$$\left(\frac{4\pi}{3} n_0\right)^{1/3} \frac{e^2}{4\pi\varepsilon_0 T_e} \ll 1. \tag{3.1}$$

Taking the 3/2 power of the left-hand side and omitting the numerical factor yields the relation

$$n_0 \lambda_D^3 \gg 1, \tag{3.2}$$

where λ_D is defined by

$$\lambda_D = (\varepsilon_0 T_e / n_0 e^2)^{1/2}. \tag{3.3}$$

This parameter has dimensions of length and is called the Debye length. The condition (3.1) or (3.2) is called the plasma condition.

As is well-known, the Coulomb potential is of long range in the sense that it extends over an infinite distance in a vacuum. In a plasma, however, the Coulomb potential of any charged particle (to be referred to as the test particle) induces an electric polarization because it attracts (or repels) the charge of the opposite (or same) sign. The polarization charge tends to cancel the charge of the test particle and restricts the effect of its Coulomb potential to a finite range. It is the Debye length that gives this range in the case of a static test charge. The restriction of the effective range of the Coulomb potential by the polarization charge is called Debye shielding.

In high-temperature plasmas, two-body Coulomb interactions or collisions are rare. The collisionality can be quantitatively estimated using the mean free path ℓ of a charged particle. When r_0 is the collision diameter, then it is found from the standard formula

$$\ell \sim \frac{1}{n_0 \pi r_0^2}. \tag{3.4}$$

The collision diameter for the electron can be estimated as the distance at which the electron orbit is significantly modified by a scatterer which may be either an electron or an ion. A strong orbit modification occurs when the electron kinetic energy $m_e v^2/2$ becomes comparable to the Coulomb energy $e^2/(4\pi\varepsilon_0 r_0)$. On average, $m_e v^2/2$ can be replaced by T_e, so that we have

$$r_0 \sim \frac{e^2}{4\pi\varepsilon_0 T_e}. \tag{3.5}$$

Substituting (3.5) into (3.4) yields

$$\ell \sim \frac{16\pi\varepsilon_0^2 T_e^2}{n_0 e^4} = 16\pi(n_0 \lambda_D^3)\lambda_D. \tag{3.6}$$

This estimation is rough, and an important effect arises from an accumulation of many small-angle scattering processes due to distant encounters. If cumulative small-angle scattering is taken into account, the electron mean free path should be written, instead of (3.6), as

$$\ell \sim \frac{16\pi(n_0 \lambda_D^3)\lambda_D}{\ln \Lambda}, \tag{3.7}$$

where Λ is of order $n_0 \lambda_D^3$. Essentially the same result can be obtained for an ion mean free path, as we can see from the mass independence of the formula.

The plasma that we shall be concerned with in this book is the high-temperature plasma in relation to thermonuclear fusion research. If a nucleus of deuterium fuses with a nucleus of tritium, an α-particle is produced and a nuetron is released. The nuclear reaction results in a reduction in total mass and a consequent release of energy in the form of a kinetic energy of the reaction products. The energy release is 17.6 MeV per reaction. In macroscopic terms, just 1 kg of deuterium and tritium mixture with equal concentrations would release 10^8 kWh of energy.

Deuterium is a plentiful resource that can be taken from the sea, but tritium is virtually nonexistent in nature and is produced by nuclear reactions, using the neutrons released by the D–T reaction and lithium,

$$\begin{cases} \mathrm{Li}^6 + \mathrm{n} \rightarrow \mathrm{T} + \mathrm{He}^4 + 4.8\,\mathrm{MeV} \\ \mathrm{Li}^7 + \mathrm{n} \rightarrow \mathrm{T} + \mathrm{He}^4 + \mathrm{n} - 2.5\,\mathrm{MeV}. \end{cases} \tag{3.8}$$

In order to induce the fusion of nuclei of deuterium and tritium, it is necessary to overcome the mutual repulsion due to their positive charges. The cross-section for D–T fusion increases with energy, reaching a maximum at 100 keV (see Fig. 9.1). This is much less than the energy released in the nuclear fusion reaction, but is a large amount of energy to supply efficiently to fuel particles.

The most promising method of supplying the energy is to heat the deuterium–tritium plasma so that the thermal velocities of the nuclei are high enough to produce the D–T reactions (see chapter 8). Fusion brought about in this way is called thermonuclear fusion. The required temperature of 10–20 keV is not as high as that corresponding to the energy of the maximum cross-section, because the D–T reactions mainly occur in the high-energy tail of the Maxwellian distribution of the fuel ions. At such high temperatures deuterium and tritium are ionized; however, the electric charge of the fuel ions is neutralized by an equal number of electrons. This kind of high-temperature plasma is required for thermonuclear fusion.

For such high temperatures that preclude confinement by material walls, magnetic fields are employed for confinement of the plasma. The tokamak and the stellarator, including the heliotron, are typical devices for magnetic confinement. In these systems, electrons and ions are confined to a toroidal region by a magnetic field (see chapter 2). In the stellarator even a single charged particle can be confined, since magnetic surfaces exist without plasma, but in the tokamak only a plasma with a current in the toroidal direction can be confined. It is possible, due to the magnetic confinement, that fuel ions may travel a distance of a million times the device size before reaction. For a reactor it would be necessary to confine a sufficiently high-density plasma for a time that would allow an adequate fraction of the fuel plasma to react. At the required ion density of around $10^{20}\,\mathrm{m}^{-3}$, this characteristic time or confinement time becomes the order of a second. This density is primarily limited by the available strength of the magnetic field, since the plasma pressure $(n_e T_e + n_D T_D + n_T T_T)$ (see section 3.3) should be less than a fraction of the magnetic field pressure $B^2/2\mu_0$. When $B \lesssim 10\mathrm{T}$ and $T_e = T_D = T_T = 10\,\mathrm{keV}$, $n_e = n_D + n_T \lesssim 10^{20}\,\mathrm{m}^{-3}$ is required for $\beta = (n_e T_e + n_D T_D + n_T T_T)/(B^2/2\mu_0) \simeq 0.1$. The magnetic field is limited by the superconducting materials available for the manufacture of magnetic coils.

Although magnetic confinement looks challenging, it has been demonstrated that the tokamak plasma can be heated to the above required temperature. The confinement time of the order of a second has been already achieved. However, both confinement and heating still require substantial research in order that they can be realized simultaneously. In this book, we explain the established results of confinement and heating in the stellarator, with an emphasis on the heliotron device, which has the potential to be a future fusion reactor based on magnetic confinement.

3.3 THE VLASOV EQUATION FOR DESCRIBING INCOMPRESSIBLE PHASE FLUID AND MOMENT EQUATIONS

First, we consider many charged particles in a given electromagnetic field with electric field $\mathbf{E}(\mathbf{x}, t)$ and magnetic field $\mathbf{B}(\mathbf{x}, t)$. When Coulomb forces between the particles are small compared with the externally applied electric and magnetic fields, we can solve the equations of motion for each particle independently to find its position and velocity. The equations for the nth changed particle are

$$\frac{d\mathbf{x}_n}{dt} = \mathbf{v}_n, \tag{3.9}$$

$$\frac{d\mathbf{v}_n}{dt} = \frac{q}{m}\left[\mathbf{E}(\mathbf{x}_n(t), t) + \mathbf{v}_n(t) \times \mathbf{B}(x_n(t), t)\right], \tag{3.10}$$

where m is a mass and q is a charge. Here only the electromagnetic force (Lorentz force) is considered. Usually, the gravitational force is negligible in problems of magnetically confined plasmas.

The motion of a single charged particle can be described in three-dimensional geometric space or real space. The particle trajectory is followed by the vector $\mathbf{x}(t) = (x(t), y(t), z(t))$. This position vector is obtained from the velocity vector $\mathbf{v}(t) = (v_x(t), v_y(t), v_z(t))$. However, the real-space description is not an effective way to follow the behavior of many particles. In order to develop theoretical tools for the description of many-particle dynamics, we need an ordered picture of particle dynamics. We represent particle trajectories in a six-dimensional space, with axes in both real space and velocity space: $(\mathbf{x}, \mathbf{v}) = (x, y, z, v_x, v_y, v_z)$. This mathematical space is called phase space. At a particular time, each particle in a plasma is represented as a single point in the phase space.

In the phase space, the trajectories of large numbers of particles have a high degree of order if the force varies smoothly. When the scale length of the force variation is long compared with the phase space distance between particles, the force is considered to be smoothly varying. Without collisions, two particles that start out with similar phase-space coordinates will always be neighbors. This follows from the fact that particles with nearly the same initial position and velocity are influenced by the same force. Another consequence is that trajectories of particles moved by the smooth force never cross in the phase space. The implication is that they are laminar in phase space. Laminar flow in phase space is the foundation for the description of the collective behavior of plasma. Incompressible laminar flow in phase space naturally leads to the Vlasov equation for describing plasma dynamics.

In plasmas the charged particles interact, and the electric and magnetic fields are determined by Maxwell's equations,

$$\frac{\partial \mathbf{B}}{\partial t} = -\nabla \times \mathbf{E} \tag{3.11}$$

$$\varepsilon_0 \frac{\partial \mathbf{E}}{\partial t} + \mathbf{J} = \frac{1}{\mu_0} \nabla \times \mathbf{B}, \tag{3.12}$$

$$\nabla \cdot \mathbf{B} = 0, \tag{3.13}$$

$$\nabla \cdot \mathbf{E} = \sigma_e / \varepsilon_0 \tag{3.14}$$

where $\mathbf{J}$ denotes the current density and σ_e denotes the charge density, which depend on position and time. Later in this section we discuss the physical meaning of $\mathbf{J}$ and σ_e for plasmas. Here we remark that they are dependent on the charged-particle dynamcis, which means that the force on a specific particle depends on the locations of all other particles in the plasma. Alternatively, each particle orbit depends on the time-dependent positions and velocities of the other particles which determine the electromagnetic fields through Maxwell's equations. Thus it is a formidable task to obtain the exact particle orbits of many charged particles.

In order to obtain self-consistent solutions of (3.9) and (3.10), we consider a stepwise approximation, or we push all particles simultaneously over a small time interval Δt. The resulting orbit predictions converge to self-consistent solutions if the forces generated by particles satisfy two conditions: (1) they vary smoothly compared with the distance between particles in the phase space; and (2) they change slowly compared with Δt.

As in any numerical calculation, the procedure involves compromises. A short time step improves the accuracy but increases the number of operations. On the other hand, a large time step leads to inaccuracies that may mask important physical processes. The goal is to extract the maximum amount of information through the minimum time of computation. Here we present a numerical method that yields sufficient accuracy for a given Δt. It is called the leapfrog method, and it gives accurate predictions of particle orbits with a few operations.

According to the leapfrog method, (3.9) and (3.10) are solved by

$$\mathbf{x}_n(\Delta t) = \mathbf{x}_n(0) + \mathbf{v}_n(\Delta t/2)\Delta t, \tag{3.15}$$

$$\begin{aligned}\mathbf{v}_n(3\Delta t/2) = \mathbf{v}_n(\Delta t/2) + \frac{q}{m} [&\mathbf{E}(\mathbf{x}_n(\Delta t), \Delta t) \\ &+ \tfrac{1}{2}(\mathbf{v}_n(3\Delta t/2) + \mathbf{v}_n(\Delta t/2)) \times \mathbf{B}]\Delta t,\end{aligned} \tag{3.16}$$

where $\mathbf{B}$ is assumed constant for simplicity. At the start of numerical calculation we usually have the set of initial particle positions and velocities, $\mathbf{x}_n(0)$ and $\mathbf{v}_n(0)$. Instead, suppose that we know the positions at $t = 0$, $\mathbf{x}_n(0)$, and the velocities at the displaced time $\Delta t/2$, $\mathbf{v}_n(\Delta t/2)$. In other words, the initial quantities are $\mathbf{x}_n(0)$ and $\mathbf{v}_n(\Delta t/2)$. Then we can find the particle positions and velocities $\mathbf{x}_n(\Delta t)$ and $\mathbf{v}_n(3\Delta t/2)$ using (3.15) and (3.16).

For a numerical simulation of the many-particle system the electric field $\mathbf{E}(\mathbf{x}_n(\Delta t), \Delta t)$ in (3.16) is determined by the other charged particles. Coulomb's law gives the electric field at the position of particle i as

$$\mathbf{E}_i = \sum_{j \neq i} \frac{q_j \mathbf{r}_{ji}}{\varepsilon_0 |\mathbf{r}_j - \mathbf{r}_i|^2}, \tag{3.17}$$

where $\mathbf{r}_{ji}$ is a unit vector pointing from particle j to particle i, while $\mathbf{r}_i$ and $\mathbf{r}_j$ are the position vectors of the two particles. When $\mathbf{x}_n(\Delta t)$ is obtained by (3.15), $\mathbf{E}(\mathbf{x}_n(\Delta t), \Delta t)$ is determined by (3.17) in principle. In realistic numerical simulations, (3.14) is solved by introduced grid points in the real space and defining the charge density σ_e at the grid points. The reason for using (3.14) is that the smoothness of the force is usually violated for the electric field given by (3.17).

When $\mathbf{x}_n(\Delta t)$ and $\mathbf{v}_n(3\Delta t/2)$ are obtained, the procedure can be iterated for $2\Delta t$, $5\Delta t/2$, $3\Delta t$, $7\Delta t/2$, etc., with the velocity and position advancing ahead of each other. It is remarked that difference approximations (3.15) and (3.16) have a time-centered form from the numerical analysis point of view, which ensures accuracy of time integration to the order of $(\Delta t)^2$.

Now we consider conservation of the phase-space volume occupied by the plasma. A phase-space volume is given by $\Delta\mathbf{x}\Delta\mathbf{v}$ at an arbitrary time t, where $\Delta\mathbf{x}$ denotes a volume element in the real space and $\Delta\mathbf{v}$ denotes a volume element in the velocity space. If the positions and velocities of relevant particles are obtained by (3.15) and (3.16) for one time step Δt, then

$$(\Delta\mathbf{x}\Delta\mathbf{v})|_t = (\Delta\mathbf{x}\Delta\mathbf{v})|_{t+\Delta t}. \tag{3.18}$$

It is easy to show (3.18) for simple cases. This means conservation of the volume element in the phase space. It is often written in the following way:

$$\frac{Df(x, y, z, v_x, v_y, v_z, t)}{Dt} = 0, \tag{3.19}$$

where D/Dt denotes the convective derivative and f is a particle distribution function in the phase space. In fluid dynamics, remember that convection means moving along with the fluid element in the real space.

Equation (3.18) also means that the density of collisionless particles is constant in a frame of reference moving with the trajectory through the phase space (Lagrangian picture). If we view the collection of particle trajectories as a fluid, (3.18) states that the fluid is locally incompressible. Often, the Lagrangian picture is not the most convenient form. Usually, we would like to describe variations of fluid quantities in a fixed frame through which the fluid moves (Eulerian picture). Eulerian equations specify the evolution of fluid quantities in a stationary frame. We shall convert (3.18) to a Eulerian form. The derivation leads to the collisionless Boltzmann equation. A form of the Boltzmann equation that includes only electromagnetic forces, the Vlasov equation, is the fundamental equation of plasma physics.

First, we consider the distribution function in two-dimensional phase space (x, v_x). For example, consider a point (x_0, v_{x0}). Suppose that the particle distribution function f has a negative spatial gradient at x_0 for $t = t_0$. Then, the phase-space density observed at the point increases with time:

$$f[x_0(t), v_{x0}(t), t] > f[x_0(t_0), v_{x0}(t_0), t_0]. \tag{3.20}$$

We shall develop a quantitative expression for the time variation of f. The distribution function $f(x, v_x, t)$ is assumed to have a negative spatial gradient at time t. At a later time, $t + \Delta t$, all particles shift to the right by a distance $\Delta x = v_x \Delta t$. Although f retains its original value at the new position of particles, the value of f at the fixed point (x, v_x) increases by

$$\Delta f_x(x, v_x) = -\Delta x \left(\frac{\partial f}{\partial x} \bigg|_{x, v_x} \right) = -v_x \Delta t \left(\frac{\partial f}{\partial x} \bigg|_{x, v_x} \right). \tag{3.21}$$

The partial derivative notation refers to a variation of f along a path in the x direction. Similarly, we can write the change of f at the point (x, v_x) that results from a gradient of f in the velocity space. Suppose that near (x, v_x) the phase-space density is higher for large v_x. In the presence of a decelerating force, the higher-velocity particles than v_x shift to occupy the region near v_x. As a result, $f(x, v_x, t)$ rises with time. The change of velocity due to an acceleration a_x acting over an internal Δt is $\Delta v_x = a_x \Delta t$. The change in f resulting from $\partial f / \partial v_x$ is written as

$$\Delta f_{v_x}(x, v_x) = -\Delta v_x \left(\frac{\partial f}{\partial v_x} \bigg|_{x, v_x} \right) = -a_x \Delta t \left(\frac{\partial f}{\partial v_x} \bigg|_{x, v_x} \right). \tag{3.22}$$

The symbol $\partial f / \partial t$ denotes the change in f with time at a constant location in the phase space. The combined effects of position and velocity gradients give

$$\frac{\partial f}{\partial t} = -v_x \frac{\partial f}{\partial x} - a_x \frac{\partial f}{\partial v_x}. \tag{3.23}$$

Equation (3.23) implies that the total change in the particle distribution function from all causes—the passage of time, the motion of the phase fluid, and the phase fluid acceleration—equals zero. The full form of (3.23) for the phase fluid motion in six dimensions becomes

$$\frac{\partial f}{\partial t} + \mathbf{v} \cdot \frac{\partial f}{\partial \mathbf{x}} + \mathbf{a} \cdot \frac{\partial f}{\partial \mathbf{v}} = 0, \tag{3.24}$$

which is the Vlasov equation. When the acceleration is given by the Lorentz force,

$$\mathbf{a} = q(\mathbf{E} + \mathbf{v} \times \mathbf{B})/m, \tag{3.25}$$

where q is an electric charge and m is a mass, (3.24) reduces to

$$\frac{\partial f}{\partial t} + \mathbf{v} \cdot \frac{\partial f}{\partial \mathbf{x}} + \frac{q}{m} (\mathbf{E} + \mathbf{v} \times \mathbf{B}) \cdot \frac{\partial f}{\partial \mathbf{v}} = 0. \tag{3.26}$$

The Vlasov equation is equivalent to the principle of the conservation of phase-space volume.

If collisions occur, the phase-space volume may change, even in the Lagrangian picture. Collisions cause a random walk of a particle trajectory through the velocity space. Usually, collisions result in a decrease of the phase-space density of the particle trajectory with time, since the particle distribution function diffuses to fill a larger phase-space volume. The effect of collisions is often added symbolically to (3.24) as a term that contributes to a small variation in f:

$$\frac{\partial f}{\partial t} + \mathbf{v} \cdot \frac{\partial f}{\partial \mathbf{x}} + \mathbf{a} \cdot \frac{\partial f}{\partial \mathbf{v}} = \left.\frac{\partial f}{\partial t}\right|_c. \tag{3.27}$$

Details of the collision term will be described in chapter 7.

For a single species of charged particle in real space, the charge density equals the electric charge of each particle, q, multiplied by the number of particles per unit volume:

$$\sigma_e(x, y, z) = qn(x, y, z). \tag{3.28}$$

With the particle distribution function the charge density is given as

$$\sigma_e(x, y, z, t) = q \int_{-\infty}^{\infty} dv_x \int_{-\infty}^{\infty} dv_y \int_{-\infty}^{\infty} dv_z f(x, y, z, v_x, v_y, v_z, t). \tag{3.29}$$

The current density $\mathbf{J}$ is the charge crossing a unit area per unit time. The current density vector points in the direction of the average flow and the unit area is normal to this flow. For a single species, the current density equals the charge density multiplied by the average flow velocity:

$$\mathbf{J}(x, y, z, t) = qn(x, y, z, t)\langle \mathbf{v}(x, y, z, t) \rangle. \tag{3.30}$$

In terms of the distribution function, it is given as

$$\mathbf{J}(x, y, z, t) = q \int_{-\infty}^{\infty} dv_x \int_{-\infty}^{\infty} dv_y \int_{-\infty}^{\infty} dv_z \mathbf{v} f(x, y, z, v_x, v_y, v_z, t). \tag{3.31}$$

Alternately, (3.30), the average velocity is defined as

$$\langle \mathbf{v}(\mathbf{x}, t) \rangle \equiv \iiint d\mathbf{v}\mathbf{v} f(\mathbf{x}, \mathbf{v}, t)/n(\mathbf{x}, t). \tag{3.32}$$

Velocity moments taken over a particle distribution function have the following general form:

$$\mathbf{M}_\ell(x, y, z, t) = \iiint d\mathbf{v}\mathbf{v}^\ell f(\mathbf{x}, \mathbf{v}, t). \tag{3.33}$$

The moment for $\ell = 0$ is a scalar quantity, or the particle density $n(x, y, z, t)$ (see (3.29)). The moment for $\ell = 1$ is the average velocity of the particles at a location multiplied by the density (see (3.31)). We can derive the moment equations by applying the operation of (3.33) to all terms in the Vlasov equation. To simplify the explanation, calculations are performed for a one-dimensional distribution function $f(x, v_x, t)$. The $\ell = 0$ moment over the Vlasov equation has the form

$$\int_{-\infty}^{\infty} dv_x \frac{\partial f}{\partial t} + \int_{-\infty}^{\infty} dv_x v_x \frac{\partial f}{\partial x} + \int_{-\infty}^{\infty} dv_x a_x \frac{\partial f}{\partial v_x} = 0. \tag{3.34}$$

By noting that a_x does not depend on v_x from (3.25) and that the distribution function must vanish as $v_x \to \pm\infty$, the $\ell = 0$ moment equation becomes

$$\frac{\partial n}{\partial t} + \frac{\partial}{\partial x}(n\langle v_x\rangle) = 0. \tag{3.35}$$

The $\ell = 1$ moment equation can be derived from

$$\int_{-\infty}^{\infty} dv_x v_x \frac{\partial f}{\partial t} + \int_{-\infty}^{\infty} dv_x v_x^2 \frac{\partial f}{\partial x} + \int_{-\infty}^{\infty} dv_x a_x v_x \frac{\partial f}{\partial v_x} = 0, \tag{3.36}$$

and it becomes

$$\frac{\partial}{\partial t}(mn\langle v_x\rangle) = -\frac{\partial}{\partial x}[mn\langle v_x\rangle^2] - \frac{\partial}{\partial x}[mn\langle(\delta v_x)^2\rangle] + mn\langle a_x\rangle. \tag{3.37}$$

We have used the fact that the velocity of a particle is expressed in terms of a deviation from the average velocity by $v_x = \langle v_x\rangle + \delta v_x$. By the definition of the average velocity, $\langle\delta v_x\rangle = 0$. Therefore $\langle v_x^2\rangle = \langle v_x\rangle^2 + \langle(\delta v_x)^2\rangle$. $\langle a_x\rangle$ is the x component of the average acceleration. Equation (3.37) expresses conservation of momentum in real space.

The LHS of (3.37) is the rate of change of momentum per unit volume, $mn\langle v_x\rangle$. The terms on the RHS contribute to the momentum change. The average momentum can be changed as follows: (i) an applied force accelerates particles in the volume element; and (ii) particles leave the volume element and are replaced by new particles with a different average momentum. The first process is familiar from single-particle motion. The second process is unique to the collective phenomena—it appears only if a system contains many particles. The first term on the RHS of (3.37) describes momentum convection by an average flow. We can understand this term by noting that the flux of momentum across a surface normal to the x direction equals the momentum per particle multiplied by the number of particles that cross the surface per unit area and unit time, $[m\langle v_x\rangle][n\langle v_x\rangle]$. The change of momentum resulting from the convection in a differential length element is proportional to the gradient of $mn\langle v_x\rangle^2$. The second term on the RHS of (3.37) represents the migration of momentum associated with spatial variations of random velocity components. This term is often represented by pressure and viscosity.

Here we discuss how velocity spreads affect the momentum balance among collisionless particles. Velocity dispersion leads to a force density that acts on the volume element. For the one-dimensional case, (3.37) shows that the force per unit volume resulting from random velocity components with respect to an average velocity is

$$-\frac{\partial}{\partial x}[mn\langle(\delta v_x)^2\rangle].$$

This implies that momentum may be transformed to or from the neighboring volume element when there is a gradient in either the particle density or the mean square velocity spread, $\langle(\delta v_x)^2\rangle$. This volumetric force is called the pressure.

The key to understanding the pressure is to remember that a volumetric force is synonymous with a rate of change of momentum per unit volume. We identify the quantity $mn\langle(\delta v_x)^2\rangle$ as the particle pressure in the x direction. The pressure notation can be extended to particles with three-dimensional variations. If particle motions are decoupled in x, y, and z, then we can define three components of the pressure; $P_x = mn\langle(\delta v_x)^2\rangle$, $P_y = mn\langle(\delta v_y)^2\rangle$, and $P_z = mn\langle(\delta v_z)^2\rangle$. When particle motions in the x, y, and z directions are not decoupled, the momentum flow becomes more complex. The momentum equation involves terms such as $mn\langle v_x v_y\rangle$ and pressure must be defined as a tensor. The cross-terms represent viscosity, which will be explained in section 3.7. Scalar pressure is useful if the distribution of the particle velocity is isotropic. Then the viscosity disappears and is defined by

$$\begin{aligned} P &= mn(\langle(\delta v_x)^2\rangle + \langle(\delta v_y)^2\rangle + \langle(\delta v_z)^2\rangle)/3 \\ &= (P_x + P_y + P_z)/3. \end{aligned} \tag{3.38}$$

A temperature T including the Boltzmann constant k_B can be defined for an isotropic Maxwellian distribution:

$$f(\mathbf{x}, \mathbf{v}, t) = n(\mathbf{x}, t)\left[\frac{m}{2\pi T(\mathbf{x}, t)}\right]^{3/2} \exp\left[-\frac{m(v_x^2 + v_y^2 + v_z^2)}{2T(\mathbf{x}, t)}\right]. \tag{3.39}$$

The temperature corresponds to the velocity dispersion. By using (3.39) the pressure given by (3.38) becomes

$$P(\mathbf{x}, t) = n(\mathbf{x}, t)T(\mathbf{x}, t). \tag{3.40}$$

The three-dimensional extension of the momentum equation (3.37) becomes

$$\frac{\partial\langle \mathbf{v}(\mathbf{x}, t)\rangle}{\partial t} + \langle \mathbf{v}(\mathbf{x}, t)\rangle \cdot \nabla\langle \mathbf{v}(\mathbf{x}, t)\rangle = \frac{\nabla P(\mathbf{x}, t)}{n(\mathbf{x}, t)m} + \frac{q}{m}(\mathbf{E} + \langle \mathbf{v}(\mathbf{x}, t)\rangle \times \mathbf{B}). \tag{3.41}$$

Next, we consider the second moment equation. We note that

$$\frac{\partial}{\partial t}\int \tfrac{1}{2} m v^2 f d\mathbf{v} = \frac{\partial}{\partial t}[\tfrac{1}{2} mn(\mathbf{x}, t)\langle \mathbf{v}(\mathbf{x}, t)\rangle^2 + \tfrac{1}{2} mn(\mathbf{x}, t)\langle(\delta \mathbf{v})^2\rangle], \tag{3.42}$$

$$\begin{aligned}\frac{\partial}{\partial x_i}\int \tfrac{1}{2} m v^2 v_i f d\mathbf{v} = \frac{\partial}{\partial x_i}[&\tfrac{1}{2} mn(\mathbf{x}, t)\langle \mathbf{v}(\mathbf{x}, t)\rangle^2\langle v_i(\mathbf{x}, t)\rangle \\ &+ \tfrac{1}{2} mn(\mathbf{x}, t)(\langle(\delta \mathbf{v})^2\rangle + \langle \delta v_i \delta v_j\rangle)\langle v_j(\mathbf{x}, t)\rangle \\ &+ \tfrac{1}{2} mn(\mathbf{x}, t)\langle(\delta \mathbf{v})^2 \delta v_i\rangle],\end{aligned} \tag{3.43}$$

and the last term of the Vlasov equation gives

$$\int \tfrac{1}{2} m v^2 \frac{q}{m}(\mathbf{E} + \mathbf{v} \times \mathbf{B}) \cdot \frac{\partial f}{\partial \mathbf{v}} d\mathbf{v} = -qn(\mathbf{x}, t)\langle \mathbf{v}(\mathbf{x}, t)\rangle \cdot \mathbf{E}. \tag{3.44}$$

We now substitute the Maxwell distribution (3.39) into the second moment equation with $\ell = 2$. We note that $\langle(\delta \mathbf{v})^2\rangle = \langle(\delta v_x)^2\langle(\delta v_x)^2\rangle + \langle(\delta v_y)^2\rangle + \langle(\delta v_z)^2\rangle = 3T/m$, $\langle \delta v_i \delta v_j\rangle = \delta_{ij} T/m$, and $\langle(\delta \mathbf{v})^2 \delta v_i\rangle = 0$, where δ_{ij} is the Kronecker δ function with $\delta_{ij} = 1$ (if $i = j$) or $\delta_{ij} = 0$ (if $i \neq j$). Then the second moment equation reduces to

$$\begin{aligned} mn\langle v_i\rangle \frac{\partial\langle v_i\rangle}{\partial t} &+ \tfrac{1}{2} mn\langle v\rangle^2 \frac{\partial n}{\partial t} + \tfrac{3}{2}\frac{\partial P}{\partial t} + mn\langle v_i\rangle\left(\langle v_j\rangle \frac{\partial\langle v_j\rangle}{\partial x_i}\right) \\ &+ \tfrac{1}{2} m\langle v\rangle^2 \frac{\partial}{\partial x_i}(n\langle v_i\rangle) + \tfrac{5}{2}\langle v_i\rangle \frac{\partial P}{\partial x_i} + \tfrac{5}{2} P \frac{\partial\langle v_i\rangle}{\partial x_i} = \hat{\mathbf{J}} \cdot \mathbf{E}, \end{aligned} \tag{3.45}$$

where $\hat{\mathbf{J}}$ denotes a current density due to single species of charged particle. We note that the second and the fifth terms cancel by virtue of the continuity equation (3.35). Similarly, in view of the equation of motion (3.41), we note that the first, fourth, two-fifth of the sixth term are equal to $\sigma_e\langle \mathbf{v}\rangle \cdot \mathbf{E} + \langle \mathbf{v}\rangle \cdot (\hat{\mathbf{J}} \times \mathbf{B})$. Now for each charged species, $\sigma_e\langle \mathbf{v}\rangle = \hat{\mathbf{J}}$; hence $\langle \mathbf{v}\rangle \cdot (\hat{\mathbf{J}} \times \mathbf{B}) = 0$. Hence, these terms cancel with $\hat{\mathbf{J}} \cdot \mathbf{E}$ on the RHS. The resultant equation of (3.45) gives

$$\frac{3}{2}\frac{dP}{dt} = \frac{5}{2}T\frac{dn}{dt}, \tag{3.46}$$

where $d/dt = \partial/\partial t + \langle \mathbf{v} \rangle \cdot \mathbf{\nabla}$. This relation holds independently for both the electron pressure P_e and the ion pressure P_i. Dividing both sides by P and using (3.40), we obtain the relation between the number density and the plasma pressure, as

$$\frac{d}{dt}\ln P = \frac{5}{3}\frac{d}{dt}\ln n. \tag{3.47}$$

Often, the coefficient is replaced with the adiabatic constant $\gamma = 5/3$. The appearance of the adiabatic relation (3.47) is the natural consequence of the Maxwellian distribution (3.39) because the absence of the third moment automatically inhibits the heat conduction.

3.4 MAGNETOHYDRODYNAMIC EQUATIONS

In order to describe a plasma with no net charge as a continuous medium in real space, we employ the moment approach shown in the previous section. Under this approximation, information about the velocity distribution function is averaged out and kinetic effects relating to wave–particle interactions, which will be discussed in chapter 8, are neglected. However, the macroscopic behavior of a magnetically confined plasma can be treated even for a three-dimensional magnetic field geometry.

From the moment with $\ell = 0$, conservation of number density or the continuity equation is given by

$$\frac{\partial n_s}{\partial t} + \mathbf{\nabla} \cdot (n_s \mathbf{v}_s) = 0, \tag{3.48}$$

where $\mathbf{v}_s$ means $\langle v_s \rangle$ in this section, and n_s and $\mathbf{v}_s$ depend on position $\mathbf{x}$ and time t. Here s denotes a particle species, and $s = e$ corresponds to electrons and $s = i$ to ions. Equation (3.48) is obtained by extending the one-dimensional continuity equation (3.41) to three-dimensional space. From the moment with $\ell = 1$, equation of motion for the particle species s is given by

$$m_s n_s \left(\frac{\partial \mathbf{v}_s}{\partial t} + (\mathbf{v}_s \cdot \mathbf{\nabla})\mathbf{v}_s \right) = -\mathbf{\nabla} P_s + n_s q_s (\mathbf{E} + \mathbf{v}_s \times \mathbf{B}), \tag{3.49}$$

which is the same as (3.38). Energy conservation equation is obtained from the moment with $\ell = 2$, which is given by

$$\frac{d}{dt}\ln P_s = \gamma \frac{d}{dt}\ln n_s. \tag{3.50}$$

Then, by noting that the charge density and the plasma current are expressed as

$$\begin{cases} \sigma_e = q_e n_e + q_i n_i \\ \mathbf{J} = n_e q_e \mathbf{v}_e + n_i q_i \mathbf{v}_i, \end{cases} \tag{3.51}$$

respectively, the moment equations can be coupled with the Maxwell euqations (3.11)–(3.14).

When the ions have a charge state specified by Z, $q_i = Ze$ and $q_e = -e$, and μ_0 and ε_0 in the Maxwell equations are the vacuum permittivity and vacuum dilectric constant, respectively. Equation (3.11) is the induction equation, (3.12) is Ampère's law determining the magnetic field due to the conduction current and the displacement current, and (3.14) is the Poisson's equation for an electric field due to space charges. Equation (3.13) means that there is no magnetic monopole. A system of equations (3.11)–(3.14), (3.48)–(3.50), and (3.51) is called two-fluid equations, where one fluid consists of pure electrons and the other pure ions. There are 16 unknown quantities, that is, n_e, n_i, P_e, P_i, $\mathbf{v}_e$, $\mathbf{v}_i$, $\mathbf{E}$, and $\mathbf{B}$, when $\mathbf{J}$ and σ_e are replaced with (3.51), and 18 equations. However it should be noted that the divergence of (3.11) corresponds to the time derivative of (3.13). It is considered that (3.11) describes the time evolution of an initial condition set by (3.13). Similarly, the divergence of (3.12) is equivalent to taking the time derivative of (3.14), when we use (3.48) to calculate the current $\mathbf{J}$ by (3.51). Thus, the Poisson equation (3.14) also specifies an initial condition. Then there remain 16 independent equations equal to the number of independent variables. This result yields that the two-fluid equations are a closed complete set, and they describe the dynamics of both an electron fluid and an ion fluid in self-consistent electromagnetic fields.

One remark in the above two-fluid equations is the lack of a damping mechanism for fluctuations such as density, velocity, pressure, electric field, and magnetic field. There are two types of dissipative mechanism in plasmas; collisional damping and Landau damping due to the wave–particle interaction. Since the two-fluid model is composed of moments and loses information about the velocity distribution function, it is not straightforward to include the latter damping mechanism. However, there are several efforts to induce the Landau damping through a fluid description (see chapter 8). The former collisonal damping mechanism is naturally included in the two-fluid model and a general form of the collisional two-fluid model is known as the Braginskii's equations.

First, we consider collisional momentum exchange between the electron fluid and the ion fluid. The frictional force for the momentum exchange is denoted by

$$\mathbf{R} = -n_e m_e(\mathbf{v}_e - \mathbf{v}_i)/\tau_{ei}, \tag{3.52}$$

where the collision time between electrons and ions is denoted by τ_{ei}. When the collisonal momentum exchange is considerable, the equation of motion for an electron fluid becomes

$$m_e n_e \left(\frac{\partial \mathbf{v}_e}{\partial t} + (\mathbf{v}_e \cdot \nabla)\mathbf{v}_e \right) = -\nabla P_e + n_e q_e (\mathbf{E} + \mathbf{v}_e \times \mathbf{B}) + \mathbf{R}, \tag{3.53}$$

and for an ion fluid,

$$m_i n_i \left(\frac{\partial \mathbf{v}_i}{\partial t} + (\mathbf{v}_i \cdot \nabla)\mathbf{v}_i \right) = -\nabla P_i + n_i q_i (\mathbf{E} + \mathbf{v}_i \times \mathbf{B}) - \mathbf{R}. \tag{3.54}$$

Here we discuss the electron–ion collison frequency $\nu_{ei} = 1/\tau_{ei}$ briefly. Suppose that an electron with a velocity v is moving toward an ion at a fixed point. When the electron reaches distance r_0 from the ion, the Coulomb potential energy at this point is $-Ze^2/(4\pi\varepsilon_0 r_0)$. r_0 is considered to be the distance at which the Coulomb potential energy becomes equal to the kinetic energy of electron $m_e v^2/2$. Thus

$$r_0 \simeq \frac{Ze^2}{2\pi\varepsilon_0 m_e v^2} \tag{3.55}$$

and the cross-section for electron–ion collisions is estimated by

$$\sigma = \pi r_0^2 \simeq \frac{Z^2 e^4}{4\pi\varepsilon_0^2 m_e^2 v^4}. \tag{3.56}$$

Then the electron–ion collision frequency is obtained by considering that the electron collides with all ions within the volume σv in a unit time,

$$\nu_{ei}(v) = n_i \sigma v = \frac{Z^2 e^4 n_i}{4\pi\varepsilon_0^2 m_e^2 v^3}. \tag{3.57}$$

When electrons obey the Maxwellian distribution, v may be replaced by the thermal velocity $v_T = \sqrt{T_e/m_e}$,

$$\nu_{ei} = \tau_{ei}^{-1} = \frac{Z^2 e^4 n_i}{4\pi\varepsilon_0^2 m_e^2 v_T^3}, \tag{3.58}$$

which is proportional to $T_3^{-3/2}$, where T_e is an electron temperature. More strictly, since the Coulomb collisions are governed by successive small-angle scattering, ν_{ei} in (3.58) is multiplied by $\ln \Lambda$. Here Λ is nearly equal to the plasma parameter $\Lambda \sim g \sim n\lambda_D^3$. For high-temperature plasmas it is approximated by $\ln \Lambda = 24 - \ln(n_e^{1/2}/T_e)$ for $T_e > 10\,\text{eV}$ and n_e in units of cm^{-3}.

The resistive one-fluid MHD equations are derived from the two-fluid equations. By taking the sum of the equations of motion (3.53) and (3.54),

$$m_e n_e \frac{d\mathbf{v}_e}{dt} + m_i n_i \frac{d\mathbf{v}_i}{dt} = -\nabla P + \mathbf{J} \times \mathbf{B} + \sigma_e \mathbf{E} \tag{3.59}$$

is given, where $P = P_e + P_i$ is a total pressure. It is noted that the mass density ρ and the average velocity $\mathbf{v}$ are approximated by

$$\begin{cases} \rho = n_i m_i + n_e m_e \simeq n_i m_i \\ \mathbf{v} = (\mathbf{v}_i n_i m_i + \mathbf{v}_e n_e m_e)/\rho \simeq \mathbf{v}_i. \end{cases} \tag{3.60}$$

In the LHS of (3.59), the electron inertia term is usually much smaller than the ion inertia term and is therefore negligible. Then we have the equation

$$\rho\left(\frac{\partial \mathbf{v}}{\partial t} + (\mathbf{v} \cdot \nabla)\mathbf{v}\right) = -\nabla P + \mathbf{J} \times \mathbf{B} + \sigma_e \mathbf{E} \tag{3.61}$$

as the equation of motion for one fluid, or the plasma is approximated by a single magneto-fluid.

Similarly, by taking the sum of the electron continuity equation and the ion continuity equation, we obtain

$$\frac{\partial \rho}{\partial t} + \nabla \cdot (\rho \mathbf{v}) = 0 \tag{3.62}$$

as the one-fluid continuity equation.

It is noted that the energy conservation equation or the adiabatic relation (3.50) can be written as

$$\frac{d}{dt}(P_s n_s^{-\gamma}) = 0. \tag{3.63}$$

After introducing charge neutrality $n_e = Zn_i = n$ and taking the sum of the adiabatic relations for the electron fluid and the ion fluid, we obtain

$$\frac{d}{dt}(P n^{-\gamma}) = 0. \tag{3.64}$$

If we use the continuity equation for the density, (3.64) can be rewritten as

$$\frac{\partial P}{\partial t} + (\mathbf{v} \cdot \nabla)P + \gamma P \nabla \cdot \mathbf{v} = 0. \tag{3.65}$$

From the equation of motion for an electron fluid (3.53), an expression for the relation between the electric field $\mathbf{E}$ and the current density $\mathbf{J}$, i.e., Ohm's law, can be derived. When the electron inertia term is negligible, Ohm's law is written as

$$\mathbf{E} + \mathbf{v} \times \mathbf{B} = \frac{1}{en} (\mathbf{J} \times \mathbf{B} - \nabla P_e + \mathbf{R}), \tag{3.66}$$

where charge neutrality is assumed. The frictional force in (3.66) is expressed as

$$\mathbf{R} = \frac{m_e}{\tau_{ei} e} \mathbf{J} = en_e \eta \mathbf{J}, \tag{3.67}$$

where Spitzer's resistivity,

$$\eta = \frac{m_e}{n_e e^2 \tau_{ei}}, \tag{3.68}$$

is introduced. When the current drift velocity $\mathbf{v}_d = -\mathbf{J}/en$ is much smaller than $\mathbf{v}$, and $|\mathbf{J} \times \mathbf{B}| \sim |\nabla P_e|$, the RHS of (3.66) is simply replaced by $\eta \mathbf{J}$. Then Ohm's law becomes

$$\mathbf{E} + \mathbf{v} \times \mathbf{B} = \eta \mathbf{J}. \tag{3.69}$$

In order to obtain the change density σ_e in the equation of motion (3.61), we take the difference of the continuity equations for the electron fluid and the ion fluid,

$$\frac{\partial \sigma_e}{\partial t} + \nabla \cdot \mathbf{J} = 0, \tag{3.70}$$

which shows a charge conservation law. When charge neutrality is imposed, $\sigma_e = 0$, (3.70) becomes $\nabla \cdot \mathbf{J} = 0$, which is often called a charge neutrality condition.

For electromagnetic fields,

$$\frac{\partial \mathbf{B}}{\partial t} = -\nabla \times \mathbf{E}, \tag{3.71}$$

$$\mu_0 \mathbf{J} = \nabla \times \mathbf{B} \tag{3.72}$$

are assumed, where the displacement current is neglected under the condition $v^2/c^2 \ll 1$. Thus, the resistive one-fluid MHD equations are closed by

$$\frac{d\rho}{dt} + \rho \nabla \cdot \mathbf{v} = 0, \tag{3.73}$$

$$\rho \frac{d\mathbf{v}}{dt} = -\nabla P + \mathbf{J} \times \mathbf{B}, \tag{3.74}$$

$$\frac{dP}{dt} + \gamma P \nabla \cdot \mathbf{v} = 0, \tag{3.75}$$

$$\mathbf{E} + \mathbf{v} \times \mathbf{B} = \eta \mathbf{J}, \tag{3.76}$$

$$\frac{\partial \mathbf{B}}{\partial t} = -\nabla \times \mathbf{E}, \tag{3.77}$$

$$\mu_0 \mathbf{J} = \nabla \times \mathbf{B}, \tag{3.78}$$

for the 14 variables $\{\rho, P, \mathbf{v}, \mathbf{B}, \mathbf{E}, \mathbf{J}\}$. As before, $\nabla \cdot \mathbf{B} = 0$ is considered as an initial condition. Here the Poisson's equation (3.59) is used to calculate the charge density $\sigma_e = \varepsilon_0 \nabla \cdot \mathbf{E}$, if necessary.

It should be noted that the resistive MHD equations are actually closed when the resistivity η is given independently. According to the Coulomb collision theory, Spitzer's resistivity is proportional to $T_e^{-3/2}$. The pressure P also includes T_e. In order to include the temperature dependence of the resistivity within the resistive MHD equations, we need an equation for the electron temperature or electron pressure evolution (see section 3.7) or a simple assumption, $T_e = T_i = P/2n$.

Here we discuss the applicability of the magneto-fluid approximation to magnetized plasmas. When a collision time between ions τ_{ii} is longer than the period of ion cyclotron motion, $\Omega_i \tau_{ii} \gg 1$, the characteristic length of charged particle motions in the perpendicular direction with respect to the magnetic field line becomes the ion Larmor radius, where $\Omega_i = q_i B/m_i$. When it is much shorter than the typical system size, such as a plasma radius or a characteristic length of inhomogeneity of the magnetic field, the magneto-fluid approximation is applicable. However, in the parallel direction with respect to the magnetic field line, the mean free path easily exceeds the characteristic length, such as a major radius of a torus. In the parallel direction the magneto-fluid approximation is also considered to be appropriate, when collisions are sufficiently frequent to keep the velocity distribution function close to Maxwellian. In collisionless plasmas the MHD equations become invalid in the direction along the magnetic field line, although they are still applicable in the perpendicular direction with respect to the magnetic field line. There is a model called a gyro-fluid, which includes effects of wave–particle interaction only in the direction along the magnetic field line (see chapter 8).

There is an extension of the above MHD model to the plasma with anisotropic pressure, described by

$$\begin{aligned}\overleftrightarrow{P} &= P_\perp \overleftrightarrow{I} + (P_\parallel - P_\perp)\mathbf{bb} \\ &= P_\perp \begin{bmatrix} 1 & 0 & 0 \\ 0 & 1 & 0 \\ 0 & 0 & 1 \end{bmatrix} + (P_\parallel - P_\perp) \begin{bmatrix} 0 & 0 & 0 \\ 0 & 0 & 0 \\ 0 & 0 & 1 \end{bmatrix},\end{aligned} \tag{3.79}$$

where $\mathbf{b} = \mathbf{B}/B$ and $B = |\mathbf{B}|$. This expression is obtained for a bi-Maxwellian distribution with $T_\perp \neq T_\parallel$, where $T_\perp$ and $T_\parallel$ are a perpendicular and parallel temperature with respect to the magnetic field line, respectively. If we take the divergence of (3.79),

$$\begin{aligned}\nabla\cdot \overleftrightarrow{P} &= \nabla P_\perp + \mathbf{b}(\mathbf{b}\cdot\nabla)(P_\parallel - P_\perp) + (P_\parallel - P_\perp)\nabla\cdot(\mathbf{bb}) \\ &= \nabla P_\perp + (P_\parallel - P_\perp)(\mathbf{b}\cdot\nabla)\mathbf{b} + (P_\parallel - P_\perp)\mathbf{b}(\nabla\cdot\mathbf{b}) \\ &\quad + \mathbf{b}(\mathbf{b}\cdot\nabla)(P_\parallel - P_\perp) \\ &= \nabla_\perp P_\perp + (P_\parallel - P_\perp)(\mathbf{b}\cdot\nabla)\mathbf{b} + \nabla_\parallel P_\parallel + (P_\parallel - P_\perp)\mathbf{b}(\nabla\cdot\mathbf{b}), \end{aligned} \tag{3.80}$$

where use is made of the identity

$$\nabla\cdot(\mathbf{bb}) = (\mathbf{b}\cdot\nabla)\mathbf{b} + \mathbf{b}(\nabla\cdot\mathbf{b}). \tag{3.81}$$

Equation (3.80) indicates that divergence of the pressure tensor $\overleftrightarrow{P}$ has components both perpendicular and parallel to the magnetic field line. The perpendicular component is

$$(\nabla\cdot \overleftrightarrow{P})_\perp = \nabla_\perp P_\perp + (P_\parallel - P_\perp)(\mathbf{b}\cdot\nabla)\mathbf{b}, \tag{3.82}$$

while the parallel component is given by

$$(\nabla\cdot \overleftrightarrow{P})_\parallel = \nabla_\parallel P_\parallel + (P_\parallel - P_\perp)\mathbf{b}(\nabla\cdot\mathbf{b}). \tag{3.83}$$

These expressions give the perpendicular component of equation of motion corresponding to (3.74),

$$\rho(d\mathbf{v}/dt)_\perp = -\nabla_\perp P_\perp - (P_\parallel - P_\perp)(\nabla\cdot\mathbf{b})\mathbf{b} + \mathbf{J}\times\mathbf{B}, \tag{3.84}$$

and the parallel component,

$$\rho(d\mathbf{v}/dt)_\parallel = -\nabla_\parallel P_\parallel - (P_\parallel - P_\perp)\mathbf{b}(\nabla\cdot\mathbf{b}). \tag{3.85}$$

From $\mathbf{b} = \mathbf{B}/B$ and $\nabla\cdot\mathbf{B} = 0$, it follows that $\nabla\cdot\mathbf{b} = -(\mathbf{B}/B^2)\cdot\nabla B$ and (3.85) becomes

$$\rho(d\mathbf{v}/dt)_\parallel = -\nabla_\parallel P_\parallel - (P_\parallel - P_\perp)(\nabla B/B)_\parallel. \tag{3.86}$$

The continuity equation (3.73), the equations of motion (3.84) and (3.86), Maxwell's equations for the electromagnetic field (3.71) and (3.72), and Ohm's law, $\mathbf{E} + \mathbf{v}\times\mathbf{B} = 0$ present an almost complete system of equations for the unknowns ρ, $\mathbf{v}$, $P_\perp$, $P_\parallel$, $\mathbf{E}$, and $\mathbf{B}$. Here we assume that the resistivity is negligible for high-temperature plasmas, which is consistent with existence of the anisotropy in pressure. We need two equations of state coupling ρ to $P_\parallel$ and

$P_\perp$. Since in the zeroth order the velocity distribution is isotropic in the perpendicular and parallel directions, the corresponding two- and one-dimensional adiabatic expressions may be used. It is noted here that the constancy of phase-space volume, $2\pi v dv V = 2\pi v_0 dv_0 V_0$, for a two-dimensional plasma gives $v^2 V =$ constant, where V is a two-dimensional volume or a surface area. On the contrary, for a one-dimensional plasma, $vV =$ constant and V becomes a line element. Thus a dimensional expression $T_\perp \propto L_\perp^{-2}$ is valid in the perpendicular direction and $T_\parallel \propto L_\parallel^{-2}$ in the parallel direction, where $L_\perp$ or $L_\parallel$ represents a characteristic length. When the motion of particles in the parallel direction is independent of the motion in the perpendicular plane, these relations satisfy $T_\perp^2 T_\parallel \propto L_\perp^{-4} L_\parallel^{-2}$, which means that $T_\perp^2 T_\parallel \propto \rho^2$ or

$$\frac{P_\perp^2 P_\parallel}{\rho^5} = \text{constant}, \tag{3.87}$$

since $\rho \propto L_\perp^{-2} L_\parallel^{-1}$. The last missing equation of state can be obtained by recognizing the connection between the magnetic field strength and the perpendicular component of the particle energy. Since the magnetic moment is conserved for the charged particle motion in the magnetic field (see section 3.6),

$$\frac{\sum_s \frac{1}{2} m_s \langle v_{\perp s}^2 \rangle}{B} = \frac{T_{\perp i} + T_{\perp e}}{B} = \frac{P_\perp}{\rho B} = \text{constant} \tag{3.88}$$

is given. The set of equations (3.73), (3.84), (3.86), (3.87), (3.88), (3.76) with $\eta = 0$, (3.77) and (3.78) is known as the Chew–Goldberger–Low (CGL) equations under the double adiabatic approximation which derives from the equations of state (3.87) and (3.88). It is noted that this approximation assumes that there is no heat flow along the magnetic field line.

3.5 MHD WAVES

Electromagnetic waves with frequencies much lower than the ion cyclotron frequency are explained here. Since, in this frequency range, the plasma behavior may be described by the magnetohydrodynamic (MHD) equations introduced in section 3.4, they are often called MHD waves. For simplicity, a uniform plasma with density n_0 in a uniform magnetic field $\mathbf{B}_0$ is considered. To study small-amplitude waves, we first linearize the ideal MHD equations by neglecting $\eta\mathbf{J}$ in Ohm's law. Using the subscript 1 for the linearized quantities we obtain

$$\frac{\partial \rho_1}{\partial t} + \nabla \cdot (\rho_0 \mathbf{v}_1) = 0, \tag{3.89}$$

$$\rho_0 \frac{\partial \mathbf{v}_1}{\partial t} = \mathbf{J}_1 \times \mathbf{B}_0 - \nabla P_1, \tag{3.90}$$

$$\frac{\partial P_1}{\partial t} + \gamma P_0 \nabla \cdot \mathbf{v}_1 = 0, \tag{3.91}$$

$$\mathbf{E}_1 + \mathbf{v}_1 \times \mathbf{B}_0 = 0, \tag{3.92}$$

$$\frac{\partial \mathbf{B}_1}{\partial t} = -\nabla \times \mathbf{E}_1, \tag{3.93}$$

$$\mu_0 \mathbf{J}_1 = \nabla \times \mathbf{B}_1, \tag{3.94}$$

$$\nabla \cdot \mathbf{B}_1 = 0. \tag{3.95}$$

Small-amplitude waves are characterized by a linear response of the plasma and thus may be treated as a Fourier mode with the amplitude $A_{\mathbf{k}}$ and the phase $(\mathbf{k} \cdot \mathbf{r} - \omega t)$ in the homogeneous plasma,

$$A(\mathbf{r}, t) = A_{\mathbf{k}} e^{-i(\omega t - \mathbf{k} \cdot \mathbf{r})} + c.c., \tag{3.96}$$

where $c.c.$ indicates the complex conjugate. The vector $\mathbf{k}$ is called the wave vector and its direction designates the direction of the wave propagation. Its magnitude $k = |\mathbf{k}|$ is called the wave number and is related to the wavelength λ through $k = 2\pi/\lambda$. ω is the angular frequency of the wave and in general is a function of the wave vector $\mathbf{k}$. ω is related to the wave frequency f, $\omega = 2\pi f$.

The nature of a small-amplitude wave can be characterized by the wave vector dependence of the wave frequency $\omega(\mathbf{k})$, which is called the dispersion relation. When the dispersion relation is given, the phase velocity $\mathbf{v}_{ph}$ of the wave is defined as $\mathbf{v}_{ph} = \omega/\mathbf{k}$ and the group velocity is $\mathbf{v}_g = \partial\omega/\partial\mathbf{k}$.

By using the Fourier amplitude expressions as given by (3.96), (3.89)–(3.95) become

$$-\omega \rho_{\mathbf{k}} + \rho_0 \mathbf{k} \cdot \mathbf{v}_{\mathbf{k}} = 0, \tag{3.89$'$}$$

$$-i\omega \rho_0 \mathbf{v}_{\mathbf{k}} = \mathbf{J}_{\mathbf{k}} \times \mathbf{B}_0 - i\mathbf{k} P_{\mathbf{k}}, \tag{3.90$'$}$$

$$-\omega P_{\mathbf{k}} + \gamma P_0 \mathbf{k} \cdot \mathbf{v}_{\mathbf{k}} = 0, \tag{3.91$'$}$$

$$\mathbf{E}_{\mathbf{k}} + \mathbf{v}_{\mathbf{k}} \times \mathbf{B}_0 = 0, \tag{3.92$'$}$$

$$\omega \mathbf{B}_{\mathbf{k}} = \mathbf{k} \times \mathbf{E}_{\mathbf{k}}, \tag{3.93$'$}$$

$$\mu_0 \mathbf{J}_{\mathbf{k}} = i\mathbf{k} \times \mathbf{B}_{\mathbf{k}}, \tag{3.94$'$}$$

$$\mathbf{k} \cdot \mathbf{B}_{\mathbf{k}} = 0. \tag{3.95$'$}$$

There are two types of waves. One is a torsional wave satisfying the incompressible condition $\nabla \cdot \mathbf{v}_1 = 0$; the other is a compressional wave, for which $\nabla \cdot \mathbf{v}_1 \neq 0$. The torsional wave is also described by vorticity in the direction of the ambient magnetic field, $\Omega(\mathbf{r}, t) \equiv (\nabla \times \mathbf{v}_1) \cdot \mathbf{B}_0/B_0$.

From (3.90$'$), the vorticity satisfies

$$-\omega \rho_0 \Omega_{\mathbf{k}} = (\mathbf{k} \cdot \mathbf{B}_0)(\mathbf{J}_{\mathbf{k}} \cdot \mathbf{B}_0) \frac{1}{B_0}. \tag{3.97}$$

From (3.92′) and (3.93′), $\mathbf{B}_\mathbf{k}$ is shown as

$$\mathbf{B}_\mathbf{k} = \frac{1}{\omega}[(\mathbf{k}\cdot\mathbf{v}_\mathbf{k})\mathbf{B}_0 - (\mathbf{k}\cdot\mathbf{B}_0)\mathbf{v}_\mathbf{k}]. \tag{3.98}$$

Substituting this equation into (3.94′) yields

$$\mathbf{J}_\mathbf{k} = \frac{i}{\mu_0\omega}[(\mathbf{k}\cdot\mathbf{v}_\mathbf{k})(\mathbf{k}\times\mathbf{B}_0) - (\mathbf{k}\cdot\mathbf{B}_0)(\mathbf{k}\times\mathbf{v}_\mathbf{k})]. \tag{3.99}$$

The dispersion relation of the torsional wave is obtained by substituting (3.99) into (3.97) and using the incompressibility $\mathbf{k}\cdot\mathbf{v}_\mathbf{k} = 0$,

$$[\mu_0\rho_0\omega^2 - (\mathbf{k}\cdot\mathbf{B}_0)^2]\Omega_\mathbf{k} = 0. \tag{3.100}$$

By introducing $k_\parallel$ for the wave vector in the direction of magnetic field, (3.100) gives the dispersion relation of the shear Alfvén wave,

$$\omega^2 = k_\parallel^2 V_A^2, \tag{3.101}$$

where

$$V_A = \frac{B_0}{\sqrt{\mu_0\rho_0}} \tag{3.102}$$

is the Alfvén velocity. Since the phase velocity of the shear Alfvén wave $\omega/k_\parallel$ is given by the square root of the ratio of the magnetic tension B_0^2/μ_0 and the mass density ρ_0, the wave may be regarded as a tensile wave of the magnetic line of force with a weight density given by ρ_0.

The dispersion relation of the compressional wave may be obtained by constructing the scalar products of $i\mathbf{k}$ and $i\mathbf{B}_0$ with the equation of motion (3.90′),

$$\omega\rho_0\mathbf{k}\cdot\mathbf{v}_\mathbf{k} = i\mathbf{k}\cdot(\mathbf{J}_\mathbf{k}\times\mathbf{B}_0) + k^2P_\mathbf{k} \tag{3.103}$$

and

$$\omega\rho_0(\mathbf{B}_0\cdot\mathbf{v}_\mathbf{k}) = (\mathbf{k}\cdot\mathbf{B}_0)P_\mathbf{k}. \tag{3.104}$$

By substituting $\mathbf{J}_\mathbf{k}$ given by (3.99) and $P_\mathbf{k}$ from (3.91′) into these equations and eliminating $(\mathbf{B}_0\cdot\mathbf{v}_\mathbf{k})$, we have the following dispersion relation:

$$\omega^4 - \omega^2k^2(V_A^2 + C_s^2) + k^2k_\parallel^2V_A^2C_s^2 = 0, \tag{3.105}$$

where

$$C_s = \left(\frac{\gamma P_0}{\rho_0}\right)^{1/2} = \left(\frac{\gamma T_0}{m_i}\right)^{1/2} \tag{3.106}$$

is the ion sound velocity, and $P_0 = n_0 T_0$ and $\rho_0 = n_0 m_i$ are used in the last expression of (3.106).

Since the dispersion relation is quadratic with respect to ω^2, we note that there are two types of compressional waves, which are coupled. Solving the dispersion relation (3.105) yields

$$\omega^2/k^2 = \tfrac{1}{2}\{(V_A^2 + C_s^2) \pm [(V_A^2 + C_s^2)^2 - 4V_A^2 C_s^2 k_\parallel^2/k^2]^{1/2}\}. \tag{3.107}$$

For perpendicular propagation with $k_\parallel = 0$, the two waves satisfy

$$\omega^2/k_\perp^2 = V_A^2 + C_s^2 \tag{3.108}$$

and

$$\omega = 0, \tag{3.109}$$

while, for parallel propagation with $k_\perp = 0$, these waves satisfy

$$\omega^2/k_\parallel^2 = V_A^2 \tag{3.110}$$

and

$$\omega^2/k_\parallel^2 = C_s^2. \tag{3.111}$$

The wave corresponding to (3.108) and (3.110) is called the fast magnetosonic wave, while that corresponding to (3.109) and (3.111) is called the slow magnetosonic wave at any angle of propagation.

3.6 THE DRIFT-KINETIC EQUATION

For plasma immersed in a strong magnetic field, the Larmor radii of the individual particles are much smaller than the characteristic length of the magnetic field inhomogeneity, L. Since $\rho_L/L \ll 1$, where ρ_L is a representative Larmor radius, we can expand the Vlasov equation for such plasmas with this small parameter. By a drift-kinetic equation we mean an approximate version, appropriate to the small Larmor radius limit, of the Vlasov equation. It is also possible to derive this equation directly from the guiding-center equations of motion rather than by expanding the Vlasov equation.

We will now turn to guiding-center drift equations for time-independent magnetic fields,

$$\frac{d\mathbf{x}}{dt} \equiv \mathbf{v}_D = v_\| \mathbf{b} + \mathbf{v}_E + \mathbf{v}_G + \frac{v_\|^2}{\Omega}(\nabla \times \mathbf{b}), \tag{3.112}$$

$$\frac{dv_\|}{dt} = \left(\frac{q}{m}\mathbf{E} - \frac{\mu}{m}\nabla B\right)\cdot\left[\mathbf{b} + \frac{v_\|}{\Omega}(\nabla \times \mathbf{b})\right], \tag{3.113}$$

$$\frac{d\mu}{dt} = 0, \tag{3.114}$$

where $\Omega = qB/m$ is the cyclotron frequency and $\mu = mv_\perp^2/(2B)$ is the magnetic moment, and

$$\mathbf{v}_E = \frac{\mathbf{E} \times \mathbf{B}}{B^2}, \tag{3.115}$$

$$\mathbf{v}_G = \frac{mv_\perp^2 \mathbf{B} \times \nabla B}{2qB^3} \tag{3.116}$$

are the $\mathbf{E} \times \mathbf{B}$ drift velocity and the gradient B drift velocity, respectively. The last term in (3.113) can be derived by using the energy conservation of a charged particle, $K = \mu B + mv_\|^2/2 + q\phi =$ constant, in the following way:

$$\begin{aligned}\frac{dv_\|}{dt} &= \frac{d}{dt}\left[\left(\frac{2}{m}\right)^{1/2}\sqrt{K - \mu B - q\phi}\right] \\ &= -\frac{1}{\sqrt{2m}}\left(\mu\frac{dB}{dt} + q\frac{d\phi}{dt}\right)(K - \mu B - q\phi)^{-1/2},\end{aligned} \tag{3.117}$$

where

$$\frac{dB}{dt} = \frac{d\mathbf{x}}{dt}\cdot\nabla B, \tag{3.118}$$

$$\frac{d\phi}{dt} = \frac{d\mathbf{x}}{dt}\cdot\nabla\phi. \tag{3.119}$$

By substituting (3.112) into $d\mathbf{x}/dt$ in (3.118) and (3.119), and noting that $(\mathbf{v}_E + \mathbf{v}_G)\cdot(\mu\nabla B + q\nabla\phi) = 0$, we obtain

$$\frac{dv_\|}{dt} = -\left(\frac{\mu}{m}\nabla B + \frac{q}{m}\nabla\phi\right)\cdot\left[\mathbf{b} + \frac{v_\|}{\Omega}(\nabla \times \mathbf{b})\right]. \tag{3.120}$$

For $\mathbf{E} = -\nabla\phi$, (3.120) is exactly equal to (3.113). A special property of this guiding-center equation is that (3.120) contains a term for the parallel drift which is of higher order in ρ_L/L, and that $v_\|$ obtained by (3.120) agrees with the usual guiding-center velocity along the magnetic field line only in the lowest order in ρ_L/L. It is noted that the last term of (3.112) is identical to the curvature drift, since $(\mathbf{b}\cdot\nabla)\mathbf{b} = -\mathbf{b}\times(\nabla\times\mathbf{b})$ for vacuum magnetic fields satisfying $\nabla\times\mathbf{B} = 0$:

$$\mathbf{v}_C = \frac{v_\parallel^2}{\Omega}\,\mathbf{b}\times(\mathbf{b}\cdot\nabla)\mathbf{b} = \frac{v_\parallel^2}{\Omega}\,(\nabla\times\mathbf{b}). \tag{3.121}$$

Here we introduce a guiding-center phase space and a guiding-center distribution function, $f(\mathbf{x}, \mu, v_\parallel, t)$. It is noted that the phase angle dependence of the gyration motion has been averaged out in this distribution function. The collisionless guiding-center drift kinetic equation is obtained from the requirement that the number of guiding centers in the phase space volume element dV be constant in time.

$$\frac{d}{dt}(fdV) = 0, \tag{3.122}$$

where

$$dV = \frac{2\pi}{m}\,Bd\mathbf{x}d\mu dv_\parallel. \tag{3.123}$$

Equation (3.122) can be transformed to yield

$$\frac{df}{dt} + \frac{1}{dV}\frac{d}{dt}(dV)f = 0. \tag{3.124}$$

By noting that the phase space volume element for guiding centers is also conserved like that for many particles, $d/dt(dV) = 0$, (3.124) is written as

$$\frac{\partial f}{\partial t} + \mathbf{v}_D\cdot\nabla f + \frac{dv_\parallel}{dt}\frac{\partial f}{\partial v_\parallel} = 0, \tag{3.125}$$

which is called the drift-kinetic equation. Since the magnetic moment is conserved or $d\mu/dt = 0$, the term containing μ does not appear in (3.125). It is noted that $v_\parallel$ is a coordinate of the guiding-center phase space, while $\mathbf{v}_\perp = \mathbf{v}_E + \mathbf{v}_G + \mathbf{v}_C$ is not.

Instead of $f(\mathbf{x}, \mu, v_\parallel, t)$, the guiding-center distribution function $f(\mathbf{x}, \mu, K, t)$ is also usable in the drift-kinetic equation. In this (3.125) is changed into

$$\frac{\partial f}{\partial t} + \mathbf{v}_D\cdot\nabla f + \frac{dK}{dt}\frac{\partial f}{\partial K} = 0, \tag{3.126}$$

where

$$\frac{dK}{dt} = q\frac{\partial\phi}{\partial t} - qv_\parallel\frac{\partial A_\parallel}{\partial t}. \tag{3.127}$$

Here the electric field generated by the magnetic induction is included through $\partial A_\parallel/\partial t$, and $A_\parallel$ is a component of vector potential given by $\mathbf{B} = \nabla \times \mathbf{A}$ parallel to the magnetic field line. Under the electrostatic approximation the last term of the RHS of (3.127) vanishes.

3.7 TRANSPORT EQUATIONS

The kinetic equation that describes the slow time scale evolution of the distribution function of species (a) in a single ion species plasma is the Vlasov equation with collision term and source term

$$\frac{\partial f_a}{\partial t} + \mathbf{v} \cdot \frac{f_a}{\partial \mathbf{x}} + \frac{e_a}{m_a}(\mathbf{E} + \mathbf{v} \times \mathbf{B}) \cdot \frac{\partial f_a}{\partial \mathbf{v}} = C_a(f_a) + S_a, \tag{3.128}$$

where $C_a(f_a)$ is the Coulomb collision operator (see chapter 7) and S_a is an external particle source term. We can calculate the moment equations as systematically, as in section 3.3. The even parity velocity moments $(1, |\mathbf{v}|^2)$ of (3.128) represent the conservation of particles and energy:

$$\frac{\partial n_a}{\partial t} + \nabla \cdot (n_a \mathbf{u}_a) = S_{na} \tag{3.129}$$

$$\frac{\partial}{\partial t}\left(\tfrac{1}{2} n_a m_a u_a^2 + \tfrac{3}{2} P_a\right) + \nabla \cdot \left[\left(\tfrac{1}{2} n_a m_a u_a^2 + \tfrac{5}{2} P_a\right)\mathbf{u}_a + \overleftrightarrow{\Pi}_a \cdot \mathbf{u}_a + \mathbf{q}_a\right]$$
$$= (e_a n_a \mathbf{E} + \mathbf{F}_{a1}) \cdot \mathbf{u}_a + Q_a + S_{Ea}. \tag{3.130}$$

Here $n_a = \int f_a d_\mathbf{v}$, $P_a = \int m_a(|\mathbf{v} - \mathbf{u}_a|^2/3) f_a d\mathbf{v}$, $\mathbf{u}_a = \int \mathbf{v} f_a d\mathbf{v}/n_a$, and $\mathbf{q}_a = \int (\mathbf{v} - \mathbf{u}_a)(m_a|\mathbf{v} - \mathbf{u}_a|^2/2) f_a d\mathbf{v}$, are the density, pressure, flow velocity, and conductive heat flux of species (a), respectively. In (3.130),

$$\overleftrightarrow{\Pi}_a = \int m_a[(\mathbf{v} - \mathbf{u}_a)(\mathbf{v} - \mathbf{u}_a) - |\mathbf{v} - \mathbf{u}_a|^2 \overleftrightarrow{I}/3] f_a d\mathbf{v} \tag{3.131}$$

is the viscosity tensor, where $(\overleftrightarrow{I})_{ij} = \delta_{ij}$, $S_{na} = \int S_a d\mathbf{v}$ and $S_{Ea} = \int (m_a v^2/2) S_a d\mathbf{v}$ are the particle and energy source (due to ionization, recombination, charge exchange, auxiliary heating, etc.), respectively; and

$$\mathbf{F}_{a1} = \int m_a \mathbf{v} C_a(f_a) d\mathbf{v}, \tag{3.132}$$

$$Q_a = \int \tfrac{1}{2} m_a |\mathbf{v} - \mathbf{u}_a|^2 C_a(f_a) d\mathbf{v} \tag{3.133}$$

are the collisional momentum and heat generation rate in species (a), respectively. The quantity $\mathbf{F}_{a1}$ will be referred to as the friction force. It should be noted that $d\mathbf{v}$ denotes a volume element in velocity space.

The two lowest-order odd velocity moments $(\mathbf{v}, v^2\mathbf{v})$ of (3.128) represent the force and heat flow balance, respectively, for species (a):

$$m_a n_a \frac{d\mathbf{u}_a}{dt} = n_a e_a(\mathbf{E} + \mathbf{u}_a \times \mathbf{B}) + \mathbf{F}_{a1} - \nabla P_a - \nabla \cdot \overleftrightarrow{\Pi}_a \tag{3.134}$$

$$\frac{\partial \mathbf{Q}_a}{\partial t} = \frac{e_a}{m_a}[\mathbf{E} \cdot (\tfrac{5}{2} P_a \overleftrightarrow{I} + \overleftrightarrow{\Pi}_a + m_a n_a \mathbf{u}_a \mathbf{u}_a) + \mathbf{Q}_a \times \mathbf{B}] + \mathbf{G}_a - \nabla \cdot \overleftrightarrow{r}_a, \tag{3.135}$$

where $d/dt \equiv \partial/\partial t + \mathbf{u}_a \cdot \nabla$,

$$\mathbf{Q}_a \equiv \int (m_a v^2/2)\mathbf{v} f_a d\mathbf{v} = (m_a n_a u_a^2/2)\mathbf{u}_a + 5P_a \mathbf{u}_a/2 + \overleftrightarrow{\Pi}_a \cdot \mathbf{u}_a + \mathbf{q}_a \tag{3.136}$$

is the total energy flux appearing in the LHS of (3.130),

$$\overleftrightarrow{r}_a \equiv \int (\tfrac{1}{2} m_a v^2 \mathbf{v}\mathbf{v}) f_a d\mathbf{v} \tag{3.137}$$

is the energy-weighted stress tensor, and

$$\mathbf{G}_a \equiv \int (\tfrac{1}{2} m_a v^2 \mathbf{v}) C_a(f_a) d\mathbf{v} = \frac{T_a}{m_a}(\tfrac{5}{2}\mathbf{F}_{a1} + \mathbf{F}_{a2}) \tag{3.138}$$

is the collisional rate of heat flux generation (or heat friction), where

$$\mathbf{F}_{a2} = \int m_a \mathbf{v}\left(\frac{m_a v^2}{2T_a} - \frac{5}{2}\right) C_a(f_a) d\mathbf{v} \tag{3.139}$$

and $T_a = P_a/n_a$. Equation (3.135) is given here for completeness of the transport equations. Contributions to momentum and heat flux generation from the external sources have been omitted in (3.134) and (3.135).

An equation for the pressure evolution may be obtained by multiplying (3.134) by $\mathbf{u}_a$ and (3.129) by $m_a u_a^2/2$, and subtracting from (3.130):

$$\frac{\partial}{\partial t}\left(\frac{3}{2}P_a\right) + \nabla \cdot \left(\mathbf{q}_a + \frac{5}{2}P_a \mathbf{u}_a\right) = -\overleftrightarrow{\Pi}_a : \nabla \mathbf{u}_a + \mathbf{u}_a \cdot \nabla P_a + Q_a + S_{pa}, \tag{3.140}$$

where $S_{pa} = S_{Ea} - (m_a u_a^2/2) S_{na}$. It is noted that $\overleftrightarrow{\Pi}_a : \nabla \mathbf{u}_a$ becomes a scalar.

The net force balance equation for the plasma is obtained by summing (3.134) over all species, noting that $\sum_a \mathbf{F}_{a1} = 0$ due to momentum conservation in Coulomb collisions and that $\sum_a n_a e_a = 0$ in a quasi-neutral plasma:

$$\sum_a m_a n_a \frac{d\mathbf{u}_a}{dt} = \mathbf{J} \times \mathbf{B} - \nabla \cdot \overleftrightarrow{P}. \tag{3.141}$$

Here $\overleftrightarrow{P} = \sum_a (P_a \overleftrightarrow{I} + \overleftrightarrow{\Pi}_a)$ is the total pressure tensor. A quasi-equilibrium is established in (3.141) on the Alfvén transit time scale $\tau_A \sim L/V_A$, where L is a typical pressure inhomogeneity scale length and $V_A = (\mu_0 \sum_a m_a n_a / B^2)^{-1/2}$ is the Alfvén velocity (see (3.102)). This equilibrium is characterized by a stationary pressure balance including the plasma flow effect:

$$\mathbf{J} \times \mathbf{B} - \nabla \cdot \left(\overleftrightarrow{P} + \sum_a m_a n_a \mathbf{u}_a \mathbf{u}_a \right) = 0. \tag{3.142}$$

For processes with characteristic time scales longer than τ_A (such as collisional diffusion), the plasma evolves temporally through a sequence of equilibria satisfying (3.142). Maxwell's equations (3.71) and (3.72), and Ohm's law (3.76) with $\nabla \cdot \mathbf{B} = 0$ at $t = 0$ are appropriate for describing the diffusive time scale behavior of the electric and magnetic fields in a magnetically confined plasma. Here neglect of the displacement current in Ampère's law (3.72) is justified for $u^2/c^2 \ll 1$. In Faraday's law (3.71), the electric field $\mathbf{E}$ should be consistent with that given by

$$\mathbf{E} = -\mathbf{u}_e \times \mathbf{B} + (\mathbf{F}_{e1} - \nabla P_e - \nabla \cdot \overleftrightarrow{\Pi}_e)/n_e e, \tag{3.143}$$

which is obtained by neglecting the electron inertia term in (3.134) for $a = e$.

In chapter 7, (3.130) is conisdered again to discuss energy transport in stellarators and heliotrons. In the transport study, (3.140) is useful and $5P_a \mathbf{u}_a/2$ is often replaced with $5P_a \boldsymbol{\Gamma}_a/(2n_a)$, where $\boldsymbol{\Gamma}_a = n_a \mathbf{u}_a$ denotes a particle flux vector.

3.8 THE AVERAGED REDUCED MHD EQUATIONS

The reduced MHD equations aim to describe plasma dynamics by keeping the essential physics of the original MHD equations, but with the minimum number of variables. Therefore, the reduced MHD equations are applicable to a subset of MHD phemonema described the the original MHD equations given in section 3.4. However, analytic or numeric studies based on the reduced MHD equations are more tractable than the original MHD equations. This approach was first successful in the tokamak MHD theory, in the study of the internal disruption, the Mirnov oscillations, and the major disruption (see chapter 8). These phenomena are basically related to the nonlinear evolution of tearing modes (see chapter 5) in tokamaks. In this section we derive reduced MHD equations describing magnetohydrodynamics in heliotrons and torsatrons. As shown later in this section, when the effects of the external helical magnetic field are neglected, our reduced MHD equations become the same as those for high-beta tokamaks.

The metrics for the toroidal coordinates (r, θ, ζ) (see Fig. 2.1) are written as

$$(d\ell)^2 = (dr)^2 + (rd\theta)^2 + \left(1 + \frac{r}{R_0} \cos\theta\right)^2 d\zeta^2, \tag{3.144}$$

where $\zeta = -R_0\phi$, and ϕ is an angle variable in the toroidal direction.

The most natural expansion parameter for toroidal plasmas is an inverse aspect ratio $\varepsilon = r/R_0$. Since a/R_0 is typically one tenth in heliotrons, $\varepsilon \ll 1$ is a reasonable assumption. Next we consider the magnitude of the helical magnetic field $\mathbf{B}_h$ generated by helical windings (see chapter 2). We assume that $\delta = |\mathbf{B}_h|/B_0 \simeq \varepsilon^{1/2}$ and we also consider δ as an expansion parameter, since $\delta < 1$. Another important parameter for toroidal plasmas is a beta value or a ratio between the plasma pressure and the magnetic pressure, $\beta = P/(B^2/2\mu_0)$. Here we assume that $\beta \simeq \delta^2 \simeq \varepsilon$. This ordering is called the "stellarator expansion ordering," and was introduced by Greene and Johnson (1961) at Princeton Plasma Physics Laboratory.

The helical magnetic fields produced by the helical windings are given by a solution of $\nabla^2\Phi = 0$ which is obtained from $\mathbf{B}_h = \nabla\Phi$ and $\nabla \cdot \mathbf{B}_h = 0$. When the toroidal effect is negligible, the magnetic potential Φ in the coordinates (r, θ, ζ) is given by

$$\Phi = \sum_{\ell-\infty}^{\infty} \sum_{p=1}^{\infty} \Phi_{\ell p} I_\ell(phr) \sin(\ell\theta - ph\zeta + \phi_{\ell p}), \tag{3.145}$$

where ℓ is a pole number of the helical winding and $h = M/R_0 \sim O(1)$. M is called a pitch number, $\Phi_{\ell p}$ is a constant coefficient, $\phi_{\ell p}$ is a phase angle, and I_ℓ is a modified Bessel function. It is noted here that $M \sim \varepsilon^{-1} \gg 1$ and, for example, $M = 19$ for the Heliotron-E device and $M = 12$ for the ATF device. Usually, $\ell = 1$, 2, and 3 are relevant to realistic devices. When the magnetic potential (3.145) is given, the total magnetic field of heliotrons is described by

$$\mathbf{B} = B_0\hat{\zeta} + \delta\nabla\Phi + \delta^2\nabla A \times \hat{\zeta} + \delta^2 B_2\hat{\zeta}, \tag{3.146}$$

with the small ordering parameter δ. The first and second terms correspond to the external toroidal magnetic field and the helical magnetic field, respectively. The third term denotes a magnetic field produced by a plasma current in the toroidal direction $\mathbf{J} = J\hat{\zeta}$, where A corresponds to the ζ component of the vector potential satisfying $\mathbf{B}_J = \nabla \times \mathbf{A} = \nabla \times (A\hat{\zeta}) = \nabla A \times \hat{\zeta}$. Since the rotational transform produced by the plasma current is expected to be less than unity, $B_\theta/B_0 \sim O(\varepsilon) \sim O(\delta^2)$ is assumed here, where B_θ denotes the poloidal magnetic field produced by the toroidal plasma current J, which is also $O(\delta^2)$. It is also remarked that the current J in the ζ direction includes the Pfirsch–Schlüter current, which is explained in chapter 4. The fourth term includes corrections of the toroidal mangetic field due to the diagmagnetic effect and toroidal curvature. When the plasma pressure is inhomogeneous in magnetized plasmas, the diamagnetic current is induced in the direction perpendicular to both $\mathbf{B}$ and ∇P, which gives a correction to the externally applied magnetic field. For finite-beta plasmas this correction affects the properties of equilibrium and stability. The corrections in the toroidal magnetic field are understood from the relation

$$B_\zeta = \frac{I}{R} = \frac{I_0 + I_1}{R_0 + r\cos\theta} \simeq \frac{I_0}{R_0}\left(1 + \frac{I_1}{I_0} - \frac{r\cos\theta}{R_0}\right). \tag{3.147}$$

Here I_0 is the toroidal coil current producing the external toroidal field $B_0 = I_0/R_0$, and I_1 denotes the diamagnetic current in the poloidal (or θ) direction. Thus $B_2 = B_0(I_1/I_0 - r\cos\theta/R_0)$. From the expression for the magnetic field (3.146) the total plasma current is obtained by using (3.78):

$$\begin{aligned}\mu_0 \mathbf{J} &= \delta^2 \nabla \times \left(\nabla A \times \hat{\zeta} + B_0 \frac{I_1}{I_0}\,\hat{\zeta}\right) \\ &= -\delta^2 \nabla_\perp^2 A\hat{\zeta} + \delta^2\,\frac{B_0}{I_0}\,\nabla_\perp I_1 \times \hat{\zeta},\end{aligned} \tag{3.148}$$

where

$$\nabla_\perp \equiv -\hat{\zeta}\,\frac{\partial}{\partial\zeta}, \tag{3.149}$$

$\nabla_\perp^2 = \nabla_\perp \cdot \nabla_\perp$ and $|\partial/\partial\zeta|/|\nabla_\perp| \sim O(\delta)$ being imposed here. It is noted that $|\partial/\partial\zeta|/|\nabla_\perp| \sim O(1)$ when $\partial/\partial\zeta$ is operated on the magnetic potential Φ.

As explained in section 3.5, there are two time scales characterized by the Alfvén wave in the MHD model. In the toroidal plasma considered here, the perpendicular wavelength is estimated by a plasma minor radius a or $k_\perp \sim 1/a$, and the parallel wavelength is estimated by a major radius or $k_\parallel \sim 1/R_0$. Thus the fast time scale is given by $\tau_f = a/V_A$, which is related to the fast magnetosonic wave, and the slow time scale is given by $\tau_s = R/V_A$, which is related to the shear Alfvén wave. The significant difference between these MHD waves comes from the compressibility. When the incompressible condition $\nabla \cdot \mathbf{v} = 0$ is assumed, the fast time scale disappears and the characteristic time becomes the transit time of the shear Alfvén wave or τ_s. Thus, we assume that $\nabla \cdot \mathbf{v} = 0$, which may be valid in toroidal plasmas immersed in the strong toroidal (or longitudinal) magnetic field. Since $\tau_s \sim O(\varepsilon^{-1})$, the ordering of the time derivative determined by the slow time scale τ_s is $|\partial/\partial t| \simeq O(\varepsilon) \simeq O(\delta^2)$.

Here we also assume that $|\mathbf{v}_\perp| \sim O(\delta^2)$ and $|\mathbf{v}_\parallel|/|\mathbf{v}_\perp| \simeq O(\varepsilon) \simeq O(\delta^2)$. The former ordering is consistent with that the time evolution is $O(\delta^2)$. From this relation $\nabla \cdot \mathbf{v} \simeq \nabla_\perp \cdot \mathbf{v}_\perp = 0$ is required up to $O(\delta^3)$, since $\nabla_\parallel \cdot \mathbf{v}_\parallel \sim O(\delta^4)$. Here the rapid variation in the ζ direction is kept in $\nabla_\parallel$.

From the induction equation (3.77) and Ohm's law (3.76),

$$\frac{d\mathbf{B}}{dt} = \frac{\partial \mathbf{B}}{\partial t} + (\mathbf{v}\cdot\nabla)\mathbf{B} = -\mathbf{B}(\nabla\cdot\mathbf{v}) + (\mathbf{B}\cdot\nabla)\mathbf{v} + \frac{\eta}{\mu_0}\,\nabla^2\mathbf{B} \tag{3.150}$$

is obtained for $\eta =$ constant. Projection of (3.150) in the ζ direction gives

$$\frac{dB_\zeta}{dt} = -B_\zeta(\mathbf{\nabla} \cdot \mathbf{v}) + (\mathbf{B} \cdot \mathbf{\nabla})v_\zeta + \frac{\eta}{\mu_0} \nabla^2 B_\zeta. \tag{3.151}$$

By substituting (3.147) into (3.151),

$$\frac{dI_1}{dt} = -I_0(\mathbf{\nabla} \cdot \mathbf{v}) + I_0 \frac{\partial v_\zeta}{\partial \zeta} + \frac{\eta}{\mu_0} \nabla^2 I_1 \tag{3.152}$$

is obtained. Here we consider the ordering of each term in (3.152); $I_1 \sim O(\delta^2)$, $d/dt \sim O(\delta^2)$, $\partial/\partial\zeta \sim O(\delta)$, $v_\zeta(\text{or } v_\parallel) \sim O(\delta^4)$, and $\eta \sim O(\delta^2)$. From the comparison between the LHS and the lst term in the RHS of (3.152), describing the resistive diffusion process, $\eta \sim O(\delta^2)$ is required. Since $|\mathbf{v}_\perp| \sim O(\delta^2)$ and $|\mathbf{v}_\parallel|/|\mathbf{v}_\perp| \sim O(\delta^2)$, $v_\zeta(\text{or } v_\parallel) \sim \delta^2 |\mathbf{v}_\perp| \sim O(\delta^4)$. From these orderings, (3.152) gives

$$dI_1/dt = 0 \tag{3.153}$$

to $O(\delta^3)$.

Now we will derive reduced MHD equations by using the orderings discussed above. First, we will consider the equation of motion (3.74). Since the inertia term is $O(\delta^4)$ (note that $|\mathbf{v}_\perp| \sim O(\delta^2)$), the RHS becomes

$$0 = -\mathbf{\nabla}_\perp P + \mathbf{J} \times B_0 \hat{\zeta}, \tag{3.154}$$

in the lowest order. Substituting $\mathbf{J}_\perp = (\mathbf{\nabla} \times \mathbf{B})_\perp/\mu_0 \simeq (\mathbf{\nabla} \times B_2 \hat{\zeta})/\mu_0$ into the second term in (3.154) gives

$$\mathbf{\nabla}_\perp \left(P + \frac{B_0^2}{\mu_0} \frac{I_1}{I_0} \right) = 0 \tag{3.155}$$

up to $O(\delta^3)$. The equilibrium state,

$$P + \frac{B_0^2}{\mu_0} \frac{I_1}{I_0} = \text{constant}, \tag{3.156}$$

is the same as that of theta pinch, which has $P(r)$ and $\mathbf{B} = (0, 0, B_z(r))$ in the cylindrical geometry. It is interesting that the lowest-order perpendicular pressure balance in heliotrons is similar to that of theta pinch.

Here we note that $\nabla \cdot \mathbf{J} = 0$ is satisfied in MHD equations. When $\mathbf{J}_\perp$ obtained from the equation of motion is substituted into $\nabla \cdot \mathbf{J} = 0$, it is shown as

$$\nabla \cdot \left\{ -\left(\rho \frac{d\mathbf{v}}{dt} + \nabla P \right) \times \mathbf{B}/B^2 + \sigma \mathbf{B} \right\} = 0. \tag{3.157}$$

Here $\mathbf{J}_\parallel$ is denoted as $\sigma\mathbf{B}$, and $\mu_0 \sigma = -\nabla_\perp^2 A/B_0 \sim O(\delta^2)$. From (3.157),

$$\mathbf{B} \cdot \nabla \times \frac{\rho}{B^2} \frac{d\mathbf{v}}{dt} = \mathbf{B} \cdot \nabla\sigma + \frac{1}{B^4} (\nabla B^2 \times \nabla P) \cdot \mathbf{B} \tag{3.158}$$

is obtained by taking the component along the magnetic field vector. Here the LHS is $O(\delta^4)$ and the RHS is $O(\delta^3)$, since $|\mathbf{B} \cdot \nabla|$ and $|\nabla B^2|$ are $O(\delta)$. The LHS of (3.158) becomes

$$\mathbf{B} \cdot \nabla \times \frac{\rho}{B^2} \frac{d\mathbf{v}}{dt} \simeq \mathbf{B} \cdot \frac{\rho}{B_0^2} \nabla \times \frac{d\mathbf{v}_\perp}{dt} = \frac{\rho}{B_0^2} \frac{d}{dt} \nabla_\perp^2 u, \tag{3.159}$$

where $\rho =$ constant $\sim O(1)$ is assumed and u is a stream function given by $\mathbf{v}_\perp = \nabla u \times \hat{\zeta}$. thus (3.158) is written as

$$-\frac{\rho}{B_0} \frac{d}{dt} \nabla_\perp^2 u = \mathbf{B} \cdot \nabla\sigma + \frac{1}{B^4} (\nabla B^2 \times \nabla P) \cdot \mathbf{B}. \tag{3.160}$$

The parallel component of the equation of motion is given by

$$\rho \frac{dv_\parallel}{dt} = -\frac{\mathbf{B}}{B} \cdot \nabla P. \tag{3.161}$$

By noting that $v_\parallel \sim O(\delta^4)$ and $d/dt \sim O(\delta^2)$, $\mathbf{B} \cdot \nabla P = 0$ is valid to $O(\delta^5)$. Here we note that

$$\begin{aligned} \frac{d}{dt} (\mathbf{B} \cdot \nabla)P &= \frac{\partial}{\partial t} (\mathbf{B} \cdot \nabla)P + (\mathbf{v} \cdot \nabla)(\mathbf{B} \cdot \nabla)P \\ &= \frac{\partial \mathbf{B}}{\partial t} \cdot \nabla P + (\mathbf{B} \cdot \nabla) \frac{\partial P}{\partial t} + \{(\mathbf{v} \cdot \nabla)\mathbf{B}\} \cdot \nabla P \\ &\quad + (\mathbf{B} \cdot \nabla)\{(\mathbf{v} \cdot \nabla)P\} - \{(\mathbf{B} \cdot \nabla)\mathbf{v}\} \cdot \nabla P \\ &= 0 \end{aligned} \tag{3.162}$$

by considering the pressure evolution equation under the incompressible condition,

$$\frac{\partial P}{\partial t} + (\mathbf{v} \cdot \nabla)P = 0, \tag{3.163}$$

and the induction equation under $\eta = 0$,

$$\frac{\partial \mathbf{B}}{\partial t} = -(\mathbf{v} \cdot \nabla)\mathbf{B} + (\mathbf{B} \cdot \nabla)\mathbf{v}. \qquad (3.164)$$

It is noted that, for $v_\parallel = 0$ and $\mathbf{B} \cdot \nabla P = 0$ at $t = 0$, $v_\parallel = 0$ is valid even for an arbitrary time $t > 0$. This also supports the ordering $v_\parallel/v_\perp \sim O(\delta^2)$. When $\mathbf{v} \simeq \mathbf{v}_\perp$, (3.163) becomes

$$\frac{\partial P}{\partial t} + (\nabla u \times \hat{\boldsymbol{\zeta}}) \cdot \nabla_\perp P = 0. \qquad (3.165)$$

The ordering of all terms in (3.165) is unified under $O(\delta^4)$, since $P \sim O(\delta^2)$, $\partial/\partial t \sim O(\delta^2)$, and $u \sim O(\delta^2)$.

Here we write the induction equation as

$$\frac{\partial}{\partial t} \nabla \times \mathbf{A} + \nabla \times \mathbf{E} = 0, \qquad (3.166)$$

which gives

$$\frac{\partial \mathbf{A}}{\partial t} + \mathbf{E} = \nabla(uB_0). \qquad (3.167)$$

The RHS of (3.167) is chosen by noting that $\mathbf{v}_\perp = \mathbf{E} \times \mathbf{B}/B^2 \simeq \mathbf{E} \times \hat{\boldsymbol{\zeta}}/B_0$ is equal to $\nabla u \times \hat{\boldsymbol{\zeta}}$. The component in the direction of $\mathbf{b} = \mathbf{B}/B \simeq \mathbf{B}/B_0$ is written as

$$\frac{\partial A}{\partial t} + E_\parallel = \mathbf{B} \cdot \nabla u. \qquad (3.168)$$

Here A denotes a parallel component of the vector potential $\mathbf{A}$. We note that the LHS of (3.168) is $O(\delta^4)$, since $\partial/\partial t \sim O(\delta^2)$, $A \sim O(\delta^2)$, and $E_\parallel = \eta J_\parallel = \eta \sigma B_0 = -(\eta/\mu_0)\nabla_\perp^2 A \sim O(\delta^4)$. However, the RHS of (3.168) is $O(\delta^3)$, since $|\mathbf{B} \cdot \nabla|$ is $O(\delta)$.

Now we may consider that (3.160), (3,165), and (3.168) are the reduced MHD equations when η is given. Here the variables are u, A, and P. For these equations, the orderings are summarized in Table 3.1.

We must recall that the lowest order of the RHS in (3.160) and (3.168) is $O(\delta^3)$, which is different from $O(\delta^4)$ of the LHS in these equations. This difference derives from the fact that two scales exist along the magnetic field line or in the toroidal direction; one is short-wavelength variation comparable to the plasma radius and the other is long-wavelength variation due to the toroidal effect. The former is produced by the external helical magnetic field $\mathbf{B}_h$, and the helical pitch length is $2\pi R_0/M$ is $O(\delta^0)$ since $M \sim O(\delta^{-2})$ and $R_0 \sim 0(\delta^{-2})$. In order to eliminate the difference of ordering between the LHS and the RHS, we apply an averaging procedure over the short-wavelength

Table 3.1 Ordering of Stellarator Expansion

Ordering	$O(1)$	$O(\delta)$	$O(\delta^2)$	$O(\delta^4)$
Toroidal field	$B_0, I_0/R_0$			
Helical field		$\mathbf{B}_h, \nabla\Phi$		
Poloidal field			$\nabla A \times \hat{\zeta}$	
Toroidal correction			$B_2, I_1/I_0$	
Beta			β	
Pressure			P	
Poloidal flux (by current)			A	
Poloidal flux (by helical field)			Ψ_h	
Stream function			u	
Perpendicular velocity			$\mathbf{v}_\perp, \nabla u \times \hat{\zeta}$	
Parallel velocity				$\mathbf{v}_\parallel$
Parallel current			$J_\parallel$	
Time derivative			$\partial/\partial t$	
Resistivity			η	
Perpendicular derivative	$\nabla_\perp$			
Derivative of Φ with respect to ζ	$\partial/\partial\zeta$			
Derivative of others with respect to ζ		$\partial/\partial\zeta$		
Derivative with respect to $\bar{\zeta}$			$\partial/\partial\bar{\zeta}$	

variation to both sides of (3.160) and (3.168). This is the origin of the term "averaging method." There are several variations in the averaging method according to the coordinate to which the averaging procedure is applied. Here we consider averaging along the toroidal or ζ direction.

Equations (3.160) and (3.168) have the form

$$\mathbf{B} \cdot \nabla F = G, \tag{3.169}$$

which is generally called a magnetic differential equation. Both sides of (3.169) have the following orderings:

$$\begin{cases} F = \delta^2 F_0 + \delta^3 F_1 + \delta^4 F_2 + \cdots \\ G = \delta^3 G_1 + \delta^4 G_2 + \cdots \end{cases} . \tag{3.170}$$

The slow and fast variations along the toroidal direction are shown by

$$\begin{cases} F = F(r, \theta, \zeta, \delta^2 \bar{\zeta}) \\ G = G(r, \theta, \zeta, \delta^2 \bar{\zeta}), \end{cases} \tag{3.171}$$

which gives the ordering $|\partial/\partial\zeta|/|\partial/\partial\bar{\zeta}| \simeq O(\delta^2)$. With the use of (3.170) and (3.171) in the magnetic differential equation (3.169),

$$B_0 \frac{\partial F_0}{\partial \zeta} = 0 \tag{3.172}$$

is obtained in the lowest order of $O(\delta^2)$. Equation (3.172) gives $F_0 = F_0(r, \theta, \delta^2 \bar{\zeta})$. To $O(\delta^3)$,

$$B_0 \frac{\partial F_1}{\partial \zeta} + \nabla \Phi \cdot \nabla F_0 = G_1 \tag{3.173}$$

is obtained, since the external helical magnetic field in (3.146) satisfies $\nabla \Phi \simeq O(\delta)$. Here we define the averaging procedure over the helical pitch length $2\pi R_0/M$ along the ζ direction:

$$\bar{f}(r, \theta, \delta^2 \bar{\zeta}) = \frac{M}{2\pi R_0} \int_0^{\frac{2\pi R_0}{M}} f(r, \theta, \zeta, \delta^2 \bar{\zeta}) d\zeta. \tag{3.174}$$

Also, the function $\langle f \rangle (r, \theta, \delta^2 \bar{\zeta})$ is defined by

$$\langle f \rangle (r, \theta, \delta^2 \bar{\zeta}) = \int_0^{\zeta} f(r, \theta, \zeta, \delta^2 \bar{\zeta}) d\zeta + C(r, \theta, \delta^2 \bar{\zeta}). \tag{3.175}$$

The integral constant $C(r, \theta, \delta^2 \bar{\zeta})$ is determined by

$$\langle \bar{f} \rangle = 0. \tag{3.176}$$

It is easy to show the following properties for $\bar{f}$ and $\langle f \rangle$:

$$\overline{\frac{\partial f}{\partial \zeta}} = 0, \tag{3.177}$$

and

$$\frac{\partial \langle f \rangle}{\partial \zeta} = \left\langle \frac{\partial f}{\partial \zeta} \right\rangle = f, \tag{3.178}$$

as shown in section 2.4. We note that $\bar{\Phi} = 0$ and $\overline{\nabla \Phi} = 0$ from (3.174). Here we apply (3.175) to (3.173), and obtain

$$F_1 = -\frac{1}{B_0} \nabla \langle \Phi \rangle \cdot \nabla F_0 + \frac{1}{B_0} \langle G_1 \rangle. \tag{3.179}$$

To $O(\delta^4)$, (3.169) gives

$$B_0 \frac{\partial F_2}{\partial \zeta} + \nabla F_1 \cdot \nabla \Phi + B_0 \frac{\partial F_0}{\partial \bar{\zeta}} + (\nabla A \times \hat{\zeta}) \cdot \nabla F_0 = G_2. \tag{3.180}$$

By noting that F_0 has only a slowly varying component in the ζ direction and applying the averaging procedure (3.174) to (3.180),

$$
\begin{aligned}
B_0 \frac{\partial F_0}{\partial \zeta} + \nabla F_0 \times \nabla A \cdot \hat{\zeta} &= \overline{G_2} - \overline{\nabla F_1 \cdot \nabla \Phi} \\
&= \overline{G_2} - \frac{1}{B_0} \overline{\nabla \langle G_1 \rangle \cdot \nabla \Phi} \\
&\quad + \frac{1}{B_0} \overline{\nabla (\nabla F_0 \cdot \nabla \langle \Phi \rangle) \cdot \nabla \Phi}
\end{aligned} \tag{3.181}
$$

is obtained, where (3.179) is substituted into the second equality.

Now we show a somewhat lengthy calculation to give

$$
\overline{\nabla (\nabla F_0 \cdot \nabla \langle \Phi \rangle) \cdot \nabla \Phi} = \tfrac{1}{2} \nabla F_0 \cdot \nabla (\overline{\nabla \langle \Phi \rangle \times \nabla \Phi} \cdot \hat{\zeta}) \times \hat{\zeta}. \tag{3.182}
$$

According to vector calculations,

$$
\begin{aligned}
\nabla (\nabla F_0 \cdot \nabla \langle \Phi \rangle) \cdot \nabla \Phi =& \nabla F_0 \cdot \{ (\nabla \Phi \cdot \nabla) \nabla \langle \Phi \rangle - (\nabla \langle \Phi \rangle \cdot \nabla) \nabla \Phi \} \\
&+ \nabla \langle \Phi \rangle \cdot \nabla (\nabla \Phi \cdot \nabla F_0)
\end{aligned} \tag{3.183}
$$

and

$$
\begin{aligned}
\nabla \times (\nabla \langle \Phi \rangle \times \nabla \Phi) =& \nabla \langle \Phi \rangle (\nabla \cdot \nabla \Phi) - \nabla \Phi (\nabla \cdot \nabla \langle \Phi \rangle) \\
&+ (\nabla \Phi \cdot \nabla) \nabla \langle \Phi \rangle - (\nabla \langle \Phi \rangle \cdot \nabla) \nabla \Phi
\end{aligned} \tag{3.184}
$$

are obtained. By noting that Φ denotes the magnetic potential and that $\nabla \cdot \nabla \Phi = 0$,

$$
\begin{aligned}
\nabla (\nabla F_0 \cdot \nabla \langle \Phi \rangle) \cdot \nabla \Phi =& \nabla F_0 \cdot \nabla \times (\nabla \langle \Phi \rangle \times \nabla \Phi) \\
&+ (\nabla F_0 \cdot \nabla \Phi)(\nabla \cdot \nabla \langle \Phi \rangle) + \nabla \langle \Phi \rangle \cdot \nabla (\nabla \Phi \cdot \nabla F_0)
\end{aligned} \tag{3.185}
$$

is given from (3.183) and (3.184). The second term of the RHS of (3.185) is written as

$$
(\nabla F_0 \cdot \nabla \Phi)(\nabla \cdot \nabla \langle \Phi \rangle) = \nabla \cdot \{ (\nabla F_0 \cdot \nabla \Phi) \nabla \langle \Phi \rangle \} - \nabla \langle \Phi \rangle \cdot \nabla (\nabla F_0 \cdot \nabla \Phi). \tag{3.186}
$$

Here we substitute $\partial / \partial \zeta \, \langle \Phi \rangle = \Phi$ into the first term of the RHS of (3.186) and note that F_0 does not include the fast variation along the ζ direction:

$$
\begin{aligned}
(\nabla F_0 \cdot \nabla \Phi)(\nabla \cdot \nabla \langle \Phi \rangle) =& \frac{\partial}{\partial \zeta} \{ \nabla \cdot [(\nabla F_0 \cdot \nabla \langle \Phi \rangle) \nabla \langle \Phi \rangle] \} \\
&- \nabla \cdot [(\nabla F_0 \cdot \nabla \langle \Phi \rangle) \frac{\partial}{\partial \zeta} \nabla \langle \Phi \rangle] - \nabla \langle \Phi \rangle \cdot \nabla (\nabla F_0 \cdot \nabla \Phi) \\
=& \frac{\partial}{\partial \zeta} \{ \nabla \cdot [(\nabla f_0 \cdot \nabla \langle \Phi \rangle) \nabla \langle \Phi \rangle] \} - \nabla \Phi \cdot \nabla (\nabla F_0 \cdot \nabla \langle \Phi \rangle) \\
&- \nabla \langle \Phi \rangle \cdot \nabla (\nabla F_0 \cdot \nabla \Phi).
\end{aligned} \tag{3.187}
$$

From (3.185) and (3.187), we obtain

$$\begin{aligned}\nabla(\nabla F_0 \cdot \nabla\langle\Phi\rangle) \cdot \nabla\Phi =& \nabla F_0 \cdot \nabla \times (\nabla\langle\Phi\rangle \times \nabla\Phi) \\ &+ \frac{\partial}{\partial\zeta}\{\nabla \cdot [(\nabla F_0 \cdot \nabla\langle\Phi\rangle)\nabla\langle\Phi\rangle]\} \\ &- \nabla\Phi \cdot \nabla(\nabla F_0 \cdot \nabla\langle\Phi\rangle),\end{aligned} \tag{3.188}$$

which gives

$$\begin{aligned}\nabla(\nabla F_0 \cdot \nabla\langle\Phi\rangle) \cdot \nabla\Phi =& \tfrac{1}{2}\nabla F_0 \cdot \nabla \times (\nabla\langle\Phi\rangle \times \nabla\Phi) \\ &+ \tfrac{1}{2}\frac{\partial}{\partial\zeta}\{\nabla \cdot [(\nabla F_0 \cdot \nabla\langle\Phi\rangle)\nabla\langle\Phi\rangle]\}.\end{aligned} \tag{3.189}$$

In the first term of the RHS of (3.189), the following relation is substituted:

$$\begin{aligned}\nabla \times (\nabla\langle\Phi\rangle \times \nabla\Phi) =& \nabla(\nabla\langle\Phi\rangle \times \nabla\Phi \cdot \hat{\zeta}) \times \hat{\zeta} \\ &+ \frac{\partial}{\partial\zeta}[\nabla \times (\Phi\nabla\langle\Phi\rangle \times \hat{\zeta})] - \nabla \times \nabla \times (\Phi^2\hat{\zeta}).\end{aligned} \tag{3.190}$$

Here we note that the RHS of (3.182), which is expected to be the final expression of the present calculations, appears in the first term of the RHS of (3.190). In order to prove the relation (3.190), we need the following equalities:

$$\begin{aligned}\nabla(\nabla\langle\Phi\rangle \times \nabla\Phi \cdot \hat{\zeta}) \times \hat{\zeta} &= \{\hat{\zeta} \times (\nabla \times (\nabla\langle\Phi\rangle \times \nabla\Phi)) + (\hat{\zeta} \cdot \nabla)(\nabla\langle\Phi\rangle \times \nabla\Phi)\} \times \hat{\zeta} \\ &= \nabla \times (\nabla\langle\Phi\rangle \times \nabla\Phi) - (\hat{\zeta} \cdot \nabla \times (\nabla\langle\Phi\rangle \times \nabla\Phi)) \\ &\quad + \frac{\partial}{\partial\zeta}(\nabla\langle\Phi\rangle \times \nabla\Phi) \times \hat{\zeta}\end{aligned} \tag{3.191}$$

and

$$\begin{aligned}\nabla \times \nabla \times (\Phi^2\hat{\zeta}) &= \nabla \times \nabla \times \left(\frac{\partial\langle\Phi\rangle}{\partial\zeta}\Phi\hat{\zeta}\right) \\ &= \frac{\partial}{\partial\zeta}[\nabla \times \nabla \times (\langle\Phi\rangle\Phi\hat{\zeta})] - \nabla \times \nabla \times \left(\langle\Phi\rangle\frac{\partial\Phi}{\partial\zeta}\hat{\zeta}\right) \\ &= \frac{\partial}{\partial\zeta}[\nabla \times (\Phi\nabla\langle\Phi\rangle \times \hat{\zeta})] + \frac{\partial}{\partial\zeta}[\nabla \times (\langle\Phi\rangle\nabla\Phi \times \hat{\zeta})] \\ &\quad - \nabla \times \nabla \times \left(\langle\Phi\rangle\frac{\partial\Phi}{\partial\zeta}\hat{\zeta}\right).\end{aligned} \tag{3.192}$$

Then the RHS of (3.190) becomes

$$\begin{aligned}&\nabla \times (\nabla\langle\Phi\rangle \times \nabla\Phi) - \hat{\zeta}(\hat{\zeta} \cdot \nabla \times (\nabla\langle\Phi\rangle \times \nabla\Phi)) \\ &\quad + \frac{\partial}{\partial\zeta}(\nabla\langle\Phi\rangle \times \nabla\Phi) \times \hat{\zeta} - \frac{\partial}{\partial\zeta}[\nabla \times (\langle\Phi\rangle\nabla\Phi \times \hat{\zeta})] \\ &\quad + \nabla \times \nabla \times \left(\langle\Phi\rangle\frac{\partial\Phi}{\partial\zeta}\hat{\zeta}\right)\end{aligned} \tag{3.193}$$

which is shown to be equal to the LHS of (3.190) by substituting the following relations into (3.193):

$$\begin{aligned}\hat{\zeta}(\hat{\zeta}\cdot\nabla\times(\nabla\langle\Phi\rangle\times\nabla\Phi)) &= \hat{\zeta}[-(\hat{\zeta}\cdot\nabla\Phi)\nabla\cdot\nabla\langle\Phi\rangle+(\nabla\Phi\cdot\nabla)(\hat{\zeta}\cdot\nabla)\langle\Phi\rangle\\ &\quad -(\nabla\langle\Phi\rangle\cdot\nabla)(\hat{\zeta}\cdot\nabla)\Phi]\\ &= -\hat{\zeta}\frac{\partial\Phi}{\partial\zeta}\nabla\cdot\nabla\langle\Phi\rangle+\hat{\zeta}(\nabla\Phi\cdot\nabla)\frac{\partial\langle\Phi\rangle}{\partial\zeta}-\hat{\zeta}(\nabla\langle\Phi\rangle\cdot\nabla)\frac{\partial\Phi}{\partial\zeta}\\ &= -\hat{\zeta}\frac{\partial\Phi}{\partial\zeta}\nabla\cdot\nabla\langle\Phi\rangle+\hat{\zeta}(\nabla\Phi\cdot\nabla\Phi)=\hat{\zeta}\left(\nabla\langle\Phi\rangle\cdot\nabla\frac{\partial\Phi}{\partial\zeta}\right),\end{aligned} \tag{3.194}$$

and

$$\begin{aligned}&\frac{\partial}{\partial\zeta}(\nabla\langle\Phi\rangle\times\nabla\Phi)\times\hat{\zeta}-\frac{\partial}{\partial\zeta}[\nabla(\langle\Phi\rangle\nabla\Phi\times\hat{\zeta})]+\nabla\times\nabla\times\left(\langle\Phi\rangle\frac{\partial\Phi}{\partial\zeta}\hat{\zeta}\right)\\ &= \nabla\frac{\partial\Phi}{\partial\zeta}(\hat{\zeta}\cdot\nabla\langle\Phi\rangle)-\nabla\langle\Phi\rangle\left(\hat{\zeta}\cdot\nabla\frac{\partial\Phi}{\partial\zeta}\right)\\ &\quad -\nabla\times(\Phi\nabla\Phi\times\hat{\zeta})+\frac{\partial\Phi}{\partial\zeta}\nabla\times(\nabla\langle\Phi\rangle\times\hat{\zeta})+\nabla\frac{\partial\Phi}{\partial\zeta}\times(\nabla\langle\Phi\rangle\times\hat{\zeta})\\ &= \Phi\nabla\frac{\partial\Phi}{\partial\zeta}-\nabla\times(\Phi\nabla\Phi\times\hat{\zeta})+\frac{\partial\Phi}{\partial\zeta}\nabla\times(\nabla\langle\Phi\rangle\times\hat{\zeta})-\hat{\zeta}\left(\nabla\frac{\partial\Phi}{\partial\zeta}\cdot\nabla\langle\Phi\rangle\right)\\ &= \hat{\zeta}\nabla\cdot(\Phi\nabla\Phi)-\frac{\partial\Phi}{\partial\zeta}\hat{\zeta}\nabla\cdot\nabla\langle\Phi\rangle-\hat{\zeta}\left(\nabla\frac{\partial\Phi}{\partial\zeta}\cdot\nabla\langle\Phi\rangle\right)\\ &= \hat{\zeta}(\nabla\Phi\cdot\nabla\Phi)-\frac{\partial\Phi}{\partial\zeta}\hat{\zeta}\nabla\cdot\nabla\langle\Phi\rangle-\hat{\zeta}\left(\nabla\langle\Phi\rangle\cdot\nabla\frac{\partial\Phi}{\partial\zeta}\right).\end{aligned} \tag{3.195}$$

Here we consider the last term in the RHS of (3.190). Expression (3.145) suggests that Φ^2 is composed of a constant term and sinusoidally oscillating terms. Thus $\nabla\times\nabla\times(\Phi^2\hat{\zeta})$ does not remain after it is averaged with respect to ζ. By noting that F_0 does not include the ζ component, from (3.189) and (3.190)

$$\begin{aligned}\nabla(\nabla F_0\cdot\overline{\nabla\langle\Phi\rangle)\cdot\nabla\Phi} &= \tfrac{1}{2}\nabla F_0\cdot\nabla\times\overline{(\nabla\langle\Phi\rangle\times\nabla\Phi)}\\ &= \tfrac{1}{2}\nabla F_0\cdot\overline{\nabla(\nabla\langle\Phi\rangle\times\nabla\Phi\cdot\hat{\zeta})}\times\hat{\zeta}\end{aligned} \tag{3.196}$$

is obtained after applying the averaging procedure (3.174). Now we go back to (3.181) and (3.182), which give

$$\bar{\mathbf{B}}\cdot\nabla F_0=\bar{G}_2-\frac{1}{B_0}\overline{\nabla\langle G_1\rangle\cdot\nabla\Phi}, \tag{3.197}$$

where

$$\bar{\mathbf{B}} \cdot \boldsymbol{\nabla} = B_0 \frac{\partial}{\partial \bar{\zeta}} + \boldsymbol{\nabla}\Psi \times \hat{\zeta} \cdot \boldsymbol{\nabla} \tag{3.198}$$

and

$$\Psi = A - \frac{1}{2B_0} \overline{\boldsymbol{\nabla}\langle\Phi\rangle \times \boldsymbol{\nabla}\Phi \cdot \hat{\zeta}}. \tag{3.199}$$

We note that the order of $|\bar{\mathbf{B}} \cdot \boldsymbol{\nabla}| \sim O(\delta^2)$.

Next, the averaged expression (3.197) for the magnetic differential equation (3.169) is applied to (3.160) and (3.168). From Ohm's law (3.168),

$$\bar{\mathbf{B}} \cdot \boldsymbol{\nabla} u = \overline{\frac{\partial A}{\partial t} + E_{\|}} = \frac{\partial A}{\partial t} + E_{\|}$$

or

$$\frac{\partial A}{\partial t} = \bar{\mathbf{B}} \cdot \boldsymbol{\nabla} u - \eta J \tag{3.200}$$

is given. From the equation of motion (3.160),

$$\begin{aligned}
\mathbf{B} \cdot \boldsymbol{\nabla}\sigma &= \frac{1}{B^4}(B \times \boldsymbol{\nabla}B^2) \cdot \boldsymbol{\nabla}P - \frac{\rho}{B_0}\frac{d}{dt}\nabla_\perp^2 u \\
&= \frac{2}{B_0^2}\hat{\zeta} \times \boldsymbol{\nabla}\left(\frac{\partial\Phi}{\partial\zeta}\right) \cdot \boldsymbol{\nabla}P_0 - \frac{2}{B_0^3}\hat{\zeta} \times \boldsymbol{\nabla}\left(\frac{\partial\Phi}{\partial\zeta}\right) \cdot \boldsymbol{\nabla}(\boldsymbol{\nabla}P_0 \cdot \boldsymbol{\nabla}_\perp\langle\Phi\rangle) \\
&\quad - \frac{3}{B_0^3}\hat{\zeta} \times \boldsymbol{\nabla}\left(\frac{\partial\Phi}{\partial\zeta}\right)^2 \cdot \boldsymbol{\nabla}P_0 - \frac{2}{B_0^3}\frac{\partial^2\Phi}{\partial\zeta^2}\hat{\zeta} \times \boldsymbol{\nabla}_\perp\Phi \cdot \boldsymbol{\nabla}P_0 \\
&\quad + \frac{1}{B_0}\hat{\zeta} \times \boldsymbol{\nabla}\left\{\left(\frac{\boldsymbol{\nabla}\Phi}{B_0}\right)^2 - 2\left(\frac{P_0}{B_0^2} + \frac{r\cos\theta}{R_0}\right)\right\} \cdot \boldsymbol{\nabla}P_0 \\
&\quad - \frac{\rho}{B_0}\frac{d}{dt}\nabla_\perp^2 u
\end{aligned} \tag{3.201}$$

is obtained by keeping terms up to $O(\delta^4)$. In the second term of the last expression, $P_1 = -\boldsymbol{\nabla}P_0 \cdot \boldsymbol{\nabla}\langle\Phi\rangle / B_0$, which is obtained from $\mathbf{B} \cdot \boldsymbol{\nabla}P = 0$ by using (3.179), is substituted. It is noted that $\boldsymbol{\nabla}P_0$ has no component parallel to the magnetic line of force. Here we have also substituted the model magnetic field

$$\mathbf{B} = B_0\left\{\hat{\zeta} + \frac{\boldsymbol{\nabla}\Phi}{B_0} + \frac{1}{B_0}\left[\boldsymbol{\nabla}A \times \hat{\zeta} - B_0\left(\frac{P_0}{B_0^2} + \frac{r\cos\theta}{R_0}\right)\hat{\zeta}\right\}\right.$$

into the first term of the RHS in the first equality of (3.201) in the following way:

$$\frac{\mathbf{B}\times\nabla B^2}{B^4} = \frac{1}{B^3}\hat{\mathbf{b}}\times\nabla B^2$$
$$= \frac{2}{B_0^2}\hat{\zeta}\times\nabla\left(\frac{\partial\Phi}{\partial\zeta}\right) - \frac{3}{B_0^3}\hat{\zeta}\times\nabla\left(\frac{\partial\Phi}{\partial\zeta}\right)^2 - \frac{2}{B_0^3}\left(\frac{\partial^2\Phi}{\partial\zeta^2}\right)\hat{\zeta}\times\nabla_\perp\Phi$$
$$+\frac{1}{B_0}\hat{\zeta}\times\nabla\left\{\left(\frac{\nabla\Phi}{B_0}\right)^2 - 2\left(\frac{P_0}{B_0^2}+\frac{r\cos\theta}{R_0}\right)\right\} + \frac{2}{B_0^3}\nabla_\perp\Phi\times\nabla_\perp\left(\frac{\partial\Phi}{\partial\zeta}\right), \tag{3.202}$$

where B^2 and B^{-3} are approximated as

$$B^2 = B_0^2\left\{1 + \frac{2}{B_0}\frac{\partial\Phi}{\partial\zeta} + \left(\frac{\nabla\Phi}{B_0}\right)^2 - 2\left(\frac{P_0}{B_0^2}+\frac{r\cos\theta}{R_0}\right)\right\}, \tag{3.203}$$

$$B^{-3} = B_0^{-3}\left(1 - \frac{3}{B_0}\frac{\partial\Phi}{\partial\zeta}\right) \tag{3.204}$$

and

$$\hat{b} = \mathbf{B}/B_0 \simeq \hat{\zeta} + \frac{\nabla_\perp\Phi}{B_0}. \tag{3.205}$$

Here the third and fifth terms of (3.202) are derived from the second term of (3.205). We also note that the fourth term in the RHS of (3.201) is written as

$$\frac{2}{B_0^3}\frac{\partial}{\partial\zeta}\left\{\left(\frac{\partial\Phi}{\partial\zeta}\right)\hat{\zeta}\times\nabla_\perp\Phi\cdot\nabla P_0\right\} - \frac{1}{B_0^3}\left\{\hat{\zeta}\times\nabla\left(\frac{\partial\Phi}{\partial\zeta}\right)^2\right\}\cdot\nabla P_0 \tag{3.206}$$

and the fifth term is written as

$$\frac{1}{B_0}\hat{\zeta}\times\nabla\left\{\frac{1}{B_0^2}\left[(\nabla_\perp\Phi)^2 + \left(\frac{\partial\Phi}{\partial\zeta}\right)^2\right] - 2\frac{r\cos\theta}{R_0}\right\}\cdot\nabla P_0. \tag{3.207}$$

Now we can express (3.201) in the following form:

$$\mathbf{B}\cdot\nabla\sigma = G_1 + G_2, \tag{3.208}$$

where

$$G_1 = \frac{2}{B_0^2}\nabla\left(\frac{\partial\Phi}{\partial\zeta}\right)\times\nabla P_0\cdot\hat{\zeta} \tag{3.209}$$

is $O(\delta^3)$,

$$G_2 = -\frac{2}{B_0^3}\nabla\left(\frac{\partial\Phi}{\partial\zeta}\right)\times\nabla(\nabla P_0\cdot\nabla_\perp\langle\Phi\rangle)\cdot\hat{\zeta}$$
$$+\frac{1}{B_0}\nabla\left\{\frac{1}{B_0^2}(\nabla_\perp\Phi)^2-\frac{1}{B_0^2}\left(\frac{\partial\Phi}{\partial\zeta}\right)^2-2\frac{r\cos\theta}{R_0}\right\}\times\nabla P_0\cdot\hat{\zeta}$$
$$-\frac{2}{B_0^3}\frac{\partial}{\partial\zeta}\left(\frac{\partial\Phi}{\partial\zeta}\nabla_\perp\Phi\times\nabla P_0\cdot\hat{\zeta}\right)-\frac{\rho}{B_0}\frac{d}{dt}\nabla_\perp^2 u \tag{3.210}$$

is $O(\delta^4)$ and the LHS of (3.208) is $O(\delta^3)$. By using the averaged form (3.197) for (3.208),

$$\overline{\mathbf{B}}\cdot\nabla\sigma = \overline{G}_2 - \frac{1}{B_0}\overline{\nabla\langle G_1\rangle\cdot\nabla\Phi} \tag{3.211}$$

is given, and is explicitly written as

$$\rho\frac{d}{dt}\nabla_\perp^2 u = -\nabla\left[2\frac{r\cos\theta}{R_0}+\frac{1}{B_0^2}\overline{(\nabla\Phi)^2}\right]\times\nabla P_0\cdot\hat{\zeta}+\frac{1}{\mu_0}\overline{\mathbf{B}}\cdot\nabla\nabla_\perp^2 A. \tag{3.212}$$

In the derivation of (3.212) we have used

$$\frac{1}{B_0}\overline{\nabla\langle G_1\rangle\cdot\nabla\Phi} = \frac{2}{B_0^3}\overline{\nabla\left\langle\nabla\left(\frac{\partial\Phi}{\partial\zeta}\right)\times\nabla P_0\cdot\hat{\zeta}\right\rangle\cdot\nabla\Phi}$$
$$= \frac{2}{B_0^3}\overline{\nabla\left(\nabla\left\langle\frac{\partial\Phi}{\partial\zeta}\right\rangle\times\nabla P_0\cdot\hat{\zeta}\right)\cdot\nabla\Phi}$$
$$= \frac{2}{B_0^3}\overline{\nabla(\nabla\Phi\times\nabla P_0\cdot\hat{\zeta})\cdot\nabla\Phi}, \tag{3.213}$$

where

$$\left\langle\frac{\partial\Phi}{\partial\zeta}\right\rangle = \frac{\partial}{\partial\zeta}\langle\Phi\rangle = \Phi$$

is substituted into the last equality. From (3.210), (3.211), and (3.213) we need to show the relation

$$\overline{\nabla\left(\frac{\partial\Phi}{\partial\zeta}\right)\times\nabla(\nabla_\perp\langle\Phi\rangle\cdot\nabla P_0)\cdot\hat{\zeta}} - \overline{\nabla(\nabla_\perp\Phi)^2}\times\nabla P_0\cdot\hat{\zeta}$$
$$+\overline{\nabla\Phi\cdot\nabla(\nabla\Phi}\times\nabla P_0\cdot\hat{\zeta}) = 0 \tag{3.214}$$

to obtain (3.212). For simplicity, we introduce $H = \nabla\Phi\times\nabla P_0\cdot\hat{\zeta}$ here. The last term of (3.214) is equal to $\overline{T} = \overline{\nabla\Phi\cdot\nabla H}$. We employ the relations

$$-\nabla \times (H\nabla\Phi \times \hat{\zeta}) \cdot \hat{\zeta} = \nabla \cdot (H\nabla\Phi) - \frac{\partial}{\partial\zeta}\left(H\,\frac{\partial\Phi}{\partial\zeta}\right) \tag{3.215}$$

and

$$\begin{aligned}
-\nabla \times (H\nabla\Phi \times \hat{\zeta}) \cdot \hat{\zeta} &= \nabla \times [(\nabla_\perp\Phi \times \hat{\zeta})(\nabla_\perp\Phi \times \hat{\zeta}) \cdot \nabla P_0] \cdot \hat{\zeta} \\
&= \nabla \times [(\nabla_\perp\Phi \times \hat{\zeta}) \times ((\nabla_\perp\Phi \times \hat{\zeta}) \times \nabla P_0) + (\nabla_\perp\Phi \times \hat{\zeta})^2 \nabla P_0] \cdot \hat{\zeta} \\
&= \nabla \times [-(\nabla_\perp\Phi \cdot \nabla P_0)\nabla_\perp\Phi + (\nabla_\perp\Phi)^2 \nabla P_0] \cdot \hat{\zeta} \\
&= \nabla \times [-(\nabla_\perp\Phi \cdot \nabla P_0)\left(\nabla\Phi - \frac{\partial\Phi}{\partial\zeta}\,\hat{\zeta}\right) + (\nabla_\perp\Phi)^2 \nabla P_0] \cdot \hat{\zeta} \\
&= \nabla(\nabla_\perp\Phi \cdot \nabla P_0) \times \nabla\Phi \cdot \hat{\zeta} + \nabla(\nabla_\perp\Phi)^2 \times \nabla P_0 \cdot \hat{\zeta},
\end{aligned} \tag{3.216}$$

where $\nabla_\perp\Phi \cdot \hat{\zeta} = 0$ is taken into account. The first term of the last equality of (3.216) is also written as

$$\begin{aligned}
&\nabla(\nabla_\perp\Phi \cdot \nabla P_0) \times \nabla\Phi \cdot \hat{\zeta} \\
&\quad = \nabla\left(\nabla_\perp\,\frac{\partial\langle\Phi\rangle}{\partial\zeta} \cdot \nabla P_0\right) \times \nabla\Phi \cdot \hat{\zeta} \\
&\quad = \frac{\partial}{\partial\zeta}\,[\nabla(\nabla_\perp\langle\Phi\rangle \cdot \nabla P_0) \times \nabla\Phi] \cdot \hat{\zeta} - \nabla(\nabla_\perp\langle\Phi\rangle \cdot \nabla P_0) \times \nabla\left(\frac{\partial\Phi}{\partial\zeta}\right) \cdot \hat{\zeta}.
\end{aligned} \tag{3.217}$$

Therefore, (3.216) and (3.217) give the relation

$$\begin{aligned}
-\nabla \times (H\nabla\Phi \times \hat{\zeta}) \cdot \hat{\zeta} &= -\frac{\partial}{\partial\zeta}\,[\nabla(\nabla_\perp\langle\Phi\rangle \cdot \nabla P_0) \times \nabla\Phi] \cdot \hat{\zeta} \\
&\quad -\nabla\left(\frac{\partial\Phi}{\partial\zeta}\right) \times \nabla(\nabla_\perp\langle\Phi\rangle \cdot \nabla P_0) \cdot \hat{\zeta} + \nabla(\nabla_\perp\Phi)^2 \times \nabla P_0 \cdot \hat{\zeta}.
\end{aligned} \tag{3.218}$$

From the relation (3.215),

$$T = -\nabla \times (H\nabla\Phi \times \hat{\zeta}) \cdot \hat{\zeta} + \frac{\partial}{\partial\zeta}\left(H\,\frac{\partial\Phi}{\partial\zeta}\right) \tag{3.219}$$

is given, and the averaging procedure (3.174) gives

$$\begin{aligned}
\overline{T} &= -\overline{\nabla \times (H\nabla\Phi} \times \hat{\zeta}) \cdot \hat{\zeta} \\
&= \overline{\nabla\Phi \cdot \nabla(\nabla\Phi{\times}}\nabla P_0 \cdot \hat{\zeta}) \\
&= \overline{-\nabla\left(\frac{\partial\Phi}{\partial\zeta}\right) \times \nabla(\nabla_\perp\langle\Phi\rangle \cdot \nabla P_0) \cdot \hat{\zeta}} + \overline{\nabla(\nabla_\perp\Phi)^2} \times \nabla P_0 \cdot \hat{\zeta}.
\end{aligned} \tag{3.220}$$

Here the relation obtained from the second and third equalities is the same as (3.214).

Finally, we note that the pressure equation becomes

$$\frac{\partial P_0}{\partial t} + (\nabla u \times \hat{\zeta}) \cdot \nabla P_0 = 0. \tag{3.221}$$

Equations (3.200), (3.212), and (3.221) comprise the reduced MHD equations for heliotron/torsatron configurations. Since the variables are $\{u, A, P_0\}$, they are often called three fields equations.

It is interesting that the reduced MHD equations for the heliotron/torsatron become those for high-beta tokamaks by assuming $\Phi = 0$ or by removing helical magnetic fields and adding a poloidal magnetic field generated by a toroidal plasma current. The reduced MHD equations for high-beta tokamakes were first derived by H. R. Strauss (1977).

From the point of view of a numeric study for MHD equilibrium and stability, the reduced MHD equations averaged over the toroidal direction assure the same level of analysis as for tokamaks. Although there is limitation for the application, they are useful in the study of MHD equilibrium and stability in heliotron configurations, as shown in chapters 4 and 5.

We present a simplified but useful model magnetic field described by

$$\Phi = 2\Phi_\ell I_\ell(hr)\sin(\ell\theta + h\zeta). \tag{3.222}$$

By substituting (3.222) into (3.199),

$$\begin{aligned}\Psi_h &\equiv -\frac{1}{2B_0}\overline{\nabla\langle\Phi\rangle \times \nabla\Phi} \cdot \hat{\zeta} \\ &= -\frac{2\ell\Phi_\ell^2}{rB_0} I_\ell(hr)I'_\ell(hr)\end{aligned} \tag{3.223}$$

is obtained. This flux function produced by the helical magnetic field gives the rotational transform as

$$\iota_h \equiv -\frac{R_0}{rB_0}\frac{d\Psi_h}{dr} = \frac{2\Phi_\ell^2 h^2 R_0}{rB_0^2} F'(hr), \tag{3.224}$$

where

$$F(hr) \equiv \frac{\ell}{hr} I_\ell(hr)I'_\ell(hr). \tag{3.225}$$

The flux function Ψ_h is also shown as

$$\Psi_h = -\frac{2\Phi_\ell^2 h}{B_0} F(hr) = -\frac{B_0 a\iota_h(a)}{hR_0}\frac{F(hr)}{F'(ha)}. \tag{3.226}$$

Next, we consider the total curvature composed of the toroidal curvature and the helical curvature with Ω:

$$\Omega \equiv \frac{2r\cos\theta}{R_0} + \frac{\overline{(\nabla\Phi)^2}}{B_0^2}. \tag{3.227}$$

By substituting (3.222) into (3.227),

$$\begin{aligned}\Omega &= \frac{2r\cos\theta}{R_0} + \frac{2\Phi_\ell^2 h^2}{B_0^2}\, G(hr)\\ &= \frac{2r\cos\theta}{R_0} + \frac{a\iota_h(a)}{R_0}\,\frac{G(hr)}{F'(ha)},\end{aligned} \tag{3.228}$$

is given, where

$$G(hr) = (I_\ell'(hr))^2 + \frac{\ell^2 + h^2r^2}{h^2r^2}\,(I_\ell(hr))^2. \tag{3.229}$$

We note that

$$\Omega = \frac{2r\cos\theta}{R_0} + \frac{h}{\ell R_0}\left(r^2\iota_h(r) + 2\int_0^r r\iota_h(r)dr\right)$$

is obtained from (3.224) and (3.228). Also, a total rotational transform is $\iota(a) = \iota_h(a) + \iota_J(a)$, where $\iota_J(a)$ is a rotational transform produced by the toroidal plasma current or the poloidal field $-dA/dr$. For currentless plasmas in heliotrons, $\iota_J(r) = 0$ is assumed.

We now give normalized expressions of the reduced MHD equations. One way is to use normalizations: $t \equiv t/(a/\varepsilon V_A)$, $r \equiv r/a$, $\zeta = \zeta/a$, $\bar{\zeta} \equiv \bar{\zeta}/R_0$, $u \equiv u/(\varepsilon a V_A)$, $A \equiv A/(\varepsilon a B_0)$, $\Phi = \Phi/(\varepsilon^{1/2} a B_0)$, and $P_0/(\varepsilon B_0^2/\mu_0)$, where $\varepsilon = a/R_0$ and V_A is the Alfvén velocity. The normalized reduced MHD equations are expressed as

$$\frac{\partial}{\partial t}\nabla_\perp^2 u = [u, \nabla_\perp^2 u] + [\nabla_\perp^2 A, \Psi] + \frac{\partial}{\partial\bar{\zeta}}\nabla_\perp^2 A + [\Omega, P], \tag{3.230}$$

$$\frac{\partial A}{\partial t} = [u, \Psi] + \frac{\partial u}{\partial\bar{\zeta}} + \bar{\eta}\nabla_\perp^2 A, \tag{3.231}$$

$$\frac{\partial P}{\partial t} = [u, P]. \tag{3.232}$$

Here the Poisson bracket is defined as $[f, g] \equiv \nabla f \times \nabla g \cdot \hat{\zeta}$ for $f(r, \theta, \bar{\zeta})$ and $g(r, \theta, \bar{\zeta})$, and

$$\Psi_h = -\frac{1}{N\varepsilon}\,\frac{F(N\varepsilon r)}{F(N\varepsilon)}\,\frac{\iota_h(1)}{\iota(1)}, \tag{3.233}$$

$$\Omega = x + \iota_h(1)\frac{G(N\varepsilon r)}{F'(N\varepsilon)}, \tag{3.234}$$

$$\bar{\eta} \equiv \frac{\eta}{\mu_0 \varepsilon a V_A} = S^{-1}, \tag{3.235}$$

where S denotes the Lundquist number (or the magntic Reynolds number). For a relatively low-temperature and high-density plasma $S \sim 10^5$, for a high-temperature plasma $S \sim 10^7$, and for a reacting plasma in a fusion reactor $S \sim 10^9$. The Lundquist number plays a role in the resistive MHD stability theory in chapter 5.

By noting that $[f, g] = \nabla \cdot (f\nabla g \times \hat{\zeta})$ and applying Gauss's theorem to a volume integral $\int[f, g]dV$,

$$\int [f, g]dV = 0 \tag{3.236}$$

is obtained by imposing the fixed boundary condition at the perfectly conducting boundary. By using the properties of Poisson brackets

$$[fg, h] = g[f, h] + f[g, h], \tag{3.237}$$

$$[f, g] = -[g, f], \tag{3.238}$$

and (3.236),

$$\int f[g, h]dV = \int g[h, f]dV = \int h[f, g]dV \tag{3.239}$$

is given. Now we multiply (3.230) by u, (3.231) by $\nabla_\perp^2 A$, and (3.232) by Ω, and sum up these three equations. Then the volume integral becomes

$$\frac{\partial}{\partial t}\int \{\tfrac{1}{2}(\nabla_\perp u)^2 + \tfrac{1}{2}(\nabla_\perp A)^2 - P\Omega\}dV = -\int \bar{\eta}(\nabla_\perp^2 A)^2 dV, \tag{3.240}$$

with the use of (3.239). The first term in the brace on the LHS is the kinetic energy density, the second term is the magnetic energy density, and the third term is the internal energy density, while the RHS denotes ohmic dissipation due to electric resistivity. When $\bar{\eta} = 0$, or for the ideal MHD model, the energy integral is conserved,

$$\int \{\tfrac{1}{2}(\nabla_\perp u)^2 + \tfrac{1}{2}(\nabla_\perp A)^2 - P\Omega\}dV = \text{constant}. \tag{3.241}$$

Finally, we note that the transformation

$$\{\Psi(r, \theta, \bar{\zeta}), u(r, \theta, \bar{\zeta}), P(r, \theta, \bar{\zeta})\} \rightarrow \{\Psi(r, -\theta, -\bar{\zeta}), -u(r, -\theta, -\bar{\zeta}), P(r, -\theta, -\bar{\zeta})\} \tag{3.242}$$

does not change the reduced MHD equations, which is considered as a spatial parity conservation. This property is derived from the fact that the transformation

$$\begin{pmatrix} \rho, \mathbf{v}, \mathbf{B} \\ P, \mathbf{J}, \mathbf{E} \end{pmatrix} \rightarrow \begin{pmatrix} \rho, \mathbf{v}, -\mathbf{B} \\ P, -\mathbf{J}, \mathbf{E} \end{pmatrix} \tag{3.243}$$

does not change the full MHD equations shown in section 3.4.

BIBLIOGRAPHY

Boozer, A. (1980). Guiding center drift equations. *Phys. Fluids*, **23**, 904.

Greene, J. M., and Johnson, J. L. (1961). Determination of hydromagnetic equilibria. *Phys. Fluids*, **4**, 875.

Hasegawa, A., and Sato, T. (1989). *Space plasma physics* 1. *Stationary processes*. Physics and Chemistry in Space 16. Springer-Verlag, Berlin.

Hasegawa, A., and Uberoi, C. (1982). *The Alfvén wave*. DOE/TIC-11197. Technical Information Center, U.S. Department of Energy.

Hirschman, S. P., and Sigmar, D. J. Neoclassical transport of impurities in tokamak plasmas. *Nucl. Fusion*, **21**, 1079.

Nishikawa, K., and Wakatani, M. (1990). *Plasma physics, basic theory with fusion applications*. Springer Series on Atoms and Plasma. Springer-Verlag, Berlin.

Shaing, K. C., and Callen, J. D. (1983). Neoclassical flows and transport in nonaxisymmetric toroidal plasmas. *Phys. Fluids*, **26**, 3315.

Strauss, H. R. (1977). Dynamics of high β tokamaks. *Phys. Fluids*, **20**, 1354.

Strauss, H. R. (1980). Stellarator equations for motion. *Plasma Phys*., **22**, 733.

Wesson, J. (1987). *Tokamaks*. Clarendon Press, Oxford.

Zakharov, L. E., and Shafranov, V. D. (1982). Equilibrium of current-carrying plasmas in toroidal configurations, in *Reviews of plasma physics*. Consultants Bureau, New York, vol. 11, p. 153.

4

THE MHD EQUILIBRIUM OF A TOROIDAL PLASMA IN THREE-DIMENSIONAL GEOMETRY

4.1 INTRODUCTION

In order to confine a plasma in a finite volume with a donut shape, macroscopic plasma motion must be suppressed and a local force balance between the expansion force due to the pressure gradient ∇P and the electromagnetic force $\mathbf{J} \times \mathbf{B}$ must be satisfied. As shown in chapter 2, a vacuum magnetic system with nested toroidal magnetic surfaces is appropriate for plasma confinement. Thus, the next question concerns the existence of nested magnetic surfaces satisfying the MHD equilibrium equations

$$\mathbf{J} \times \mathbf{B} = \nabla P, \tag{4.1}$$

$$\mu_0 \mathbf{J} = \nabla \times \mathbf{B}, \tag{4.2}$$

$$\nabla \cdot \mathbf{B} = 0. \tag{4.3}$$

It is well- known that, when the system has a geometrical symmetry, such as axisymmetry or helical symmetry, the existence of nested magnetic surfaces can be proved. The tokamak, the Reversed Field Pinch (RFP), and the Spheromak are axisymmetric toroidal systems. In order to produce the rotational transform in these devices, a plasma current in the toroidal direction is required. The helically symmetric straight plasma satisfying the MHD equilibrium equations (4.1)–(4.3) also has nested magnetic surfaces. However, in order to eliminate end-loss of the straight plasma column, the bending of the helically symmetric straight system into a toroidal stellarator or heliotron is a natural development from the point of view of high-temperature plasma confinement. Then, the resultant toroidal stellarator or heliotron loses helical symmetry. One significant characteristic of the toroidal heliotron is the possibility of plasma confinement without a net toroidal current as in tokamaks.

According to (4.2), the plasma current is obtained by integrating the poloidal magnetic field B_p in the poloidal direction along the curve closed C encircling the magnetic axis:

$$\oint_c B_p d\ell = \mu_0 I_p. \tag{4.4}$$

The rotational transform of the stellarator and the helotron is finite, although $I_p = 0$. The mechanism by which the helical magnetic field in the stellarator and heliotron produces the rotational transform can be understood by a schematic picture of magnetic field lines on the magnetic surface, shown in Fig. 4.1. The angles θ and ϕ denote the poloidal and toroidal angles, respectively. The magnetic field line moves back and forth in the θ direction when ϕ is increased; however, an averaged behavior shows movement in one direction. when B_p or the projection of $\mathbf{B}$ in the θ direction is weaker for $d\theta/d\phi > 0$ than for $d\theta/d\phi < 0$, $\oint B_p d\ell = 0$ may become possible, since the length is longer for $d\theta/d\phi > 0$ than for $d\theta/d\phi < 0$. It should be noted that the local plasma current is not zero in order to satisfy (4.1), even if the net plasma current inside the magnetic surface is zero. Therefore, the terminology of "currentless or current-free" MHD equilibrium is sometimes misleading. For example, the plasma current to maintain the toroidal MHD equilibrium, called the Pfirsch–Schlüter current, exists in both the stellarator and the heliotron. However, the Pfirsch–Schlüter currrent is pressure-driven and is kept finite without an external electric field, if the pressure is sustained by the external plasma heating in present experiments or by alpha particle heating in a future fusion reactor. Thus continuous operation of a fusion reactor is principally possible in the stellarator and the heliotron, when the magnetic field is generated by superconducting coils.

For the axisymmetric tokamak plasma or straight helically symmetric plasma, MHD equilibria are studied on the basis of the Grad–Shafranov equation. When the existence of a magnetic surface is assumed, we show in section 4.2 that the Grad–Shafranov type equilibrium equation can be derived even for

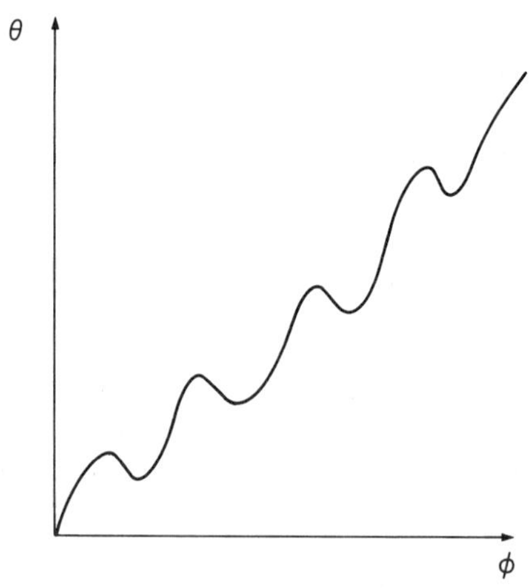

Fig. 4.1 The behavior of a magnetic field line on the $\theta - \phi$ plane.

three-dimensional stellarators and heliotrons. For heliotrons, since they have fairly large pitch numbers of $M \geq 8$ (e.g., the Heliotron D device at Kyoto University had the largest M, $M = 25$, and the CHS at the National Institute for Fusion Science has the lowest M, $M = 8$), MHD equilibrium averaged over the toroidal angle is extremely useful. When this averaging procedure is combined with a small parameter expansion to analyze the equilibrium, the resultant equilibrium equation becomes similar to the Grad–Shafranov equation for the axisymmetric tokamak plasma, although both an averaged curvature term related to the vacuum helical magnetic field produced by helical windings and a poloidal flux also generated by helical windings are newly included in the equilibrium equation. The small parameter is usually an inverse aspect ratio $\epsilon = r/R_0$ for the large-aspect ratio heliotron or b_h/B_0 for the low-aspect ratio helical system, where $b_h = |\mathbf{b}_h|$ is the helical magnetic field. We explain the averaged MHD equilibrium equation and its properties in section 4.3.

It is expected that the "currentless" MHD equilibrium is natural for a steady state stellarator or heliotron plasma. However, if plasma heating is sufficiently rapid, flux-conserving MHD equilibrium or FCT (Flux-Conserving Torus) equilibrium become possible. Although it is maintained only for a period of the order of the energy confinement time, it is useful to understand the MHD equilibrium of a finite-beta plasma in the stellarator and heliotron. We discuss the properties of FCT equilibrium in section 4.3. One significant result obained by assuming FCT equilibrium is the appearance of a toroidal plasma current or an FCT current without a toroidal electric field, which arises from the constraint that the rotational transform profile is kept constant during the rapid heating. Although the FCT equilibrium was first discussed in tokamaks to realize high-beta plasma confinement, it has an interesting implication for the stellarator and heliotron.

It is known that the ideal MHD equations conserve energy for an isolated system separated by a conducting wall, even if a vacuum region exists between the plasma column and the wall. By defining a potential energy for a plasma without a macroscopic plasma flow, the MHD equilibrium is obtained as the minimum energy state of the potential energy. This suggests a method for solving the MHD equilibrium based on a variational principle, which we explain in section 4.4.

As explained in chapter 2, a recent trend in stellarator research is directed toward the study of helical magnetic axis systems. For the spatial axis stellarators, the MHD equilibrium is studied by using the Solov'ev–Shafranov equation. This equation solves the shift of the magnetic axis due to finite beta effects. In a specific limit for an axisymmetric toroidal plasma, the Solov'ev–Shafranov equation derived in section 4.5 gives the Shafranov shift of the circular cross-section tokamak.

In the final section 4.6, we discuss the Pfirch–Schlüter current generally. Then one of the sets of magnetic coordinates, called the Boozer coordinates, which are often used to study equilibrium properties in the stellarator and heliotron, are introduced. The details of dissipative MHD equilibrium, including resistivity, are discussed particularly in the presence of rational magnetic surfaces. Finally, we explain the relations between two well-known sets of magnetic coordinates, the Boozer and Hamada coordinates.

4.2 THE GENERALIZED GRAD–SHAFRANOV EQUATION

From the MHD equilibrium equation (4.1), $\mathbf{B}\cdot\nabla P=0$ and $\mathbf{J}\cdot\nabla P=0$ are obtained automatically. Here we assume that perfect nested magnetic surfaces exist in the three-dimensional toroidal geometry corresponding to the stellarator and heliotron for simplicity. Then the plasma pressure P becomes constant on the magnetic surface and the vectors $\mathbf{B}$ and $\mathbf{J}$ are located on the magnetic surface.

Now we introduce a scalar function $a(\mathbf{r})$ denoting the magnetic surface by $a(\mathbf{r})=\text{constant}$ to consider geometric properties, where $\mathbf{r}$ is a position vector. The function $a(\mathbf{r})$ satisfies

$$\mathbf{B}\cdot\nabla a=0, \tag{4.5}$$

$$\mathbf{J}\cdot\nabla a=\nabla\cdot(\mathbf{B}\times\nabla a)=0, \tag{4.6}$$

and

$$\nabla\cdot\mathbf{B}=0. \tag{4.7}$$

Since $\mathbf{B}$ and $\mathbf{J}$ are located on the magnetic surfaces, (4.6) and (4.7) include derivatives with respect to the poloidal and toroidal angles but no radial derivative. When we use the poloidal magnetic flux function $\Psi(\mathbf{r})$ to define the magnetic surface, with $\Psi(\mathbf{r})=\text{constant}$, (4.5) is equivalent to

$$\mathbf{B}\cdot\nabla\Psi=0 \tag{4.8}$$

and (4.6) becomes

$$\mathbf{J}\cdot\nabla\Psi=0. \tag{4.9}$$

In order to obtain $\Psi(\mathbf{r})$, we take a projection of the MHD equilibrium equation $\mathbf{J}\times\mathbf{B}=\nabla P$ in the direction of $\nabla\Psi$:

$$(\nabla\times\mathbf{B})\times\mathbf{B}\cdot\nabla\Psi=\mu_0 P'(\Psi)|\nabla\Psi|^2. \tag{4.10}$$

Here we consider an equilibrium with a longitudinal plasma current as a specific case. Then the magnetic field will be given by

$$\mathbf{B}=\nabla\Psi\times\mathbf{b}_\theta+F\mathbf{b}_\zeta, \tag{4.11}$$

where $\mathbf{b}_\theta$ and $\mathbf{b}_\zeta$ are vectors on the magnetic surface, and F is a poloidal current flux function. Here $\mathbf{b}_\theta$ and $\mathbf{b}_\zeta$ should be determined consistently with the condition of MHD equilibrium and they are functions of Ψ in the three-dimensional configuration. The plasma current is given by

$$\mu_0 \mathbf{J} = \nabla \times (\nabla\Psi \times \mathbf{b}_\theta) + F\nabla \times \mathbf{b}_\zeta + \nabla F \times \mathbf{b}_\zeta. \tag{4.12}$$

By substituting (4.11) and (4.12) into (4.7), (4.8), and (4.9), we obtain

$$\nabla \cdot \mathbf{b}_\zeta = 0, \tag{4.13}$$

$$\mathbf{b}_\zeta \cdot \nabla\Psi = 0, \tag{4.14}$$

$$\nabla\Psi \cdot \nabla \times \mathbf{b}_\zeta = 0, \tag{4.15}$$

$$\nabla \cdot (|\nabla\Psi|^2 \mathbf{b}_\theta) = 0, \tag{4.16}$$

$$\mathbf{b}_\theta \cdot \nabla\Psi = 0, \tag{4.17}$$

$$\nabla\Psi \cdot \nabla \times \mathbf{b}_\theta = 0. \tag{4.18}$$

Now $\mathbf{b}_\theta$ and $\mathbf{b}_\zeta$ are chosen to satisfy (4.13)–(4.18). We introduce two scalar functions $h(\mathbf{r})$ and $\lambda(\mathbf{r})$, as

$$\mathbf{b}_\theta = \mathbf{b}_\zeta + (\nabla\Psi \times (\nabla h \times \nabla\Psi))/|\nabla\Psi|^2 \tag{4.19}$$

and

$$\mathbf{b}_\zeta = \nabla\Psi \times \nabla\lambda. \tag{4.20}$$

It is easy to see that (4.13), (4.14), and (4.17) are satisfied automatically. Use of

$$\mathbf{b}_\theta \times \nabla\Psi = \mathbf{b}_\zeta \times \nabla\Psi + \nabla h \times \nabla\Psi \tag{4.21}$$

obtained from (4.19) yields the relation

$$\begin{aligned} \nabla\Psi \cdot \nabla \times \mathbf{b}_\theta &= \nabla \cdot (\mathbf{b}_\theta \times \nabla\Psi) = \nabla \cdot (\mathbf{b}_\zeta \times \nabla\Psi) \\ &= \nabla\Psi \cdot \nabla \times \mathbf{b}_\zeta. \end{aligned} \tag{4.22}$$

By using (4.15), (4.20), and (4.22),

$$\nabla \cdot ((\nabla\Psi \times \nabla\lambda) \times \nabla\Psi) = 0 \tag{4.23}$$

is obtained to determine the function λ. Also, (4.16) gives

$$\nabla \cdot (\nabla\Psi \times (\nabla h \times \nabla\Psi)) + (\nabla\Psi \times \nabla\lambda) \cdot \nabla|\nabla\Psi|^2 = 0 \tag{4.24}$$

with the use of (4.19) and (4.20). Equation (4.24) is considered to determine the function h. Thus the expression of the magnetic field vector given by (4.11) is consistent with the MHD equilibrium. The vectors $\mathbf{b}_\theta$ and $\mathbf{b}_\zeta$ are given by (4.19) and (4.20) after the differential equations (4.23) and (4.24) are solved to obtain λ and h. By noting (4.10) yields

$$\mathbf{J}\cdot(\mathbf{B}\times\nabla\Psi) = P'(\Psi)|\nabla\Psi|^2, \tag{4.25}$$

and by replacing $\mathbf{B}$ with (4.11),

$$\mathbf{J}\cdot\left(\mathbf{b}_\theta + F\,\frac{\mathbf{b}_\zeta\times\nabla\Psi}{|\nabla\Psi|^2}\right) = P'(\Psi) \tag{4.26}$$

is obtained. By substituting (4.12) into (4.26), we find

$$\begin{aligned}
&-\nabla\cdot[(\mathbf{b}_\theta)^2\nabla\Psi] + \mathbf{b}_\theta\times(\nabla\times\mathbf{b}_\theta)\cdot\nabla\Psi \\
&-\tfrac{1}{2}(\mathbf{b}_\zeta)^2\,\frac{dF^2}{d\Psi} + F\mathbf{b}_\theta\cdot\nabla\times\mathbf{b}_\zeta - (\mathbf{b}_\theta\times\mathbf{b}_\zeta)\cdot\nabla\Psi\,\frac{dF}{d\Psi} \\
&-F\,\frac{\mathbf{b}_\zeta\times\nabla\Psi}{|\nabla\Psi|^2}\cdot\nabla\times(\mathbf{b}_\theta\times\nabla\Psi) + F^2\,\frac{\mathbf{b}_\zeta\times\nabla\Psi}{|\nabla\Psi|^2}\cdot\nabla\times\mathbf{b}_\zeta = \mu_0\,\frac{dP}{d\Psi}.
\end{aligned} \tag{4.27}$$

Here we have used the fact that the poloidal current function F depends on Ψ only. The differential equation (4.27) is interpreted as a generalized Grad–Shafranov equation, applicable to any toroidal plasma with the longitudinal plasma current. The essential assumption to derive (4.27) is that P and F are surface functions. One significant different between (4.27) and the ordinary Grad–Shafranov equation for tokamaks (see (4.28)) is the requirement of solving differential equations (4.23) and (4.24) simultaneously with the genealized Grad–Shafranov equation. We now assume axisymmetry for the generalized Grad–Shafranov equation. Then $\mathbf{b}_\theta$ and $\mathbf{b}_\zeta$ become independent of the poloidal magnetic flux function Ψ. Here $\mathbf{b}_\theta = \mathbf{b}_\zeta = \nabla\phi$ is a natural choice, where ϕ is a toroidal angle variable. By noting that $|\nabla\phi| = 1/R$ and substituting $\mathbf{b}_\theta = \mathbf{b}_\zeta = \nabla\phi$ into (4.27), we obtain the Grad–Shafranov equation for axisymmetric toroidal plasmas:

$$\nabla\cdot\left[\frac{1}{R^2}\,\nabla\Psi\right] + \frac{1}{2R^2}\,\frac{dF^2}{d\Psi} = -\mu_0\,\frac{dP}{d\Psi}. \tag{4.28}$$

Here the sixth term in (4.27) disappears due to the fact that $\nabla\times(\nabla\phi\times\nabla\Psi)$ has components parallel to $\nabla\phi$ or $\nabla\Psi$ only. For tokamaks, when $P(\Psi)$ and $F(\Psi)$ are given definitely, the solution of (4.28) gives nested magnetic surfaces by $\Psi = \text{constant}$. We also note that the current density (4.12) becomes $\mu_0\mathbf{J} = \nabla\times(\nabla\Psi\times\nabla\phi) + \nabla F\times\nabla\phi$ and the toroidal component of $\mathbf{J}$ is given by

$$\mu_0 J_\phi = \frac{1}{R}\,\nabla^2\Psi + \nabla\Psi\cdot\nabla(1/R^2) = R\nabla\cdot\left[\frac{1}{R^2}\,\nabla\Psi\right].$$

Thus a well-known relation,

$$\mu_0 J_\phi = \mu_0 R \frac{dP}{d\Psi} + \frac{1}{2R}\frac{dF^2}{d\Psi}, \tag{4.29}$$

is obtained. There are several established numerical schemes to solve (4.28). Usually, the institute or laboratory investigating thermonuclear fusion has several numerical codes to study MHD equilibrium quantitatively.

In the cylindrical coordinates (R, ϕ, Z), the Grad–Shafranov equation (4.28) becomes

$$R\frac{\partial}{\partial R}\left(\frac{1}{R}\frac{\partial \Psi}{\partial R}\right) + \frac{\partial^2 \Psi}{\partial Z^2} = -\mu_0 R^2 \frac{dP}{d\Psi} - F\frac{dF}{d\Psi}. \tag{4.30}$$

Here we are interested in solutions corresponding to nested magnetic surfaces after the free functions $P(\Psi)$ and $F(\Psi)$ are given. The simplest choice that still has a physical meaning is $\mu_0 P' = -C$ and $FF' = A$, with constants A and C. This class of equilibrium was investigated by Solovév (1968). It is noted that $-AZ^2/2 + CR^4/8$ is a particular solution of (4.30), and constant number, R^2 and $(R^4 - 4R^2Z^2)$ are homogeneous solutions of (4.30). Then the solution is written in a convenient form:

$$\Psi(R, Z) = \frac{C\gamma}{8}\left[(R^2 - R_a^2)^2 - R_b^4\right] + \frac{C}{2}\left[(1-\gamma)R^2 - \frac{A}{C}\right]Z^2, \tag{4.31}$$

where γ, R_a, and R_b are constants. The flux function has been normalized so that $\Psi(R, Z) = 0$ on the plasma boundary.

The quantities R_a and R_b are identified by examining Ψ along the midplane $Z = 0$, as illustrated in Fig. 4.2. The point $(R_a, 0)$ corresponds to the magnetic axis. The outer edge of the plasma is located at $((R_a^2 + R_b^2)^{1/2}, 0)$, which in terms of the standard tokamak notation corresponds to $(R_a^2 + R_b^2)^{1/2} = R_0 + a$. Similarly, the inner edge is located at $((R_a^2 - R_b^2)^{1/2}, 0)$, where $(R_a^2 - R_b^2)^{1/2} = R_0 - a$. These two relations can be solved simultaneously for R_a and R_b in terms of R_0 and a. The result is $R_a^2 = R_0^2(1 + \epsilon_a^2)$ and $R_b^2 = 2R_0^2\epsilon_a$, where $\epsilon_a = a/R_0$. The constant γ is related to the elongation defined by $\kappa = S/\pi a^2$, where S is the area of the cross-section. κ is closely related but not exactly equal to the height h in Fig. 4.2. The relationship between γ and κ can be found analytically, but even with the simple case under consideration, the algebra becomes cumbersome. The MHD equilibrium solution (4.31) is applicable to the spherical tokamak, which is a recently proposed concept, since it is an exact solution and it is valid for any aspect ratio ϵ_a. The spherical tokamak is an attempt to obtain the highest beta limit through the aspect ratio dependence of the beta limit inherent in the tokamak configuration (see section 5.5). By its nature, the tight aspect ratio of the spherical tokamak implies a compact design of fusion reactor.

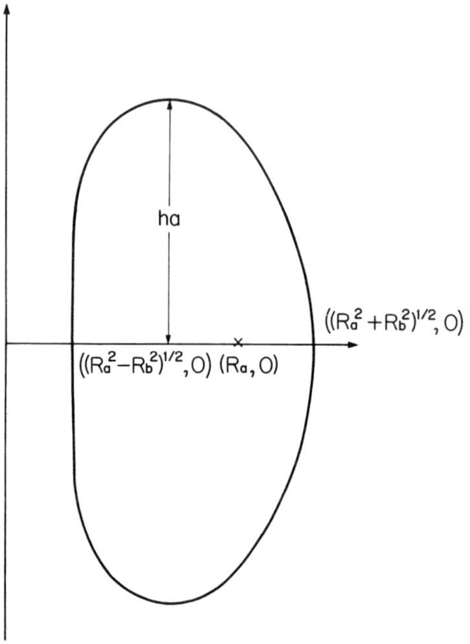

Fig. 4.2 The noncircular cross-section of Solov'ev equilibrium.

4.3 THE AVERAGED MHD EQUILIBRIUM EQUATION

Even if we assume the existence of magnetic surfaces for heliotrons, it is not easy to solve the generalized Grad–Shafranov equation describing three-dimensional MHD equilibrium. Instead of solving (4.27), a more realistic approximation is to employ the averaging procedure in the toroidal direction, as explained in section 3.7. When this approximation is allowed, the lowest-order MHD equilibrium averaged over the toroidal direction becomes axisymmetric, like the tokamak equilibrium. According to the stellarator expansion, we may calculate the first-order correction to the lowest-order equilibrium. It describes the three-dimensional structure of the MHD equilibrium approximately. Although the stellarator expansion gives higher-order corrections systematically, the calculations soon become tedious. Usually, the higher-order corrections are assumed to be small, and the lowest-order equilibrium is calculated for the study of the finite-beta currentless MHD equilibrium of the heliotron. It yields a rotational transform, a magnetic well or hill, and a Shafranov shift. When we need to know the shapes of the magnetic surfaces in the three-dimensional geometry, we can calculate the first-order correction. To show the accuracy of the stellarator expansion based on the averaging procedure in the toroidal direction, we will compare the averaged MHD equilibria with the three-dimensional MHD equilibria obtained by a three-dimensional MHD equilibrium code called VMEC (see section 4.4).

First, we note the following equation:

$$\frac{d}{dt}(\bar{\mathbf{B}}\cdot\nabla f)-\bar{\mathbf{B}}\cdot\nabla\frac{df}{dt}=\frac{\partial}{\partial t}[\Psi,f]-\left[\Psi,\frac{\partial f}{\partial t}\right]+\left[u,\frac{\partial f}{\partial\zeta}\right]$$
$$-\frac{\partial}{\partial\zeta}[u,f]+[u,[\Psi,f]]-[\Psi,[u,f]], \tag{4.32}$$

expressed by using the Poisson bracket and $\bar{\mathbf{B}}\cdot\nabla f=[\Psi,f]+\partial f/\partial\zeta$ (see (3.198)). For physical quantities f, g, and h the relation

$$[[f,g],h]+[[g,h],f]+[[h,f],g]=0 \tag{4.33}$$

is obtained, which is called the Jacob identity. Applying it to (4.32), the RHS becomes

$$\left[\frac{\partial A}{\partial t}-\frac{\partial u}{\partial\zeta}-[u,\Psi],f\right] \tag{4.34}$$

since $\partial\Psi_h/\partial t=0$ (see (3.223)). When resistivity is neglected in Ohm's law (3.231), (4.34) becomes zero automatically and

$$\frac{d}{dt}(\bar{\mathbf{B}}\cdot\nabla f)=\bar{\mathbf{B}}\cdot\nabla\frac{df}{dt} \tag{4.35}$$

is obtained. By substituting pressure P into f, (4.35) gives

$$\frac{d}{dt}(\bar{\mathbf{B}}\cdot\nabla P)=0, \tag{4.36}$$

since the pressure evolution follows (3.232). This means that, when $\bar{\mathbf{B}}\cdot\nabla P=0$ initially, it is valid for $t>0$.

Here we assume that the derivatives with respect to ζ vanish in the reduced MHD equations (3.230), (3.231), and (3.232):

$$\frac{\partial}{\partial t}\nabla_\perp^2 u=[u,\nabla_\perp^2 u]+[\nabla_\perp^2 A,\Psi]+[\Omega,P], \tag{4.37}$$

$$\frac{\partial A}{\partial t}=[u,\Psi], \tag{4.38}$$

$$\frac{\partial P}{\partial t}=[u,P], \tag{4.39}$$

where the resistivity is also neglected. Equations for the static MHD equilibrium are obtained under the conditions that $u=0$ and that the LHS of these equations vanishes. For this case (4.36) becomes $\bar{\mathbf{B}}\cdot\nabla P=[\Psi,P]=0$, which is equivalent to $P=P(\Psi)$. Also, from (4.37),

$$\left[\nabla_\perp^2 A+\Omega\frac{dP}{d\Psi},\Psi\right]=0 \tag{4.40}$$

is given. Then the equilibrium equation becomes

$$\nabla_\perp^2 A = -\Omega \frac{dP}{d\Psi} + g(\Psi), \tag{4.41}$$

where g is an arbitrary function of Ψ corresponding to an equilibrium plasma current. By noting that the magnetic surface average of $\mu_0 J_\zeta = \nabla_\perp^2 A$ vanishes under the currentless constraint,

$$\langle \nabla_\perp^2 A \rangle = -\langle \Omega \rangle \frac{dP}{d\Psi} + g(\Psi) = 0, \tag{4.42}$$

where

$$\langle f \rangle = \int \frac{f}{|\nabla\Psi|} \, d\ell \Big/ \int \frac{d\ell}{|\nabla\Psi|}. \tag{4.43}$$

Here $d\ell$ is a line element on the poloidal cross-section of the magnetic surface $\Psi =$ constant. Then the equilibrium equation (4.41) is written as

$$\nabla_\perp^2 A = (\langle \Omega \rangle - \Omega) \frac{dP}{d\Psi}. \tag{4.44}$$

One method for obtaining MHD equilibrium is to follow the time evolution of a configuration with vacuum magnetic surfaces described by $\Psi_h =$ constant and an assumed pressure profile $P(\Psi_h)$ initially by adding a viscosity term $\bar{\mu}\nabla_\perp^4 u$ in (4.37) and a resistivity term $\bar{\eta}\nabla_\perp^2 A$ in (4.38). Here $\bar{\mu}$ and $\bar{\eta}$ are constant for simplicity, which does not change the results. One way to remove the kinetic energy is to put $u = 0$ after the increase in the kinetic energy has almost stopped. Then we calculate the time evolution again. By iterating these procedures the increase in the kinetic energy finally disappears, which corresponds to the resistive MHD equilibrium for a sufficiently small $\bar{\eta}$. This state also satisfies the currentless constraint (4.42).

We write (4.41) in the following form:

$$\frac{1}{R_0} \nabla_\perp^2 (\Psi - \Psi_h) = -R_0 \Omega_0 \frac{dP}{d\Psi} - J_\zeta(\Psi), \tag{4.45}$$

where $\Psi - \Psi_h = R_0^2 A$ is substituted in the LHS of (4.41) and $\Omega_0 = \overline{(\nabla\Phi)^2}/B_0^2$. By noting that the toroidal plasma current is described by $J_\zeta(\Psi) = R_0 P'(\Psi) + FF'(\Psi)/R_0$ (see (4.29)), (4.45) can be written as

$$\nabla \cdot \frac{1}{R_0^2} \nabla(\Psi - \Psi_h) = -P'(\Psi)(1 + \Omega_0) - FF'(\Psi)/R_0^2, \tag{4.46}$$

where F is given in (4.11). This expression for the averaged MHD equilibrium of the stellarator and heliotron suggests the following equation, when a toroidal correction is taken into account with $R = R_0 + r\cos\theta$:

$$\nabla \cdot \frac{1}{R^2}\,\nabla(\Psi - \Psi_h) = -P(\Psi)(1 + \Omega_0) - FF'(\Psi)/R^2. \tag{4.47}$$

The form of the Grad–Shafranov type equation (4.47) may be derived more rigorously, as shown by Pustovitov and Shafranov (1990).

Here we given an approximate solution of the MHD equilibrium equation (4.41). From the expression of Ψ_h (3.226),

$$\Psi_h(r) \simeq -\frac{\iota_h(a)}{2(\ell-1)R_0}\, r^{2(\ell-1)}\left(1 + \frac{h^2 r^2}{2\ell}\right) \tag{4.48}$$

is given by assuming that $hr \gg 1$. Also, $B_0 = 1$ and $a = 1$ are assumed. The vacuum rotational transform is also given by (3.224),

$$\iota_h(r) \simeq \iota_h(a)\left(\frac{2(\ell-1) + h^2 r^2}{2(\ell-1) + h^2}\right) r^{2(\ell-2)}, \tag{4.49}$$

for $hr \ll 1$. For a low-beta plasma with $\beta \ll 1$, the flux function Ψ may have the form

$$\Psi(r, \theta) = \Psi_h(r) + A_1(r, \theta) + \cdots, \tag{4.50}$$

where $A_1(r, \theta)$ belongs to $O(\beta)$ and satisfies

$$\nabla_\perp^2 A_1 + \frac{2r}{R_0}\cos\theta P'(\Psi_h) = 0, \tag{4.51}$$

which is obtained from (4.41). Here the first term in the expansion (4.50) satisfies

$$-\Omega_0 P'(\Psi_h) + g(\Psi_h) = 0. \tag{4.52}$$

By assuming $P'(\Psi_h) = P_0/\Psi_h(a)$ for simplicity, the solution of (4.51) is given by

$$A_1(r, \theta) = -\frac{P_0}{4\Psi_h(a)R_0}\,(r^3 - r)\cos\theta. \tag{4.53}$$

The position of the magnetic axis is given by $\nabla(\Psi_h + A_1) = 0$. When it is shown by $r = r_m$ and $\theta = 0$, this gives

$$r_m^{2\ell-3} \simeq \frac{P_0 R_0}{2(\iota_h(a))^2}\,(\ell - 1). \tag{4.54}$$

For the $\ell = 2$ case, such as the Heliotron E, the ATF, and the CHS,

$$r_m = \frac{P_0 R_0}{2(\iota_h(a))^2}, \tag{4.55}$$

which predicts that the position of the magnetic axis changes linearly with respect to $\beta(0)$ (central beta value) $\propto P_0$. Figure 4.3 shows the magnetic axis position versus the central beta value for the currentless equilibria of the Heliotron E, which are obtained by several numerical codes under the fixed boundary condition where the conducting wall is assumed to be on the outermost magnetic surface. Here the STEP(EQ) code is used to solve the Grad–Shafranov type equation (4.44) and the VMEC code is the three-dimensional MHD equilibrium code based on the variational principle, assuming the existence of magnetic surfaces (see section 4.4). Figure 4.3 shows that the magnetic axis shift is proportional to the central beta as shown by (4.55) for the low-beta region; however, it has a tendency to saturate in the higher-beta region. This kind of universal behavior is known for various stellarators and heliotrons from numerical studies of currentless equilibrium. Here Δ is defined as the difference in the magnetic axis position between $\beta(0) = 0$ and $\beta(0) \neq 0$. Usually, the beta value corresponding to $\Delta/a \simeq 0.5$ is considered to be the *equilibrium beta limit*, since the confinement property may degrade due to the steepening of the pressure gradient in the outer region of the toroidal plasma.

Here we note that the equation for $A_1(r,\theta)$ (4.51) is valid only near the magnetic axis. From (4.41) a more general equation is given for $A_1(r,\theta) = \tilde{A}(r)\cos\theta$,

$$\frac{d^2\tilde{A}}{dr^2} + \frac{1}{r}\frac{d\tilde{A}}{dr} + \left(\frac{dJ_0/dr}{d\Psi_h/dr} - \frac{dP_0/dr}{(d\Psi_h/dr)^2}\frac{d\Omega_0}{dr} - \frac{1}{r^2}\right)\tilde{A} = -\frac{2r}{R_0}\frac{dP_0/dr}{d\Psi_h/dr}, \tag{4.56}$$

where J_0 denotes a toroidal plasma current producing the rotational transform ι_J. When the third and fourth terms in the LHS are neglected, (4.56) becomes equivalent to (4.51) under the assumption of $A_1(r,\theta) = \tilde{A}(r)\cos\theta$. Then the toroidal shift of the magnetic surface $\Psi(r,\theta) =$ constant is given by

$$\xi(r) = \frac{\tilde{A}(r)}{R_0 r(\iota_h(r) + \iota_J(r))}. \tag{4.57}$$

In order to obtain a complete solution, (4.56) must be supplemented by boundary conditions.

For a free boundary equilibrium case, $\tilde{A}$ vanishes at $r = 0$, and $\tilde{A}$ and $d\tilde{A}/dr$ are connected at the plasma–vacuum boundary at $r = a_0$. In the vacuum region $\tilde{A}$ must vanish at a conducting wall, or at infinity if there is no such wall. We consider that the plasma column of radius a_0 is surrounded by a conducting

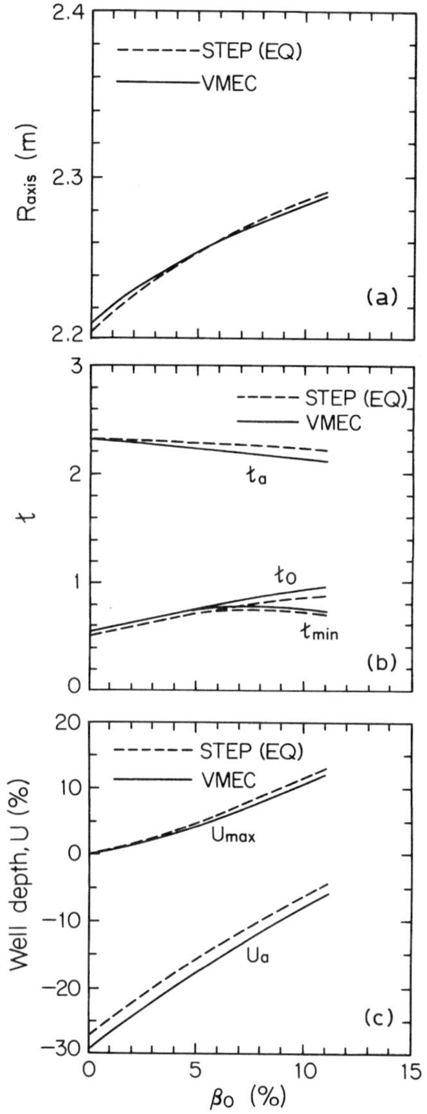

Fig. 4.3 Pressure dependence for (a) a shift of magnetic axis, (b) a change of rotational transform characterized by the central value t_0, edge value t_a, and minimum value t_{min}, and (c) the well depth characterized by U_{max} and U_a, where $U = (V'(0) - V'(\Psi))/V'(0)$ and $V' = dV/d\Phi$, obtained by the VMEC and STEP(EQ) codes for the Heliotron E. In (a), R_{axis} denotes the major radius of the magnetic axis. The pressure profile is assumed to be $P(\Psi) = P_0(1 - \Psi)^2$.

wall of radius $a_1 (> a_0)$, and that the plasma pressure vanishes at $r \geq a_0$. Then a solution for $r \leq a$ satisfying the boundary condition is

$$\tilde{A} = C_1 y_1 + y_2 \int_0^r \frac{y_1 F}{W} dr - y_1 \int_0^r \frac{y_2 F}{W} dr, \tag{4.58}$$

where y_1 and y_2 constitute two independent solutions of the homogeneous equation given by (4.56) and $y_1(0) = 0$. We note that y_2 goes to infinity at $r = 0$. C_1 is a constant found from the boundary conditions,

$$C_1 = \left. \frac{(MQ - rM')}{(y_1' r - Q y_1)} \right|_{r=a_0}, \tag{4.59}$$

where

$$M = y_2 \int_0^r \frac{y_1 F}{W} dr - y_1 \int_0^r \frac{y_2 F}{W} dr,$$

$$W = y_1 y_2' - y_1' y_2, \quad F = -\frac{2r}{R_0} \frac{dP_0/dr}{d\Psi_h/dr},$$

$$Q = -\frac{1 + (a_0/a_1)^2}{1 - (a_0/a_1)^2}.$$

We can see that at $(ry_1')|_{r=a_0} = y_1|_{r=a_0} Q$, C_1 and thus $\xi(r)$ becomes infinite. This occurs for $(ry_1'/y_1)|_{r=a_0} < 0$, since $Q < 0$. This means that the free boundary MHD equilibrium is lost at a certain critical beta value, since $(ry_1'/y_1)|_{r=a_0}$ may become negative for a large toroidal shift case.

According to the numerical results from both the STEP(EQ) and VMEC codes for currentless equilibria in $\ell = 2$ heliotrons, the rotational transform at the magnetic axis generally increases with an increase of $\beta(0)$ (see Fig. 4.4), which is different from the vacuum rotational transform (4.49) for $\ell = 2$. Contrary to the increase of $\iota(0)$, $\iota(a)$ decreases with the increase of $\beta(0)$ under the currentless condition. This can be understood by solving the Grad–Shafranov type equilibrium equation

$$\nabla_\perp^2 \Psi = K - C \frac{r}{R_0} \cos\theta + \nabla_\perp^2 \left(\frac{\iota_h r^2}{2R_0} \right), \tag{4.60}$$

which is obtained from (4.46), where $FF'(\Psi) = -K = \text{constant}$, $R_0^2 P'(\Psi) = C/2 = \text{constant}$, and $\Psi_h = \iota_h r^2 / 2R_0$ are assumed (see (4.48)). Also, the toroidal curvature is assumed to be larger than that due to the external helical magnetic field or $\varepsilon > \delta^2$, since $\Omega_0 \simeq \delta^2$ and $\delta = |\mathbf{B}_h|/B_0$ (see (3.228)). In (4.60) the averaged magnetic surface over the pitch length in the toroidal direction is considered to be circular and the plasma boundary is assumed at $r = a$. These are crude approximations of a realistic heliotron con-

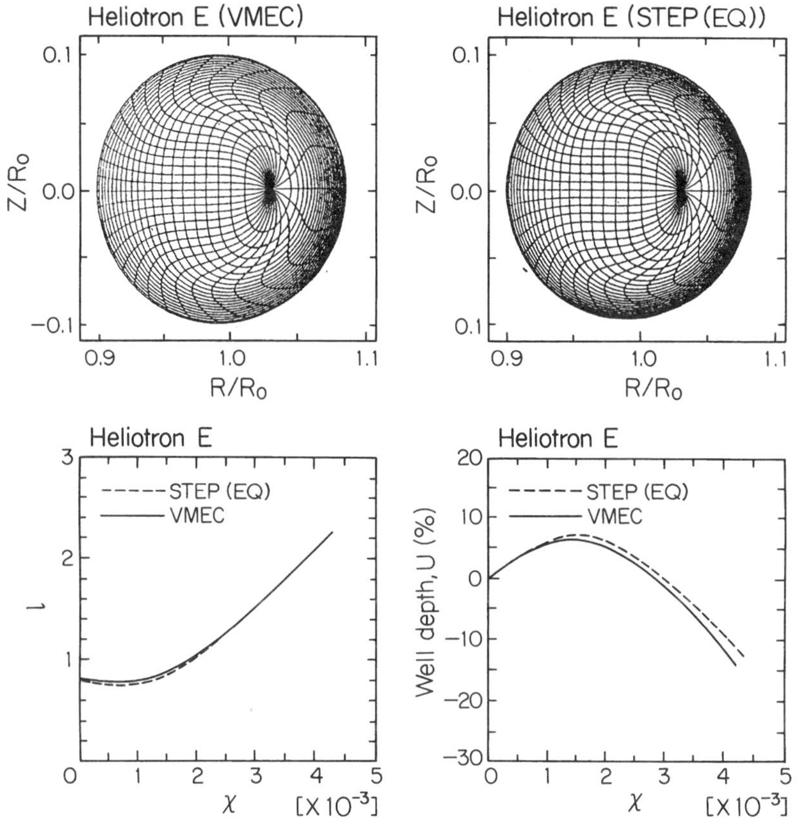

Fig. 4.4 The averaged magnetic surfaces, rotational transform profiles and well depth profile given by $U = (V'(0) - V'(\Psi))/V'(0)$ obtained by the VMEC and STEP(EQ) codes for $\beta(0) = 7\%$ in the Heliotron where $V' = dV/d\Phi$. The pressure profile is assumed to be $P(\Psi) = P_0(1 - \Psi)^2$. In the lower figure χ corresponds to Ψ.

figuration; however, we may understand finite-beta effects on MHD equilibrium qualitiatively.

By imposing the fixed boundary condition $\Psi(a, \theta) = 0$, the solution of (4.60) becomes

$$\Psi(\hat{r}, \theta) = \frac{a^2}{2R_0} (\hat{K} + \iota_h - \hat{C}\hat{r} \cos\theta)(\hat{r}^2 - 1), \tag{4.61}$$

where $\hat{r} = r/a$, $\hat{K} = KR_0/2$, and $\hat{C} = Ca/4$. From (4.29) the toroidal plasma current density J is given by

$$J = K - C \frac{r}{R_0} \cos\theta. \tag{4.62}$$

When a total plasma current is denoted by I_0,

$$\hat{K} = \iota_J = \frac{RI_0}{2\pi a^2} \tag{4.63}$$

is the rotational transform produced by the plasma current. It should be noted that $B_0 = 1$ is already assumed. Since $\bar{\beta}$ is defined as

$$\bar{\beta} = \frac{2}{\pi a^2} \int P r dr d\theta, \tag{4.64}$$

and $P(\Psi) = -C\Psi/2$,

$$\hat{C} = \frac{\bar{\beta}}{\varepsilon_a(\iota_h + \iota_J)} \tag{4.65}$$

is obtained by substituting (4.61) into (4.64), where $\varepsilon_a = a/R_0$. It is noted that the rotational transform and the shift of the magnetic axis can be obtained from the magnetic flux function (4.61).

The definition of the rotational transform,

$$\iota = 2\pi R_0 \left[\int_0^{2\pi} \left(\frac{1}{r} \frac{d\Psi}{dr} \right)^{-1} d\theta \right]^{-1}, \tag{4.66}$$

gives

$$\iota_a = (\iota_h + \iota_J)(1 - \nu^2)^{1/2}, \tag{4.67}$$

where

$$\nu = \frac{\bar{\beta}}{\varepsilon_a(\iota_h + \iota_J)^2}. \tag{4.68}$$

The magnetic axis satisfies $\partial\Psi/\partial r|_{\theta=0} = 0$, which gives the shift of the magnetic axis, or the Shafranov shift:

$$\frac{\Delta}{a} = \frac{\nu}{1 + (1 + 3\nu^2)^{1/2}}. \tag{4.69}$$

Here we consider the constraints for the toroidal plasma current in the stellarataor and heliotron. In the stationary state it is expected that the net toroidal plasma current is zero, since there is no toroidal electric field to drive the current. In chapter 7 we discuss the bootstrap current driven by the pressure gradient in the context of neoclassical transport theory. When the plasma temperature increases and collisionality decreases to generate particles trapped in the local mirror field of the stellarator and heliotron, the bootstrap current will contribute to the MHD equilibrium of heliotron devices such as the LHD. The existence of the bootstrap current has already been shown, and it exceeds several hundreds of kA in large tokamaks, such as TFTR, JT-60, and JET. Also, the bootstrap current can be observed in the stellarators and heliotrons such as W7-AS, ATF, and CHS, although the magnitude is a few kA at present. The effect of the bootstrap current on the Shafranov shift can be included through ι_J in (4.68) and (4.69). However, since currentless MHD equilibrium is appropriate in the context of the ideal MHD theory, we assume that $\iota_J = 0$ at first. For extremely low-beta plasmas, the rotational transform given by (4.67) becomes $\iota_a = \iota_h$. With an increase in plasma pressure, $\nu \to 1$, $\iota_a \to 0$. In other words, the rotational transform at the plasma surface decreases with an increase in $\bar{\beta}$. When the rotational transform disappears, the MHD equilibrium in the toroidal geometry will be lost. In other words, the beta limit appears at $\nu = 1$ or

$$\bar{\beta} = \varepsilon_a \iota_h^2. \tag{4.70}$$

In this case, the shift in the magnetic axis is given by

$$\frac{\Delta}{a} = \frac{1}{3}. \tag{4.71}$$

This is smaller than the usual estimation, $\Delta/a \simeq 1/2$, for the equilibrium beta limit. Expression (4.70) suggests that the equilibrium beta limit is $\bar{\beta} \simeq \varepsilon_a$ for a heliotron with $\iota_h \simeq 1$. Usually, $\varepsilon_a = 0.1$–0.2 in heliotrons, and the equilibrium beta limit is sufficiently high for a fusion reactor design.

An example of the rotational transform for finite-beta plasmas of the Heliotron E under the currentless condition is shown in Fig. 4.4. Here the pressure profile is assumed to be $P = P_0(1 - \Psi)^2$. According to the above dicussion, the rotational transform at the plasma surface ι_a decreases with an increase in $\beta(0)$ (see Fig. 4.3(b)). It is also noted that the rotational transform at the magnetic axis ι_0 increases. This tendency depends on the shape of the magnetic surface. When the averaged magnetic surface over the toroidal angle near the magnetic axis is close to circular, the increase of ι_0 due to the finite beta effect is general. This property is related to the existence of magnetic shear in heliotron devices. Here we will write the rotational transform without a plasma as

$$\iota^v(r) = \iota_0^v + (\iota_a^v - \iota_0^v)(r^2/a^2) \tag{4.72}$$

for simplicity. Here r is a radius measured from the magnetic axis in the vacuum configuration. Also, ι_0^v and ι_a^v are the vacuum rotational transforms at $r = 0$ and $r = a$, respectively. The heliotron configuration is characterized by $\iota_a^v > \iota_0^v$. When the plasma pressure becomes finite, the magnetic axis shifts toward the outside of the torus with Δ. Roughly by substituting $r \approx \Delta$

in (4.72), the rotational transform at the magnetic axis becomes $\iota_0^v + \alpha(\iota_a^v - \iota_0^v)(\Delta^2/a^2)$, which is larger than ι_0^v, where α is a coefficient of order unity. When the beta value increases substantially, ι_0 can be larger than ι_a in some heliotron devices.

Now we consider a flux-conserving stellarator. The concept of the flux-conserving tokamak was introduced by Clark and Sigma (1977) and its approach is recognized to be useful in obtaining high-beta troidal plasmas. When a toroidal plasma is heated rapdily, the ratio between the poloidal and toroidal fluxes is conserved during the increase in the plasma pressure. This means that the safety factor or the rotational transform profile is conserved during the rapid heating. This kind of configuration is called a flux-conserving tokamak. The same idea is applicable to stellarators. In stellarators the steady state equilibrium is expected to be currentless; however, the flux-conserving stellarator might be realized during the heating phasse. Expressions for the rotational transform (4.67) and the Shafranov shift (4.69) are usable in the flux-conserving stellarator. In this case ι_a is fixed at the vacuum rotational transform of ι_h even for a finite-pressure plasma obtained by a rapid heating, and

$$\nu^2 = 1 - \left(\frac{\iota_h}{\iota_J + \iota_h}\right)^2 \tag{4.73}$$

is given. Since ν is related to $\bar{\beta}$ through (4.68), (4.73) yields

$$\iota_J/\iota_h = \{\tfrac{1}{2}[1 + (1 + 4H^2)^{1/2}]\}^{1/2} - 1, \tag{4.74}$$

where $H = \bar{\beta}/(\varepsilon_a \iota_h^2)$. Also with H, the Shafranov shift is expressed as

$$\begin{cases} \dfrac{\Delta}{a} = \dfrac{\nu}{1 + (1 + 3\nu^2)^{1/2}} \\ \nu = \dfrac{2H}{1 + (1 + 4H^2)^{1/2}}. \end{cases} \tag{4.75}$$

The result shown by (4.74) means that the net toroidal plasma current appears under the flux-conserving constraint in stellarators. This current increases according to the plasma pressure and $\iota_J > \iota_h$ is possible for $H \gg 1$. Furthermore, the relations (4.75) mean that for high-beta plasmas with $\bar{\beta} > \varepsilon_a \iota_h^2$ or $H > 1$, the Shafranov shift is less than the maximum value in the case of currentless equilibrium (4.71). Therefore, there is no equilibrium beta limit in the flux-conserving stellarator.

The toroidal current driven by the increase in the plasma pressure decays according to Ohm's law under the finite resistivity; and the equilibrium state approaches the currentless resistive equilibrium slowly.

4.4 A THREE-DIMENSIONAL MHD EQUILIBRIUM CALCULATION BASED ON THE VARIATIONAL PRINCIPLE

One of the basic problems of plasma physics is the calculation of general three-dimensional MHD equilibria and the assessment of their stability. This is essential for the design of magnetic confinement devices without an ignorable coordinate, such as the stellarator and heliotron with helical windings. This problem has been solved in the following two ways:

(i) an expansion method with a small parameter, the lowest-order equilibrium being the averaged one in the toroidal direction;
(ii) a variational method to obtain the equilibrium state iteratively in the three-dimensional geometry.

The latter approach is described here. Starting from given distributions of the plasma pressure and magnetic field, the distributions are adjusted under the magnetic flux, mass, and entropy conservation in such a way that the potential energy

$$W = \int \left(\frac{B^2}{2\mu_0} + \frac{P}{\gamma - 1} \right) dV \tag{4.76}$$

is decreased monotonically, where dV is a volume element. A state of minimum energy thus obtained represents an MHD equilibrium. The minimum energy state is related to an allowed degree of freedom in the plasma motion. If all degrees of freedom are considered, the obtained MHD equilibrium should be stable for all perturbations.

An MHD equilibrium characterized as a stationary state of the potential energy W is given by

$$\delta W = 0 \tag{4.77}$$

for an infinitesimal displacement vector $\xi(\mathbf{x}, t)$ that satisfies the constraints of mass density and magnetic flux conservation,

$$\tilde{\rho} = -\nabla \cdot (\rho \boldsymbol{\xi}), \tag{4.78}$$

$$\tilde{\mathbf{B}} = \nabla \times (\boldsymbol{\xi} \times \mathbf{B}). \tag{4.79}$$

These are obtained from the linearized equations

$$\frac{\partial \tilde{\rho}}{\partial t} + \nabla \cdot (\rho \tilde{\mathbf{v}}) = 0$$

and

$$\frac{\partial \tilde{\mathbf{B}}}{\partial t} = \nabla \times (\tilde{\mathbf{v}} \times \mathbf{B}),$$

by assuming that $\tilde{\mathbf{v}} = \partial \boldsymbol{\xi}/\partial t$ and integrating them under the initial condition $\boldsymbol{\xi}(\mathbf{x}, 0) = 0$. We assume that the change of plasma state is adiabatic and that there is a relation between the pressure P and mass density ρ,

$$P = \rho^{\gamma}, \tag{4.80}$$

where γ is a specific heat ratio. In this case

$$\tilde{P} = -(\boldsymbol{\xi} \cdot \nabla)P - \gamma P(\nabla \cdot \boldsymbol{\xi}) \tag{4.81}$$

is obtained for infinitesimal $\boldsymbol{\xi}$. We can derive the MHD equilibrium equation from condition (4.77). For simplicity, we assume that the plasma extends to the conducting wall of the toroidal chamber without a vacuum region in between. Then the boundary condition at the wall is assumed to be $\boldsymbol{\xi} \cdot \mathbf{n} = 0$. On this wall the perturbed magnetic field also satisfies $\tilde{\mathbf{B}} \cdot \mathbf{n} = 0$, or the normal component of the magnetic field does not appear, where $\mathbf{n}$ is a unit vector normal to the conducting wall. By using the constraints (4.79) and (4.81), the variation of the potential energy can be written as

$$\delta W = -\int \mathbf{F} \cdot \boldsymbol{\xi} dV, \tag{4.82}$$

where $\mathbf{F}$ represents the local electromagnetic force:

$$\mathbf{F} = \frac{1}{\mu_0}(\nabla \times \mathbf{B}) \times \mathbf{B} - \nabla P. \tag{4.83}$$

the Euler–Lagrange equation to satisfy $\delta W = 0$ is thus the MHD equilibrium condition $\mathbf{F} = 0$.

Next, we consider the method of finding the energy minimum of the heliotron and stellarator configuration. In order to apply it to a three-dimensional MHD equilibrium calculation, we consider the vacuum magnetic field $\mathbf{B}$ possessing the magnetic surfaces $\Psi(\mathbf{x}) = \text{constant}$, where Ψ is a flux function. The mass density is assumed arbitrarily as $\rho(\Psi)$, which gives $P(\Psi)$ by (4.80). These $\mathbf{B}$ and P do not satisfy $\mathbf{F} = 0$. The path of the displacement to the energy minimum that satisfies $\mathbf{F} = 0$ is described by $\boldsymbol{\xi}(\mathbf{x}, t)$. Here t is considered as an artificial time. By considering that $\boldsymbol{\xi}$ is an infinitesimal displacement,

$$\frac{dW}{dt} = -\int \mathbf{F} \cdot \frac{\partial \boldsymbol{\xi}}{\partial t} dV \tag{4.84}$$

is given from (4.82). Since W is bounded from below due to magnetic flux and mass conservation and is positive definite for $\gamma > 1$, the equilibrium corresponding to the minimum energy state exists. Thus, by finding the path along which $|dW/dt|$ decreases monotonically, the equilibrium will be reached. To reduce $|dW/dt|$, we note the relation

$$\left| \int \mathbf{F} \cdot \frac{\partial \xi}{\partial t}\, dV \right|^2 \leq \int |\mathbf{F}|^2 dV \int \left| \frac{\partial \xi}{\partial t} \right|^2 dV \tag{4.85}$$

which gives the equality if and only if

$$\partial \xi / \partial t = \alpha \mathbf{F}, \tag{4.86}$$

where $\alpha > 0$ is an arbitrary real constant. Here we choose $\alpha = 1$. Then the decent path is given by

$$\partial \xi / \partial t = \mathbf{F}, \tag{4.87}$$

and the maximum rate of decrease in W along this path is given by

$$dW/dt = -\int |\mathbf{F}|^2 dV. \tag{4.88}$$

Equation (4.87) comprises the descent path equation for relaxing W to its minimum. We also note that $dW/dt = 0$ if and only if $\mathbf{F} = 0$.

If we solve for the decent path by applying a difference scheme to (4.87), the convergence to an equilibrium solution keeping the numerical stability is usually slow. In order to accelerate the convergence, (4.86) is converted from a parabolic differential equation to a hyperbolic one by adding a new term, while retaining an explicit form for the force vector $\mathbf{F}$,

$$\frac{\partial^2 \xi}{\partial t^2} + \frac{1}{\alpha} \frac{\partial \xi}{\partial t} = \mathbf{F}, \tag{4.89}$$

which is called the second-order Richardson scheme. The parameter α is chosen to enhance the decay rate of the potential energy and the optimum value is estimated as

$$\frac{1}{\alpha} \simeq -\frac{d}{dt} \left(\ln \int |\mathbf{F}|^2 dV \right). \tag{4.90}$$

Multiplying (4.89) by $\partial \xi / \partial t$ and integrating over the plasma region yields

$$\frac{\partial}{\partial t} (W_k + W) = -\frac{2}{\alpha} W_k, \tag{4.91}$$

where $W_k = \int (1/2)|\partial\xi/\partial t|^2 dV$ is considered to be the kinetic energy. Thus, for $\alpha > 0$, the sum of the kinetic and potential energies, which is bounded from below, decays monotonically until $W_k = 0$, and the MHD equilibrium is attained.

Bauer, Betancourt, and Garabedian (1984) first developed the three-dimensional MHD equilibrium and stability code called BETA, on the basis of the variational principle, assuming the existence of magnetic surfaces. In this code the MHD equilibrium is obtained for one pitch length by assuming periodicity from the pitch number M of the stellarator and heliotron. Here the pressure profile is assigned as a function of the magnetic surface.

As for MHD stability analysis, the second minimization of the potential energy is examined, in order to find the minimum energy state for a full torus with M periods after a perturbation is added for the toroidal equilibrium given by the first minimization. When the minimum energy of the second minimization is lower (higher) than that corresponding to the first minimization, the equilibrium is unstable (stable) to the assumed perturbation.

The most widely used three-dimensional MHD equilibrium code is VMEC, developed by Hirshman and Whitson (1983) at the Oak Ridge National Laboratory. Here Fourier expansions are employed for angular variables on the magnetic surfaces. Also, the angular variable in the poloidal direction is chosen to minimize the number of Fourier modes. In numerical codes for calculating MHD equilibrium based on the variational principle, the slowest convergence occurs at the magnetic axis. Usually, the axis region is treated separately to increase the numerical accuracy.

4.5 THE SOLOV'EV–SHAFRANOV EQUATION

In this section we discuss the MHD equilibrium of a stellarator with an helical magnetic axis. The helical axis configuration is different from that of heliotrons; however, it is possible to produce a helical magnetic axis even in heliotrons, when the pitch number M is even. In this case two helical coils may have different coil currents. This difference in coil currents can generate an $\ell = 1$ helical magnetic field, which produces a helical magnetic axis.

In the straight approximation, or by neglecting toroidal effects, the magnetic axis is shown in Fig. 2.21. Here we will try to retain the toroidal effects in the derivation of an equilibrium equation for the helical axis stellarator. We employ special magnetic coordinates related to the helical magnetic axis. Along the helical magnetic axis we introduce a coordinate s. A radial coordinate ρ is given to specify the magnetic surface, or $\rho =$ constant corresponds to the magnetic surface. A poloidal angle is defined by

$$\begin{cases} \omega = \theta + \alpha(s) \\ \alpha(s) = \int_0^s \kappa ds, \end{cases} \tag{4.92}$$

where θ is an angle measured from the direction of the normal curvature vector and κ is the torsion of the helical magnetic axis. It is noted that the coordinates (ρ, ω, s) constitute orthogonal coordinates, and the line element is given by

$$(d\ell)^2 = (d\rho)^2 + (\rho d\omega)^2 + (1 - k\rho\cos\theta)^2(ds)^2, \tag{4.93}$$

where k is the curvature of the magnetic axis. When the magnetic axis has no torsion, or $\kappa = 0$ and $\omega = \theta$, the coordinates become the standard toroidal coordinates (see Fig. 2.1).

Here we show expressions for the curvature and the torsion when the helical magnetic axis is described by

$$\begin{cases} R(\phi) = R_0 + r_0\cos(N\phi) \\ Z(\phi) = r_0\sin(N\phi), \end{cases} \tag{4.94}$$

where (R, ϕ, Z) correspnds to the cylindrical coordinates. Here we introduce local orthogonal coordinates (X_L, Y_L, Z_L) on the helical magnetic axis at $\phi = \phi_0$. It is assumed that the (X_L, Y_L) plane is perpendicular to the axis and the Z_L-coordinate is along the tangential direction to the axis. Then, the unit vector in the Z_L-direction, $\mathbf{e}_{ZL}$, is given by

$$\mathbf{e}_{ZL} \equiv \left.\frac{d\mathbf{r}}{d\phi}\right|_{\phi=\phi_0}, \tag{4.95}$$

where $\mathbf{r} = (R(\phi)\cos\phi, R(\phi)\sin\phi, Z(\phi))$ in Cartesian coordinates, with the same Z-axis as in (R, ϕ, Z). Definition (4.95) gives

$$\begin{cases} (\mathbf{e}_{ZL})_X = -[r_0 N\sin(N\phi_0)\cos\phi_0 + R(\phi)\sin\phi_0]/L_Z \\ (\mathbf{e}_{ZL})_Y = [-r_0 N\sin(N\phi_0)\sin\phi_0 + R(\phi)\cos\phi_0]/L_Z \\ (\mathbf{e}_{ZL})_Z = r_0 N\cos(N\phi_0)/L_Z, \end{cases} \tag{4.96}$$

where (X, Y, Z) denote three directions in Cartesian coordinates from the helical magnetic axis at $\phi = \phi_0$. Here

$$L_Z = [r_0^2 N^2 + (R(\phi))^2]^{1/2}. \tag{4.97}$$

We take the X_L-coordinate parallel to the principal normal direction of the helical magnetic axis, and its unit vector, $\mathbf{e}_{XL}$, is given from Frenet's formula,

$$\mathbf{e}_{XL} = \left.\frac{1}{k}\frac{d\mathbf{e}_{ZL}}{d\phi}\right|_{\phi=\phi_0}, \tag{4.98}$$

where k is the curvature of the axis. Substituting (4.96) into the definition (4.98) yields

$$\begin{cases} (\mathbf{e}_{XL})_X = \dfrac{1}{kL_Z}\left[(\mathbf{e}_{ZL})_X \dfrac{dL_Z}{d\phi_0} - R(\phi_0)\cos\phi_0 \right. \\ \qquad \left. - r_0 N^2 \cos(N\phi_0)\cos\phi_0 + 2r_0 N \sin(N\phi_0)\sin\phi_0\right] \\ (\mathbf{e}_{XL})_Y = \dfrac{1}{kL_Z}\left[(\mathbf{e}_{ZL})_Y \dfrac{dL_Z}{d\phi_0} - R(\phi_0)\sin\phi_0 \right. \\ \qquad \left. - r_0 N^2 \cos(N\phi_0)\sin\phi_0 - 2r_0 N \sin(N\phi_0)\cos\phi_0\right] \\ (\mathbf{e}_{XL})_Z = \dfrac{1}{kL_Z}\left[(\mathbf{e}_{ZL})_Z \dfrac{dL_Z}{d\phi_0} - r_0 N^2 \sin(N\phi_0)\right], \end{cases} \tag{4.99}$$

where

$$dL_Z/d\phi_0 = -R(\phi_0) N r_0 \sin(N\phi_0)/L_Z. \tag{4.100}$$

From (4.99), the curvature $k(\phi_0)$ is written as

$$k(\phi_0) = \frac{1}{L_Z^2}\left\{ r_0^2 N^4 + 2r_0^2 N^2 + 2r_0^2 N^2 \sin^2(N\phi_0) \right. \\ \left. + 2r_0 R_0 N^2 \cos(N\phi_0) - (R(\phi_0))^2 \left(1 - \frac{r_0^2 N^2}{L_Z^2} \sin^2(\phi_0)\right)\right\}^{1/2}. \tag{4.101}$$

The remaining unit vector in the Y_L-direction is given by $\mathbf{e}_{YL} = \mathbf{e}_{ZL} \times \mathbf{e}_{XL}$.

$$\begin{cases} (\mathbf{e}_{YL})_X = \dfrac{1}{kL_z^2}\left[r_0^2 N^3 \sin\phi_0 + r_0^2 N^2 \sin(N\phi_0)\cos\phi_0 \right. \\ \qquad \left. + R(\phi_0) r_0 \{N\cos(N\phi_0)\sin\phi_0 - N^2 \sin(2N\phi_0)\cos\phi_0\}\right] \\ (\mathbf{e}_{YL})_Y = \dfrac{1}{kL_z^2}\left[-r_0^2 N^3 \cos\phi_0 + r_0^2 N^2 \sin(2N\phi_0)\sin\phi_0\right. \\ (\mathbf{e}_{YL})_Z = \dfrac{1}{kL_z^2}\left[2r_0^2 N^2 \sin^2(N\phi_0) + (R(\phi_0))^2 + R(\phi_0) r_0 N^2 \cos(N\phi_0)\right]. \end{cases} \tag{4.102}$$

Again from Frenet's formula,

$$\left.\frac{d\mathbf{e}_{YL}}{d\phi}\right|_{\phi=\phi_0} = -\kappa \mathbf{e}_{XL}, \tag{4.103}$$

which gives the torsion of the helical magnetic axis κ as

$$\kappa(\phi_0) = \left\{ -\frac{r_0^2}{N}(R_0 - 2r_0\cos(N\phi_0)) + \frac{1}{N^3}\left[\left(3r_0^3 + \frac{r_0 L_Z^2}{N^2} - R_0^2 r_0\right)\cos(N\phi_0) \right.\right.$$
$$\left.\left. + 2R_0 r_0^2 \cos 2(N\phi_0) - r_0^3\cos^3(N\phi_0)\right]\right\} \times$$
$$\times \left\{ r_0^4 + \frac{1}{N^2}\left[r_0^2 R_0^2 + \left(r_0^2 + \frac{L_Z^2}{N^2}\right)^2 + 2r_0\left(R_0 r_0^2 - r_0^3 + \frac{R_0 L_Z^2}{N^2}\right)\cos(N\phi_0) \right.\right.$$
$$\left.\left. - r_0^2\left(2r_0^2 + \frac{L_Z^2}{N^2}\right)\cos^2(N\phi_0)\right]\right\}^{-1}. \tag{4.104}$$

Also, we note that the length of the helical magnetic axis for one period is

$$L = 2\pi r_0 \left[1 + \left(\frac{R_0}{N r_0}\right)^2\right]^{1/2}, \tag{4.105}$$

where $R_0 \gg r_0$ is assumed. There is a relation

$$N\phi = 2\pi s/L \tag{4.106}$$

between ϕ and s.

In order to study the MHD equilibrium of the helical axis stellarator analytically, we expand the plasma pressure, plasma current, and magnetic field by assuming $k \ll 1$. In the lowest order of $k \to 0$, a cylindrical plasma model appears. In this case the pressure, the current, and the magnetic field depend on the radius ρ only, and they are described by $P_0(\rho)$, $\mathbf{B}_0 = (0, B_{\omega 0}(\rho), B_{s0}(\rho))$, and $\mathbf{J}_0 = (0, J_{\omega 0}(\rho), J_{s0}(\rho))$. When the first-order quantities with respect to k are included,

$$\begin{cases} P = P_0(\rho) + P_1(\rho, \omega, s) \\ \mathbf{B} = \mathbf{B}_0(\rho) + \mathbf{B}_1(\rho, \omega, s) \\ \mathbf{J} = \mathbf{J}_0(\rho) + \mathbf{J}_1(\rho, \omega, s). \end{cases} \tag{4.107}$$

The first-order quantities can be expanded with respect to the angle $\theta = \omega - \alpha(s)$ as

$$\begin{cases} P_1 = \sum_n \tilde{P}_n \cos(\omega - \kappa_n s) \\ \mathbf{B}_1 = \sum_n \{(\tilde{\mathbf{B}}_n - \tilde{B}_{\rho n}\mathbf{e}_\rho)\cos(\omega - \kappa_n s) + \tilde{B}_{\rho n}\mathbf{e}_\rho \sin(\omega - \kappa_n s)\} \\ \mathbf{J}_1 = \sum_n \{(\tilde{\mathbf{J}}_n - \tilde{J}_{\rho n}\mathbf{e}_\rho)\cos(\omega - \kappa_n s) + \tilde{J}_{\rho n}\mathbf{e}_\rho \sin(\omega - \kappa_n s)\}, \end{cases} \tag{4.108}$$

where

$$\kappa_n = \kappa_0 + 2\pi n/L \tag{4.109}$$

and

$$\kappa_0 = \int_0^L \frac{\kappa}{L}\, ds. \tag{4.110}$$

Here we note that L is the length of the helical magnetic axis for one field period and that the direction of axis winding is anticlockwise. Under the first-order approximation in the expansion with respect to the curvature, the cross-sections of the magnetic surfaces remain as circles, but with displaced centers. The magnetic surface with a radius ρ', at which the plasma pressure is $P_0(\rho')$, is related to ρ and ξ in the following way:

$$\rho = \rho' + \xi(\rho', \omega, s) = \rho' + \sum_n \xi_n(\rho')\cos(\omega - \kappa_n s). \tag{4.111}$$

From the definition, the plasma pressure on this surface has a relation $P_0(\rho') = P(\rho, \omega, s)$. By expanding the pressure P in the form

$$P(\rho' + \xi, \omega, s) = P_0(\rho') + P_1(\rho', \omega, s) + \frac{dP_0}{d\rho}\,\xi(\rho', \omega, s) + \cdots, \tag{4.112}$$

we can find the relation between the displacement ξ and the pressure correction P_1 in the linear approximation,

$$P_1(\rho, \omega, s) = -\frac{dP_0}{d\rho}\,\xi(\rho, \omega, s), \tag{4.113}$$

where the difference between ρ' and ρ is negligible.

In order to solve the MHD equilibrium we use the radial components of $\mathbf{J} \times \mathbf{B} = \nabla P$, $\mathbf{B} \cdot \nabla P = 0$, and $\mathbf{J} \cdot \nabla P = 0$. They are supplemented with $\mu_0 \mathbf{J} = \nabla \times \mathbf{B}$ and $\nabla \cdot \mathbf{B} = 0$. The radial component of $\mathbf{J} \times \mathbf{B} = \nabla P$ gives

$$d\tilde{P}_n/d\rho = J_{\omega 0}\tilde{B}_{sn} - J_{s0}\tilde{B}_{\omega n} + \tilde{J}_{\omega n}B_{s0} - \tilde{J}_{sn}B_{\omega 0} \tag{4.114}$$

for the perturbations denoted by (4.108). $\mathbf{B} \cdot \nabla P = 0$ gives

$$\tilde{B}_{\rho n} = -\frac{\xi_n}{\rho}\,(B_{\omega 0} - \kappa_n \rho B_{s0}) \tag{4.115}$$

and $\mathbf{J} \cdot \nabla P = 0$ gives

$$\tilde{J}_{\rho n} = -\frac{\xi_n}{\rho}(J_{\omega 0} - \kappa_n \rho J_{s0}). \tag{4.116}$$

From $\mu_0 \mathbf{J} = \nabla \times \mathbf{B}$,

$$\mu_0 J_\rho = \frac{1}{1 - k\rho\cos\theta} \left\{ \frac{1}{\rho} \frac{\partial}{\partial \omega} [(1 - k\rho\cos\theta)B_s] - \frac{\partial B_\omega}{\partial s} \right\} \tag{4.117}$$

$$\mu_0 J_\omega = \frac{1}{1 - k\rho\cos\theta} \left\{ \frac{\partial B_\rho}{\partial s} - \frac{\partial}{\partial \rho} [(1 - k\rho\cos\theta)B_s] \right\} \tag{4.118}$$

$$\mu_0 J_s = \frac{1}{\rho} \frac{\partial}{\partial \rho} (\rho B_\omega) - \frac{1}{\rho} \frac{\partial B_\rho}{\partial \omega} \tag{4.119}$$

are obtained. Noting that

$$k(s)\cos\theta = \sum_n k_n \cos(\omega - \kappa_n s), \tag{4.120}$$

where

$$k_n = \frac{1}{L} \int_0^L k(s) \cos(\kappa_n s - \int_0^s \kappa(s')\, ds')\, ds, \tag{4.121}$$

we obtain

$$\mu_0 \tilde{J}_{\rho n} = -\kappa_n \tilde{B}_{\omega n} - \frac{\tilde{B}_{sn}}{\rho} + k_n B_{s0} \tag{4.122}$$

and

$$\mu_0 \tilde{J}_{sn} = \frac{1}{\rho} \frac{d}{d\rho} (\rho \tilde{B}_{\omega n}) - \frac{1}{\rho} \tilde{B}_{\rho n} \tag{4.123}$$

from (4.117) and (4.119), respectively. Instead of (4.118) we use $\nabla \cdot \mathbf{J} = 0$, which is required from $\mu_0 \mathbf{J} = \nabla \times \mathbf{B}$,

$$\frac{1}{\rho(1 - k\rho\cos\theta)} \left\{ \frac{\partial}{\partial \rho} [(\rho - k\rho^2 \cos\theta) J_p] + \frac{\partial}{\partial \omega} [(1 - k\rho\cos\theta) J_\omega] + \rho \frac{\partial J_s}{\partial s} \right\} = 0, \tag{4.124}$$

which gives

$$\frac{d}{d\rho} (\rho \tilde{J}_{\rho n}) - (\tilde{J}_{\omega n} - \kappa_n \rho \tilde{J}_{sn}) + k_n \rho J_{\omega 0} = 0. \tag{4.125}$$

$\nabla \cdot \mathbf{B} = 0$ also gives

$$\frac{d}{d\rho} (\rho \tilde{B}_{\rho n}) - (\tilde{B}_{\omega n} - \kappa_n \rho \tilde{B}_{sn}) + k_n \rho B_{\omega 0} = 0. \tag{4.126}$$

From (4.115), (4.116), (4.122), and (4.126),

$$\tilde{B}_{\omega n} = -\mu_0 J_{s0}\xi_n + \frac{1}{1+\kappa_n^2\rho^2}\left\{\left(k_n\rho + \frac{\xi_n}{\rho}\right)(B_{\omega 0} + \kappa_n\rho B_{s0}) - (B_{\omega 0} - \kappa_n\rho B_{s0})\frac{d\xi_n}{d\rho}\right\} \tag{4.127}$$

and

$$\tilde{B}_{sn} = \mu_0 J_{\omega 0}\xi_n + \frac{1}{1+\kappa_n^2\rho^2}\left\{k_n\rho(B_{s0} - \kappa_n\rho B_{\omega 0}) - \kappa_n\xi_n(B_{\omega 0} + \kappa_n\rho B_{s0}) + \kappa_n\rho(B_{\omega 0} - \kappa_n\rho B_{s0})\frac{d\xi_n}{d\rho}\right\} \tag{4.128}$$

are obtained. Also, from (4.114), (4.116), and (4.125),

$$\tilde{J}_{sn} = -\xi_n\frac{dJ_{s0}}{d\rho} + \frac{\kappa_n\rho}{1+\kappa_n^2\rho^2}(J_{\omega 0} - \kappa_n\rho J_{s0})\frac{d\xi_n}{d\rho} - \frac{k_n}{\kappa_n}J_{\omega 0} - \kappa_n\frac{(\xi_n - k_n/\kappa_n^2)}{1+\kappa_n^2\rho^2}\frac{(J_{\omega 0} - \kappa_n\rho J_{s0})(B_{\omega 0} + \kappa_n\rho B_{s0})}{b_{\omega 0} - \kappa_n\rho B_{s0}} \tag{4.129}$$

and

$$\tilde{J}_{\omega n} = \xi_n\left(\kappa_n J_{s0} - \frac{dJ_{\omega 0}}{d\rho}\right) - \frac{J_{\omega 0} - \kappa_n\rho J_{s0}}{1+\kappa_n^2\rho^2}\frac{d\xi_n}{d\rho} - \kappa_n^2\rho\frac{(\xi_n - k_n/\kappa_n^2)}{1+\kappa_n^2\rho^2}\frac{(J_{\omega 0} - \kappa_n\rho J_{s0})(B_{\omega 0} + \kappa_n\rho B_{s0})}{B_{\omega 0} - \kappa_n\rho B_{s0}} \tag{4.130}$$

are obtained. Now, by substituting $\tilde{B}_{\rho n}$, $\tilde{B}_{\omega n}$, and $\tilde{J}_{sn}$ into (4.123), we obtain the following ordinary differential equation for the displacement ξ_n:

$$\frac{1}{\rho}\frac{d}{d\rho}\left[\rho\frac{(B_{\omega 0} - \kappa_n\rho B_{s0})^2}{1+\kappa_n^2\rho^2}\frac{d\xi_n}{d\rho}\right] + \frac{\kappa_n^2\xi_n - k_n}{1+\kappa_n^2\rho^2}\times \times\left[2\frac{B_{\omega 0}^2 - \kappa_n^2\rho^2 B_{s0}^2}{1+\kappa_n^2\rho^2} - (B_{\omega 0} - \kappa_n\rho B_{s0})^2 - 2\mu_0\rho\frac{dP_0}{d\rho}\right] = 0. \tag{4.131}$$

This equation is similar to the Euler equation describing the radial displacement ξ_r at the marginal stability under the ideal MHD model in the analysis of stability of a cylindrical plasma (see section 5.4). If we assume that the azimuthal mode number is $m = 1$ and the longitudinal wave number $k = \kappa_n$ in the Euler equation (see (5.168)), it becomes (4.131) when $k_n = 0$. The curvature term proportional to k_n plays the role of an external force. It is natural that the coefficients in the expansion of the relative curvature k_n fall off rapidly with index n and that the dominant toroidal corrections are derived from terms characterized by low values of n. For this situation, $|\kappa_n\rho| \ll 1$ and $|\kappa_n^2\xi_n| \ll 1$ may be assumed to solve (4.131). Then, by integrating this equation, we have

$$\xi_n = k_n \int_{\rho_1}^{\rho_2} \rho \frac{G}{D}\, d\rho, \tag{4.132}$$

which corresponds to the relative displacement between the magnetic surface with radius ρ_1 and that with ρ_2. In the expression of (4.132),

$$G = 2\mu_0[\langle P_0\rangle_\rho - P_0(\rho)] + \tfrac{1}{2}[\langle B_{\omega 0}^2\rangle_\rho + 2\langle \kappa_n\rho B_{\omega 0}B_{s0}\rangle_\rho - 3\langle \kappa_n^2\rho^2 B_{s0}^2\rangle_\rho], \tag{4.133}$$

$$D = [B_{\omega 0}(\rho) - \kappa_n\rho B_{s0}]^2, \tag{4.134}$$

and the bracket $\langle f\rangle_\rho$ denotes the surface average, defined by

$$\langle f\rangle_\rho = \frac{1}{\pi\rho^2}\int_0^{2\pi} d\omega \int_0^{\rho} f(\rho, \omega, s)\rho d\rho. \tag{4.135}$$

When we take $\rho_1 = 0$ and $\rho_2 = a$ in (4.132), ξ_n becomes the shift of the magnetic axis under the fixed boundary condition. On the other hand, we can estimate the shift of the center of the plasma column by taking $\rho_1 = b$ and $\rho_2 = a$, where b is the radius of the circular conducting wall. Here we can assume

$$\left\{\begin{aligned} P_0 &= 0 \\ B_{\omega 0}(\rho) &= aB_{\omega 0}(a)/\rho \\ B_{s0} &= B_{s0}(a) = \text{constant} \end{aligned}\right. \tag{4.136}$$

for $a \le \rho_0 \le b$. Then we have the expression

$$\begin{aligned} \xi_n = &-\frac{3}{8}k_n b^2\left(1 - \frac{a^2}{b^2}\right) \\ &+ \frac{k_n ab}{2[B_{\omega 0}(a) - \kappa_n aB_{s0}][B_{\omega 0}(b) - \kappa_n bB_{s0}]} \times \\ &\times \left\{\left(2\mu_0\langle P\rangle_a + \frac{\langle B_\omega^2\rangle_a}{2} + \langle \kappa_n\rho B_\omega B_{s0}\rangle_a\right)\left(1 - \frac{a^2}{b^2}\right)\right. \\ &+ B_{\omega 0}^2(a)\left[\ln\frac{b}{a} + \frac{1}{4}\left(1 - \frac{a^2}{b^2}\right)\right] \\ &\left. - \kappa_n aB_{s0}B_{\omega 0}(a)\left[\ln\frac{b}{a} + 1 - \frac{b^2}{a^2}\right]\right\} \end{aligned} \tag{4.137}$$

from (4.132); $\langle f\rangle_\rho$ in G is replaced with $\langle f\rangle_a$. For the currentless stellarator with an helical magnetic axis, which satisfies $B_{\omega 0} = 0$, we obtain

$$\xi_n = -\frac{3}{8}k_n b^2\left(1-\frac{a^2}{b^2}\right) + \frac{k_n\bar{\beta}}{2\kappa_n^2}\left(1-\frac{a^2}{b^2}\right), \tag{4.138}$$

where $\bar{\beta} = 2\mu_0\langle P\rangle_a/B_{s0}^2$. Here we note that (4.137) is also applicable to the Shafranov shift in a tokamak. Using $k_n = 0$ for $n \neq 0$ and $k_0 = 1/R$, we obtain a familiar expression for the shift of the plasma column in a circular tokamak under a large aspect ratio approximation:

$$\xi_0 = \frac{b^2}{2R}\left\{\ln\frac{b}{a} + \left(1-\frac{a^2}{b^2}\right)\left(\beta_p + \frac{\ell_i - 1}{2}\right)\right\}, \tag{4.139}$$

where $\beta_p = 2\mu_0\langle P\rangle_a/B_{\omega 0}^2(a)$ and $\ell_i = \langle B_\omega^2\rangle_a/B_{\omega 0}^2(a)$ is an internal inductance related to a profile of toroidal plasma current.

Here we show an expression for the specific volume $V'(\Phi)$ for the helical axis stellarator

$$V'(\Phi) = \frac{dV}{d\Phi} = \oint \frac{1-\langle k\rho\cos\theta\rangle}{\langle B_s\rangle}\,ds, \tag{4.140}$$

where the line integral is carried out over one field period. Since B_s is approximated as

$$B_s = \frac{B_{s0}}{1-k\rho\cos\theta}, \tag{4.141}$$

the specific volume becomes

$$V'(\Phi) = \frac{1}{B_{s0}}\oint (1 - 2\langle k\rho\cos\theta\rangle - \langle k^2\rho^2\cos^2\theta\rangle - \langle k\rho\cos\theta\rangle^2)\,ds. \tag{4.142}$$

As an example of an helical axis stellarator, we show a finite-beta MHD equilibrium of Asperator NP-4 with an average major radius of $R_0 = 152.4$ cm, a minor radius of $a = 9.5$ cm, and a radius of helical magnetic axis of $r_0 = 19.05$ cm in Fig. 4.5. Here the number of magnetic field periods is $N = 8$ and $\bar{\beta} = 4.4\%$ for the pressure profile $P = P_0(1-\rho^2/a^2)^2$. Figure 4.5, showing magnetic surfaces at every quarter period is obtained using the BETA code. Figure 4.6 shows the profile of the rotational transform ι per one period for several beta values in Asperator NP-4. The profile at almost zero beta is nearly flat and the shear is fairly weak. As the beta value increases, the rotational transform increases at the magnetic axis and decreases at the edge under the constraint of currentless equilibrium. This tendency is similar to that in the currentless finite-beta equilibrium of heliotrons.

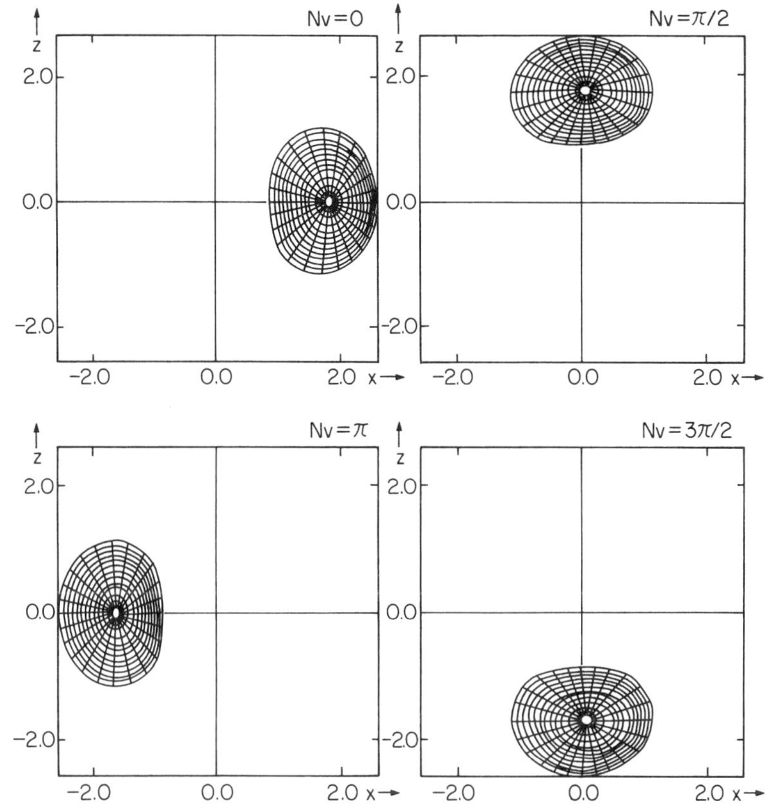

Fig. 4.5 Magnetic surfaces at different toroidal positions of the Asperator NP-4, obtained viewing the BETA code at $\bar{\beta} = 4.4\%$ for a pressure profile of $P = P_0(1 - \rho^2/a^2)^2$.

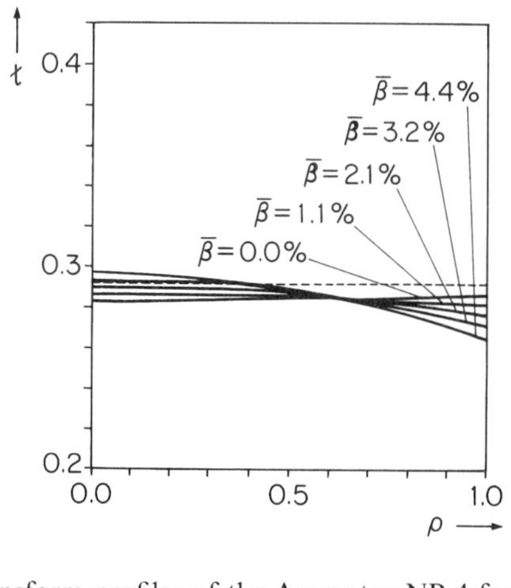

Fig. 4.6 Rotational transform profiles of the Asperator NP-4 for various beta values, obtained using the BETA code for a pressure profile of $P = P_0(1 - \rho^2/a^2)^2$.

In the helical axis stellarator, the rotational transform ι is produced by the torsion. The value near the magnetic axis per one field period is given by

$$\iota = 1 - \int_0^L \kappa(s)\, ds. \tag{4.143}$$

In the approximation of the straight helical magnetic axis, the curvature and the torsion are given by

$$\bar{k} = \frac{1}{r_0\left[1 + \left(\dfrac{R_0}{Nr_0}\right)^2\right]}, \quad \bar{\kappa} = \frac{(R_0/Nr_0)}{r_0\left[1 + \left(\dfrac{R_0}{Nr_0}\right)^2\right]}. \tag{4.144}$$

From (4.143) and (4.144), the rotational transform is given by

$$\iota = 1 - \frac{\bar{\kappa}L}{2\pi}. \tag{4.145}$$

The dashed line in Fig. 4.6 shows the value given by (4.145) for the case of Asperator NP-4.

From the results of the BETA code for the finite-beta currentless MHD equilibria of Asperator NP-4, the helical shift of the magnetic axis and the toroidal shift of the magnetic axis are plotted as a function of $\bar{\beta}$ in Fig. 4.7. Here circles correspond to the BETA code results. Using (4.132) for the pressure profile $P = P_0(1 - \rho^2/a^2)^2$ and the currentless constraint or $B_{\omega 0} = 0$, we find the relative shift $\xi_n(\rho_0)$ between the magnetic axis and the magnetic surface with radius ρ_0,

$$\xi_n(\rho_0) = -\tfrac{3}{8}\, k_n \rho_0^2 = \frac{k_n \bar{\beta}}{2\kappa_n^2} \frac{\rho_0^2}{a^2}\left(3 - \frac{\rho_0^2}{a^2}\right), \tag{4.146}$$

where we have used $\rho_1 = 0$ and $\rho_2 = \rho_0$. We need explicit expressions for k_n and κ_n in order to evaluate $\xi_n(\rho_0)$.

Here we use the expressions for the curvature (4.101) and the torsion (4.104) for Asperator NP-4. By using small parameters of the Asperator NP-4, $\varepsilon_R = r_0/R_0 \sim \varepsilon_N = N \ll 1$ and $I_R = R_0/(r_0 N) \sim 1$, we epand (4.101) and (4.104), and keep terms up to the first order in ε_N or ε_R:

$$\begin{cases} k(\phi) = \bar{k}\left[1 + \dfrac{I_R^3 - I_R}{1 + I_R^2}\, \varepsilon_N \cos(N\phi)\right] \\ \kappa(\phi) = -\bar{\kappa}\left\{1 - \dfrac{1}{1 + I_R^2}\, [2(1 + I_R^2)\varepsilon_R + I_R(3 + I_R^2)\varepsilon_N] \cos(N\phi)\right\}, \end{cases} \tag{4.147}$$

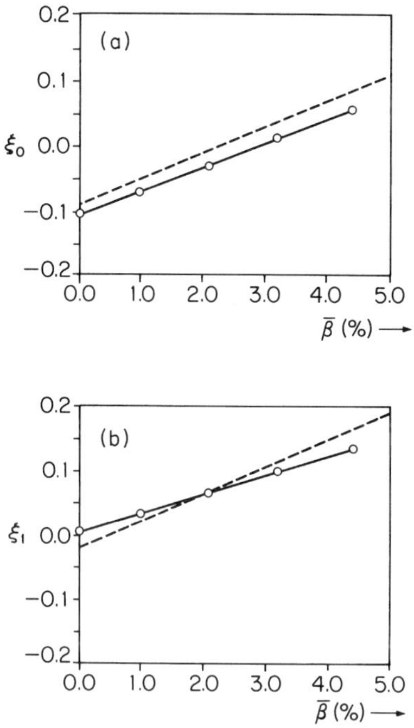

Fig. 4.7 The helical magnetic axis shift ξ_0 and toroidal magnetic axis shift ξ_1 as a function of average beta $\bar{\beta}$ in the Asperator NP-4.

with $\bar{k}$ and $\bar{\kappa}$ given in (4.144). Noting the relation (4.106), we obtain the following expressions for the curvature and the torsion of helical magnetic axis in the Asperator NP-4,

$$\begin{cases} k(s) = \bar{k} + \tilde{k}\cos\dfrac{2\pi s}{L} \\ \kappa(s) = -\bar{\kappa} + \tilde{\kappa}\cos\dfrac{2\pi s}{L}. \end{cases} \tag{4.148}$$

By noting (4.109) and (4.110), we obtain

$$\kappa_n = -\bar{\kappa} + \frac{2\pi n}{L}. \tag{4.149}$$

Substituting (4.148) and (4.149) into (4.121) yields the expression

$$k_n = \bar{k}J_n\left(\frac{\tilde{\kappa}L}{2\pi}\right) + \frac{1}{2}\tilde{k}\left[J_{n+1}\left(\frac{\tilde{\kappa}L}{2\pi}\right) + J_{n-1}\left(\frac{\tilde{\kappa}L}{2\pi}\right)\right], \tag{4.150}$$

where J_n is a Bessel function of the nth order. Since $\tilde{k} \ll \bar{k}$ for Asperator NP-4, the first term is dominant in (4.150). From (4.149) and (4.150), k_n becomes smaller as $|n|$ increases, and $k_{-1} \sim k_1$ and $|\kappa_{-1}| > |\kappa_1|$. Thus we have

$$\frac{k_{-1}}{2\kappa_{-1}^2} \ll \frac{k_0}{2\kappa_0^2} \sim \frac{k_1}{2\kappa_1^2}. \tag{4.151}$$

These relations suggest that (4.111) is approximated as

$$\begin{aligned}\rho &\sim \rho_0 + \xi_0 \cos(\omega - \kappa_0 s) + \xi_1 \cos(\omega - \kappa_1 s) \\ &\sim \rho_0 + \xi_0 \cos\theta + \xi_1 \cos(\theta - 2\pi s/L).\end{aligned} \tag{4.152}$$

Here, ξ_0 and ξ_1 in the expression (4.152) correspond to the helical shift and the toroidal shift of the magnetic axis, respectively, when $\rho_0 = a$ is substituted into (4.146). After normalizing all quantities by a, they are written as

$$\xi_0 = -\frac{3}{8}k_0 + \frac{k_0\bar{\beta}}{\kappa_0^2}, \tag{4.153}$$

where

$$k_0 = \bar{k}J_0\left(\frac{\tilde{\kappa}L}{2\pi}\right), \qquad \kappa_0 = -\bar{\kappa}, \tag{4.154}$$

and

$$\xi_1 = -\frac{3}{8}k_1 + \frac{k_1\bar{\beta}}{\kappa_1^2}, \tag{4.155}$$

where

$$k_1 = \bar{k}J_1\left(\frac{\tilde{\kappa}L}{2\pi}\right), \qquad \kappa_1 = \frac{2\pi\iota}{L}. \tag{4.156}$$

When the toroidal correction of the torsion vanishes or $\tilde{\kappa} = 0$, (4.153) reduces to the helical shift in the straight helical configuration with $k_0 = \bar{k}$, and the toroidal shift (4.155) disappears due to $k_1 = 0$. This result shows that the toroidal shift proportional to $\bar{\beta}$ in (4.155) is caused by the toroidal correction of the torsion $\tilde{\kappa}$ and the average curvature $\bar{k}$. In Fig. 4.7 we plot ξ_0 and ξ_1 given by (4.153) and (4.155), respectively, with dotted lines for Asperator NP-4. It is noted that the agreement is reasonable and that the toroidal shift is roughly comparable to the helical shift.

Since we have obtained the expressions for the shift of the plasma column, we can estimate the magnetic well from (4.142). The specific volume is expressed as

$$V'(\Phi) = \frac{1}{B_{s0}} \int_0^L \left[1 - k\rho_0 \left\{ \left(2\,\frac{\xi_0}{\rho_0} + \frac{d\xi_0}{d\rho_0} \right) + \left(2\,\frac{\xi_1}{\rho_0} + \frac{d\xi_1}{d\rho_0} \right) \cos\frac{2\pi s}{L} \right\} - \tfrac{1}{2}\, k^2 \rho_0^2 \right] ds. \tag{4.157}$$

By substituting ξ_n in (4.146) with $n = 0$ and $n = 1$ into expression (4.157), we obtain

$$V'(\Phi) = \frac{L}{B_{s0}} \left[1 + \frac{\Phi}{\pi B_{s0}} \left(k_0^2 - 6\,\frac{k_0^2}{\kappa_0^2}\,\bar{\beta} \right) \right], \tag{4.158}$$

where $\Phi = \pi\rho_0^2 B_{s0}$. Expression (4.158) shows that the magnetic hill at $\bar{\beta} = 0$ changes into a magnetic well or $V'' < 0$ for $\bar{\beta} > \kappa_0^2/6$. This is an example of magnetic well formation due to the finite-beta effect.

4.6 THE PFIRSCH–SCHLÜTER CURRENT, AND EQUILIBRIUM WITH RATIONAL MAGNETIC SURFACES

The Pfirsch–Schlüter current that appears in finite-beta toroidal plasmas has recently gained renewed interest, because there is a new trend toward designing a stellarator that minimizes the Pfirsch–Schlüter current. Also, there is strong interest in tokamaks with a substantial bootstrap current. As will be shown in chapter 7, the definition of the bootstrap current is closely related to the definition of the Pfirsch–Schlüter current. First, general expressions for the parallel current in a nonaxisymmetric toroidal plasma are derived by using Hamada coordinates. Then, we present an expression for the Pfirsch–Schlüter current.

Within the framework of MHD, the equilibrium state is characterized by a magnetic field **B** that is assumed to form nested toroidal magnetic surfaces characterized by the volume enclosed, V, a current density, **J**, and a scalar pressure $P(V)$. The current density fulfills the equilibrium equation

$$\mathbf{J} \times \mathbf{B} = \nabla P \tag{4.159}$$

and

$$\nabla \cdot \mathbf{J} = 0. \tag{4.160}$$

From (4.159), the current perpendicular to the magnetic field lines is given by

$$\mathbf{J}_\perp = P'[(\mathbf{B} \times \nabla V)/B^2], \tag{4.161}$$

where $P' = dP/dV$ and $\mathbf{J} = \mathbf{J}_\perp + \mathbf{J}_\parallel$, with $\mathbf{J}_\parallel$ being the current flowing along the magnetic field lines. For $\mathbf{J}_\perp$ given by (4.161), (4.160) becomes an equation for $\mathbf{J}_\parallel$, and by writing $\mathbf{J}_\parallel = \sigma\mathbf{B}$ it changes into a magnetic differential equation:

$$\mathbf{B}\cdot\nabla\sigma = -P'\nabla\cdot[(\mathbf{B}\times\nabla V)/B^2]. \tag{4.162}$$

From now on, we use the magnetic coordinates (V, θ, ζ), with a radial coordinate defined by the volume enclosed by thge magnetic surface, and θ and ζ are angular variables in the poloidal and toroidal directions, respectively. Here $B^V = \mathbf{B}\cdot\nabla V = 0$, $B^\theta = \mathbf{B}\cdot\nabla\theta$, and $B^\zeta = \mathbf{B}\cdot\nabla\zeta$ and the contravariant components of $\mathbf{B}$. We can write the magnetic field vector as

$$\mathbf{B} = B^\theta\mathbf{e}_\theta + B^\zeta\mathbf{e}_\zeta, \tag{4.163}$$

or as

$$\mathbf{B} = B_V\nabla V + B_\theta\nabla\theta + B_\zeta\nabla\zeta \tag{4.164}$$

in terms of the covariant components $B_V = \mathbf{B}\cdot\mathbf{e}_V$, $B_\theta = \mathbf{B}\cdot\mathbf{e}_\theta$, and $B_\zeta = \mathbf{B}\cdot\mathbf{e}_\zeta$, where $\mathbf{e}_\theta = J\nabla\zeta\times\nabla V$ and $\mathbf{e}_\zeta = J\nabla V\times\nabla\theta$. Here J denotes the Jacobian of the coordinate transformation. After substituting (4.164) into the RHS of (4.162) and using $\nabla\cdot(\mathbf{e}_\theta/J) = \nabla\cdot(\mathbf{e}_\zeta/J) = 0$, we obtain

$$\nabla\cdot\left(\mathbf{B}\times\frac{\nabla V}{B^2}\right) = J^{-1}\left[\mathbf{e}_\theta\cdot\nabla\left(\frac{B_\zeta}{B^2}\right) - \mathbf{e}_\zeta\cdot\nabla\left(\frac{B_\theta}{B^2}\right)\right]. \tag{4.165}$$

By noting that $B^2 = B_\theta B_\theta + B_\zeta B^\zeta$, the RHS of (4.165) becomes

$$\nabla\cdot\left(\mathbf{B}\times\frac{\nabla V}{B^2}\right) = -\mathbf{B}\cdot\nabla\left(\frac{B_\theta}{JB^\zeta B^2}\right) + B^\theta\mathbf{e}_\theta\cdot\nabla\left(\frac{J}{\Psi_T'\Psi_P'}\right) \tag{4.166}$$

or

$$\nabla\cdot\left(\mathbf{B}\times\frac{\nabla V}{B^2}\right) = \mathbf{B}\cdot\nabla\left(\frac{B_\zeta}{JB^\theta B^2}\right) - B^\zeta\mathbf{e}_\zeta\cdot\nabla\left(\frac{J}{\psi_T'\Psi_P'}\right). \tag{4.167}$$

Here we have used the fact that $JB^\theta = \Psi_P'(V)$ and $JB^\zeta = \Psi_T'(V)$, where Ψ_P and Ψ_T are surface quantities equal, respectively, to the poloidal and toroidal magnetic flux enclosed by the magnetic surface V.

Since there are two expressions (4.166) and (5.167) for $\nabla\cdot(\mathbf{B}\times\nabla V/B^2)$, (4.162) can be written as

$$\mathbf{B}\cdot\nabla\left[\sigma - P'\left(\frac{\alpha_1}{\alpha_1+\alpha_2}\frac{B_\theta}{JB^\zeta B^2} - \frac{\alpha_2}{\alpha_1+\alpha_2}\frac{B_\zeta}{JB^\theta B^2}\right)\right]$$
$$= -\frac{P'}{\Psi_T'\Psi_P'}\left(\frac{\alpha_1}{\alpha_1+\alpha_2}B^\theta\mathbf{e}_\theta - \frac{\alpha_2}{\alpha_1+\alpha_2}B^\zeta\mathbf{e}_\zeta\right)\cdot\nabla J, \tag{4.168}$$

where α_1 and α_2 are weighting factors free to be chosen with $\alpha_1 + \alpha_2 \neq 0$. In the case of Hamada coordinates with $J = 1$, the RHS of (4.168) vanishes and the parallel plasma current is obtained as a compensating current which results from the condition $\nabla \cdot \mathbf{J} = 0$, where $\mathbf{J}_\perp$ is given by (4.161). This part of $\mathbf{J}_\parallel$ is called Pfirsch–Schlüter current. From (4.168) we obtain

$$\mathbf{J}_\parallel = \sigma \mathbf{B} = P' \left[\frac{\alpha_1}{\alpha_1 + \alpha_2} \frac{B_\theta}{B^\zeta} - \frac{\alpha_2}{\alpha_1 + \alpha_2} \frac{B_\zeta}{B^\theta} \right] \frac{\mathbf{B}}{B^2} + \gamma(V)\mathbf{B}, \tag{4.169}$$

which is also written as

$$\mathbf{J}_\parallel = P' \frac{B_\theta}{B^\zeta} \frac{\mathbf{B}}{B^2} + \gamma_1(V)\mathbf{B}, \tag{4.170}$$

where $\gamma_1(V) = \gamma(V) + [\alpha_2/((\alpha_1 + \alpha_2)B^\theta B^\zeta)]$. The term $\gamma(V)\mathbf{B}$ in (4.169) is to be determined by an additional condition. As we will discuss in chapter 7, this term contains the bootstrap current, which is driven in the presence of trapped particles.

Here we will write $\gamma_1(V)$ in terms of the total toroidal current flowing inside a magnetic surface,

$$I_T(V) = \int_0^V \int_0^{2\pi} \mathbf{J} \cdot \nabla \zeta \, d\theta \, dV, \tag{4.171}$$

which also gives

$$I_T'(V) = J^\zeta(V). \tag{4.172}$$

Thus $\mathbf{J} \cdot \nabla \zeta = J^\zeta(V)$ is a surface quantity in Hamada coordinates. Since $\mathbf{J}_\perp$ and $\mathbf{J}_\parallel$ are expressed as (4.161) and (4.169), the poloidal and toroidal current densities are given by

$$J^\theta = (\mathbf{J}_\perp + \mathbf{J}_\parallel) \cdot \nabla \theta = \frac{P'}{B^\zeta} \left(\frac{\alpha_1}{\alpha_1 + \alpha_2} \right) + \gamma(V) B^\theta \tag{4.173}$$

and

$$J^\zeta = (\mathbf{J}_\perp + \mathbf{J}_\parallel) \cdot \nabla \zeta = -\frac{P'}{B^\theta} \left(\frac{\alpha_2}{\alpha_2 + \alpha_2} \right) + \gamma(V) B^\zeta. \tag{4.174}$$

From (4.172) and (4.174), $\gamma(V)$ is expressed in terms of α_1, α_2, and I_T':

$$\gamma(V) = \frac{I_T'}{B_\zeta} + \frac{P'}{B^\theta B^\zeta} \left(\frac{\alpha_2}{\alpha_1 + \alpha_2} \right). \tag{4.175}$$

Then (4.173) with (4.175) gives

$$J^{\theta} = \frac{P'}{B^{\theta}} + \frac{B^{\zeta}}{B^{\zeta}} I'_T. \tag{4.176}$$

Also, by substituting (4.175) into (4.169), we obtain

$$\mathbf{J}_{\parallel} = \frac{P'}{B^2} \frac{B_{\theta}}{B^{\zeta}} \mathbf{B} + \frac{I'_T}{B^{\zeta}} \mathbf{B}, \tag{4.177}$$

which gives $\gamma_1(V) = I'_T/B^{\zeta}$ by comparing (4.177) with (4.170). In the currentless stellarator and heliotron, the total toroidal current flowing inside each magnetic surface disappears, or $I_T(V) = 0$. Then the parallel current (4.177) corresponds to the Pfirsch–Schlüter current:

$$\mathbf{J}_{ps} = \frac{P'}{B^2} \frac{B_{\theta}}{B^{\zeta}} \mathbf{B}. \tag{4.178}$$

Here we analyze the Pfirsch–Schlüter current in a large aspect ratio tokamak with the standard tokamak model field,

$$\mathbf{B} = \bar{B}\left[\frac{\varepsilon}{q} \mathbf{e}_{\vartheta} + (1 - \varepsilon \cos \vartheta)\mathbf{e}_{\phi}\right], \tag{4.179}$$

where (r, ϑ, ϕ) are the usual toroidal coordinates. The magnetic surfaces are defined by $r =$ constant and $\varepsilon = r/R_0$, where R_0 is the major radius, and q is the safety factor. The magnitude of the magnetic field (4.179) is approximately given by

$$B = \bar{B}(1 - \varepsilon \cos \vartheta), \tag{4.180}$$

since $\varepsilon \ll 1$ and $q \sim O(1)$. The Hamada coordinates (V, θ, ζ) for this standard tokamak model is given analytically with $\theta = (\vartheta + \varepsilon \sin \vartheta)/2\pi$ and $\zeta = (\phi + 2q\varepsilon \sin \vartheta)/2\pi$. In these coordinates the covariant components are $B_V = -q\bar{B} \sin \vartheta[1 + (r/q)(dq/dr)]/2\pi^2 r R_0$ and

$$\begin{cases} B_{\theta} = -4\pi r q \bar{B} \cos \vartheta \\ B_{\zeta} = 2\pi R_0 \bar{B} \end{cases} \tag{4.181}$$

and the contravariant components are $B^V = 0$ and

$$\begin{cases} B^{\theta} = \bar{B}/2\pi q R_0, \\ B^{\zeta} = \bar{B}/2\pi R_0. \end{cases} \tag{4.182}$$

The volume enclosed by a magnetic surface is $V = 2\pi^2 R_0 r^2$ and the derivative given by the prime is $d/dV = (4\pi^2 R_0 r)^{-1} d/dr$. Then (4.178) yields

$$\mathbf{J}_{ps} = -\frac{2q}{\bar{B}} \left(\frac{dP}{dr}\right) \cos \theta \mathbf{e}_{\zeta}, \tag{4.183}$$

which is the lowest-order expression of the Pfirsch–Schlüter current.

Solenoidal vectors such as the magnetic field can be written in the Clebsch representation,

$$\mathbf{B} = \nabla\Psi \times \nabla\theta_0, \tag{4.184}$$

where a magnetic line of force is defined by a cross-line between $\Psi =$ constant and $\theta_0 =$ constant surfaces. The function Ψ can be chosen such that $2\pi\Psi$ is equal to the toroidal magnetic flux inside the $\Psi =$ constant surface (see (4.190)). Then θ_0 becomes an angle-like variable.

In addition to the Clebsch or contravariant representation, the magnetic field in the finite-pressure plasma can be written in the covariant form

$$\mathbf{B} = \nabla\chi + \beta\nabla\Psi. \tag{4.185}$$

It is noted that there is arbitrariness for θ_0, χ, and β in the representation of $\mathbf{B}$, after Ψ is defined. If θ_0, χ, and β represent $\mathbf{B}$, then the following $\bar{\theta}_0$, $\bar{\chi}$, and $\bar{\beta}$ also give a representation of $\mathbf{B}$:

$$\bar{\theta}_0 = \theta_0 + \theta_*(\Psi), \quad \bar{\chi} = \chi + \chi_*(\Psi), \quad \bar{\beta} = \beta - \frac{d\chi_*(\Psi)}{d\Psi}, \tag{4.186}$$

where the functions θ_* and χ_* are arbitrary functions of Ψ.

We follow a magnetic field line from $\chi(0)$ and $\theta_0(0)$ in a toroidal system. In general, χ and θ_0 will not return to their original values $\chi(0)$ and $\theta_0(0)$. After one toroidal circuit, they are shown as

$$\chi = \chi(0) + 2\pi g(\Psi), \quad \theta_0 = \theta_0(0) - 2\pi\iota(\Psi). \tag{4.187}$$

Since both $(\chi(0), \theta_0(0))$ and (4.187) are representations at the same toroidal position, g and ι must be functions of Ψ alone. Here $\iota(\Psi)$ is the rotational transform. Also, after one poloidal circuit,

$$\chi = \chi(0) + 2\pi I(\Psi), \quad \theta_0 = \theta_0(0) + 2\pi\sigma(\Psi). \tag{4.188}$$

Here I and σ must also be functions of Ψ alone. Thus the periodicities of the torus are satisfied by defining the poloidal angle θ and the toroidal angle ϕ as

$$\begin{cases} \theta_0 = \sigma\theta - \iota\phi \\ \chi = g\phi + I\theta. \end{cases} \tag{4.189}$$

Hereafter, we consider (Ψ, θ, ϕ) as Boozer coordinates.

To show that $\sigma = 1$, we use the fact that $2\pi\Psi$ equals the magnetic flux inside a magnetic surface, or

$$2\pi\Psi = \int \mathbf{B} \cdot d\mathbf{S}_t. \tag{4.190}$$

The surface area element $d\mathbf{S}_t$ in (Ψ, θ, ϕ) cordinates is

$$d\mathbf{S}_t = \frac{\nabla\phi}{\nabla\phi \cdot (\nabla\Psi \times \nabla\theta)} d\theta d\Psi. \tag{4.191}$$

By using θ_0 in (4.189) for the expression (4.184), $\mathbf{B} = \sigma(\Psi)\nabla\Psi \times \nabla\theta + \iota(\Psi)\nabla\phi \times \nabla\Psi$ is given and (4.190) yields

$$\Psi = \int \sigma d\Psi. \tag{4.192}$$

This means that $\sigma = 1$, and the contravariant form of $\mathbf{B}$ is shown as

$$\mathbf{B} = \nabla\Psi \times \nabla\theta + \iota(\Psi)\nabla\phi \times \Psi, \tag{4.193}$$

while the covariant form is

$$\mathbf{B} = g(\Psi)\nabla\phi + I(\Psi)\nabla\theta + \beta_* \nabla\Psi, \tag{4.194}$$

where $\beta_* = \beta + (dg/d\Psi)\phi + (dI/d\Psi)\theta$. The total toroidal current density inside a flux surface is

$$\int \mathbf{J} \cdot d\mathbf{S}_t = \frac{1}{\mu_0} \int \mathbf{B} \cdot d\boldsymbol{\ell}_p = \frac{1}{\mu_0} \int \mathbf{B} \cdot \frac{\nabla\phi \times \nabla\Psi}{\nabla\phi \cdot (\nabla\Psi \times \nabla\theta)} d\theta = \frac{2\pi}{\mu_0} I(\Psi), \tag{4.195}$$

where $d\boldsymbol{\ell}_p$ is the line element vector in the poloidal direction. The total poloidal current outside a flux surface can similarly be shown to be $2\pi g(\Psi)/\mu_0$.

The covariant representation of $\mathbf{B}$, given in (4.194), yields a simple expression for the current density

$$\mathbf{J} = \frac{1}{\mu_0} \nabla \times \mathbf{B} = \frac{1}{\mu_0} \left((\nabla\phi \times \nabla\Psi) \frac{\partial\beta}{\partial\phi} - (\nabla\Psi \times \nabla\theta) \frac{\partial\beta}{\partial\theta} \right). \tag{4.196}$$

The cross-product of this expression with the covariant representation of $\mathbf{B}$ (4.193) yields an equation for β:

$$\nabla P = \mathbf{J} \times \mathbf{B} = \frac{1}{\mu_0} \nabla\phi \cdot (\nabla\Psi \times \nabla\theta) \left(\frac{\partial\beta}{\partial\phi} + \iota \frac{\partial\beta}{\partial\theta} \right) \nabla\Psi. \tag{4.197}$$

We note that the Jacobian can be found from the inner product between the covariant and contravariant representation of **B**,

$$J^{-1} = \nabla\phi \cdot (\nabla\Psi \times \nabla\theta) = \frac{B^2}{g + \iota I}. \tag{4.198}$$

The equation for β is then

$$\frac{\partial\beta}{\partial\phi} + \iota\frac{\partial\beta}{\partial\theta} = \frac{\mu_0}{B^2}(g + \iota I)\frac{dP}{d\Psi}. \tag{4.199}$$

The parallel component for **J** can be found from the inner product between the covariant expression for **B** (4.194), and **J** (4.196),

$$\frac{\mu_0 J_\parallel}{B} = \frac{1}{g + \iota I}\left(I\frac{\partial\beta}{\partial\phi} - g\frac{\partial\beta}{\partial\theta}\right). \tag{4.200}$$

Let us now solve the equations for β and $J_\parallel/B$. The function β need not be periodic in θ and ϕ; however, $J_\parallel/B$ must be. First, one finds an homogeneous solution for β

$$\gamma_h(\Psi)(\iota\phi - \theta), \tag{4.201}$$

from (4.199), with γ_h an arbitrary function of Ψ. To this solution the inhomogeneous solution of (4.199) must be added. To find the inhomogeneous solution, let

$$\frac{1}{B^2} = \frac{1}{\bar{B}^2}\left(1 + \sum_{m,n}{}' \delta_{m,n}(\Psi)\cos(n\phi - m\theta + \lambda_{m,n})\right), \tag{4.202}$$

with the prime on the sum implying that the term with $n = 0$ and $m = 0$ is eliminated. Here we have assumed that the field strength on each magnetic surface is known, and that it can be appropriately expanded in a Fourier series. Then one finds

$$\beta = \gamma_h(\Psi)(\iota\phi - \theta) + \frac{\mu_0}{\bar{B}^2}(g + \iota I)\frac{dP}{d\Psi}\phi + \beta_*, \tag{4.203}$$

with β_*, which is the same as that in (4.194), equal to

$$\beta_* = \frac{\mu_0}{\bar{B}^2}\frac{dP}{d\Psi}\sum_{m,n}{}' \frac{g + \iota I}{n - \iota m}\delta_{mn}\sin(n\phi - m\theta + \lambda_{mn}). \tag{4.204}$$

The expressions become simpler if the force-free current part of β is singled out by defining

$$\gamma = \gamma_h + \frac{\mu_0}{\bar{B}^2}\frac{dP}{d\psi}I; \tag{4.205}$$

then

$$\beta = \gamma(\Psi)(\iota\phi - \theta) + \frac{\mu_0}{\bar{B}^2}\frac{dP}{d\Psi}(g\phi + I\theta) + \beta_*. \tag{4.206}$$

Thus the parallel current (4.200), is given by

$$\frac{\mu_0 J_\parallel}{B} = \gamma(\Psi) + \frac{\mu_0}{\bar{B}^2}\frac{dP}{d\Psi}\sum_{m,n}{}' \frac{nI + mg}{n - \iota m}\delta_{mn}\cos(n\phi - m\theta + \lambda_{mn}). \tag{4.207}$$

The first term on the RHS of this equation represents the force-free current and the second term the Pfirsch–Schlüter current. The poloidal and toroidal currents can be evaluated using the expression of β_* given below (4.194) and (4.206),

$$\frac{dI}{d\Psi} = \gamma - \left(\frac{\mu_0}{\bar{B}^2}\frac{dP}{d\Psi}\right)I, \tag{4.208}$$

$$\frac{dg}{d\Psi} = -\iota\gamma - \left(\frac{\mu_0}{\bar{B}^2}\frac{dP}{d\Psi}\right)g. \tag{4.209}$$

Now we consider Ohm's law,

$$\mathbf{E} + \mathbf{v} \times \mathbf{B} = \eta\mathbf{J}, \tag{4.210}$$

which allows us to evaluate a collisional diffusion coefficient. An electrostatic potential part of the electric field can be separated. Let us introduce

$$\mathbf{E} = -\nabla\Phi + \mathbf{E}_A. \tag{4.211}$$

The perpendicular part of $\mathbf{E}_A$ can be written in terms of a velocity $\mathbf{u}$:

$$\mathbf{E}_{A\perp} + \mathbf{u} \times \mathbf{B} = 0. \tag{4.212}$$

Herre $\mathbf{u}$ represents a pinch velocity of the magnetic field and the plasma. Then Ohm's law can be written as

$$-\nabla\Phi + \mathbf{E}_{A\parallel} + (\mathbf{v} - \mathbf{u}) \times \mathbf{B} = \eta_\perp\mathbf{J}_\perp + \eta_\parallel\mathbf{J}_\parallel. \tag{4.213}$$

With expressions (4.193) and (4.198), the parallel component of this equation in the coordinates (Ψ, θ, ϕ) gives

$$\frac{\partial\Phi}{\partial\phi} + \iota\frac{\partial\Phi}{\partial\theta} = (g + \iota I)\left(\frac{E_{A\parallel}}{B} - \eta_\parallel\frac{J_\parallel}{B}\right). \tag{4.214}$$

This equation, with the expression for $J_\parallel/B$ (4.207), implies the results

$$E_{A\parallel} = \frac{\eta_\parallel}{\mu_0}\gamma(\Psi)B \tag{4.215}$$

and

$$\Phi = \Phi_0(\Psi) - \eta_\parallel \frac{(g+\iota I)}{\bar{B}^2}\frac{dP}{d\Psi}\sum_{m,n}{}' \frac{nI+mg}{(n-\iota m)^2}\delta_{mn}(\Psi)\times$$
$$\times \sin(n\phi - m\theta + \lambda_{mn}), \tag{4.216}$$

with Φ_0 an arbitrary function of Ψ.

To understand the meaning of the velocity **u**, we will consider Faraday's law:

$$\partial \mathbf{B}/\partial t = -\nabla \times \mathbf{E} = \nabla \times (\mathbf{u} \times \mathbf{B}) - \nabla \times \mathbf{E}_{A\parallel}. \tag{4.217}$$

Evaluating $\mathbf{u} \times \mathbf{B}$ with the contravariant expression for $\mathbf{B} = \nabla \times (\Psi\nabla\theta) - \nabla \times (\Psi_p \nabla\phi)$, one finds that

$$\frac{\partial \Psi(\mathrm{x}, t)}{\partial t} + \mathbf{u}\cdot\nabla\Psi = -\left(\frac{\eta_\parallel}{\mu_0}\right)\gamma I, \tag{4.218}$$

$$\frac{\partial \Psi_p(\mathrm{x}, t)}{\partial t} + \mathbf{u}\cdot\nabla\Psi_p = \left(\frac{\eta_\parallel}{\mu_0}\right)\gamma g, \tag{4.219}$$

where the poloidal magnetic flux is Ψ_p and the relation $d\Psi_p/d\Psi = \iota$ has been used. Here we have used the covariant form of **B** (4.194) to obtain the RHS of the above equations. One can define the plasma loop voltage $V_L(\Psi, t)$ by

$$V_L = \frac{\partial}{\partial \Psi}\left(\frac{1}{2\pi}\int \mathbf{E}\cdot\mathbf{B}d\mathbf{r}\right) \tag{4.220}$$

which has the usual approximation in a large aspect ratio tokamak, $V_L \simeq 2\pi R E_\parallel$, where $d\mathbf{r}$ is a volume element. Evaluating (4.220) by using (4.215) for $\mathbf{E}_\parallel$, one finds that

$$\gamma(\Psi) = \frac{2\pi\mu_0 V_L}{(g+\iota I)\eta_\parallel}. \tag{4.221}$$

In the steady state, the loop voltage becomes a constant V_{L0} and (4.218) gives

$$\mathbf{u}\cdot\nabla\Psi = -\frac{2\pi I}{g+\iota I}V_{L0}, \tag{4.222}$$

and the particle flux due to **u** with the mass density $\rho(\Psi)$ is

$$\Gamma_p = \int \rho\mathbf{u}\cdot d\mathbf{S}_\psi = -(2\pi)^3\frac{\rho I}{B^2}V_{L0}, \tag{4.223}$$

where $d\mathbf{S}_\psi = J\nabla\Psi d\theta d\phi$. This is the classical pinch effect. In stellarators and heliotrons usually $V_{L0} = 0$ and there is no such pinch effect.

The diffusive particle flux can be evaluated using Ohm's law (4.213). First, we operate $\mathbf{B}\times$ on (4.213), take components in the $\nabla\Psi$ direction, and use the covariant form (4.194) and the electrostatic potential (4.216). Finally, we obtain

$$\begin{aligned}(\mathbf{v}-\mathbf{u})\cdot\nabla\Psi &= \frac{1}{g+\iota I}\left(I\frac{\partial\Phi}{\partial\phi}-g\frac{\partial\Phi}{\partial\theta}\right)-\eta_\perp\frac{\nabla P}{B^2}\cdot\nabla\Psi\\ &= \frac{\eta_\parallel}{\bar{B}^2}\frac{dP}{d\Psi}\sum_{m,n}{}'\left(\frac{nI+mg}{n-\iota m}\right)^2\delta_{mn}(\Psi)\cos(n\phi-m\theta+\lambda_{mn})\\ &\quad -\eta_\perp\frac{dP}{d\Psi}\frac{|\nabla\Psi|^2}{B^2}.\end{aligned} \tag{4.224}$$

The total diffusive particle flux crossing a magnetic surface is

$$\begin{aligned}\Gamma_d &= \int\rho(\mathbf{v}-\mathbf{u})\cdot\frac{\nabla\Psi}{\nabla\phi\cdot(\nabla\Psi\times\nabla\theta)}\,d\theta d\phi\\ &= (g+\iota I)\rho\int(\mathbf{v}-\mathbf{u})\cdot\nabla\Psi\,\frac{d\theta d\phi}{B^2}.\end{aligned} \tag{4.225}$$

The expression for the total diffusive flux can be rewritten by using (4.224):

$$\Gamma_d = -(D_\parallel + D_\perp)\frac{dP}{d\Psi}, \tag{4.226}$$

with

$$D_\parallel = 2\pi^2\eta_\parallel\frac{(g+\iota I)}{\bar{B}^4}\sum_{n,m}{}'\left(\frac{nI+mg}{n-\iota m}\right)^2\delta_{mn}^2(\Psi), \tag{4.227}$$

$$D_\perp = \eta_\perp\int\frac{\nabla\Psi}{B^2}\cdot\frac{\nabla\Psi}{\nabla\phi\cdot(\nabla\Psi\times\nabla\theta)}\,d\theta d\phi. \tag{4.228}$$

When the plasma is in a steady state, particle conservation implies that

$$\nabla\cdot n\mathbf{v} = S_n, \tag{4.229}$$

with S_n the source of particle per unit volume. The total flux of particles $\Gamma = \Gamma_p + \Gamma_d$ obeys

$$\frac{d\Gamma}{d\Psi} = \int S_n\frac{d\theta d\phi}{\nabla\phi\cdot(\nabla\Psi\times\nabla\theta)} = S_n(\Psi), \tag{4.230}$$

with $S_n(\Psi)d\Psi$ the number of particles added between two differentially separated flux surfaces. The parallel current-driven diffusion coefficient (4.227) is singular at each rational surface; that is, near the rational surface,

$$D_\parallel \propto \frac{\delta_{mn}^2}{(\Psi-\Psi_{mn})^2}, \tag{4.231}$$

with Ψ_{mn} the value of Ψ at $n=\iota m$. The total particle flux $\Gamma(\Psi)$ must be smoothly varying in Ψ, and near the rational surface the pressure gradient is given by

$$\frac{dP}{d\Psi} \propto \frac{(\Psi - \Psi_{mn})^2}{\delta_{mn}^2}; \tag{4.232}$$

consequently, unless $|\delta_{mn}| = 0$ at $\Psi = \Psi_{mn}$, $dP/d\Psi$ will vanish at rational surfaces. The Pfirsch–Schlüter or pressure-driven part of the parallel current near the rational surface is shown as

$$(J_\parallel)_{ps} \propto \frac{\delta_{mn}}{(\Psi - \Psi_{mn})} \frac{dP}{d\Psi} \cos(n\phi - m\theta + \lambda_{mn}) \propto (\Psi - \Psi_{mn}), \tag{4.233}$$

from the expression (4.207). Consequently, the Pfirsch–Schlüter part of $J_\parallel$ vanishes at each rational surface rather than being singular. Interestingly, the electrostatic potential (4.216) retains a finite variation on rational magnetic surfaces even though $dP/d\Psi$ vanishes on these surfaces.

An application of the expression for $D_\parallel$ is to derive the Pfirsch–Schlüter diffusion coefficient for a stellarator. This is done by assuming that the field strength has the standard form

$$\frac{1}{B^2} \simeq \frac{1}{\bar{B}^2}[1 - 2\epsilon_t \cos\theta - 2\epsilon_h \cos(M\phi - \ell\theta)]. \tag{4.234}$$

The nonvanishing terms in δ_{mn} are $\delta_{10} = 2\epsilon_t$ and $\delta_{\ell M} = 2\epsilon_h$. We assume that the plasma has no net toroidal current, $I = 0$, and that the toroidal field dominates in (4.234), which gives $g = R_0\bar{B}$. The diffusion coefficient $D_\parallel^*$ is equal to $D_\parallel$ divided by the area of the averaged magnetic surface $(2\pi r) \times (2\pi R_0)$ and also divided by $d\Psi/dr = r\bar{B}$, since $D_\parallel^*$ multiplies dP/dr rather than $dP/d\Psi$ to obtain the particle flux. From (4.227),

$$D_\parallel^* = \frac{2}{\iota^2}\frac{\eta_\parallel}{\bar{B}^2}\sum_{m,n}{}' \left(\frac{\delta_{mn}}{2\epsilon_t}\right)^2 \left(\frac{\iota}{\iota - (n/m)}\right)^2 \tag{4.235}$$

is obtained with $\epsilon_t = r/R_0$. The magnetic field component with $n = 0$, $m = 1$ corresponding to $\delta_{10} = 2\epsilon_t$ gives a coefficient

$$D_{\parallel 01}^* = \frac{2}{\iota^2}\frac{\eta_\parallel}{\bar{B}^2}, \tag{4.236}$$

which is the Pfirsch–Schlüter diffusion coefficient for axisymmetric toroidal plasmas. The magnetic component with $n = M$, $m = \ell$ corresponding to $\delta_{\ell M} = 2\epsilon_h$ gives a coefficient

$$D_{\parallel M\ell}^* = \frac{2}{\iota^2}\frac{\eta_\parallel}{\bar{B}^2}\left(\frac{\epsilon_h}{\epsilon_t}\right)^2 \left(\frac{\iota}{\iota - M/\ell}\right)^2. \tag{4.237}$$

Usually, heliotrons are characterized by $\epsilon_h > \epsilon_t$ but $M/\ell \gg \iota$. Thus the Pfirsch–Schlüter diffusion coefficient is approximately given by (4.236), since the contribution from (4.237) becomes small. However, if the resonant perturbation with $\iota = n/m$ is finite, $D^*_{\parallel mn}$ contributes significantly.

Here we note the relationship between the Boozer coordinates and other magnetic coordinates. Let $(\Psi_m, \theta_m, \phi_m)$ be any set of magnetic coordinates. By magnetic coordinates we mean that

$$\mathbf{B} \cdot \nabla\Psi_m = 0, \quad \mathbf{B} \cdot \nabla(\theta_m - \iota\phi_m) = 0. \tag{4.238}$$

Naturally, θ_m and ϕ_m must preserve the periodicities of the torus. By the expressions for the alternative magnetic coordinates, where Ψ_m is a function of Ψ alone, the angular variables are written as

$$\theta_m = \theta + \iota G(\Psi, \theta, \phi), \quad \phi_m = \phi + G(\Psi, \theta, \phi), \tag{4.239}$$

where G is an arbitrary function periodic in θ and ϕ. An important characteristic of the coordinate system is described by its Jacobian J,

$$\frac{1}{J_m} \equiv (\nabla\Psi_m \times \nabla\theta_m) \cdot \nabla\phi_m = \frac{d\Psi_m}{d\Psi}\left(1 + \frac{\partial G}{\partial\phi} + \iota\frac{\partial G}{\partial\theta}\right)\frac{B^2}{g + \iota I}. \tag{4.240}$$

It is noted that there is freedom in the choice of the Jacobian.

The well-known magnetic coordinates are the Hamada coordinates, which will be denoted by a subscript H here. In these coordinates the Jacobian is a constant. For the Hamada coordinates,

$$\frac{\partial G_H}{\partial\phi} + \iota\frac{\partial G_H}{\partial\theta} = \frac{1}{J_H}\frac{d\Psi}{d\Psi_H}\frac{g + \iota I}{B^2} - 1 \tag{4.241}$$

is given from (4.240). Since G_H is to be periodic, $\Psi_H(\Psi)$ must be chosen so that

$$\frac{d\Psi_H}{d\Psi} = \frac{1}{J_H}\frac{g + \iota I}{\bar{B}^2}. \tag{4.242}$$

Then, the solution of (4.241) is shown by

$$G_H(\Psi, \theta, \phi) = {\sum_{m,n}}' \frac{\delta_{mn}}{n - \iota m}\sin(n\phi - m\theta + \lambda_{mn}), \tag{4.243}$$

with the expression (4.202) for $1/B^2$. Equations (4.204) and (4.243) imply that

$$\beta_* = (g + \iota I)\frac{\mu_0}{\bar{B}^2}\frac{dP}{d\Psi}G_H. \tag{4.244}$$

Using (4.206), (4.208), and (4.209), and expressions for the Hamada angles, $\theta_H = \theta + \iota G_H$ and $\phi_H = \phi + G_H$, one finds that

$$\beta = -\frac{dg}{d\Psi}\phi_H - \frac{dI}{d\Psi}\theta_H. \tag{4.245}$$

BIBLIOGRAPHY

Bauer, F., Betancourt, O., and Garabedian, P. (1984). *Magnetohydrodynamic equilibrium and stability of stellarators*. Springer-Verlag, New York.

Bauer, F., Betancourt, O., Garabedian, P., and Wakatani, M. (1987). *The beta equilibrium, stability, and transport codes*. Academic Press, San Diego.

Bickerton, R. J., Connor, J. W., and Taylor, J. B. (1971). Diffusion driven plasma currents and bootstrap tokamak. *Nature, Physical Science*, **229**, 110.

Boozer, A. H. (1981). Plasma equilibrium with rational magnetic surfaces. *Phys. Fluids*, **24**, 1999.

Callen, J. D., and Dory, R. A. (1972). Magnetohydrodynamic equilibrium in sharply curved axisymmetric devices. *Phys. Fluid*, **15**, 1523.

Clark, J. F., and Sigma, D. J. (1977). High–pressure flux-conserving tokamak equilibria. *Phys. Rev. Letters*, **38**, 70.

Freidberg, J. P. (1987). *Ideal Magnetohydrodynamics*. Plenum, New York.

Greene, J. M., Johnson, J. L., and Weimer, K. E. (1966). Critical pressure for equilibrium in a toroidal system. *Plasma Phys.*, **8**, 145.

Hamada, S. (1961). Hydromagnetic equilibria and their proper coordinates. *Nucl. Fusion*, **2**, 23.

Hirshman, S. P., and Whitson, J. C. (1983). Steepest–descent moment method for three-dimensional magnetohydrodynamic equilibria. *Phys. Fluids*, **26**, 3553.

Ichiguchi, K., and Wakatani, M. (1988). Analysis of a spatial axis stellarator by numerical and analytical techniques. *Nucl. Fusion*, **28**, 411.

Kikuchi, M. and Azumi, M. (1995). Experimental evidence for the bootstrap current in a tokamak. *Plasm Phys. Contr. Fusion*, **37**, 1215.

Nakamura, Y., Wakatani, M., and Ichiguchi, K. (1993). Low-n mode stability analysis for $\ell = 2$ heliotron/torsatron by VMEC–STEP code. *J. Plasma Fusion Res.*, **69**, 41.

Nishikawa, K., and Wakatani, M. (1993). *Plasma physics, basic theory with fusion applications*. Springer-Verlag, Berlin.

Pustovitov, V. D., and Shafranov, V. D. (1990). Equilibrium and stability of plasmas in stellarators, in *Reviews of plasma physics*. Consultant Bureau, New York, vol. 15, p. 163.

Solovév, L. S. (1968). The theory of hydromagnetic stability of toroidal plasma configurations. *Sov. Phys.—JETP*, **26**, 400.

Solovév, L. S., and Shafranov, V. D. (1970). Plasma confinement in a closed magnetic system, in *Reviews of Plasma Physics*. Consultants Bureau, New York, vol. 5, p. 1.

Strauss, H. R., (1980). Stellarator equations of motion. *Plasma Phys.*, **22**, 733.

5

MHD INSTABILITIES IN HELIOTRONS

5.1 INTRODUCTION

Stability studies with the MHD model have progressed significantly since the 1950s, when magnetic confinement research began. For tokamaks, it is considered that MHD stability theory was almost completed by the development of ballooning mode theory in the latter half of the 1970s. MHD studies for stellarators also began in the 1950s. However, progress was slow because it was not easy to obtain three-dimensional MHD equilibrium of finite-beta plasma. For toroidal stellarator MHD equilibrium, there are two problems as discussed in chapter 4: the existence of nested magnetic surfaces, and the viability of currentless equilibrium. Regarding the first point, the existence of general three-dimensional equilibrium has been disproved. However, since the existence of equilibrium in the straight stellarator configuration is guaranteed by the helical symmetry, we often study the equilibrium assuming that nested magnetic surfaces would exist if the toroidal effects were sufficiently small. In other words, it is accepted in practice that the magnetic surfaces exist when no magnetic islands or no stochastic magnetic field line regions appear in the calculation of the vacuum magnetic surfaces by line tracing.

Currentless equilibrium means that the toroidal current averaged over each magnetic surface is zero, although the local current density is nonzero. The significant property of currentless equilibrium is that the rotational transform profile at finite beta is different from that in the vacuum magnetic field. In heliotrons the value of the rotational transform at the magnetic axis increases, while it decreases at the edge. In currentless equilibrium, there are two factors degrading stability in the resistive time scales, including Coulomb collisions. One is the appearance of two resonant surfaces with the same mode numbers satisfying $\iota = n/m$, where ι is a rotational transform and m and n are the poloidal and toroidal mode numbers, respectively. The other is that a weak shear region with $d\iota/d\Psi \simeq 0$ appears, where $\Psi = \text{constant}$ denotes the magnetic surface.

According to neoclassical transport theory, there appears to be a bootstrap current in stellarators (see chapter 7) as well as in tokamaks. Experiments to confirm this current have already been carried out. Analysis of the MHD equilibrium in heliotrons consistently including the bootstrap current has been already attempted. A sufficient condition to suppress current-driven

MHD instabilities is that the rotational transform produced by the plasma current, $\iota_J(r)$, is less than $\iota_h(r) - \iota_h(0)$, where $\iota_h(r)$ is the rotational transform produced by the external stellarator field. When the current is localized near the plasma surface, this criterion is easily satisfied.

With regard to MHD stability in currentless stellarator plasmas, the ideal and resistive pressure-driven instabilities are important in evaluating the critical beta for stellarator and heliotron devices. Interchange and ballooning modes are typically pressure-driven. For ideal interchange modes, the Mercier criterion, $D_I > 0$, becomes a good indication of the whole stability properties; in particular, of the beta limit. It is the beta value corresponding to the marginal state of the ideal pressure-driven mode. Ballooning modes also play a role in the beta limit; however, a negative shear (or $q' = dq/d\Psi < 0$, where q is a safety factor) in heliotrons has a stabilizing effect on the ballooning mode, which implies that the interchange mode is more dangerous in the negative shear region. For currentless finite-beta equilibrium, when the Shafranov shift is large, the rotational transform usually decreases from the magnetic axis and a $q' > 0$ region appears. In this case, the ballooning mode may become more unstable than the interchange mode.

One of the characteristics of heliotrons is that the magnetic hill region cannot be removed completely from the edge plasma region. The magnetic hill easily destabilizes the ideal interchange mode, while the magnetic shear has a stabilizing effect. In the resistive MHD model, the resistive interchange mode easily becomes unstable at a beta value lower than the ideal beta limit, because the magnetic shear stabilization is not effective for the resistive mode. There is an extension of the Mercier criterion for the ideal interchange mode to the resistive interchange mode shown by $D_R > 0$ for the stability. It is noted that D_I is always larger than D_R, although in this book D_R is explained only for cylindrical geometry. While in the stability analysis of D_I and D_R it is assumed that the instability has a localized mode structure at the rational surface, the low-n ideal or resistive interchange instability has a broad radial mode structure around the rational surface when the growth rate is substantial. However, as the beta value approaches the critical value, the low-n ideal interchange instability tends to become localized at the rational surface and, therefore, it may be considered that the critical beta is equal to the beta value given by $D_I = 0$. Similarly, it is suggested that $D_R = 0$ also gives the critical beta due to the low-n resistive interchange mode. Since $|D_R|$ is small but D_R is negative in the magnetic hill region of heliotrons, even for $\beta \simeq 0$, the most crucial instability in heliotrons is the resistive interchange mode, which seems consistent with experimental results. In tokamaks, although the tearing and rippling modes are studied as important resistive instabilities, they are not relevant to heliotrons because the toroidal plasma current is small enough or the MHD equilibrium is currentless.

In section 5.2, we summarize the linearized MHD equations and linear MHD stability. First, we derive the ballooning mode equation for stellarators, and the localized stability criterion of the ideal interchange mode D_I is given in section 5.3. We derive the growth rate of the resistive interchange mode in section 5.4. Numerical codes for studying ideal MHD stability are explained briefly, and we discuss the relation between low-n modes and high-n modes in section 5.5. In section 5.6, we compare the pressure-driven modes observed in heliotrons with theoretical results based on the resistive interchange modes. In section 5.7, current-driven instabilities are summarized, and we give a comparison between theory and experiment. In the final section 5.8 we discuss the relation between magnetic surface destruction in finite-beta plasmas of heliotron and the stability criterion for the resistive interchange mode.

5.2 LINEAR MHD STABILITY AND THE ENERGY PRINCIPLE

The problem of stability can be stated qualitatively as follows. The existence of an MHD equilibrium implies a state in which the sum of the forces acting on the plasma is exactly zero. If the plasma is perturbed from this state, the resulting perturbed forces either restore the plasma to its original equilibirum or cause a further enhancement of the initial disturbance. The former corresponds to stability and the latter to instability. The question of ideal MHD stability is most important, since plasmas are considered to suffer catastrophic termination as a consequence of such instabilities. It is the first priority to obtain ideal MHD stability for a fusion reactor. In order to avoid ideal MHD instabilities, somewhat unusual but ingenious magnetic configurations are investigated in magnetic fusion research, particularly in stellarator and heliotron research.

In MHD stability, efforts are devoted toward linear stability analysis. Ideal MHD instabilities are so virulent that it is more important in practice to avoid them on the basis of linear stability studies rather than to investigate the nonlinear evolution of the unstable state. Physical intuition is easily acquired when analytic treatment is possible. Thus the discussion in this chapter is concerned primarily with analytic linear stability theory. However, when we assess the merits of actual magnetic configurations, we need quantitative results of numerical studies of linear stability. Sometimes the approximations used in the analytic theories are not accurate enough for realistic designs of experimental devices.

First, the definition of an ideal MHD instability is given here. We assume that all quantities of interest are linearized about their equilibrium values:

$$G(\mathbf{r}, t) = G_0(\mathbf{r}) + \tilde{G}(\mathbf{r}, t). \tag{5.1}$$

Here $G_0(\mathbf{r})$ is the equilibrium value and $|\tilde{G}(\mathbf{r}, t)|/|G_0(\mathbf{r})| \ll 1$ is considered to be a small first-order perturbation. Since the equilibrium is time independent, the perturbation can be written as

$$\tilde{G}(\mathbf{r}, t) = G_1(\mathbf{r})e^{-i\omega t}. \tag{5.2}$$

The stability definition is such that if any of the eigenfrequencies ω corresponds to exponential growth, the system is exponentially unstable. If there is no such eigenfrequency, it is considered to be exponentially stable. In other words, in (5.2),

$$\begin{cases} \mathrm{Im}(\omega) > 0\text{: exponential instability,} \\ \mathrm{Im}(\omega) \leq 0\text{: exponential stability.} \end{cases} \tag{5.3}$$

Here we assume implicitly that modes satisfying given boundary conditions are discrete with respect to eigenfrequencies. It is the unstable part of the complex ω plane that is of interest in this chapter, although usually a continuous spectrum exists in the stable part, as we shall discuss later.

It is convenient to express all perturbed quantities in terms of a vector $\tilde{\xi}$, defined by

$$\tilde{\mathbf{v}} = \partial\tilde{\xi}/\partial t. \tag{5.4}$$

The vector $\tilde{\xi}$ represents the displacement of the plasma away from its equilibrium position. First, we aim to express all perturbed quantities in terms of $\tilde{\xi}$ and then we obtain a single equation describing the time evolution of $\tilde{\xi}$.

In an initial value formulation, one needs to specify appropriate initial data. A very convenient choice of initial data for stability problems is as follows:

$$\begin{cases} \tilde{\xi}(\mathbf{r}, 0) = \tilde{\mathbf{B}}(\mathbf{r}, 0) = \tilde{\rho}(\mathbf{r}, 0) = \tilde{P}(\mathbf{r}, 0) = 0, \\ \left.\dfrac{\partial\tilde{\xi}}{\partial t}\right|_{t=0} = \tilde{\mathbf{v}}(\mathbf{r}, 0) \neq 0. \end{cases} \tag{5.5}$$

This corresponds to the situation in which at $t = 0$, the plasma is in its exact equilibrium position but is moving away with a small velocity $\tilde{\mathbf{v}}(\mathbf{r}, 0)$. Under these conditions the linearized form of the equation of continuity (3.73), the pressure equation (3.75), and Faraday's law (3.77) can be integrated with respect to time, and we obtain

$$\tilde{\rho} = -\nabla \cdot (\rho_0\tilde{\xi}), \tag{5.6}$$

$$\tilde{P} = -\tilde{\xi} \cdot \nabla P_0 - \gamma P_0 \nabla \cdot \tilde{\xi}, \tag{5.7}$$

$$\tilde{\mathbf{B}} = \nabla \times (\tilde{\xi} \times \mathbf{B}_0). \tag{5.8}$$

It is noted that (5.8) gives $\nabla \cdot \tilde{\mathbf{B}} = 0$. These quantities given by (5.6)–(5.8) can be substituted into the equation of motion, leading to a single equation for the displacement vector $\tilde{\xi}$,

$$\rho_0 \frac{\partial^2\tilde{\xi}}{\partial t^2} = \mathbf{F}(\tilde{\xi}), \tag{5.9}$$

where the force operator is written as

$$\mathbf{F}(\tilde{\xi}) = \frac{1}{\mu_0}(\nabla \times \mathbf{B}_0) \times \tilde{\mathbf{B}} + \frac{1}{\mu_0}(\nabla \times \tilde{\mathbf{B}}) \times \mathbf{B}_0 + \nabla(\tilde{\xi} \cdot \nabla P_0 + \gamma P_0 \nabla \cdot \tilde{\xi}). \tag{5.10}$$

Equation (5.10) plus appropriate boundary conditions at the magnetic axis and the plasma boundary, in addition to conditions (5.5), constitute the formulation of linear stability for three-dimensional equilibria as an initial value problem. The initial value approach has the advantage of directly determining the actual time evolution of a given initial perturbation. It is also easy to extend the numerical formulation to the nonlinear problem.

The normal-mode formulation is very efficient in investigating the linear stability. This can be done by letting all perturbed quantities vary as (5.2).

Then (5.6)–(5.8) reduce to

$$\rho_1 = -\nabla \cdot (\rho_0 \xi), \tag{5.11}$$

$$P_1 = -\xi \cdot \nabla P_0 - \gamma P_0 \nabla \cdot \xi, \tag{5.12}$$

$$\mathbf{B}_1 = \nabla \times (\xi \times \mathbf{B}_0). \tag{5.13}$$

It is noted that $\xi(\mathbf{r})$ is different from $\tilde{\xi}(\mathbf{r}, t)$. We also obtain

$$-\rho_0 \omega^2 \xi = \mathbf{F}(\xi) \tag{5.14}$$

from (5.9), where

$$\mathbf{F}(\xi) = \frac{1}{\mu_0}(\nabla \times \mathbf{B}_0) \times \mathbf{B}_1 + \frac{1}{\mu_0}(\nabla \times \mathbf{B}_1) \times \mathbf{B}_0 + \nabla(\xi \cdot \nabla P_0 + \gamma P_0 \nabla \cdot \xi). \tag{5.15}$$

Equation (5.14) represents the normal-mode formulation of the linear stability problem for three-dimensional equilibria. In this approach appropriate boundary conditions on ξ at the magnetic axis and at the plasma boundary are required. Equation (5.14) is usually solved as an eigenvalue problem for the eigenvalue ω^2. Within the context of the ideal MHD model, the normal-mode approach is quite efficient with respect to numerical computations. However, it is difficult to exploit the advantage in the nonlinear formulation. We note that the usefulness of the normal-mode method is strongly coupled to the assumption that the eigenvalues are discrete, so that the concept of exponential stability is valid. This is indeed true for the unstable part of the spectrum of ω^2.

The force operator **F** possesses an important mathematical property; that is, **F** is a self-adjoint operator. To demonstrate this property it is necessary to show that, for any two arbitrary vectors ξ and $\boldsymbol{\eta}$ satisfying appropriate boundary conditions, the following relation holds:

$$\int \boldsymbol{\eta} \cdot \mathbf{F}(\xi) dV = \int \xi \cdot \mathbf{F}(\boldsymbol{\eta}) dV. \tag{5.16}$$

Here we assume that $\mathbf{n} \cdot \xi = \mathbf{n} \cdot \boldsymbol{\eta} = 0$ on the plasma boundary, where $\mathbf{n}$ is a unit vector normal to the boundary. This corresponds to the existence of a perfectly conducting wall at the plasma boundary.

The integrand in (5.16) can be written as

$$\boldsymbol{\eta} \cdot \mathbf{F}(\xi) = \boldsymbol{\eta} \cdot \left[\frac{1}{\mu_0} (\nabla \times \mathbf{B}) \times \mathbf{Q} + \frac{1}{\mu_0} (\nabla \times \mathbf{Q}) \times \mathbf{B} + \nabla(\xi \cdot \nabla P + \gamma P \nabla \cdot \xi) \right], \tag{5.17}$$

where $\mathbf{Q} = \mathbf{B}_1$ and the zero subscript has been dropped from all equilibrium quantities. The last term is integrated by parts and $\boldsymbol{\eta} \cdot \nabla(\gamma P \nabla \cdot \xi)$ becomes $\gamma P(\nabla \cdot \xi)(\nabla \cdot \boldsymbol{\eta})$. We now write $\xi = \xi_\perp + \zeta_\parallel \mathbf{b}$ and $\boldsymbol{\eta} = \boldsymbol{\eta}_\perp + \eta_\parallel \mathbf{b}$. Then parallel components of the first and third terms in (5.17) are written as

$$\begin{aligned} \frac{1}{\mu_0} \mathbf{B} \cdot (\nabla \times \mathbf{B}) \times \mathbf{Q} &= -\mathbf{Q} \cdot \mathbf{J} \times \mathbf{B} = -\mathbf{Q} \cdot \nabla P \\ &= \nabla \cdot [\nabla P \times (\xi \times \mathbf{B})] = -\nabla \cdot [(\xi \cdot \nabla P)\mathbf{B}], \end{aligned} \tag{5.18}$$

$$\mathbf{B} \cdot \nabla(\xi \cdot \nabla P) = \nabla \cdot [(\xi \cdot \nabla P)\mathbf{B}]. \tag{5.19}$$

By considering (5.18) and (5.19), it is seen that there is no parallel component in the first three terms in (5.17). Consequently, we find that

$$\boldsymbol{\eta} \cdot \mathbf{F}(\xi) = -\gamma P(\nabla \cdot \xi)(\nabla \cdot \boldsymbol{\eta}) + I_1, \tag{5.20}$$

where I_1 is a function of only the perpendicular components of ξ and $\boldsymbol{\eta}$:

$$I_1 = \boldsymbol{\eta}_\perp \cdot \left[\frac{1}{\mu_0} (\nabla \times \mathbf{B}) \times \mathbf{Q} + \frac{1}{\mu_0} (\nabla \times \mathbf{Q}) \times \mathbf{B} + \nabla(\xi_\perp \cdot \nabla P) \right]. \tag{5.21}$$

The last term in (5.21) is integrated by parts again and the first two terms are rewritten using standard vector identities:

$$I_1 = \frac{1}{\mu_0} \boldsymbol{\eta}_\perp \cdot [\mathbf{Q} \cdot \nabla \mathbf{B} + \mathbf{B} \cdot \nabla \mathbf{Q} - \nabla(\mathbf{B} \cdot \mathbf{Q})] - (\xi_\perp \cdot \nabla P) \nabla \cdot \boldsymbol{\eta}_\perp. \tag{5.22}$$

The three terms inside the square brackets in (5.22) are shown as follows:

$$\begin{aligned} \boldsymbol{\eta}_\perp \cdot (\mathbf{Q} \cdot \nabla \mathbf{B}) = \boldsymbol{\eta}_\perp \cdot [(\mathbf{B} \cdot \nabla \xi_\perp) \cdot \nabla \mathbf{B} - (\xi_\perp \cdot \nabla \mathbf{B}) \cdot \nabla \mathbf{B}] \\ - B^2 (\boldsymbol{\eta}_\perp \cdot \boldsymbol{\kappa}) \nabla \cdot \xi_\perp, \end{aligned} \tag{5.23}$$

$$\begin{aligned}
\boldsymbol{\eta}_\perp \cdot (\mathbf{B} \cdot \nabla \mathbf{Q}) &= \mathbf{B} \cdot \nabla(\boldsymbol{\eta}_\perp \cdot \mathbf{Q}) - \mathbf{Q} \cdot (\mathbf{B} \cdot \nabla \boldsymbol{\eta}_\perp) \\
&= \nabla \cdot [(\boldsymbol{\eta}_\perp \cdot \mathbf{Q})\mathbf{B}] - (\mathbf{B} \cdot \nabla \boldsymbol{\xi}_\perp - \boldsymbol{\xi}_\perp \cdot \nabla \mathbf{B} \\
&\quad - \mathbf{B}\nabla \cdot \boldsymbol{\xi}_\perp) \cdot (\mathbf{B} \cdot \nabla \boldsymbol{\eta}_\perp) \\
&= -(\mathbf{B} \cdot \nabla \boldsymbol{\xi}_\perp) \cdot (\mathbf{B} \cdot \nabla \boldsymbol{\eta}_\perp) + (\boldsymbol{\xi}_\perp \cdot \nabla \mathbf{B}) \cdot (\mathbf{B} \cdot \nabla \boldsymbol{\eta}_\perp) \\
&\quad - B^2(\boldsymbol{\eta}_\perp \cdot \boldsymbol{\kappa})\nabla \cdot \boldsymbol{\xi}_\perp, \qquad (5.24)
\end{aligned}$$

$$\begin{aligned}
-\boldsymbol{\eta}_\perp \cdot \nabla(\mathbf{B} \cdot \mathbf{Q}) &= -\nabla \cdot [(\mathbf{B} \cdot \mathbf{Q})\boldsymbol{\eta}_\perp] + (\mathbf{B} \cdot \mathbf{Q})\nabla \cdot \boldsymbol{\eta}_\perp \\
&= -B^2(\nabla \cdot \boldsymbol{\xi}_\perp)(\nabla \cdot \boldsymbol{\eta}_\perp) \\
&\quad - \left[\boldsymbol{\xi}_\perp \cdot \nabla \frac{B^2}{2} + B^2(\boldsymbol{\xi}_\perp \cdot \boldsymbol{\kappa})\right]\nabla \cdot \boldsymbol{\eta}_\perp, \qquad (5.25)
\end{aligned}$$

where $\boldsymbol{\kappa} = (\mathbf{B}/B) \cdot \nabla(\mathbf{B}/B)$. In (5.23) and (5.24) the full divergence contributions have been dropped, since they integrate to zero. From (5.20), (5.22), (5.23), (5.24), and (5.25), we find that

$$\begin{aligned}
\boldsymbol{\eta} \cdot \mathbf{F}(\boldsymbol{\xi}) &= -\frac{B^2}{\mu_0}(\nabla \cdot \boldsymbol{\xi}_\perp)(\nabla \cdot \boldsymbol{\eta}_\perp) - \frac{1}{\mu_0}(\mathbf{B} \cdot \nabla \boldsymbol{\xi}_\perp) \cdot (\mathbf{B} \cdot \nabla \boldsymbol{\eta}_\perp) \\
&\quad - \gamma P(\nabla \cdot \boldsymbol{\xi})(\nabla \cdot \boldsymbol{\eta}) \\
&\quad - \left[\boldsymbol{\xi}_\perp \cdot \nabla\left(P + \frac{B^2}{2\mu_0}\right) + \frac{B^2}{\mu_0}\boldsymbol{\xi}_\perp \cdot \boldsymbol{\kappa}\right](\nabla \cdot \boldsymbol{\eta}_\perp) \\
&\quad - 2\frac{B^2}{\mu_0}(\boldsymbol{\eta}_\perp \cdot \boldsymbol{\kappa})\nabla \cdot \boldsymbol{\xi}_\perp + I_2, \qquad (5.26)
\end{aligned}$$

where

$$\begin{aligned}
I_2 = \frac{1}{\mu_0}&[\boldsymbol{\eta}_\perp \cdot [(\mathbf{B} \cdot \nabla \boldsymbol{\xi}_\perp) \cdot \nabla \mathbf{B} - (\boldsymbol{\xi}_\perp \cdot \nabla \mathbf{B}) \cdot \nabla \mathbf{B}] \\
&+ (\boldsymbol{\xi}_\perp \cdot \nabla \mathbf{B}) \cdot (\mathbf{B} \cdot \nabla \boldsymbol{\eta}_\perp)]. \qquad (5.27)
\end{aligned}$$

Here we will use two identities to rewrite I_2:

$$\begin{aligned}
\nabla \cdot \{[\boldsymbol{\eta}_\perp \cdot (\boldsymbol{\xi}_\perp \cdot \nabla \mathbf{B})]\mathbf{B}\} &= (\mathbf{B} \cdot \nabla \boldsymbol{\eta}_\perp) \cdot (\boldsymbol{\xi}_\perp \cdot \nabla \mathbf{B}) \\
&\quad + \boldsymbol{\eta}_\perp \cdot (\mathbf{B} \cdot \nabla \boldsymbol{\xi}_\perp) \cdot \nabla \mathbf{B} + \boldsymbol{\eta}_\perp \cdot (\mathbf{B}\boldsymbol{\xi}_\perp : \nabla\nabla)\mathbf{B}, \qquad (5.28)
\end{aligned}$$

$$\begin{aligned}
\boldsymbol{\eta}_\perp \cdot (\boldsymbol{\xi}_\perp \cdot \nabla)(\mathbf{B} \cdot \nabla \mathbf{B}) &= \boldsymbol{\eta}_\perp \cdot (\boldsymbol{\xi}_\perp \cdot \nabla \mathbf{B}) \cdot \nabla \mathbf{B} \\
&\quad + \boldsymbol{\eta}_\perp \cdot (\mathbf{B}\boldsymbol{\xi}_\perp : \nabla\nabla)\mathbf{B}. \qquad (5.29)
\end{aligned}$$

By noting that the divergence term integrates to zero, I_2 is given by

$$\begin{aligned}
I_2 &= -\frac{1}{\mu_0}\boldsymbol{\eta}_\perp \cdot (\boldsymbol{\xi}_\perp \cdot \nabla)(\mathbf{B} \cdot \nabla \mathbf{B}) \\
&= -(\boldsymbol{\eta}_\perp \boldsymbol{\xi}_\perp : \nabla\nabla)\left(P + \frac{B^2}{2\mu_0}\right), \qquad (5.30)
\end{aligned}$$

where the relation obtained from $\mu_0 \nabla P = (\nabla \times \mathbf{B}) \times \mathbf{B}$,

$$\nabla_{\perp}\left(P + \frac{B^2}{2\mu_0}\right) = \frac{1}{\mu_0}\mathbf{B}\cdot\nabla\mathbf{B}, \tag{5.31}$$

has been used. We also note that the RHS of (5.31) is rewritten as $(B^2/\mu_0)\boldsymbol{\kappa}$. The final result for the volume integral of $\boldsymbol{\eta}\cdot\mathbf{F}(\xi)$ becomes

$$\begin{aligned}\int \boldsymbol{\eta}\cdot\mathbf{F}(\xi)dV = -\int dV\Bigg[&\frac{1}{\mu_0}(\mathbf{B}\cdot\nabla\xi_{\perp})\cdot(\mathbf{B}\cdot\nabla\boldsymbol{\eta}_{\perp}) + \gamma P(\nabla\cdot\xi)(\nabla\cdot\boldsymbol{\eta}) \\ &+\frac{B^2}{\mu_0}(\nabla\cdot\xi_{\perp} + 2\xi_{\perp}\cdot\boldsymbol{\kappa})(\nabla\cdot\boldsymbol{\eta}_{\perp} + 2\boldsymbol{\eta}_{\perp}\cdot\boldsymbol{\kappa}) \\ &-\frac{4B^2}{\mu_0}(\xi_{\perp}\cdot\boldsymbol{\kappa})(\boldsymbol{\eta}_{\perp}\cdot\boldsymbol{\kappa}) + (\boldsymbol{\eta}_{\perp}\xi_{\perp}:\nabla\nabla)\left(P + \frac{B^2}{2\mu_0}\right)\Bigg],\end{aligned} \tag{5.32}$$

which clearly shows a self-adjoint form. This expression was given by Freidberg (1987).

By making use of the self-adjointness of $\mathbf{F}$, it is straightforward to show that for any discrete normal mode, the corresponding eigenvalue ω^2 is purely real. This is shown by forming the dot product of (5.14) with $\xi^*(\mathbf{r})$ and integrating over the plasma volume,

$$\omega^2 \int \rho|\xi|^2 dV = -\int \xi^*\cdot\mathbf{F}(\xi)dV. \tag{5.33}$$

The same procedure is then applied to the complex conjugate of (5.14) with $\xi(\mathbf{r})$. Using the self-adjointness of $\mathbf{F}$ it follows that

$$(\omega^2 - \omega^{*2})\int \rho|\xi|^2 dV = 0, \tag{5.34}$$

which gives that $\omega^2 = \omega^{*2}$ is purely real. In terms of the definition of exponential stability, a normal mode with $\omega^2 > 0$ corresponds to a pure oscillation and hence would be considered stable. On the contrary, for $\omega^2 < 0$, there is a branch which grows exponentially, which corresponds to instability. Clearly, the transition from stability to instability occurs at $\omega^2 = 0$.

The spectrum properties of $\mathbf{F}$ follow from an examination of the operator $(\mathbf{F}/\rho - \lambda)^{-1}$ for all complex λ. If this operator exists and is bounded for a given λ, the linearized MHD equations with the following form,

$$(\mathbf{F}/\rho - \lambda)\xi = \mathbf{a} \tag{5.35}$$

obtained by applying Laplace transforms to (5.9) for the initial value method, can be inverted, yielding

$$\xi = (\mathbf{F}/\rho - \lambda)^{-1}\mathbf{a}. \tag{5.36}$$

The spectrum of $\mathbf{F}$ consists of those values of λ for which the operator $(\mathbf{F}/\rho - \lambda)^{-1}$ *cannot* be inverted. There are two important cases. First, when λ is such that $(\mathbf{F}/\rho - \lambda)\xi = 0$ possesses a nontrivial solution, these values of λ correspond to the *discrete* spectra of $\mathbf{F}$.

The second case is such that $(\mathbf{F}/\rho - \lambda)^{-1}$ exists, but is unbounded on a magnetic surface in the plasma column. For example, in a cylindrical plasma with an inhomogeneous plasma density one finds that

$$\mathbf{F}/\rho - \lambda = k_{\parallel}^2 V_A^2(r) - \lambda, \tag{5.37}$$

where $V_A(r)$ is the local Alfvén velocity. In these situation there is a continuous range of λ over which the operator is ill-behaved. This point can be understood by setting $\lambda = k_{\parallel}^2 V_A^2(r)$ and varying r over the range $0 < r < a$. These values of λ constitute the *continuous* spectra of $\mathbf{F}$. Thus the spectrum of the force operator contains both discrete and continuous eigenvalues. However, usually continuous ones exist only for $\omega^2 \geq 0$ in the ideal MHD stability problem. This result provides us with a motivation for examining MHD stability by the normal-mode approach by restricting attention only to the question of whether or not an exponentially growing mode exists. It is noted that a somewhat complicated situation appears as $\omega^2 \to 0$, since unstable modes can either accumulate or make smooth transitions from instability to stability.

Equation (5.14) represents the normal-mode formulation of the three-dimensional linearized MHD stability problem. Because of the self-adjointness of $\mathbf{F}$, the problem can be formulated in a variational principle. The dot product of (5.14) with ξ^* is formed and then integrated over the plasma volume, which yields

$$\omega^2 = \frac{\delta W(\xi^*, \xi)}{K(\xi^*, \xi)}, \tag{5.38}$$

where

$$\delta W(\xi, \xi^*) = -\tfrac{1}{2}\int \xi^* \cdot \mathbf{F}(\xi) dV, \tag{5.39}$$

$$K(\xi^*, \xi) = \tfrac{1}{2}\int \rho|\xi|^2 dV. \tag{5.40}$$

The variational principle states that any allowable function ξ for which ω^2 becomes an extremum is an eigenfunction of the ideal MHD normal-mode equations with the eigenvalue ω^2. It is seen by letting $\xi \to \xi + \delta\xi$, $\omega^2 \to \omega^2 + \delta\omega^2$, and setting $\delta\omega^2 = 0$, since ω^2 is an extremum. We find that

$$\delta\omega^2 = \frac{\delta W(\delta\xi^*, \xi) + \delta W(\xi^*, \delta\xi) - \omega^2[K(\delta\xi^*, \xi) + K(\xi^*, \delta\xi)]}{K(\xi^*, \xi)} \tag{5.41}$$

by assuming that $\delta W(\delta\xi^*, \delta\xi)$ and $K(\delta\xi^*, \delta\xi)$ are negligibly small. Using the self-adjointness of $\mathbf{F}$ and setting $\delta\omega^2 = 0$, we obtain

$$\int dV\{\delta\xi^* \cdot [\mathbf{F}(\xi) + \rho\omega^2\xi] + \delta\xi \cdot [\mathbf{F}(\xi^*) + \rho\omega^2\xi^*]\} = 0. \tag{5.42}$$

Since $\delta\xi$ is arbitrary, (5.42) implies (5.14), and the variational principle (5.38) is equivalent to the normal mode approach for the linear MHD stability.

It is often useful to determine whether a given system is stable or unstable without calculating growth rates ω^2 and eigenfunctions $\xi(\mathbf{r})$. It is far more important to determine the conditions for avoiding instabilities than to calculate precise growth rates. For such problems, the variational formulation can be simplified further, leading to the *energy principle* which exactly determines the stability boundaries but only estimates the growth rates.

The physical basis of the energy principle is the fact that energy is conserved in the ideal MHD model. As a consequence, when the system is unstable, the most negative eigenvalue of ω^2 represents a minimum of the potential energy δW. This implies that the question of stability (or instability) can be determined by analyzing only the sign of δW and not the full variational problem. Thus, the energy principle states that an equilibrium is stable if and only if

$$\delta W(\xi^*, \xi) \geq 0 \tag{5.43}$$

for all ξ which are bounded in energy and satisfy appropriate boundary conditions. On the other hand, if it is negative for any displacement, the equilibrium is unstable.

An elegant proof of the energy principle has been given by Laval, Mercier, and Pellat (1965). Here we consider a displacement vector $\tilde{\xi}(\mathbf{r}, t)$ again. Then the energy $H(t)$ is given by

$$\begin{aligned} H(t) &= K\left(\frac{\partial\tilde{\xi}}{\partial t}, \frac{\partial\tilde{\xi}}{\partial t}\right) + \delta W(\tilde{\xi}, \tilde{\xi}) \\ &= \tfrac{1}{2}\int dV\left[\rho\left(\frac{\partial\tilde{\xi}}{\partial t}\right)^2 - \tilde{\xi} \cdot \mathbf{F}(\tilde{\xi})\right]. \end{aligned} \tag{5.44}$$

Using the self-adjointness of $\mathbf{F}$ yields

$$\frac{dH}{dt} = \int dV\left[\rho\,\frac{\partial^2\tilde{\xi}}{\partial t^2} - \mathbf{F}(\tilde{\xi})\right] \cdot \frac{\partial\tilde{\xi}}{\partial t} = 0, \tag{5.45}$$

which shows $H(t) = H_0 =$ constant and corresponds to the energy conservation.

First, we assume that $\delta W > 0$ for all allowable $\tilde{\xi}$, which is sufficient for stability. The energy conservation is written as

$$\delta W = H_0 - K. \tag{5.46}$$

This implies that unbounded growth of K (e.g., by exponential instability) violates the assumption. To show the necessity of the energy principle we assume that a perturbation $\tilde{\boldsymbol{\eta}}(\mathbf{r})$ exists such that $\delta W(\tilde{\boldsymbol{\eta}}, \tilde{\boldsymbol{\eta}}) < 0$. We consider a displacement satisfying the initial conditions

$$\begin{cases} \tilde{\xi}(\mathbf{r}, 0) = \tilde{\boldsymbol{\eta}}(\mathbf{r}), \\ \dfrac{\partial \tilde{\xi}}{\partial t}(\mathbf{r}, 0) = 0. \end{cases} \tag{5.47}$$

From energy conservation,

$$H_0 = (K + \delta W)|_{t=0} = \delta W(\tilde{\boldsymbol{\eta}}, \tilde{\boldsymbol{\eta}}) < 0. \tag{5.48}$$

Now we calculate $d^2 I/dt^2$, where

$$I(t) = K(\tilde{\xi}, \tilde{\xi}) = \tfrac{1}{2}\int dV \rho(\tilde{\xi})^2, \tag{5.49}$$

$$\begin{aligned} \frac{d^2 I}{dt^2} &= \int dV \rho\left[\left(\frac{\partial \tilde{\xi}}{\partial t}\right)^2 + \tilde{\xi} \cdot \frac{\partial^2 \tilde{\xi}}{\partial t^2}\right] \\ &= \int dV \left[\rho\left(\frac{\partial \tilde{\xi}}{\partial t}\right)^2 + \tilde{\xi} \cdot \mathbf{F}(\tilde{\xi})\right] \\ &= 2K\left(\frac{\partial \tilde{\xi}}{\partial t}, \frac{\partial \tilde{\xi}}{\partial t}\right) - 2\delta W(\tilde{\xi}, \tilde{\xi}). \end{aligned} \tag{5.50}$$

From (5.46), we obtain

$$\frac{d^2 I}{dt^2} = 4K\left(\frac{\partial \tilde{\xi}}{\partial t}, \frac{\partial \tilde{\xi}}{\partial t}\right) - 2H_0 > -2H_0 > 0. \tag{5.51}$$

This implies that I grows without bound as $t \to \infty$, which indicates that $\tilde{\xi}$ increases at least as fast as t. Thus, if any perturbation $\tilde{\boldsymbol{\eta}}$ exists that makes $\delta W < 0$, then the equilibrium is unstable and $\tilde{\xi}$ grows without bound. Consequently, $\delta W > 0$ for all allowable displacements is a necessary condition for stability.

The energy principle is valid if the plasma is either directly surrounded by a perfectly conducting wall or isolated from the wall by a vacuum region. In the first case, application of the energy principle is straightfoward since the boundary condition to be satisfied is

$$\mathbf{n} \cdot \xi_\perp|_{\mathbf{r}=\mathbf{r}_W} = 0, \tag{5.52}$$

where $\mathbf{r}_W$ denotes the wall position.

When a vacuum region exists, the situation is more complicated, because the vacuum magnetic fields must be included through the boundary conditions at the plasma surface.

From (5.17) and (5.39), by integration by parts of $\xi^* \cdot \nabla(\gamma P \nabla \cdot \xi)$ and $\xi^* \cdot (\nabla \times \mathbf{Q}) \times \mathbf{B}$, we obtain

$$\delta W = \frac{1}{2}\int dV \left\{ \frac{|Q|^2}{\mu_0} + \gamma P |\nabla \cdot \xi|^2 - \xi^* \cdot [\mathbf{J} \times \mathbf{Q} + \nabla(\xi \cdot \nabla P)] \right\}$$
$$- \frac{1}{2}\int_S dS (\mathbf{n} \cdot \xi^*) \left(\gamma P \nabla \cdot \xi - \frac{\mathbf{B} \cdot \mathbf{B}_1}{\mu_0} \right), \tag{5.53}$$

where the second term denotes that the contribution from the plasma surface and $\mathbf{B}_1$ instead of $\mathbf{Q}$ has been used here.

Next, writing the displacement vector as $\xi = \xi_\perp + \xi_\parallel \mathbf{b}$, $\xi_\parallel^* \mathbf{b} \cdot [\mathbf{J} \times \mathbf{Q} + \nabla(\xi \cdot \nabla P)] = 0$ from (5.18) and (5.19). The term $\xi_\perp^* \cdot \nabla(\xi \cdot \nabla P)$ in (5.53) is also integrated by parts. The result is

$$\delta W = \frac{1}{2}\int_P dV \left[\frac{|Q|^2}{\mu_0} - \xi_\perp^* \cdot (\mathbf{J} \times \mathbf{Q}) + \gamma P |\nabla \cdot \xi|^2 + (\xi_\perp \cdot \nabla P) \nabla \cdot \xi_\perp^* \right]$$
$$+ \frac{1}{2}\int_S dS (\mathbf{n} \cdot \xi_\perp^*) \left(\frac{\mathbf{B} \cdot \mathbf{B}_1}{\mu_0} - \gamma P \nabla \cdot \xi - \xi_\perp \cdot \nabla P \right), \tag{5.54}$$

where we refer to the first term as δW_F in the following. Here the labels P and S for the integrals correspond to the unperturbed plasma volume and plasma surface, respectively.

Here we consider that a jump in the pressure and tangential magnetic field is sustained, if surface currents are allowed to flow. Then the jump condition at the plasma surface,

$$\left[\left[P + \frac{B^2}{2\mu_0} \right]\right] = 0, \tag{5.55}$$

is obtained from the MHD equilibrium equation (5.31), where $[[Q]] \equiv \hat{Q} - Q$, and Q and $\hat{Q}$ denote a variable in the plasma and vacuum region, respectively. By expanding the condition (5.55) about the perturbed surface $\mathbf{r}_P + \xi$, where $\mathbf{r}_P$ is the unperturbed surface and $\xi = \xi(\mathbf{r}_P)$, we obtain

$$\left[P_1 + \frac{\mathbf{B} \cdot \mathbf{B}_1}{\mu_0} + \xi \cdot \nabla \left(P + \frac{B^2}{2\mu_0} \right) \right]\Bigg|_{\mathbf{r}=\mathbf{r}_P} = \left[\frac{\hat{\mathbf{B}} \cdot \hat{\mathbf{B}}_1}{\mu_0} + \xi \cdot \nabla \frac{\hat{B}^2}{2\mu_0} \right]\Bigg|_{\mathbf{r}=\mathbf{r}_P}. \tag{5.56}$$

Substituting (5.56) with (5.12) into the second integral in (5.54) yields

$$\frac{1}{2}\int_S dS|\mathbf{n}\cdot\xi_\perp|^2\mathbf{n}\cdot\left[\left[\nabla\left(P+\frac{B^2}{2\mu_0}\right)\right]\right]+\frac{1}{2}\int_S dS(\mathbf{n}\cdot\xi_\perp^*)\,\frac{\hat{\mathbf{B}}\cdot\hat{\mathbf{B}}_1}{\mu_0}. \tag{5.57}$$

It should be noted that the first term showing the surface contribution δW_S vanishes unless surface currents flow on the plasma–vacuum boundary.

The second term in (5.57) is identical to the perturbed magnetic energy in the vacuum region:

$$\delta W_V=\frac{1}{2}\int_V dV\,\frac{|\hat{\mathbf{B}}_1|^2}{\mu_0}. \tag{5.58}$$

Here $\hat{\mathbf{B}}_1$ is written as $\hat{\mathbf{B}}_1=\nabla\times\hat{\mathbf{A}}_1$, with $\nabla\times\hat{\mathbf{B}}_1=0$, and $|\hat{\mathbf{B}}_1|^2=\hat{\mathbf{B}}_1\cdot\nabla\times\hat{\mathbf{A}}_1^*$. Integrating (5.58) by parts yields a surface integral

$$\delta W_V=-\frac{1}{2}\int_S dS\,\frac{\mathbf{n}\cdot(\hat{\mathbf{A}}_1^*\times\hat{\mathbf{B}}_1)}{\mu_0}. \tag{5.59}$$

It is noted that the contribution on the outer conducting wall vanishes by virtue of $\mathbf{n}\cdot\hat{\mathbf{B}}_1|_{\mathbf{r}=\mathbf{r}_W}=0$. We use the relation

$$\mathbf{n}\cdot\hat{\mathbf{B}}_1=\mathbf{n}\cdot\nabla\times(\xi_\perp\times\hat{\mathbf{B}}) \tag{5.60}$$

and $\mathbf{n}\cdot\hat{\mathbf{B}}_1=\mathbf{n}\cdot\nabla\times\hat{\mathbf{A}}_1$, which give $\hat{\mathbf{A}}_1=\xi_\perp\times\hat{\mathbf{B}}+\nabla f$. If f is chosen so that the gauge condition is $\hat{\mathbf{B}}_1\cdot(\mathbf{n}\times\nabla f)=0$, then (5.59) reduces to

$$\delta W_V=\frac{1}{2}\int_S dS(\mathbf{n}\cdot\xi_\perp^*)\left(\frac{\hat{\mathbf{B}}\cdot\hat{\mathbf{B}}_1}{\mu_0}\right), \tag{5.61}$$

by noting $\mathbf{n}\cdot\hat{\mathbf{B}}=0$ on the plasma–vacuum boundary. Equation (5.61) is identical to the second term of (5.57). Thus the potential energy δW in the energy principles is shown as

$$\delta W=\delta W_F+\delta W_S+\delta W_V, \tag{5.62}$$

where the fluid, surface, and vacuum contributions are given bythe first volume integral of (5.54), and the first surface integral of (5.57) and (5.58), respectively. Here we evaluate the potential energy for the fluid δW_F for a cylindrical tokamak model. The relevant equilibrium quantities are $P(r)$ and $\mathbf{B}=B_\theta(r)\mathbf{e}_\theta+B_z(\mathbf{r})\mathbf{e}_z$, which satisfy

$$\left(P+\frac{B_\theta^2+B_z^2}{2\mu_0}\right)'+\frac{B_\theta^2}{\mu_0 r}=0 \tag{5.63}$$

for MHD equilibrium. Although two field components are present, the system possesses (θ, z) symmetry, so that the perturbations can be Fourier analyzed in these two coordinates, and the expression

$$\boldsymbol{\xi}(\mathbf{r}) = \boldsymbol{\xi}(r)\exp[i(m\theta + kz)] \tag{5.64}$$

is relevant to describe the perturbation. It is this dual symmetry that is responsible for the algebraic minimization of the potential energy with respect to two components of $\boldsymbol{\xi}$.

In the first integral of (5.54), the terms in the integrand are rearranged as follows. The first two terms are separated into two components,

$$|\mathbf{Q}|^2 = |\mathbf{Q}_\perp|^2 + |\mathbf{Q}_\parallel|^2 \tag{5.65}$$

and

$$\boldsymbol{\xi}_\perp^* \cdot (\mathbf{J} \times \mathbf{Q}) = J_\parallel(\boldsymbol{\xi}_\perp^* \times \mathbf{b}) \cdot \mathbf{Q}_\perp + Q_\parallel \boldsymbol{\xi}_\perp^* \cdot (\mathbf{J}_\perp \times \mathbf{b}), \tag{5.66}$$

where $\mathbf{J}_\perp$ and $Q_\parallel$ can be written as

$$\mathbf{J}_\perp = \frac{\mathbf{b} \times \nabla P}{B}, \tag{5.67}$$

$$\begin{aligned} Q_\parallel &= \mathbf{b} \cdot \nabla \times (\boldsymbol{\xi}_\perp \times \mathbf{B}) \\ &= \mathbf{b} \cdot (\mathbf{B} \cdot \nabla \boldsymbol{\xi}_\perp - \boldsymbol{\xi}_\perp \cdot \nabla \mathbf{B} - \mathbf{B} \nabla \cdot \boldsymbol{\xi}_\perp) \\ &= -B(\nabla \cdot \boldsymbol{\xi}_\perp + 2\boldsymbol{\xi}_\perp \cdot \boldsymbol{\kappa}) + \frac{\mu_0}{B}\, \boldsymbol{\xi}_\perp \cdot \nabla P. \end{aligned} \tag{5.68}$$

Here we have used (5.31) to introduce $\boldsymbol{\kappa} = \mathbf{b} \cdot \nabla \mathbf{b}$. By using (5.65) and (5.66) with expressions (5.67) and (5.68), the first integral of (5.54) becomes

$$\begin{aligned} \delta W_F = \tfrac{1}{2} \int_P dV \bigg[& \frac{|\mathbf{Q}_\perp|^2}{\mu_0} + \frac{B^2}{\mu_0} |\nabla \cdot \boldsymbol{\xi}_\perp + 2\boldsymbol{\xi}_\perp \cdot \boldsymbol{\kappa}|^2 \\ & + \gamma P |\nabla \cdot \boldsymbol{\xi}|^2 - 2(\boldsymbol{\xi}_\perp \cdot \nabla P)(\boldsymbol{\kappa} \cdot \boldsymbol{\xi}_\perp^*) - J_\parallel (\boldsymbol{\xi}_\perp^* \times \mathbf{b}) \cdot \mathbf{Q}_\perp \bigg]. \end{aligned} \tag{5.69}$$

Here we note that the $|\mathbf{Q}_\perp|^2$ term represents the energy required to bend magnetic field lines and contributes to the shear Alfvén wave. The second term corresponds to the energy necessary to compress the magnetic field and contributes to the compressional Alfvén wave. The $\gamma P|\nabla \cdot \boldsymbol{\xi}|^2$ term represents the energy required to compress the plasma and contributes to the sound wave. Each of the contributions described above is stabilizing. The remaining two terms can be positive or negative and thus can drive instabilities. For the cylindrical tokamak model, the plasma displacement is written as

$$\boldsymbol{\xi}_\perp = \xi \mathbf{e}_r + \eta \mathbf{e}_\eta, \tag{5.70}$$

with

$$\eta = (\xi_\theta B_z - \xi_z B_\theta)/B,$$
$$\mathbf{e}_\eta = (B_z \mathbf{e}_\theta - B_\theta \mathbf{e}_z)/B.$$

The individual terms in the integrand of (5.69) are evaluated as

$$\mathbf{Q}_\perp = [\nabla \times (\xi_\perp \times \mathbf{B})]_\perp = iF\xi \mathbf{e}_r + \left\{ iF\eta + \xi \left[\frac{B_z' B_\theta}{B} - \frac{rB_z}{B} \left(\frac{B_\theta}{r} \right)' \right] \right\} \mathbf{e}_\eta, \tag{5.71}$$

$$\boldsymbol{\kappa} = \mathbf{b} \cdot \nabla \mathbf{b} = -\frac{B_\theta^2}{rB^2} \mathbf{e}_r, \tag{5.72}$$

$$\nabla \cdot \xi_\perp + 2\xi_\perp \cdot \boldsymbol{\kappa} = \frac{(r\xi)'}{r} - 2 \frac{B_\theta^2}{rB^2} \xi + i \frac{G}{B} \eta, \tag{5.73}$$

$$\mu_0 J_\parallel = \mu_0 \mathbf{J} \cdot \mathbf{b} = \frac{B_z}{rB} (rB_\theta)' - \frac{B_\theta B_z'}{B}, \tag{5.74}$$

where $F = mB_\theta/r + kB_z$ and $G = -kB_\theta + mB_z/r$. By substituting (5.71)–(5.74) into the integral (5.69), we obtain

$$\begin{aligned} \frac{\delta W_F}{2\pi R_0} = \frac{\pi}{\mu_0} \int_0^a r dr \Bigg[& F^2 |\xi|^2 + \left| iF\eta + \xi \left(\frac{B_z' B_\theta}{B} - \frac{rB_z}{B} \left(\frac{B_\theta}{r} \right)' \right) \right|^2 \\ & + B^2 \left| \frac{(r\xi)'}{r} - \frac{2B_\theta^2}{rB^2} \xi + i \frac{G}{B} \eta \right|^2 + \frac{2\mu_0 P' B_\theta^2}{rB^2} |\xi|^2 \\ & - \mu_0 J_\parallel \left\{ iF(\xi\eta^* - \xi^* \eta) - |\xi|^2 \left(\frac{B_z' B_\theta}{B} - \frac{rB_z}{B} \left(\frac{B_\theta}{r} \right)' \right) \right\} \Bigg], \end{aligned} \tag{5.75}$$

where a and $2\pi R_0$ are a plasma radius and a plasma length of cylindrical plasma, respectively. It is noted that $\nabla \cdot \xi = 0$ is assumed to eliminate the positive definite term. Here we note that η appears only algebraically. The terms explicitly containing η can be grouped together as follows:

$$\begin{aligned} & k_0^2 B^2 |\eta|^2 + 2 \frac{ikBB_\theta}{r} (\eta\xi^* - \eta^* \xi) + \frac{iGB}{r} (\eta (r\xi^*)' - \eta^* (r\xi)') \\ & = \left| ik_0 B\eta + \frac{2kB_\theta}{rk_0} \xi + \frac{G}{rk_0} (r\xi)' \right|^2 - \left| \frac{2kB_\theta}{rk_0} \xi + \frac{G}{rk_0} (r\xi)' \right|^2, \end{aligned} \tag{5.76}$$

where $k_0^2 = k^2 + m^2/r^2$. Since η appears only in a positive term, δW_F is minimized by choosing η to make the first term of the RHS of (5.76) vanish. With this choice for η, the integrand of (5.75) can be written as

$$r(A_1 \xi'^2 + 2A_2 \xi\xi' + A_3 \xi^2), \tag{5.77}$$

where

$$A_1 = \frac{F^2}{k_0^2} \tag{5.78}$$

$$A_2 = \frac{1}{rk_0^2}\left(k^2 B_z^2 - \frac{m^2 B_\theta^2}{r^2}\right) \tag{5.79}$$

$$A_3 = F^2 + \frac{2\mu_0 P' B_\theta^2}{rB^2} + \frac{B^2}{r^2}\left(1 - 2\frac{B_\theta^2}{B^2}\right)^2 - \frac{1}{r^2 k_0^2}(G + 2kB_\theta)^2 + \frac{2B_\theta B_z}{rB}\left[\frac{B_\theta B_z'}{B} - \frac{rB_z}{B}\left(\frac{B_\theta}{r}\right)'\right]. \tag{5.80}$$

It is noted that ξ in (5.77) can be real. The middle term of (5.77) can be integrated by parts. This leads to the following form of δW_F:

$$\frac{\mu_0 \delta W_F}{2\pi^2 R_0} = \int_0^a (f\xi'^2 + g\xi^2)dr + \left[\frac{k^2 r^2 B_z^2 - m^2 B_\theta^2}{k_0^2 r^2}\right]_{r=a} \xi^2(a), \tag{5.81}$$

where

$$f = rA_1 = \frac{rF^2}{k_0^2}, \tag{5.82}$$

$$g = rA_3 - (rA_2)'. \tag{5.83}$$

By using the equilibrium relation (5.63) and after a lengthy calculation, g can be written as

$$g = \frac{2k^2}{k_0^2}(\mu_0 P)' + \left(\frac{k_0^2 r^2 - 1}{k_0^2 r^2}\right) rF^2 + \frac{2k^2}{rk_0^4}\left(kB_z - \frac{mB_\theta}{r}\right)F. \tag{5.84}$$

We will now give a classification of MHD instabilities, which are usually discussed from the points of view such as mode structure and main driving source. First, we distinguish whether a given instability is predominantly an internal or external mode. Second, each instability can be characterized by its dominant driving source. Thus, a given instability is either pressure-driven or current-driven.

Consider a magnetic configuration in which the plasma is surrounded by a vacuum. Instabilities the mode structure of which does not require any motion of the plasma–vacuum interface away from its equilibrium position are called internal or fixed boundary modes. Often, the internal mode appears in the neighborhood of a resonant or rational surface (where the operator $\mathbf{B} \cdot \nabla$ vanishes) inside the plasma column. The boundary condition of the internal mode is $\mathbf{n} \cdot \xi|_{\mathbf{r}=\mathbf{r}_p} = 0$ or is equivalent to moving the conducting wall to the plasma surface, and the potential energy of fluid δW_F is only minimized. On the contrary, for external or free-boundary modes, the plasma–vacuum interface moves from its equilibrium position. The external mode often has a resonant or rational surface in the vacuum region. For such instabilities, δW_S and δW_V must be evaluated as well as δW_F, since $\mathbf{n} \cdot \xi|_{\mathbf{r}=\mathbf{r}_p} \neq 0$.

The potential energy (5.69) shows that there are two possible sources of MHD instability, one proportional to ∇P and the other to $J_{\parallel}$. Instabilities in which the dominant destabilizing term is that proportional to ∇P are called pressure-driven modes. However, it is possible to consider that they are driven by perpendicular currents, since ∇P is directly related to diamagnetic currents through the MHD equilibrium condition. Usually, the most unstable pressure-driven instabilities are internal modes, with short wavelengths perpendicular to the magnetic field but long wavelengths parallel to the magnetic field. The pressure-driven modes are usually subdivided into two categories: interchange and ballooning modes.

Interchange instabilities are very similar in nature to the Rayleigh–Taylor instability in fluid dynamics, wherre the gravitational force plays a role. In magnetically confined plasmas, magnetic field line curvature plays the same role as gravity. The interchange perturbation can lead to instabiity depending on the relative sign of the magnetic field line curvature with respect to the pressure gradient. If the magnetic field lines are concave toward the plasma, their tension tends to make them shorten. The plasma pressure, on the other hand, has a tendency to expand outward. In this case a perturbation that interchanges two flux tubes at different radii leads to an enhancement of the natural tendencies, or to a lowering of the potential energy. Since the perturbation is constant along the magnetic field line to avoid line bending, this type of mode is sometimes called a flute mode. Perpendicular to the magnetic field line the unstable perturbation has very rapid oscillations. When the magnetic field lines are convex to the plasma, the system is stable to interchange perturbations. These are considered as localized interchanges, which leads to stability conditions expressed in terms of local values of the equilibrium quantities. Examples of such conditions are the Suydam criterion (in a cylindrical plasma) and the Mercier criterion (in a toroidal plasma).

We need several properties of the magnetic configuration to stabilize interchange instabilities. First, if there are magnetic shear or rotational transform changes from one magnetic surface to another, it is impossible to interchange two neighboring flux tubes without some line bending. A second method is to make a magnetic well or "average-favorable curvature." As a given magnetic field line encircles the torus, it passes through alternating regions of favorable and unfavorable curvature to the interchange. By a careful design of the magnetic configuration, the "average curvature" can be made favorable.

Ballooning modes are internal pressure-driven instabilities that appear in two- or three-dimensional toroidal configurations. The term "ballooning" refers to the mode structure. Usually, the curvature of the magnetic field line alternates between favorable and unfavorable regions in toroidal configurations. Thus, a perturbation that is not constant but varying slowly along a magnetic field line in such a way that the mode is localized in the unfavorable curvature region can lead to more unstable situations than the simple interchange perturbation. The "ballooning" nature of the perturbation in the unfavorable curvature region increases the pressure-driven destabilizing contribution in δW_F, which is not overcome by the line bending stabilization when the plasma pressure increases.

Like interchange modes, ballooning modes are amenable to analysis. The most unstable perturbation has a long wavelength along the magnetic field line and a short wavelength perpendicular to it. By exploiting the short perpendicular wavelength nature of the instabilities, the general multidimensional (two-dimensional in tokamaks and three-dimensional in stellarators) stability problem reduces to the solution of a one-dimensional differential equation on each flux surface. Substantial progress has been made since the "ballooning representation" has been introduced. One property of the ballooning mode equation is that, in a special limit, it leads to the localized interchange stability criterion, or Mercier criterion, as shown in section 5.3.

A current-driven mode is one in which the dominant driving source of instability is proportional to $J_{\parallel}$ in the potential energy δW_F (see 5.69)). Current-driven instabilities are often known as kink modes or helical modes, and can be excited even in a zero-pressure force-free plasma. In general, the most unstable perturbation has a long parallel wavelength and a perpendicular wavelength that is comparable to a plasma radius. Current-driven modes can have the mode structure of either internal or external perturbations, depending upon the location of the resonant or rational surface.

The external kink mode is the most dangerous instability in tokamaks. The main destabilizing term for higher-m modes is the radial gradient of the parallel current, where m is a poloidal mode number. For low-m modes, and for $m = 1$ in particular, the current profile is not as important as the total current itself. The basic form of the perturbation is such that the plasma surface kinks with a helix structure. The unstable modes occur for a long parallel wavelength in order to minimize line bending.

There are several ways in which the stability of a given configuration to external kink modes can be improved. For a given configuration there is usually a critical parallel current indicating the onset of instability. Stability can be achieved by keeping the parallel current below this value. In other words, for a fixed parallel current, a toroidal device with a lower aspect ratio or a larger a/R (R is the major radius of the torus) will suppress the formation of long-wavelength kinks. Second, higher-m kink modes can be stabilized by peaking the $J_{\parallel}$ profile with a fixed total current. Third, since low-m kink modes have a broad radial mode structure, a perfectly conducting wall has a stabilizing effect when it is placed close to the plasma surface.

The internal kink mode with properties similar to those of the external kink mode is in general a weaker instability. In low-β tokamaks, usually only $m = 1$ mode becomes unstable, when a $q = 1$ surface exists inside the plasma column. Sometimes a resistive effect is crucial for the internal kink mode, since the topology of the magnetic surface is no longer conserved and the formation of a magnetic island is allowed. Thus the $m = 1$ resistive internal kink mode is considered to play a role in the sawtooth oscillation or internal disruption commonly observed in tokamaks. When a plasma current is induced in stellarators and heliotrons and an $\iota = \iota_h + \iota_J = 1$ surface appears in the central region, the $m = 1$ internal kink mode will be destabilized when the parallel current is substantial. Here ι_h and ι_J are the rotational transform produced by the external helical magnetic field and the plasma current, respectively.

5.3 THE BALLOONING MODE EQUATION

First, we express the magnetic and current density perturbations $\mathbf{B}_1$ and $\mathbf{J}_1 = \nabla \times \mathbf{B}_1/\mu_0$ appearing in the linearized equation of motion,

$$\rho \frac{\partial \mathbf{v}_1}{\partial t} = \mathbf{J} \times \mathbf{B}_1 + \mathbf{J}_1 \times \mathbf{B} - \nabla P_1, \tag{5.85}$$

in terms of the vector $\mathbf{T}$, which is defined as

$$\mathbf{T} = \nabla \times (\tilde{\xi} \times \mathbf{B}) + \frac{(\mu_0 \mathbf{J} \times \nabla a)}{|\nabla a|^2} (\tilde{\xi} \cdot \nabla a). \tag{5.86}$$

Here, as in section 4.2, a is the label for a magnetic surface. After substituting $\mathbf{v}_1$, P_1, $\mathbf{J}_1$, and $\mathbf{B}_1$ expressed with the displacement vector $\tilde{\xi}$ into (5.85), we obtain

$$\begin{aligned} \rho \frac{\partial^2 \tilde{\xi}}{\partial t^2} = & \frac{1}{\mu_0} (\nabla \times \mathbf{T}) \times \mathbf{B} + \mathbf{J} \times \left[\mathbf{T} - (\mathbf{T} \cdot \nabla a) \frac{\nabla a}{|\nabla a|^2} \right] + \nabla(\gamma P \nabla \cdot \tilde{\xi}) \\ & + (\tilde{\xi} \cdot \nabla a) \left[\nabla P' + \frac{\mu_0 \mathbf{J}^2}{|\nabla a|^2} \nabla a - \left(\nabla \times \frac{(\mathbf{J} \times \nabla a)}{|\nabla a|^2} \right) \times \mathbf{B} \right], \end{aligned} \tag{5.87}$$

where we have used the MHD equilibrium equations (4.1)–(4.3), and

$$\mathbf{T} \cdot \nabla a = \mathbf{B}_1 \cdot \nabla a = \mathbf{B} \cdot \nabla (\tilde{\xi} \cdot \nabla a), \tag{5.88}$$

which is a consequence of the definition (5.86). Here the prime denotes a derivative with respect to a. In order to show that (5.85) is equivalent to (5.87), it is easier to substitute (5.86) into (5.87) and derive (5.85). Expanding the vector $\mathbf{T}$ in terms of the three orthogonal vectors ∇a, $\mathbf{B}$, and $(\mathbf{B} \times \nabla a)$, as

$$\mathbf{T} = T_1 \nabla a + T_2 \frac{(\mathbf{B} \times \nabla a)}{|\nabla a|^2} - T_3 \mathbf{B}, \tag{5.89}$$

we transform the first two terms on the RHS of (5.87) to

$$\frac{1}{\mu_0}(\nabla\times\mathbf{T})\times\mathbf{B}+\mathbf{J}\times\left[\mathbf{T}-(\mathbf{T}\cdot\nabla a)\frac{\nabla a}{|\nabla a|^2}\right]$$
$$=\nabla a\left[\frac{\mathbf{B}}{\mu_0}\cdot\nabla T_1-\frac{\mathbf{J}\cdot\mathbf{B}}{|\nabla a|^2}T_2-2P'T_3+\frac{\mathbf{B}^2(\nabla a\cdot\nabla T_3)}{|\nabla a|^2}\right]$$
$$+\frac{(\mathbf{B}\times\nabla a)}{\mu_0|\nabla a|^2}[\mathbf{B}\cdot\nabla T_2+(\mathbf{B}\times\nabla a)\cdot\nabla T_3]-T_2\frac{\mathbf{B}}{\mu_0}\times\left[\nabla\times\frac{(\mathbf{B}\times\nabla a)}{|\nabla a|^2}\right]. \quad (5.90)$$

Here we write the plasma current as

$$\mathbf{J}=\sigma\mathbf{B}+(\mathbf{B}\times\nabla P)/B^2, \quad (5.91)$$

where $\sigma\equiv\mathbf{J}\cdot\mathbf{B}/B^2$ and $\mathbf{J}\times\mathbf{B}=\nabla P$ is satisfied with (5.91). By noting that $\nabla\cdot\mathbf{J}=0$ gives

$$\mathbf{B}\cdot\nabla\sigma=-P'(\mathbf{B}\times\nabla a)\cdot\nabla(1/B^2), \quad (5.92)$$

we obtain

$$\nabla\times\left(\frac{\mathbf{J}\times\nabla a}{|\nabla a|^2}\right)=\mathbf{B}\frac{\nabla a\cdot\nabla\sigma}{|\nabla a|^2}+\sigma\nabla\times\frac{(\mathbf{B}\times\nabla a)}{|\nabla a|^2}-\mu_0\mathbf{J}\frac{P'}{B^2}$$
$$+\frac{(\mathbf{B}\times\nabla a)}{|\nabla a|^2}\left(\nabla a\cdot\nabla\frac{P'}{B^2}\right). \quad (5.93)$$

Taking the projections on the three vectors introduced in (5.89) yields

$$\nabla\times\left(\frac{\mathbf{B}\times\nabla a}{|\nabla a|^2}\right)=\frac{\mathbf{B}}{B^2}\left[\nabla\cdot\left(B^2\frac{\nabla a}{|\nabla a|^2}\right)+\mu_0P'\right]+\frac{(\mathbf{B}\times\nabla a)}{B^2}S, \quad (5.94)$$

where

$$S=\frac{(\mathbf{B}\times\nabla a)}{|\nabla a|^2}\cdot\nabla\times\frac{(\mathbf{B}\times\nabla a)}{|\nabla a|^2}, \quad (5.95)$$

which is related to a *local magnetic shear*. This plays an important role in the ballooning mode stability. Here we substitute (5.90), (5.93), and (5.94) into (5.87) and finally obtain

$$\rho\frac{\partial^2\tilde{\xi}}{\partial t^2}=\nabla T_0+\nabla a\left[K\tilde{\xi}\cdot\nabla a+\frac{\mathbf{B}}{\mu_0}\cdot\nabla T_1+T_2\left(\frac{S}{\mu_0}-\frac{\mathbf{J}\cdot\mathbf{B}}{|\nabla a|^2}\right)\right.$$
$$\left.-2P'T_3+\frac{B^2}{\mu_0}\frac{\nabla a\cdot\nabla T_3}{|\nabla a|^2}\right]$$
$$+\frac{(\mathbf{B}\times\nabla a)}{\mu_0|\nabla a|^2}\nabla\cdot[T_2\mathbf{B}+T_3(\mathbf{B}\times\nabla a)], \quad (5.96)$$

where $T_0 = \gamma P \nabla \cdot \tilde{\xi}$ and

$$K = \frac{\mu_0 \mathbf{J}^2}{|\nabla a|^2} + \frac{P'}{B^2}\left(\mu_0 P' + \frac{\nabla a \times \nabla B^2}{|\nabla a|^2}\right) - \sigma S. \tag{5.97}$$

In the following we write the perturbation $\tilde{\xi}$ as

$$\tilde{\xi} = \xi \frac{\nabla a}{|\nabla a|^2} + \eta \frac{(\mathbf{B} \times \nabla a)}{B^2} + \tau \frac{\mathbf{B}}{B^2}. \tag{5.98}$$

Substituting this expression into (5.86) yields

$$T_1 = \mathbf{B} \cdot \nabla \xi / |\nabla a|^2, \tag{5.99}$$

$$T_2 = \frac{|\nabla a|^2}{B^2}\left[\mathbf{B} \cdot \nabla \eta + \left(\frac{\mathbf{J} \cdot \mathbf{B}}{|\nabla a|^2} - \frac{S}{\mu_0}\right)\xi\right], \tag{5.100}$$

$$T_3 = \frac{\nabla \xi \cdot \nabla a}{|\nabla a|^2} + \frac{(\mathbf{B} \times \nabla a)}{B^2} \cdot \nabla \eta + \frac{\xi}{B^2}\nabla \cdot \left(B^2 \frac{\nabla a}{|\nabla a|^2}\right) + \frac{2\mu_0 \xi P'}{B^2}. \tag{5.101}$$

First, we consider the projection of the linearized equation of motion (5.96) on the vector **B**,

$$\rho \frac{\partial^2 \tau}{\partial t^2} = \mathbf{B} \cdot \nabla T_0. \tag{5.102}$$

It follows from this equation that a longitudinal component of $\tilde{\xi}$ along **B** can develop only as a result of the compressibility of the plasma or $\nabla \cdot \tilde{\xi} \neq 0$ and only if $\nabla \cdot \tilde{\xi} \neq f(a)$, where $f(a)$ is a function of a only. From the projection of (5.96) on the vector $(\mathbf{B} \times \nabla a)$,

$$\rho \frac{|\nabla a|^2}{B^2}\frac{\partial^2 \eta}{\partial t^2} = \frac{(\mathbf{B} \times \nabla a)}{B^2} \cdot \nabla T_0 + [\mathbf{B} \cdot \nabla T_2 + (\mathbf{B} \times \nabla a) \cdot \nabla T_3]/\mu_0. \tag{5.103}$$

The last projection of (5.96) on the vector ∇a yields

$$\frac{1}{B^2 |\nabla a|^2}\rho \frac{\partial^2 \xi}{\partial t^2} = \frac{\nabla a \cdot \nabla T_3}{\mu_0 |\nabla a|^2} + \frac{1}{B^2}\left[\frac{\nabla a \cdot \nabla T_0}{|\nabla a|^2} - 2P' T_3 + T_2\left(\frac{S}{\mu_0} - \frac{\mathbf{J} \cdot \mathbf{B}}{|\nabla a|^2}\right) + \frac{\mathbf{B}}{\mu_0} \cdot \nabla T_1 + K\xi\right]. \tag{5.104}$$

Here we will write (5.104) in another form. The equation (5.96) can be written as

$$\rho \frac{\partial^2 \tilde{\xi}}{\partial t^2} = \nabla T_0 + B^2 U \nabla a + \frac{(\mathbf{B} \times \nabla a)}{\mu_0 |\nabla a|^2} (\mathbf{B} \cdot \nabla T_2) + \frac{B^2}{\mu_0} \nabla T_3 - \frac{1}{\mu_0} (\mathbf{B} \cdot \nabla T_3)\mathbf{B}, \tag{5.105}$$

where

$$B^2 U \equiv K(\tilde{\xi} \cdot \nabla a) + \frac{\mathbf{B}}{\mu_0} \cdot \nabla T_1 + T_2 \left(\frac{S}{\mu_0} - \frac{\mathbf{J} \cdot \mathbf{B}}{|\nabla a|^2} \right) - 2P' T_3. \tag{5.106}$$

Taking the vector product of (5.105) and $\mathbf{B}/B^2$, and then taking the divergence of the resulting equation, we obtain

$$\frac{\partial^2}{\partial t^2} \nabla \cdot \left(\rho \frac{\tilde{\xi} \times \mathbf{B}}{B^2} \right) = -\nabla T_0 \cdot \nabla \times \left(\frac{\mathbf{B}}{B^2} \right) - \mathbf{J} \cdot \nabla T_3 - (\mathbf{B} \times \nabla a) \cdot \nabla U + \frac{1}{\mu_0} \nabla \cdot \left(\frac{\mathbf{B} \cdot \nabla T_2}{|\nabla a|^2} \nabla a \right). \tag{5.107}$$

From the RHS of (5.107), we collect the terms with T_3 and transform them as

$$-\mathbf{J} \cdot \nabla T_3 + (\mathbf{B} \times \nabla a) \cdot \nabla \frac{2P' T_3}{B^2} = -\sigma \mathbf{B} \cdot \nabla T_3 + P' \frac{(\mathbf{B} \times \nabla a)}{B^2} \cdot \nabla T_3 + \frac{4P' T_3}{B^2} (\mathbf{B} \times \boldsymbol{\kappa}) \cdot \nabla a, \tag{5.108}$$

where the curvature of magnetic field line $\boldsymbol{\kappa} = (\mathbf{B}/B) \cdot \nabla(\mathbf{B}/B) = \nabla(2\mu_0 P + B^2)/(2B^2) - (\mathbf{B}/2B^4)(\mathbf{B} \cdot \nabla)B^2$ has been substituted in the last term of the RHS of (5.108). We also transform the last term of (5.107):

$$\nabla \cdot \left[(\mathbf{B} \cdot \nabla T_2) \frac{\nabla a}{|\nabla a|^2} \right] = \frac{\nabla a}{|\nabla a|^2} \cdot \nabla (\mathbf{B} \cdot \nabla T_2) + (\mathbf{B} \cdot \nabla T_2) \nabla \cdot \frac{\nabla a}{|\nabla a|^2} = \mathbf{B} \cdot \nabla \left(\frac{\nabla a \cdot \nabla T_2}{|\nabla a|^2} \right) + \nabla T_2 \cdot \nabla \times \left(\frac{\mathbf{B} \times \nabla a}{|\nabla a|^2} \right). \tag{5.109}$$

Using the relations (5.106), (5.108), and (5.109), (5.107) takes the form

$$\frac{\partial^2}{\partial t^2}\nabla\cdot\left(\rho\frac{\tilde{\xi}\times\mathbf{B}}{B^2}\right)=-\nabla T_0\cdot\nabla\times\left(\frac{\mathbf{B}}{B^2}\right)$$
$$+\left[-\sigma\mathbf{B}\cdot\nabla T_3+P'\frac{\mathbf{B}\times\nabla a}{B^2}\cdot\nabla T_3+\frac{4P'T_3}{B^2}(\mathbf{B}\times\boldsymbol{\kappa})\cdot\nabla a\right]$$
$$-(\mathbf{B}\times\nabla a)\cdot\nabla\left[\frac{K\tilde{\xi}\cdot\nabla a}{B^2}+\frac{\mathbf{B}\cdot\nabla T_1}{\mu_0 B^2}+\frac{T_2}{B_2}\left(\frac{S}{\mu_0}-\frac{\mathbf{J}\cdot\mathbf{B}}{|\nabla a|^2}\right)\right]$$
$$+\frac{\mathbf{B}}{\mu_0}\cdot\nabla\left(\frac{\nabla a\cdot\nabla T_2}{|\nabla a|^2}\right)+\frac{\nabla T_2}{\mu_0}\cdot\nabla\times\left(\frac{\mathbf{B}\times\nabla a}{|\nabla a|^2}\right). \tag{5.110}$$

Using the relation (5.94) and

$$\mathbf{B}\times\left(\frac{\mathbf{B}}{B^2}\right)=\mu_0\frac{\sigma\mathbf{B}}{B^2}+2\frac{(\mathbf{B}\times\boldsymbol{\kappa})}{B^2}-\mu_0\frac{P'}{B^2}\frac{(\mathbf{B}\times\nabla a)}{B^2}, \tag{5.111}$$

we obtain the transformation

$$-\mathbf{B}T_0\cdot\nabla\times\left(\frac{\mathbf{B}}{B^2}\right)+\nabla T_2\cdot\nabla\times\left(\frac{\mathbf{B}\times\nabla a}{|\nabla a|^2}\right)$$
$$=-\frac{\sigma\mathbf{B}\cdot\nabla T_0}{B^2}-2\frac{\mathbf{B}\times\boldsymbol{\kappa}}{B^2}\cdot\nabla T_0+\frac{S}{\mu_0}\frac{\mathbf{B}\times\nabla a}{B^2}\cdot\nabla T_2$$
$$+\frac{\mathbf{B}\cdot\nabla T_2}{B^2}\nabla\cdot\left(B^2\frac{\nabla a}{|\nabla a|^2}\right)+\frac{P'}{B^2}\left[\mu_0\frac{|\nabla a|^2}{B^2}\rho\frac{\partial^2\eta}{\partial t^2}-(\mathbf{B}\times\nabla a)\cdot\nabla T_3\right], \tag{5.112}$$

where (5.103) has been used for the last term in the RHS of (5.112). By substituting (5.112) into (5.110), we finally obtain

$$\frac{\partial^2}{\partial t^2}\left[\nabla\cdot\left(\rho\frac{\tilde{\xi}\times\mathbf{B}}{B^2}\right)-\mu_0\frac{P'}{B^2}\frac{|\nabla a|^2}{B^2}\rho\eta\right]=\frac{\mathbf{B}}{\mu_0}\cdot\nabla\left(\frac{\nabla a\cdot\nabla T_2}{|\nabla a|^2}\right)$$
$$+\frac{\mathbf{J}\cdot\mathbf{B}}{|\nabla a|^2}\frac{(\mathbf{B}\times\nabla a)}{B^2}\cdot\nabla T_2-(\mathbf{B}\times\nabla a)\cdot\nabla\left(\frac{\mu_0 K(\xi\cdot\nabla a)+\mathbf{B}\cdot\nabla T_1}{\mu_0 B^2}\right)$$
$$+T_2(\mathbf{B}\times\nabla a)\cdot\nabla\left(\frac{\mathbf{J}\cdot\mathbf{B}/|\nabla a|^2-S/\mu_0}{B^2}\right)$$
$$-\sigma\mathbf{B}\cdot\nabla T_3+\frac{4P'T_3}{B^2}\nabla a\cdot(\mathbf{B}\times\boldsymbol{\kappa})$$
$$-\mu_0\frac{\sigma\mathbf{B}\cdot\nabla T_0}{B^2}-2\frac{(\mathbf{B}\times\boldsymbol{\kappa})}{B^2}\cdot\nabla T_0+\frac{\mathbf{B}\cdot\nabla T_2}{\mu_0 B^2}\nabla\cdot\left(\frac{B^2\nabla a}{|\nabla a|^2}\right). \tag{5.113}$$

We note that (5.113) has been obtained without any assumptions about the form of the displacement.

Here we consider ballooning-type perturbations that have short wavelengths perpendicular to **B** and long wavelengths in the parallel direction, but not almost constant along the magnetic field line as for interchange modes. These ballooning perturbations satisfy the conditions

$$\frac{|\nabla a\cdot\nabla x|}{|\nabla a|},\quad \frac{|(\mathbf{B}\times\nabla a)\cdot\nabla x|}{B|\nabla a|}\gg\frac{|x|}{L},\quad \frac{|\mathbf{B}\cdot\nabla x|}{B}, \tag{5.114}$$

where x represents ξ and η, and L is a characteristic transverse dimension of the plasma column. On the contrary, the interchange perturbations are characterized by

$$\frac{|\nabla a\cdot\nabla x|}{|\nabla a|}\gg\frac{|(\mathbf{B}\times\nabla a)\cdot\nabla x|}{B|\nabla a|}\gg\frac{|x|}{L}\gg\frac{|\mathbf{B}\cdot\nabla x|}{B}. \tag{5.115}$$

It is convenient to describe the ballooning mode in an eikonal form, shown by

$$\tilde{\xi}=\hat{\xi}(\mathbf{r})\exp\left[\frac{i}{\varepsilon}\hat{S}(\mathbf{r})-i\omega t\right], \tag{5.116}$$

where

$$\mathbf{k}_\perp=\nabla\hat{S}, \tag{5.117}$$

which gives $\mathbf{B}\cdot\nabla\hat{S}=0$. Here $\varepsilon\ll 1$ and $L|\nabla\hat{S}|\sim O(1)$ are assumed.

The smallness of ε makes it possible to simplify the linearized equation (5.102), (5.103), and (5.113). After ξ in the form (5.116) is substituted in $T_0=\gamma P\nabla\cdot\tilde{\xi}$ and (5.99)–(5.101), T_1, T_2, and T_3 in these equations are shown in the form $T_j=\hat{T}_j\exp[(i/\varepsilon)\hat{S}-i\omega t](j=1,2,3)$. For example, $T_0=\gamma P\nabla\cdot\tilde{\xi}$ gives

$$\hat{T}_0=\gamma P\left(\nabla\cdot\hat{\xi}+\frac{i}{\varepsilon}\mathbf{k}_\perp\cdot\hat{\xi}\right), \tag{5.118}$$

$T_1=\mathbf{B}\cdot\nabla\xi/|\nabla a|^2$ gives

$$\hat{T}_1=\mathbf{B}\cdot\nabla\hat{\xi}/|\nabla a|^2, \tag{5.119}$$

and $T_2=|\nabla a|^2[\mathbf{B}\cdot\nabla\eta+(\mathbf{J}\cdot\mathbf{B}/|\nabla a|^2-S/\mu_0)\xi]/B^2$ gives

$$\hat{T}_2=|\nabla a|^2[\mathbf{B}\cdot\nabla\hat{\eta}+(\mathbf{J}\cdot\mathbf{B}/|\nabla a|^2-S/\mu_0)\hat{\xi}]/B^2. \tag{5.120}$$

For $\hat{T}_3$ we use the relation (5.101), which can be modified to include T_0 with $T_0/\gamma P=\nabla\cdot\tilde{\xi}=\nabla\xi\cdot\nabla a/|\nabla a|^2+\xi\nabla\cdot(\nabla a/|\nabla a|^2)+\nabla\eta\cdot(\mathbf{B}\times\nabla a)/B^2+$

$\eta\nabla\cdot(\mathbf{B}\times\nabla a/B^2)+\nabla\cdot(\tau\mathbf{B}/B^2)$ and $\boldsymbol{\kappa}=\nabla(2\mu_0 P+B^2)/2B^2-(\mathbf{B}/2B^4)(\mathbf{B}\cdot\nabla)B^2$. Then

$$\hat{T}_3 = \frac{\hat{T}_0}{\gamma P}-\nabla\cdot\left(\hat{\tau}\frac{\mathbf{B}}{B^2}\right)+2\hat{\boldsymbol{\xi}}\cdot\boldsymbol{\kappa} \tag{5.121}$$

is obtained. Here, by substituting the eikonal form (5.116) into (5.103), we obtain

$$-\rho\omega^2 \frac{|\nabla a|^2}{B^2}\hat{\eta} = \frac{i}{\varepsilon}k_{\perp B}\left(\hat{T}_0+\frac{B^2}{\mu_0}\hat{T}_3\right)+\frac{(\mathbf{B}\times\nabla a)}{B^2}\cdot\nabla\hat{T}_0$$
$$+\frac{\mathbf{B}}{\mu_0}\cdot\nabla\hat{T}_2+\frac{1}{\mu_0}(\mathbf{B}\times\nabla a)\cdot\nabla\hat{T}_3, \tag{5.122}$$

where we have used the notation

$$k_{\perp B} \equiv \frac{\mathbf{B}\times\nabla a}{B^2}\cdot\mathbf{k}_\perp. \tag{5.123}$$

When we keep the lowest-order quantities only in (5.122), we find that

$$\hat{T}_0+\frac{B^2}{\mu_0}\hat{T}_3 = 0. \tag{5.124}$$

Substituting (5.124) into (5.121) yields

$$\hat{T}_0 = \frac{\gamma P B^2}{\gamma\mu_0 P+B^2}\left[\nabla\cdot\left(\hat{\tau}\frac{\mathbf{B}}{B^2}\right)-2\hat{\boldsymbol{\xi}}\cdot\boldsymbol{\kappa}\right]. \tag{5.125}$$

By comparing this expression of $\hat{T}_0$ with (5.118),

$$\mathbf{k}_\perp\cdot\hat{\boldsymbol{\xi}} = 0 \tag{5.126}$$

in the lowest order of ε. By noting that

$$\hat{\boldsymbol{\xi}}\cdot\boldsymbol{\kappa} = \left(\hat{\boldsymbol{\xi}}_\perp+\hat{\tau}\frac{\mathbf{B}}{B^2}\right)\cdot\boldsymbol{\kappa}$$
$$= \hat{\xi}\frac{(\mathbf{k}_\perp\times\mathbf{B})}{B^2 k_{\perp B}}\cdot\boldsymbol{\kappa} = \hat{\xi}\frac{(\mathbf{B}\times\boldsymbol{\kappa})\cdot\mathbf{k}_\perp}{B^2 k_{\perp B}}, \tag{5.127}$$

we obtain

$$\hat{T}_0 = \frac{\gamma P B^2}{\gamma\mu_0 P+B^2}\left[\nabla\cdot\left(\hat{\tau}\frac{\mathbf{B}}{B^2}\right)-2\hat{\xi}\frac{(\mathbf{B}\times\boldsymbol{\kappa})\cdot\mathbf{k}_\perp}{B^2 k_{\perp B}}\right]. \tag{5.128}$$

In (5.127), we note the fact that $\boldsymbol{\kappa}$ is directed along the normal to the magnetic field line. Also, $\hat{\boldsymbol{\xi}}_\perp = \hat{\xi}(\mathbf{k}_\perp\times\mathbf{B})/(B^2 k_{\perp B})$ has been used as a consequence of (5.126). Then $\hat{T}_0$ is expressed in terms of two components of the vector $\hat{\xi}$, $\hat{\xi}$ and $\hat{\tau}$.

Since we have obtained expressions for $\hat{T}_0$ and $\hat{T}_3$, it is now possible to simplify (5.113). Here we note that $\hat{T}_0$, $\hat{T}_1$, $\hat{T}_2$, and $\hat{T}_3$ are of the same order in ε, $O(\varepsilon^0)$. By noting the inequalities (5.114), we retain only those terms including the $O(\varepsilon^{-1})$ contribution coming from the derivatives of T_0, T_1, T_2, and ξ in the direction perpendicular to $\mathbf{B}$. Thus (5.113) becomes

$$
\begin{aligned}
-\frac{\partial^2}{\partial t^2}\nabla\cdot\left[\frac{\rho}{B^2}(\mathbf{B}\times\tilde{\xi})\right] &= -\frac{\mathbf{B}}{\mu_0}\cdot\nabla\left(\frac{\nabla a\cdot\nabla T_2}{|\nabla a|^2}\right) \\
&+\frac{\mathbf{J}\cdot\mathbf{B}}{|\nabla a|^2}\frac{(\mathbf{B}\times\nabla a)}{B^2}\cdot\nabla T_2-(\mathbf{B}\times\nabla a)\cdot\nabla\left(\frac{\mu_0 K\xi+\mathbf{B}\cdot\nabla T_1}{\mu_0 B^2}\right) \\
&-2\frac{(\mathbf{B}\times\boldsymbol{\kappa})}{B^2}\cdot\nabla T_0. \qquad (5.129)
\end{aligned}
$$

By noting that T_0, T_1, T_2, and ξ include the phase factor $\exp(i\hat{S}/\varepsilon - i\omega t)$, here we consider a quantity $(ik_{\perp B}/\varepsilon)\exp(i\hat{S}/\varepsilon - i\omega t)$. The LHS of (5.129) is rewritten as

$$
\begin{aligned}
-\frac{\partial^2}{\partial t^2}\nabla\cdot\left(\frac{\rho\mathbf{B}}{B^2}\times\tilde{\xi}\right) &= \frac{ik_{\perp B}}{\varepsilon}\exp\left(\frac{i}{\varepsilon}\hat{S}-i\omega t\right)\rho\omega^2\frac{(\mathbf{k}_\perp\times\mathbf{B})}{k_{\perp B}B^2}\cdot\hat{\xi} \\
&= \frac{ik_{\perp B}}{\varepsilon}\exp\left(\frac{i\hat{S}}{\varepsilon}-i\omega t\right)\rho\omega^2\frac{|\mathbf{k}_\perp|^2\hat{\xi}}{k_{\perp B}^2 B^2}. \qquad (5.130)
\end{aligned}
$$

For the RHS of (5.129), we retain the terms including $\mathbf{k}_\perp$ only, and obtain

$$
\begin{aligned}
\rho\omega^2\frac{|\mathbf{k}_\perp|^2}{B^2k_{\perp B}^2}\hat{\xi} &= \frac{1}{\mu_0 k_{\perp B}}\mathbf{B}\cdot\nabla\left(\frac{\mathbf{k}_\perp\cdot\nabla a}{|\nabla a|^2}\hat{T}_2\right) \\
&+\frac{\mathbf{J}\cdot\mathbf{B}}{|\nabla a|^2}\hat{T}_2-K\hat{\xi}-\frac{\mathbf{B}}{\mu_0}\cdot\nabla\left(\frac{\mathbf{B}\cdot\nabla\hat{\xi}}{|\nabla a|^2}\right)-2\frac{(\mathbf{B}\times\boldsymbol{\kappa})\cdot\mathbf{k}_\perp}{B^2k_{\perp B}}\hat{T}_0 \qquad (5.131)
\end{aligned}
$$

after dividing by $(ik_{\perp B}/\varepsilon)\exp(i\hat{S}/\varepsilon - i\omega t)$.

Here we will introduce several equations to rewrite (5.131). By noting that $\mathbf{B}\cdot\nabla\hat{S} = 0$ and $\mathbf{B}\cdot\nabla a = 0$, we find that

$$
\mathbf{B}\cdot\nabla k_{\perp B} = \mathbf{B}\cdot\nabla\left[\frac{(\mathbf{B}\times\nabla a)}{B^2}\cdot\nabla\hat{S}\right] = 0. \qquad (5.132)
$$

Similarly, we obtain

$$
\begin{aligned}
\mathbf{B}\cdot\nabla\left(\frac{\mathbf{k}_\perp\cdot\nabla a}{|\nabla a|^2}\right) &= \mathbf{B}\cdot\nabla\left(\frac{\nabla a}{|\nabla a|^2}\cdot\nabla\hat{S}\right)\\
&= \mathbf{B}\cdot\left[\left(\frac{\nabla a}{|\nabla a|^2}\cdot\nabla\right)\nabla\hat{S}+(\nabla\hat{S}\cdot\nabla)\frac{\nabla a}{|\nabla a|^2}\right]\\
&= \frac{\nabla a}{|\nabla a|^2}\cdot(\mathbf{B}\cdot\nabla)\nabla\hat{S}+\nabla\hat{S}\cdot(\mathbf{B}\times\nabla)\frac{\nabla a}{|\nabla a|^2}\\
&= \frac{\nabla a}{|\nabla a|^2}\cdot\nabla(\mathbf{B}\cdot\nabla\hat{S})+\left[-\left(\frac{\nabla a}{|\nabla a|^2}\cdot\nabla\right)\mathbf{B}+(\mathbf{B}\cdot\nabla)\frac{\nabla a}{|\nabla a|^2}\right]\cdot\nabla\hat{S}\\
&= \frac{\nabla a}{|\nabla a|^2}\cdot\nabla(\mathbf{B}\cdot\nabla\hat{S})+\left[\left(\nabla\cdot\frac{\nabla a}{|\nabla a|^2}\right)\mathbf{B}-(\nabla\cdot\mathbf{B})\frac{\nabla a}{|\nabla a|^2}\right.\\
&\left.-\nabla\times\left(\mathbf{B}\times\frac{\nabla a}{|\nabla a|^2}\right)\right]\cdot\nabla\hat{S}. \qquad (5.133)
\end{aligned}
$$

Then, with the use of (5.94) and $\mathbf{B}\cdot\nabla\hat{S}=0$, (5.133) becomes

$$
\mathbf{B}\cdot\nabla\left(\frac{\mathbf{k}_\perp\cdot\nabla a}{|\nabla a|^2}\right) = -S\,\frac{\mathbf{B}\times\nabla a}{B^2}\cdot\mathbf{k}_\perp = -Sk_{\perp B}, \qquad (5.134)
$$

where S is given by (5.95). As a consequence of (5.126), from

$$
\hat{\xi}_\perp\times\nabla a = \hat{\xi}\,\frac{\mathbf{k}_\perp\cdot\nabla a}{B^2k_{\perp B}}\,\mathbf{B} \qquad (5.135)
$$

and

$$
\hat{\xi}_\perp\times\nabla a = -\hat{\eta}\,\frac{|\nabla a|^2}{B^2}\,\mathbf{B}, \qquad (5.136)
$$

a relation between $\hat{\xi}$ and $\hat{\eta}$ is given as

$$
\hat{\eta} = -\hat{\xi}\,\frac{\mathbf{k}_\perp\cdot\nabla a}{|\nabla a|^2k_{\perp B}}. \qquad (5.137)
$$

From (5.120) and (5.137),

$$
\begin{aligned}
\hat{T}_2 &= \frac{|\nabla a|^2}{B^2}\left[\mathbf{B}\cdot\nabla\left(-\hat{\xi}\,\frac{\mathbf{k}_\perp\cdot\nabla a}{|\nabla a|^2k_{\perp B}}\right)+\left(\frac{\mathbf{J}\cdot\mathbf{B}}{|\nabla a|^2}-\frac{S}{\mu_0}\right)\hat{\xi}\right]\\
&= -\frac{\mathbf{k}_\perp\cdot\nabla a}{B^2k_{\perp B}}(\mathbf{B}\cdot\nabla\hat{\xi})-\hat{\xi}\,\frac{|\nabla a|^2}{B^2}\,\mathbf{B}\cdot\nabla\left(\frac{\mathbf{k}_\perp\cdot\nabla a}{|\nabla a|^2k_{\perp B}}\right)+\hat{\xi}\,\frac{|\nabla a|^2}{B^2}\left(\frac{\mathbf{J}\cdot\mathbf{B}}{|\nabla a|^2}-\frac{S}{\mu_0}\right)\\
&= -\frac{\mathbf{k}_\perp\cdot\nabla a}{B^2k_{\perp B}}(\mathbf{B}\cdot\nabla\hat{\xi})+\frac{\mathbf{J}\cdot\mathbf{B}}{B^2}\,\hat{\xi} \qquad (5.138)
\end{aligned}
$$

is obtained. Here, in the last equality (5.132) and (5.134) have been used. By using (5.10), (5.132), (5.137), and (5.138), the first term of the RHS of (5.131) is written as

$$
\begin{aligned}
&\frac{1}{k_{\perp B}}\mathbf{B}\cdot\nabla\left(\frac{\mathbf{k}_\perp\cdot\nabla a}{|\nabla a|^2}\frac{\mathbf{J}\cdot\mathbf{B}}{B^2}\hat{\xi}\right)\\
&=\frac{\mathbf{k}_\perp\cdot\nabla a}{|\nabla a|^2 k_{\perp B}}\hat{\xi}\mathbf{B}\cdot\nabla\left(\frac{\mathbf{J}\cdot\mathbf{B}}{B^2}\right)-\frac{\mathbf{J}\cdot\mathbf{B}}{B^2}\mathbf{B}\cdot\nabla\hat{\eta}\\
&=2\frac{\mathbf{k}_\perp\cdot\nabla a}{|\nabla a|^2 k_{\perp B}}\hat{\xi}P'\frac{(\mathbf{B}\times\nabla a)\cdot\boldsymbol{\kappa}}{B^2}-\frac{\mathbf{J}\cdot\mathbf{B}}{B^2}\left[\frac{B^2}{|\nabla a|^2}\hat{T}_2-\left(\frac{\mathbf{J}\cdot\mathbf{B}}{|\nabla a|^2}-\frac{S}{\mu_0}\right)\hat{\xi}\right]\\
&=-\frac{\mathbf{J}\cdot\mathbf{B}}{|\nabla a|^2}\hat{T}_2+\frac{\mathbf{J}\cdot\mathbf{B}}{B^2}\left(\frac{\mathbf{J}\cdot\mathbf{B}}{|\nabla a|^2}-\frac{S}{\mu_0}\right)\hat{\xi}+2P'\hat{\xi}\frac{(\mathbf{B}\times\nabla a)\cdot\boldsymbol{\kappa}}{B^2}\frac{\mathbf{k}_\perp\cdot\nabla a}{|\nabla a|^2 k_{\perp B}}.
\end{aligned}
\tag{5.139}
$$

Here, the equation

$$
\mathbf{B}\cdot\nabla\left(\frac{\mathbf{J}\cdot\mathbf{B}}{B^2}\right)=2P'\frac{(\mathbf{B}\times\nabla a)}{B^2}\boldsymbol{\kappa}
\tag{5.140}
$$

obtained from the MHD equilibrium equations $\mathbf{J}\times\mathbf{B}=\nabla P$ and $\nabla\times(\mathbf{J}\times\mathbf{B})=(\mathbf{B}\cdot\nabla)\mathbf{J}-(\mathbf{J}\cdot\nabla)\mathbf{B}=0$ has been used.

Now, from (5.131) we obtain

$$
\begin{aligned}
\rho\omega^2\frac{|\mathbf{k}_\perp|^2}{B^2k_{\perp B}^2}\hat{\xi}&=\frac{1}{k_{\perp B}}\mathbf{B}\cdot\nabla\left[\frac{\mathbf{k}_\perp\cdot\nabla a}{|\nabla a|^2}\left\{\frac{\mathbf{J}\cdot\mathbf{B}}{B^2}\hat{\xi}-\frac{(\mathbf{k}_\perp\cdot\nabla a)}{\mu_0 B^2 k_{\perp B}}(\mathbf{B}\cdot\nabla\hat{\xi})\right\}\right]\\
&\quad+\frac{\mathbf{J}\cdot\mathbf{B}}{|\nabla a|^2}\left[\frac{\mu_0\mathbf{J}\cdot\mathbf{B}}{B^2}\hat{\xi}-\frac{(\mathbf{k}_\perp\cdot\nabla a)}{B^2k_{\perp B}}(\mathbf{B}\cdot\nabla\hat{\xi})\right]\\
&\quad-\left[\frac{\mu_0J^2}{|\nabla a|^2}+\frac{P'}{B^2}\left(\mu_0P'+\frac{\nabla a\cdot\nabla B^2}{|\nabla a|^2}\right)-\sigma S\right]\hat{\xi}\\
&\quad-\frac{\mathbf{B}}{\mu_0}\cdot\nabla\left(\frac{\mathbf{B}\cdot\nabla\hat{\xi}}{|\nabla a|^2}\right)-2\frac{(\mathbf{B}\times\boldsymbol{\kappa})\cdot\mathbf{k}_\perp}{B^2k_{\perp B}}\hat{T}_0\\
&=2P'\hat{\xi}\frac{\mathbf{B}\cdot(\mathbf{k}_\perp\times\boldsymbol{\kappa})}{B^2k_{\perp B}}-2\frac{(\mathbf{B}\times\boldsymbol{\kappa})\cdot\mathbf{k}_\perp}{B^2k_{\perp B}}\hat{T}_0\\
&\quad-\frac{\mathbf{B}}{\mu_0}\cdot\nabla\left(\frac{|\mathbf{k}_\perp|^2}{B^2k_{\perp B}^2}\mathbf{B}\cdot\nabla\hat{\xi}\right),
\end{aligned}
\tag{5.141}
$$

where (5.139) has been substituted into the first term in the RHS of the first equality. Also, from (5.102) it follows that

$$
\rho\omega^2\hat{\tau}=(\mathbf{B}\cdot\nabla)\hat{T}_0,
\tag{5.142}
$$

where $\mathbf{B}\cdot\mathbf{k}_\perp=0$ has been used.

Here, by substituting the value of $\hat{T}_0$ given by (5.128) into (5.141) and (5.142), we finally obtain two ballooning mode equations for $\hat{\xi}$ and $\hat{\tau}$:

$$\rho\omega^2 \frac{|\mathbf{k}_\perp|^2}{B^2 k_B^2} \hat{\xi} = -\frac{\mathbf{B}}{\mu_0} \cdot \nabla \left[\frac{|\mathbf{k}_\perp|^2}{B^2 k_{\perp B}^2} (\mathbf{B} \cdot \nabla)\hat{\xi} \right] - 2\frac{(\mathbf{B} \times \boldsymbol{\kappa}) \cdot \mathbf{k}_\perp}{B^2 k_{\perp B}} \left[P'\hat{\xi} + \frac{\gamma P B^2}{\gamma \mu_0 P + B^2} \times \times \left\{ \nabla \cdot \left(\hat{\tau} \frac{\mathbf{B}}{B^2} \right) - 2\hat{\xi} \frac{(\mathbf{B} \times \boldsymbol{\kappa}) \cdot \mathbf{k}_\perp}{B^2 k_{\perp B}} \right\} \right] \tag{5.143}$$

and

$$-\rho\omega^2 \hat{\tau} = \mathbf{B} \cdot \nabla \left[\frac{\gamma P B^2}{\gamma \mu_0 P + B^2} \left\{ \nabla \cdot \left(\hat{\tau} \frac{\mathbf{B}}{B^2} \right) - 2\hat{\xi} \frac{(\mathbf{B} \times \boldsymbol{\kappa}) \cdot \mathbf{k}_\perp}{B^2 k_{\perp B}} \right\} \right]. \tag{5.144}$$

The ballooning mode equation for the marginal stability of stellarators is often shown by

$$\mathbf{B} \cdot \nabla \left[\frac{|\mathbf{k}_\perp|^2}{B^2 k_{\perp B}^2} (\mathbf{B} \cdot \nabla)\hat{\xi} \right] + 2(1 - \Gamma) \frac{(\mathbf{B} \times \boldsymbol{\kappa}) \cdot \mathbf{k}_\perp \mu_0 P'}{B^2 k_{\perp B}} \hat{\xi} = 0, \tag{5.145}$$

where Γ is considered to be the eigenvalue to evaluate the growth rate for $0 < \Gamma < 1$ and $\Gamma = 0$ corresponds to marginal state.

This equation is most effectively solved in coordinate systems in which the $\mathbf{B} \cdot \nabla$ operator acquires its most simple representation. One particular choice is the magnetic coordinates (Ψ, χ, ϕ), where Ψ identifies the magnetic surface, χ is the poloidal angle-like variable that straightens the magnetic field lines, and ϕ is the toroidal angle. Then the magnetic field can be written as $\mathbf{B} = \nabla\alpha \times \nabla\Psi$, where $\alpha = \phi - q(\Psi)\chi$ is an angular variable that labels the magnetic field lines and $q(\Psi)$ is the safety factor. In these coordinates the wave vector can be expressed as

$$\frac{\mathbf{k}_\perp}{k_{\perp B}} = \nabla\alpha + \chi_k \nabla q = \nabla\phi - q(\Psi)\nabla\chi - \frac{dq}{d\Psi}(\chi - \chi_k)\nabla\Psi. \tag{5.146}$$

From the first equality in (5.146), we can identify χ_k as the radial wave number.

It should be noted that $\hat{\xi}$ in the ballooning mode equation (5.145) describes perturbation in the ballooning space $-\infty < \chi < \infty$. The basic idea of the ballooning representation is as follows. For simplicity, we consider the plasma displacement vector $\tilde{\xi}(\Psi, \chi, \phi)$ for axisymmetric tori as $\xi = \xi(\Psi, \chi)\exp(-in\phi)$, where n is a toroidal mode number. Physically acceptable displacements must satisfy the poloidal periodicity constraint $\xi(\Psi, \chi) = \xi(\Psi, \chi + 2\pi)$. Connor, Hastie, and Taylor (1978) pointed out that a periodic $\xi(\Psi, \chi)$ can be constructed from a series of nonperiodic functions, $\bar{\xi}(\Psi, \chi + 2\pi\ell)$, called quasimodes, defined over the infinite domain $-\infty < \chi < \infty$, where ℓ is an integer. A particular form is given by

$$\xi(\Psi, \chi) = \sum_{\ell=-\infty}^{\infty} \bar{\xi}(\Psi, \chi + 2\pi\ell). \tag{5.147}$$

The function $\xi(\Psi, \chi)$ is periodic with respect to χ if $\bar{\xi}$ satisfies appropriate convergence properties as $\chi \to \pm\infty$ and $\bar{\xi}(\Psi, \chi \to \pm\infty) \to 0$. The important property of the ballooning representation is that $\bar{\xi}$ satisfies the same equation as ξ. Here an unstable solution of the ballooning mode equation obtained numerically for a heliotron is shown in Fig. 5.1. Here $\hat{\chi}_k = 0$ is assumed for simplicity. The magnitude of B^2 and the eigenfunction $\hat{\xi}$ are plotted as a function of the poloidal angle θ instead of χ. This solution is fairly localized at $-\pi \leq \theta \leq \pi$, although (5.145) has been solved in the ballooning space.

5.4 THE MERCIER CRITERION

As explained in section 5.2, the interchange instability is similar to the Rayleigh–Taylor instability that appeared in the equilibrium state, in that a heavy liquid is supported by a light liquid in the presence of a gravitational force. In magnetically confined plasmas the gravity is replaced by the average

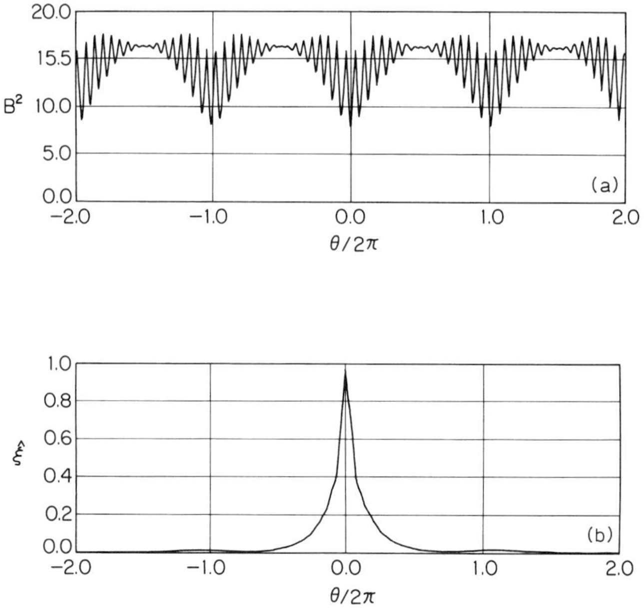

Fig. 5.1 (a) The oscillatory behavior of B^2 as a function of poloidal angle θ, and (b) a solution of the ballooning mode equation $\hat{\xi}$ in the ballooning space $-\infty < \theta < \infty$.

magnetic curvature. In the heliotron and torsatron it is given by $d\Omega/dr$ in the large aspect ratio limit (see (3.228)). When $d\Omega/dr$ is positive, the magnetic curvature is unfavorable and the interchange modes are destabilized for $(d\Omega/dr)(dP/dr) < 0$.

In tokamaks the magnetic curvature is governed by the toroidal effect. For a circular cross-section tokamak, the boundary between the favorable and unfavorable curvature to the interchange modes is given by $q = 1$. The interchange instabilities have two versions. One type is destabilized within the context of the ideal MHD model, and is called *ideal interchange modes*. In this case, shear stabilization can be expected to suppress the interchange modes. The balance between the driving force proportional to the unfavorable magnetic curvature and the shear stabilization gives a beta limit or an upper limit to the stably confined plasma pressure.

On the other hand, the interchange instability appears even in a finite-beta plasma below the beta limit, when the resistive MHD model governs the plasma dynamics. In this case, the shear stabilization disappears due to the effect of resistivity, and the plasma with unfavorable curvature is vulnerable to the interchange perturbation; this is referred to as *resistive interchange modes*. For both types of interchange modes, the magnetic well has a stabilizing effect, since it corresponds to a favorable magnetic curvature.

In this section, the stability criterion for the ideal interchange modes will be derived. There are several ways to obtain the stability criterion for toroidal plasmas, which is called the Mercier criterion. One interesting method to obtain it is to use the ballooning mode equation (5.145). Here, for simplicity, we show a derivation of the Mercier criterion for axisymmetric toroidal plasmas. It can be generalized to nonaxisymmetric toroidal plasmas by retaining the dependence on the toroidal angle in equilibrium quantities. In these cases the ballooning mode equation (5.145) becomes an ordinary differential equation along the magnetic field line and the eigenvalue Γ depends on the location of the magnetic field line on the magnetic surface.

The magnetic coordinates (Ψ, χ, ϕ) of an axisymmetric torus represent a transformation of the poloidal variables in the cylindrical coordinates (R, ϕ, Z) as

$$\begin{cases} \Psi = \Psi(R, Z) \\ \chi = \chi(R, Z) \\ \phi = \phi, \end{cases} \tag{5.148}$$

where Ψ is the flux function satisfying the Grad–Shafranov equation and χ is an orthogonal poloidal angle variable. Once Ψ is given, either analytically or numerically, one can then evaluate the magnetic fields and the curvature vector in the coordinates (R, ϕ, Z) as

$$\mathbf{B} = B_\phi \mathbf{e}_\phi + B_p \mathbf{b}_p, \tag{5.149}$$

$$\boldsymbol{\kappa} = (\mathbf{b} \cdot \nabla)\mathbf{b}, \tag{5.150}$$

where $B_\phi = F(\Psi)/R$, $\mathbf{B}_p = (\nabla\Psi/R) \times \mathbf{e}_\phi$, $B_p = |\nabla\Psi|/R$, $\mathbf{b}_p = \mathbf{B}_p/B_p$ and $\mathbf{b} = \mathbf{B}/B$. The poloidal angle variable χ is determined from the orthgonality condition $\nabla\Psi \cdot \nabla\chi = 0$, where χ is normalized such that one complete poloidal rotation corresponds to a change in χ of 2π.

The transformation of the volume element in axisymmetric tori is given by

$$d\mathbf{r} = 2\pi R dR dZ = 2\pi J d\Psi d\chi, \tag{5.151}$$

where the Jacobian J satisfies

$$\frac{R}{J} = (\nabla\chi \times \nabla\Psi) \cdot \mathbf{e}_\phi. \tag{5.152}$$

Since $R\mathbf{B}_p \cdot \nabla\chi = (\nabla\chi \times \nabla\Psi) \cdot \mathbf{e}_\phi$, the Jacobian has the form

$$JB_p = 1/(\mathbf{b}_p \cdot \nabla\chi). \tag{5.153}$$

It is convenient to introduce a set of locally orthogonal unit vectors by $(\mathbf{n}, \mathbf{t}, \mathbf{b})$, where

$$\begin{cases} \mathbf{n} = \dfrac{\nabla\Psi}{|\nabla\Psi|} \\ \mathbf{t} = \dfrac{B_\phi}{B}\mathbf{b}_p - \dfrac{B_p}{B}\mathbf{e}_\phi \\ \mathbf{b} = \dfrac{B_p}{B}\mathbf{b}_p + \dfrac{B_\phi}{B}\mathbf{e}_\phi. \end{cases} \tag{5.154}$$

It is noted that $\mathbf{n}$ is a normal vector with respect to magnetic surface, while $\mathbf{b}$ and $\mathbf{t}$ are tangential vectors lying in the magnetic surface parallel and perpendicular to the equilibrium magnetic field, respectively. Using these definitions one can write $\mathbf{k}$ as follows:

$$\begin{cases} \mathbf{k} = k_n\mathbf{n} + k_t\mathbf{t}, \\ k_n = \mathbf{n} \cdot \nabla\hat{S} = (\mathbf{n} \cdot \nabla)\dfrac{\partial\hat{S}}{\partial\Psi} \\ k_t = \mathbf{t} \cdot \nabla\hat{S} = (\mathbf{t} \cdot \nabla\chi)\dfrac{\partial\hat{S}}{\partial\chi} + (\mathbf{t} \cdot \mathbf{e}_\phi)\dfrac{1}{R}\dfrac{\partial\hat{S}}{\partial\phi}. \end{cases} \tag{5.155}$$

Similarly, one can obtain

$$\mathbf{b} \times \mathbf{k}_\perp = k_n\mathbf{n} - k_n\mathbf{t}. \tag{5.156}$$

The curvature can be decomposed as

$$\boldsymbol{\kappa} = \kappa_n\mathbf{n} + \kappa_t\mathbf{t}, \tag{5.157}$$

where $\kappa_n = \mathbf{n} \cdot \boldsymbol{\kappa}$ is the normal curvature and $\kappa_t = \mathbf{t} \cdot \boldsymbol{\kappa}$ is the geodesic curvature.

Since $\mathbf{k}_\perp = \nabla \hat{S}$, from (5.146) and (5.155), it is considered that

$$\begin{aligned} \frac{k_n}{k_{\perp B}} &= -\frac{dq}{d\Psi}(\chi - \chi_k)|\nabla\Psi| \\ &= -RB_p \int_{\chi_k}^{\chi} \frac{\partial}{\partial\Psi}\left(\frac{JB_\phi}{R}\right) d\chi' \equiv RB_p \hat{\chi}, \end{aligned} \tag{5.158}$$

$$\begin{aligned} \frac{k_t}{k_{\perp B}} &= -\frac{B_p}{RB} - \frac{B_\phi}{B}(\mathbf{b}_p \cdot \nabla\chi) q \\ &= -\frac{B}{RB_p} \end{aligned} \tag{5.159}$$

in the case of orthogonal coordinates (Ψ, χ, ϕ) for an axisymmetric torus. It should be noted that the safety factor $q(\Psi)$ is shown as

$$\begin{aligned} q(\Psi) &= \frac{1}{2\pi}\int_0^{2\pi} \hat{q}(\Psi, \chi) d\chi \\ &= \frac{1}{2\pi}\int_0^{2\pi} \frac{JB_\phi}{R} d\chi, \end{aligned} \tag{5.160}$$

since $J = 1/(\mathbf{B}_p \cdot \nabla\chi) \to r/B_\theta$ in the cylindrical model. The second equality of (5.158) implies a generalization of the first equality. In (5.159), q is replaced with $\hat{q} = JB_\phi/R$ by considering (5.153).

By noting that

$$\mathbf{b} \cdot \nabla X = \frac{1}{JB}\frac{\partial X}{\partial \chi} \tag{5.161}$$

for $X(\Psi, \chi)$, we can rewrite the ballooning mode equation (5.145) with $\Gamma = 0$ in the following form:

$$\frac{\partial}{\partial\chi}\left(f\frac{\partial\hat{\xi}}{\partial\chi}\right) - g\hat{\xi} = 0, \tag{5.162}$$

where

$$\begin{aligned} f &= \frac{k_n^2 + k_t^2}{JB^2 k_{\perp B}^2}, \\ g &= -\frac{2\mu_0 JRB_p}{B^2 k_{\perp B}^2}\frac{dP}{d\Psi}(k_t^2\kappa_n - k_t k_n \kappa_t). \end{aligned} \tag{5.163}$$

Here $P' = dP/d\Psi$ and $k_t/k_{\perp B} = -B/(RB_p)$ are substituted in the expression for g. It is easy to understand that secular algebraic terms proportional to χ^α as

$\chi \to \infty$ appear through the coefficient k_n from the first equality in (5.158), where α is a constant.

The algebraic and periodic behavior of f and g are separated by rewriting (5.162) as

$$\frac{\partial}{\partial \chi}\left[(a_0 \hat{\chi}^2 + a_1)\frac{\partial \hat{\xi}}{\partial \chi}\right] + \left(\frac{\partial c_0}{\partial \chi}\hat{\chi} + c_1\right)\hat{\xi} = 0, \tag{5.164}$$

where

$$\begin{cases} a_0 = \dfrac{R^2 B_p^2}{JB^2}, \\ a_1 = \dfrac{1}{JR^2 B_p^2}, \\ c_0 = \dfrac{\mu_0 (dP/d\Psi) F}{B^2}, \\ c_1 = \dfrac{2\mu_0 J (dP/d\Psi)\kappa_n}{RB_p}. \end{cases} \tag{5.165}$$

In deriving the coefficient $\partial c_0/\partial \chi$ the following relation has been used:

$$\begin{aligned} \kappa_t &= \mathbf{t} \cdot (\mathbf{b} \cdot \nabla)\mathbf{b} = -B\mathbf{t} \cdot \nabla(1/B), \\ &= (\mathbf{t} \cdot \nabla \chi)\left(\frac{1}{B}\frac{\partial B}{\partial \chi}\right) = \frac{F}{JRB_p B^2}\left(\frac{\partial B}{\partial \chi}\right), \end{aligned} \tag{5.166}$$

where we note $\mathbf{t} \cdot \nabla P = 0$ and the relation (5.153).

The goal now is to calculate the asymptotic solution for $\hat{\xi}$ as $\chi \to \infty$ and to determine the threshold for oscillatory behavior. The appropriate expansion for $\hat{\xi}$ can be written as

$$\hat{\chi} = \hat{\chi}^{\alpha}\left[\hat{\xi}_0 + \frac{\hat{\xi}_1}{\hat{\chi}} + \frac{\hat{\xi}_2}{\hat{\chi}^2} + \cdots\right], \tag{5.167}$$

where $\hat{\xi}_j(\Psi, \chi)$ are assumed to be periodic functions of χ with the same period as the equilibrium (2π). The parameter α is the indicial coefficient to be determined from (5.164). When α is imaginary or complex, $\hat{\xi}$ oscillates for large χ. If α is real, no oscillatory solutions exist. Thus the stability boundary defining the Mercier criterion corresponds to the transition value of α separating oscillatory and nonoscillatory behavior. This stability theorem has been given by Newcomb (1958).

For simplicity, we consider a cylindrical plasma column described by the equilibrium equation (5.63). The ideal MHD stability of the cylindrical plasma can be analyzed from a knowledge of the radial structure of $\xi(r)$, which is determined from the Euler–Lagrange equation,

$$\frac{d}{dr}\left(f\frac{d\xi}{dr}\right) - g\xi = 0, \tag{5.168}$$

for minimizing the potential energy δW_F given by (5.81). Here f and g are shown by (5.82) and (5.83). Equation (5.168) can be used to derive a set of necessary and sufficient conditions for the stability of arbitrary internal modes.

If the equilibrium magnetic surface has shear, then $F = \mathbf{k} \cdot \mathbf{B}$ is exactly zero on the resonant surface $r = r_s$, and no longer zero away from r_s. Here F is approximated as $F \simeq F'(r_s)(r - r_s)$. We now assume now that $\xi(r)$ is localized around the resonant surface. The leading order contributions to f and g are given by

$$f \simeq \left[\frac{rF'^2}{k_0^2}\right]\bigg|_{r=r_s} x^2 \tag{5.169}$$

$$g \simeq \left[\frac{2\mu_0 k^2 P'}{k_0^2}\right]\bigg|_{r=r_s}, \tag{5.170}$$

where $x = r - r_s$. Then the function ξ that minimizes δW_F satisfies the Euler–Lagrange equation

$$\frac{d}{dx}\left(x^2\frac{d\xi}{dx}\right) + D_s\xi = 0, \tag{5.171}$$

which has a solution

$$\xi = c_1 x^{s_1} + c_2 x^{s_2}, \tag{5.172}$$

where

$$D_s = -\left[\frac{2\mu_0 k^2 P'}{rF'^2}\right]\bigg|_{r=r_s} \tag{5.173}$$

and

$$s_{1,2} = -\tfrac{1}{2} \pm \tfrac{1}{2}(1 - 4D_s)^{1/2}. \tag{5.174}$$

Since at least one solution for ξ is always singular, one cannot simply use the solution (5.172) as a trial function to minimize δW_F. If one solves the eigenvalue problem including the inertia term for unstable modes as shown in section 5.6, the singularity will disappear.

The usefulness of the energy principle is that a simpler problem can be solved to study the stability instead of analyzing the full eigenvalue problem. Here we want to know whether or not ξ can be modified very near $x = 0$ so that ξ becomes a well-behaved and allowable trial function for $\delta W_F < 0$ (i.e., unstable).

For $1 - 4D_s < 0$ the roots are complex and the solution (5.172) is shown as

$$\xi = \frac{1}{|x|^{1/2}} [c_1 \sin(k_r \ell n|x|) + c_2 \cos(k_r \ell n|x|)], \tag{5.175}$$

where $k_r = (4D_s - 1)^{1/2}/2$. The solution as $x \to 0$ oscillates infinitely rapidly with a diverging envelope proportional to $1/|x|^{1/2}$. Here we show that it is possible to construct a well-behaved $\xi(x)$ such that $\delta W_F < 0$, because of the oscillatory behavior as shown in Fig. 5.2. In regions I and V, $\xi = 0$, so that $\delta W_F(I) = \delta W_F(V) = 0$. In regions II and IV, $\xi(x)$ satisfies (5.171). The corresponding contributions to δW_F are estimated by multiplying (5.171) by ξ and integrating over each region. Since either ξ or ξ' is zero at the end-points of regions II and IV, we find that

$$\begin{cases} \delta W_F(II) = \displaystyle\int_{x_1}^{x_2} (x^2 \xi'^2 - D_s \xi^2) dx = x^2 \xi \xi' \Big|_{x_1}^{x_2} = 0 \\ \delta W_F(IV) = \displaystyle\int_{x_3}^{x_4} (x^2 \xi'^2 - D_s \xi^2) dx = x^2 \xi \xi' \Big|_{x_3}^{x_4} = 0. \end{cases} \tag{5.176}$$

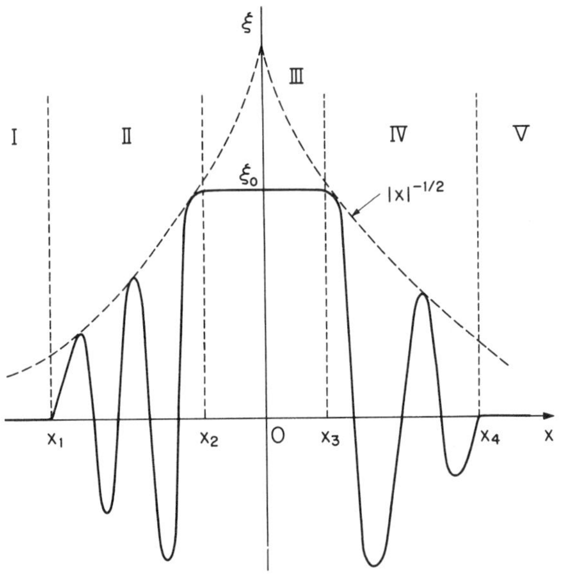

Fig. 5.2 The trial function for the unstable interchange mode in a cylindrical plasma.

The total contribution to δW_F comes from the region III where $\xi = \xi_0 = \text{constant}$,

$$\delta W_F = \int_{x_2}^{x_3} (x^2 \xi'^2 - D_s \xi^2) dx = -D_s \xi_0^2 (x_3 - x_2). \tag{5.177}$$

By the assumption of $D_s > 1/4$, in (5.177) $\delta W_F < 0$ and the system becomes unstable. When $1 - 4D_s > 0$ the roots of the indicial equation (5.174) are real and at least one solution is singular. In this case, oscillatory solutions do not exist and a localized, well-behaved trial function cannot be constructed. Therefore, if $1 - 4D_s > 0$, the equilibrium is stable to localized interchange perturbations.

If one uses

$$k = -\left[\frac{mB_\theta}{rB_z}\right]\Bigg|_{r=r_s}, \tag{5.178}$$

at the resonant surface of a cylindrical plasma, then an expression for D_s only in terms of the equilibrium quantities is given by

$$D_s = -\left[\frac{2\mu_0 P'}{rB_z^2}\left(\frac{q}{q'}\right)^2\right]\Bigg|_{r=r_s}, \tag{5.179}$$

where $q = rB_z/(R_0 B_\theta)$ is a safety factor and R_0 is the major radius of the equivalent torus. By using (5.178) and (5.179), the usual form of the Suydam criterion is given from $D_s < 1/4$:

$$\frac{1}{4}\left(\frac{q'}{q}\right)^2 + \frac{2\mu_0 P'}{rB_z^2} > 0. \tag{5.180}$$

The Suydam criterion indicates that the stability boundary is determined by two competing effects: the interchange drive term from the combination of a negative pressure gradient and an unfavorable curvature of the B_θ field, and the shear stabilization term proportional to $(q')^2$, which represents the work done in bending the field line when interchanging two flux tubes.

To determine the value of α for the solution of (5.167), it is substituted into (5.164). The resulting equation has the form of a polynomial series in $\hat{\chi}$. By equating the coefficients of descending powers of $\hat{\chi}$ to zero and making use of appropriate periodicity requirements, one obtains a solution for α and $\hat{\xi}_j$.

The zeroth-order equation corresponding to the highest power of $\hat{\chi}^{\alpha+2}$ is given by

$$\frac{\partial}{\partial \chi}\left(a_0 \frac{\partial \hat{\xi}_0}{\partial \chi}\right) = 0, \tag{5.181}$$

which has a solution $\hat{\xi}_0 =$ constant or $\hat{\xi}_0 = 1$. The first-order equation corresponding to $\hat{\chi}^{\alpha+1}$ can be written as

$$\frac{\partial}{\partial \chi}\left[a_0\left(\frac{\partial \hat{\xi}_1}{\partial \chi} + \alpha \frac{\partial \hat{q}}{\partial \Psi}\right)\right] + \frac{\partial c_0}{\partial \chi} = 0, \tag{5.182}$$

which yields

$$\frac{\partial \hat{\xi}_1}{\partial \chi} + \alpha \frac{\partial \hat{q}}{\partial \Psi} = \frac{K - c_0}{a_0} \tag{5.183}$$

by integration with respect to χ, where $K(\Psi)$ is a free integration function. This function K is determined by requiring that $\hat{\xi}_1$ is periodic with respect to χ. Thus, integration of (5.183) over one period in χ gives

$$K(\Psi) = \langle c_0 \rangle + \alpha \frac{dq}{d\Psi} \langle a_0 \rangle, \tag{5.184}$$

where the averaging procedure is defined by

$$\langle Q \rangle \equiv \int_0^{2\pi} \frac{Q d\chi}{a_0} \Big/ \int_0^{2\pi} \frac{d\chi}{a_0} \tag{5.185}$$

and

$$2\pi = \langle a_0 \rangle \int_0^{2\pi} \frac{d\chi}{a_0}. \tag{5.186}$$

Next, the second-order equation corresponding to $\hat{\chi}^{\alpha}$ is obtained as

$$\frac{\partial}{\partial \chi}\left\{a_0\left[\frac{\partial \hat{\xi}_2}{\partial \chi} + (\alpha - 1)\frac{\partial \hat{q}}{\partial \Psi}\hat{\xi}_1\right]\right\} + (\alpha + 1)a_0 \frac{\partial \hat{q}}{\partial \Psi}\left(\frac{\partial \hat{\xi}_1}{\partial \chi} + \alpha \frac{\partial \hat{q}}{\partial \Psi}\right) + c_1 + \frac{\partial c_0}{\partial \chi}\hat{\xi}_1 = 0. \tag{5.187}$$

Here it is noted that the solution of $\hat{\xi}_2$ is not explicitly required. The indicial coefficient α is determined as a periodicity constraint by integrating (5.187) over one period in χ. By noting that the contribution from the first term in (5.187) vanishes if both $\hat{\xi}_1$ and $\hat{\xi}_2$ are periodic, we obtain

$$\int_0^{2\pi}\left[(\alpha + 1)a_0 \frac{\partial \hat{q}}{\partial \chi}\left(\frac{\partial \hat{\xi}_1}{\partial \chi} + \alpha \frac{\partial \hat{q}}{\partial \Psi}\right) + c_1 - c_0 \frac{\partial \hat{\xi}_1}{\partial \chi}\right] d\chi = 0. \tag{5.188}$$

Substituting $\partial \hat{\xi}_1 / \partial \chi$ from (5.183) and using (5.184) leads to the indicial equation

$$\alpha^2 + \alpha + D_M = 0, \tag{5.189}$$

where

$$D_M = \left[\langle a_0 c_1\rangle + \langle c_0^2\rangle - \left\langle a_0 c_0 \frac{\partial\hat{q}}{\partial\Psi}\right\rangle - \langle c_0\rangle\left(\langle c_0\rangle - \left\langle a_0 \frac{\partial\hat{q}}{\partial\Psi}\right\rangle\right)\right] \Big/ \left(\langle a_0\rangle \frac{dq}{d\Psi}\right)^2. \tag{5.190}$$

We note that $\langle a_0 \partial\hat{q}/\partial\Psi\rangle = (dq/d\Psi)\langle a_0\rangle$. The root of (5.189) is

$$\alpha = -\tfrac{1}{2} \pm \tfrac{1}{2}(1 - 4D_M)^{1/2}. \tag{5.191}$$

The transition from oscillatory to nonoscillatory behavior occurs for $D_M = 1/4$. Consequently, the Mercier criterion for interchange modes is given by

$$D_M < 1/4 \tag{5.192}$$

for stability. When we use the physical variables, expression (5.190) becomes

$$D_M = \left[\mu_0 \frac{dP}{d\Psi} \Big/ (dq/d\Psi)^2\right] \times$$
$$\times \left[2\left\langle\frac{RB_p\kappa_n}{B^2}\right\rangle + \left\langle\frac{G}{B^4}\right\rangle - \left\langle\frac{1}{B^2}\right\rangle\left\langle\frac{G}{B^2}\right\rangle\right] \Big/ \left(\left\langle\frac{R^2B_p^2}{JB^2}\right\rangle\right)^2, \tag{5.193}$$

where

$$G = F\left(\mu_0 \frac{dP}{d\Psi} F - \frac{R^2 B_p^2}{J} \frac{\partial\hat{q}}{\partial\Psi}\right) \tag{5.194}$$

and

$$\langle Q\rangle = \int_0^{2\pi} \left(\frac{QB^2}{R^2B_p^2}\right) J d\chi \Big/ \int_0^{2\pi} \left(\frac{B^2}{R^2B_p^2}\right) J d\chi. \tag{5.195}$$

The Mercier criterion is a function only of the equilibrium quantities and it can be tested immediately after the MHD equilibrium is obtained from the Grad–Shafranov equation. To guarantee the ideal interchange stability, it must be satisfied on each flux surface. Since $dP/d\Psi < 0$ for normal pressure profiles, $\langle RB_p\kappa_n/B^2\rangle > 0$ is desirable to suppress the interchange instability, which corresponds to the good normal curvature of the magnetic field line. In Fig. 5.3 is shown the Mercier unstable region (the shaded region) in the $(\beta_0, \bar{r}/\bar{a})$ plane for Heliotron DR, where β_0 is a central beta value, $\bar{r}$ is an average radius of the noncircular magnetic surface $\Psi = \text{constant}$, and $\bar{a}$ corresponds to the outermost magnetic surface. Here the pressure profile is assumed to be

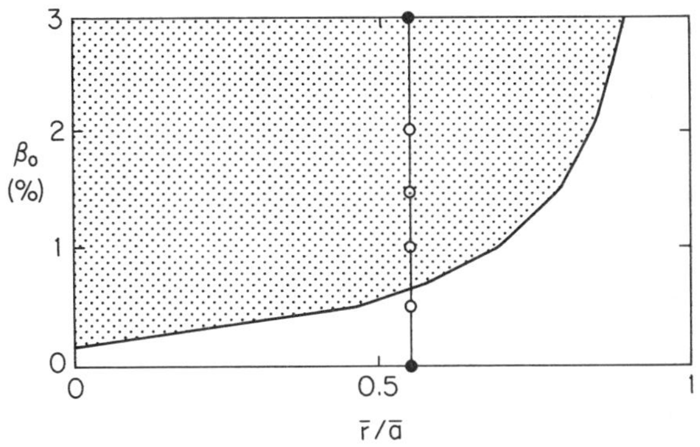

Fig. 5.3 The Mercier criterion in the $(\beta_0, \bar{r}/\bar{a})$ plane for the Heliotron DR. The $\iota = 1$ position is shown by circles. The shaded part is the Mercier unstable region of the Heliotron DR.

$P = P_0(1 - \Psi)^2$. Since the favorable magnetic curvature region is limited to being only near the magnetic axis, first the inner region becomes unstable with an increase in β_0. However, the edge region retains its stability, since the magnetic shear in heliotrons is sufficiently high.

5.5 THE RESISTIVE INTERCHANGE INSTABILITY

Here, we consider the resistive MHD model in the limit of small resistivity, $\eta \to 0$. Since our interest is in unstable modes with slow growth rates, we consider time variations such that $\partial/\partial t \sim \eta^{1/3}$. Ideal modes are independent of resistivity when it is small. Resistive diffusion becomes negligible in this limit. The reason for choosing $\eta^{1/3}$ is shown in the following.

In studying perturbations around the MHD equilibrium, we consider the displacement vector $\boldsymbol{\xi}$ and the perturbed magnetic field $\mathbf{B}_1$. Our coordinates in this section are the cylindrical ones (r, θ, z). The equilibrium quantities are independent of t, θ, and z, and we can introduce

$$\boldsymbol{\xi}(r, \theta, z, t) = \boldsymbol{\xi}(r) \exp\left(\gamma_g t + im\theta - i\frac{n}{R}z\right). \tag{5.196}$$

Here γ_g denotes the growth rate (which should not be confused with the ratio of specific heats). $2\pi R$ corresponds to a periodic length of cylindrical plasma. The linearized MHD equation for the perturbations can be written as

$$\rho_0 \gamma_g^2 \boldsymbol{\xi} = \frac{1}{\mu_0}(\nabla \times \mathbf{B}_1) \times \mathbf{B}_0 + \mathbf{J}_0 \times \mathbf{B}_1 + \nabla[\gamma P_0(\nabla \cdot \boldsymbol{\xi}) + \boldsymbol{\xi} \cdot \nabla P_0], \tag{5.197}$$

$$\mathbf{B}_1 - \frac{\eta}{\gamma_g \mu_0} \nabla \times (\nabla \times \mathbf{B}_1) = \nabla \times (\boldsymbol{\xi} \times \mathbf{B}_0). \tag{5.198}$$

When the resistivity η vanishes, these reduce to the equations for ideal MHD modes. Now we consider the boundary layer in the neighborhood of rational surface $r = r_s$, which is defined by requiring that all three terms in (5.198) should be comparable for the length scale $(r - r_s)$ and the growth rate γ_g or $\gamma_g(r - r_s)^2 \sim \eta$. Here r_s denotes the rational surface in the cylindrical model. This condition corresponds to the fact that the boundary layer thickness is about a resistive skin depth on the time scale of the growth rate. We need another relation between the time scale and the length scale. In MHD stability analyses, it is determined from the shear Alfvén wave dispersion relation, $\omega^2 = k_\parallel^2 V_A^2$, which suggests $\rho_0\omega^2 \simeq \mathbf{B}_0 \cdot \nabla(\mathbf{B}_0 \cdot \nabla) \sim (F')^2(r - r_s)^2$ in (5.197), where $F \equiv mB_{0\theta}/r - nB_{0z}/R$ and $F' = -B_{0\theta}nq'/r_s$ for the cylindrical equilibrium with $\mathbf{B}_0 = (0, B_{0\theta}(r), B_{0z}(r))$ and $B_{0\theta} \ll B_{0z}$. The prime denotes a derivative with respect to r. Then we find that (5.198) gives

$$(r - r_s) \sim \gamma_g \sim \eta^{1/3} \sim \varepsilon \ll 1 \tag{5.199}$$

for $\omega \sim \gamma_g$, where ε is an ordering parameter. The vector perturbed quantities are projected along three directions in the following way,

$$\xi = (\xi_r^{(1)} + \cdots)\mathbf{e}_r + \frac{\mathbf{B}_0 \times \mathbf{e}_r}{B_0}(\xi_\perp^{(0)} + \xi_\perp^{(1)} + \cdots) + \frac{\mathbf{B}_0}{B_0}(\xi_b^{(0)} + \xi_b^{(1)} + \cdots), \tag{5.200}$$

$$\mathbf{B}_1 = (B_{1r}^{(2)} + \cdots)\mathbf{e}_r + \frac{\mathbf{B}_0 \times \mathbf{e}_r}{B_0}(B_{1\perp}^{(1)} + B_{1\perp}^{(2)} + \cdots) + \frac{\mathbf{B}_0}{B_0}(B_{1b}^{(1)} + B_{1b}^{(2)} + \cdots), \tag{5.201}$$

by retaining the low-order components with respect to the ordering parameter ε. Furthermore, we assume that

$$\nabla \cdot \xi = \varepsilon(\nabla \cdot \xi)^{(1)} + \cdots. \tag{5.202}$$

In order to understand the orderings in the expressions (5.200) and (5.201), we consider the divergence-free displacement vector satisfying (5.202):

$$\begin{aligned}(\nabla \cdot \xi)^{(0)} &= \frac{1}{r}\frac{\partial}{\partial r}(r\xi_r^{(1)}) + \frac{i}{rB_0}\left(mB_{0z} + \frac{n}{R}rB_{0\theta}\right)\xi_\perp^{(0)} + \frac{i}{rB_0}\left(mB_{0\theta} - \frac{n}{R}rB_{0z}\right)\xi_b^{(0)} \\ &\simeq \frac{\partial\xi_r^{(1)}}{\partial r} + \frac{i}{r_sB_0}\left(mB_{0z} + \frac{n}{R}r_sB_{0\theta}\right)\xi_\perp^{(0)} = 0,\end{aligned} \tag{5.203}$$

since $\partial/\partial r \sim O(\varepsilon^{-1})$ and $mB_{0\theta} - nrB_{0z}/R = 0$ at $r = r_s$. This relation explains the relative ordering between $\xi_r^{(1)}$ and $\xi_\perp^{(0)}$. We also note that the lowest order of ξ_b is comparable to that of $\xi_\perp$ or $\xi_b^{(0)} \sim \xi_\perp^{(0)}$, since ξ_θ and ξ_z are compsed with $\xi_\perp$ and ξ_b. The condition $\nabla \cdot \mathbf{B}_1 = 0$ yields a similar relation to (5.203) between $B_{1r}^{(2)}$ and $B_{1\perp}^{(1)}$. Furthermore, the lowest order of B_{1b} satisfies $B_{1b}^{(1)} \sim B_{1\perp}^{(1)}$. The radial component of (5.198) is written as

$$B_{1r}^{(2)} - \frac{\eta}{\gamma_g \mu_0} \frac{\partial^2}{\partial r^2} B_{1r}^{(2)} = \mathbf{B}_0 \cdot \nabla \xi_r^{(1)}$$
$$= -i \frac{B_{0\theta}}{r_s} nq'(r - r_s)\xi_r^{(1)} \tag{5.204}$$

in the lowest order with $O(\varepsilon^2)$ by noting the ordering given by (5.199). Equation (5.204) justifies the claim that the lowest order of B_{1r} is $O(\varepsilon^2)$.

Now we consider the equation of motion (5.197). We assume that the pressure term γP_0 is negligible. This is allowed for the low-beta plasmas in the cylindrical model. However, dropping this term is not always justifiable. First, we take the radial component of (5.197). We can see that the term $\partial(\boldsymbol{\xi} \cdot \nabla P_0)/\partial r$ is much larger than the inertial term. Both the first and second terms on the RHS of (5.197) are written as

$$(\nabla \times \mathbf{B}_1) \times \mathbf{B}_0 + \mu_0 \mathbf{J}_0 \times \mathbf{B}_1 = \nabla \cdot (\mathbf{B}_0\mathbf{B}_1 + \mathbf{B}_1\mathbf{B}_0 - \mathbf{B}_0\mathbf{B}_1 \overleftrightarrow{I}) \simeq \varepsilon \nabla \left(B_0 B_{1b}^{(1)} \right). \tag{5.205}$$

Thus, to the lowest order, the equation of motion gives

$$\frac{1}{\mu_0} B_0 B_{1b}^{(1)} = \xi_r^{(1)} \frac{dP_0}{dr}. \tag{5.206}$$

This corresponds to the fact that the perturbed fluid and magnetic pressures approximately balance in the direction perpendicular to the field lines. If this is violated, fast magnetosonic waves will be excited.

In order to obtain information from the equation of motion, we must consider higher-order quantities. A convenient way to do this is to use an operator that annihilates the lowest-order quantities in the full equation of motion. The most convenient form of an annihilator is $\nabla \cdot (\mathbf{B}_0/B_0^2)\times$ and each term is written as

$$\nabla \cdot \frac{\mathbf{B}_0}{B_0^2} \times (\rho_0 \gamma_g^2 \boldsymbol{\xi}) \simeq -\frac{1}{r}\frac{d}{dr}[\rho_0 \gamma_g^2 r \xi_\perp^{(0)}/B_0]$$
$$\simeq -\frac{\rho_0 \gamma_g^2}{B_0} \frac{\partial}{\partial r} \xi_\perp^{(0)} \simeq \frac{i r_s}{m B_{0z} + n r_s B_{0\theta}/R} \rho_0 \gamma_g^2 \frac{\partial^2 \xi_r^{(1)}}{\partial r^2}, \tag{5.207}$$

$$\nabla \cdot \frac{\mathbf{B}_0}{B_0^2} \times [(\nabla \times \mathbf{B}_1) \times \mathbf{B}_0 + \mu_0 \mathbf{J}_0 \times \mathbf{B}_1]$$
$$= \mu_0 \mathbf{J}_0 \cdot \nabla \frac{B_{1b}^{(1)}}{B_0} - \mu_0 \mathbf{B}_1 \cdot \nabla \frac{\mathbf{J}_0 \cdot \mathbf{B}_0}{B_0^2} - \mathbf{B}_0 \cdot \nabla \left(\frac{(\nabla \times \mathbf{B}_1) \cdot \mathbf{B}_0}{B_0^2} \right), \tag{5.208}$$

and

$$\nabla \cdot \left[\frac{\mathbf{B}_0}{B_0^2} \times \nabla(\xi \cdot \nabla P_0) \right] = \frac{\mu_0}{B_0^2} \mathbf{J}_0 \cdot \nabla(\xi \cdot \nabla P_0) + \frac{\mathbf{B}_0 \times \nabla B_0^2}{B_0^4} \cdot \nabla(\xi \cdot \nabla P_0). \tag{5.209}$$

In the last expression of (5.207), (5.203) has been substituted. The lowest order of the first term of (5.208) combines with (5.209) to yield

$$\frac{1}{B_0}\left(2\mathbf{J}_0 + \frac{\mathbf{B}_0 \times \nabla B_0^2}{\mu_0 B_0^2}\right) \cdot \nabla B_{1b}^{(1)} \simeq \frac{\mathbf{B}_0}{B_0^3} \times \nabla\left(2P_0 + \frac{B_0^2}{\mu_0}\right) \cdot \nabla B_{1b}^{(1)}. \qquad (5.210)$$

In deriving this expression we have used (5.206) to express every term with $B_{1b}^{(1)}$, and we have expressed the current density as

$$\mathbf{J}_0 = \frac{\mathbf{B}_0 \times \nabla P_0}{B_0^2} + \frac{\mathbf{J}_0 \cdot \mathbf{B}_0}{B_0^2}\mathbf{B}_0 \qquad (5.211)$$

and dropped the gradient parallel to the field line since it is sufficiently small. The important fact in (5.210) is that the quantity $\nabla(2P_0 + B_0^2/\mu_0)$ is proportional to the curvature of the magnetic field lines. It represents the destabilizing force in terms of field line curvature when the curvature is unfavorable.

The second term in (5.208) is negligible in the expansion with respect to ε by noting that $B_{1r}^{(2)} \ll B_{1b}^{(1)}$, while the third term yields

$$\begin{aligned}\mathbf{B}_0 \cdot \nabla\left(\frac{\mathbf{B}_0 \cdot (\nabla \times \mathbf{B}_1)}{B_0^2}\right) &\simeq \mathbf{B}_0 \cdot \nabla\left[\frac{\mathbf{B}_0}{B_0^2} \cdot \left(\mathbf{e}_r \times \frac{\mathbf{B}_0 \times \mathbf{e}_r}{B_0}\right)\frac{\partial B_{1\perp}^{(1)}}{\partial r}\right] \\ &\simeq \mathbf{B}_0 \cdot \nabla\left[-\frac{ir_s}{mB_{0z} + nr_s B_{0\theta}/R}\frac{\partial^2 B_{1r}^{(2)}}{\partial r^2}\right],\end{aligned} \qquad (5.212)$$

where an expression similar to (5.203), obtained from $\nabla \cdot \mathbf{B}_1 = 0$, has been used. Also, we note that only the term with the radial derivative will be sufficiently large to make a significant contribution. Combining (5.207), (5.210) and (5.212) yields

$$\begin{aligned}\rho_0 \gamma_g^2 \frac{\partial^2 \xi_r^{(1)}}{\partial r^2} &= \frac{2}{\mu_0}\frac{B_{0\theta}^2}{r_s^3 B_0^3}\left(mB_{0z} + \frac{nr_s}{R}B_{0\theta}\right)^2 B_{1b}^{(1)} \\ &\quad - i\frac{B_{0\theta}}{r_s}nq'(r - r_s)\frac{\partial^2 B_{1r}^{(2)}}{\partial r^2} \\ &= \frac{2m^2 B_{0\theta}^2}{\mu_0 r_s^3 B_{0z}}B_{1b}^{(1)} - i\frac{B_{0\theta}}{r_s}nq'(r - r_s)\frac{\partial^2 B_{1r}^{(2)}}{\partial r^2},\end{aligned} \qquad (5.213)$$

where we have used the relation of $B_0 \sim B_{0z} \gg B_{0\theta}$ which is required to realize the rational surface satisfying $q = m/n$ at r_s. Here we have also used the equilibrium relation $(P_0 + B_0^2/2\mu_0)' = -B_{0\theta}^2/(\mu_0 r)$ in the cylindrical plasma.

Now (5.204), (5.206), and (5.213) are closed for variables $\xi_r^{(1)}$, $B_{1r}^{(2)}$, and $B_{1b}^{(1)}$. It is noted again that the effect of the γP_0 term in (5.197) is assumed to be negligible.

Before analyzing these closed equations, we will introduce a lot of dimensionless parameters. First, we define a dimensionless growth rate Q as

$$\gamma_g \equiv \left(\frac{\eta n^2 q'^2 B_{0\theta}^2}{\mu_0 \rho_0 r_s^2}\right)^{1/3} Q \tag{5.214}$$

and a dimensionless length x as

$$r - r_s = L_R x, \quad L_R \equiv \left(\frac{\eta^2 \rho_0 r_s^2}{\mu_0^2 n^2 q'^2 B_{0\theta}^2}\right)^{1/6}. \tag{5.215}$$

Here it is convenient to scale the perturbed magnetic field:

$$B_{1r}^{(2)} = -i\frac{L_R n q' B_{0\theta}}{r_s}\Psi. \tag{5.216}$$

By using these dimensionless quantities, (5.204) and (5.213) become

$$\Psi'' = Q(\Psi - x\xi), \tag{5.217}$$

$$Q^2\xi'' = -D_s\xi - x\Psi'', \tag{5.218}$$

or

$$Q^2\xi'' = -D_s\xi + Qx^2\xi - Qx\Psi, \tag{5.219}$$

where $\xi \equiv \xi_r^{(1)}$, a prime denotes a derivative with respect to x, and D_s, given by (5.179), is a measure of the magnetic field line curvature and comes from the RHS of (5.210).

The most important approximation is that the resistive layer is localized, i.e., $L_R \ll a$, where a is the radius of the cylindrical plasma column. In fact, L_R turns out to be of the order of millimeters in tokamaks and stellarators. This is indeed very localized compared to $a \simeq 20 - 100\,\text{cm}$. However, a situation arises in which the fluid equations become invalid, when the resistive layer becomes less than the ion Larmor radius. Here we will pay attention to the solutions of this set of equations under the fluid approximation. A very useful technique for resistive MHD stability theory has been the use of the Fourier transform. Note that the equations are of fourth order in d/dx but only of second order in x. Thus, by introducing

$$\xi = \int_{-\infty}^{\infty} d\mu \zeta(\mu) \exp(i\mu x), \tag{5.220}$$

(5.217) and (5.218) reduce to

$$\frac{d}{d\mu}\frac{\mu_2}{\mu^2 + Q}\frac{d\zeta}{d\mu} - Q\left(\mu^2 - \frac{D_s}{Q^2}\right)\zeta = 0. \tag{5.221}$$

This equation is singular at $\mu \to 0$, where it behaves as

$$\begin{cases} \zeta \propto \mu^s \\ \zeta \propto \mu^{-1-s}, \end{cases} \tag{5.222}$$

with s given by (5.174) for the plus sign case. Equation (5.221) is also singular for $\mu \to \infty$, with the behavior

$$\zeta \propto \exp\left(\pm\tfrac{1}{2}Q^{1/2}\mu^2\right). \tag{5.223}$$

Here it is pointed out that there is a sequence of exact solutions of (5.221),

$$\zeta = \mu^s \exp(-\tfrac{1}{2}Q^{1/2}\mu^2)\sum_{j=0}^{\ell} a_j\mu^{2j}, \tag{5.224}$$

where the summation is trancated at $j = \ell$. By imposing the condition that ζ does not diverge at $\mu \to \infty$, a well posed eigenvalue problem is given, and the eigenvalues are given by

$$Q^{3/2} = \frac{D_s}{s + 2\ell + \frac{1}{2} + [4s\ell + (2\ell + \frac{1}{2})^2]^{1/2}}. \tag{5.225}$$

Substituting (5.224) into (5.221) yields

$$\begin{aligned} \sum_{j=0}^{\ell}\Bigg[&\mu^4\left\{Q^{1/2}(2s+4j+1) - Q^2 + \frac{D_s}{Q}\right\} \\ &+ \mu^2\{(s+2j)(s+2j-1) - Q^{3/2}(2s+4j+3) - Q^3 + 2D_s\} \\ &+ \{Q(s+2j)(s+2j+1) + QD_s\}\Bigg]a_j\mu^{s+2j}\exp(-\tfrac{1}{2}Q^{1/2}\mu^2) = 0. \end{aligned} \tag{5.226}$$

We note that the term proportional to μ^s (or the $j = 0$ term) identically vanishes, since s satisfies $s(s+1) + D_s = 0$, which gives (5.174). We note that there are $(\ell + 3)$ equations from the coefficients for terms with $\mu^{s+2m}(m = 0, 1, \cdots, \ell + 2)$. In order to obtain the coefficients $a_j(j = 0, 1, \cdots, \ell)$ the coefficient of the leading term proportional to $\mu^{s+2\ell+4}$ should vanish,

$$Q^{-3}D_s - Q^{-3/2}(2s + 4\ell + 1) - 1 = 0, \tag{5.227}$$

which gives the dispersion relation (5.225). Then the number of equations becomes equal to $(\ell + 1)$, which is the number of coefficients $a_j(j = 0, 1, \cdots, \ell)$. Here ℓ is an integer that depends on the number of terms in the truncated series. This procedure works when D_s is positive so that $Q^{1/2}$ is positive and the solution decays at $\mu \to \infty$. This shows an infinite sequence of unstable resistive interchange modes. Here the instability criterion for the resistive interchange mode is $D_s > 0$.

By examining the spatial parity of (5.217) and (5.218) (or (5.219)) it is found that there exist even and odd solutions (see Fig. 5.4). It is also noted that the eigenvalues given by (5.225) are degenerate, and valid for both even and odd solutions.

Now we will discuss the matching conditions in resistive mode theory. By treating the matching conditions, we will obtain a more precise understanding of the resistive stability criterion.

The overall behavior of the perturbation can be sketched as shown in Fig. 5.5. Region I is the outer region closest to the magnetic axis. Generally, when treating boundary layers, the boundary layer is called the inner region, and the rest is called the outer region. Here, however, we have two outer regions, one of which is inside the rational surface. We simply call it region I. Region II is the resistive boundary layer, and region III is the other outer region. The solution is region I near the boundary layer approaches the form

$$\xi \simeq A_I \left|\frac{r - r_s}{a}\right|^s + B_I \left|\frac{r - r_s}{a}\right|^{-1-s}, \tag{5.228}$$

where the length scale a has been introduced to make the constants A_I and B_I dimensionless and s is given by (5.174) for the plus sign case. In ideal MHD stability theory, the first term is called the small solution and the second term the large solution. Similarly, in region II the solution approaches, for large negative x,

$$\xi \simeq A_{IIL}|x|^s + B_{IIL}|x|^{-1-s}, \tag{5.229}$$

where $x \equiv (r - r_s)/L_R$ and L_R is defined by (5.215), and the suffix $L(R)$

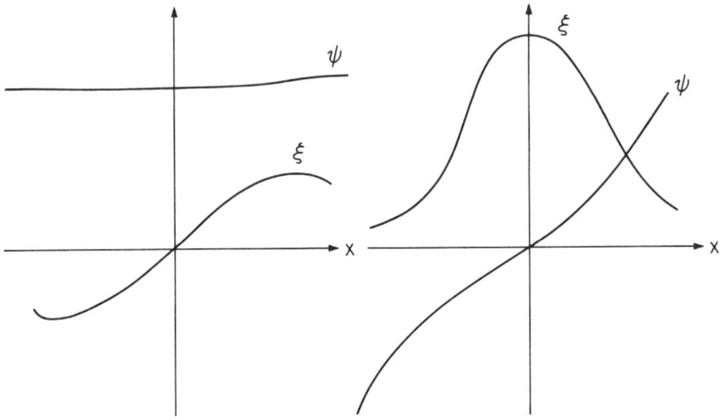

Fig. 5.4 The left figure shows the odd mode defined by $\xi(x) = -\xi(-x)$ and the right figure shows the even mode defined by $\xi(x) = \xi(-x)$. It is noted that Ψ is even (odd) when ξ is odd (even).

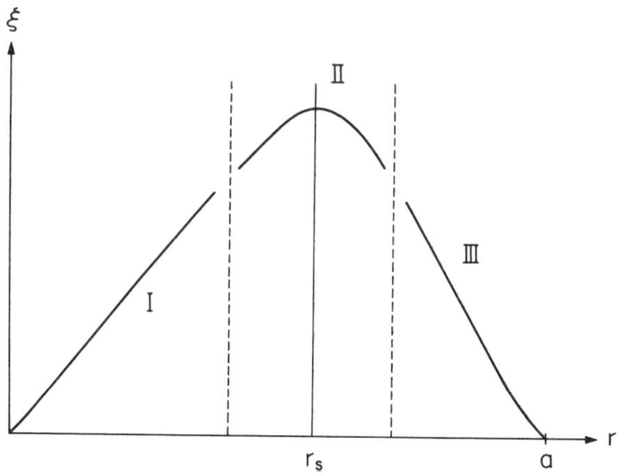

Fig. 5.5 The behavior of the solution $\xi(r)$ in the cylindrical plasma with the rational surface $r = r_s$. Here a is a radius of a conducting wall. Regions I and III are governed by the ideal MHD model and region II is a resistive layer governed by the resisti:'e MHD model.

denotes the left (right) side with respect to r_s. The matching conditions for the boundary between regions I and II are

$$\begin{cases} A_I = (a/L_R)^s A_{IIL} \\ B_I = (a/L_R)^{-1-s} B_{IIL}. \end{cases} \tag{5.230}$$

Since the equations and the boundary conditions are all homogeneous, the magnitude of ξ can be scaled in each region, and complete matching can be accomplished if

$$\frac{A_I}{B_I} = \left(\frac{a}{L_R}\right)^{2s+1} \frac{A_{IIL}}{B_{IIL}}, \tag{5.231}$$

wherre $2s + 1 = (1 - 4D_s)^{1/2}$. We expect that, for most cases, A_I/B_I will be some finite number. It depends on the parameters of the outside region I, and not on the growth rate. To obtain the matching, the growth rate must be adjusted in the resistive layer II so that $A_{IIL} \simeq 0$, since $a/L_R \gg 1$. The exceptional case occurs when $B_I \simeq 0$. This situation corresponds to the vanishing of the large solution, and the stability is marginal within the context of ideal MHD. When a system is close to a state that is unstable in ideal MHD theory, it will have a strongly modified resistive instability. A typical example is the resistive kink mode with $(m, n) = (1, 1)$ destabilized at the $q = 1$ surface in the cylindrical tokamak plasma. Similarly, there is a matching condition between regions II and III of the form

$$\frac{A_{III}}{B_{III}} = \left(\frac{a}{L_R}\right)^{2s+1} \frac{A_{IIR}}{B_{IIR}}. \tag{5.232}$$

From now on, we will consider that Q and D_s are both small. In this limit the growth rates become affected by the matching condition. We look at the behavior of the exponential solutions of the resistive layer equations (5.217) and (5.218), as shown by (5.223). Thus, as Q becomes small, we must consider the length scale given by μ^{-1}, which becomes small as $Q^{1/4}$. Otherwise, the exponential in (5.223) would grow infinitely. That is, there is an inner layer determined by the factor of $Q^{1/4}$ which is thinner than the resistive skin depth. From the dispersion relation (5.225), it is also assumed that D_s becomes small as $Q^{3/2}$ in the limit of small Q. There appear to be two kinds of solution in this limit, $\Psi \sim x\xi$ and $\Psi \sim Q^{3/2}x\xi$. These two solutions of Ψ correspond to the even and odd modes, respectively, shown in Fig. 5.4. Here, these solutions do not degenerate. It turns out that the faster growing modes are the even ones, and we will concentrate on them. Under these approximations, the leading term in (5.217) is

$$\Psi^{(0)''} \simeq 0, \tag{5.233}$$

which has a solution

$$\begin{aligned}\Psi^{(0)} &= B_{II}^{(0)} + A_{II}^{(0)}x \\ &= B_{II}^{(0)} + A_{II}^{(0)} \frac{a}{L_R} \frac{(r - r_s)}{a}.\end{aligned} \tag{5.234}$$

Since D_s is small, s approximately vanishes in (5.228), which gives agreement with (5.234) by $\Psi \sim x\xi$.

In the lowest order of (5.234), Ψ is approximated as a constant, $\Psi^{(0)} \simeq B_{II}^{(0)}$. This is the origin of the so-called "constant Ψ" approximation. Using this approximation for Ψ, (5.219) can be solved for ξ by the lowest-order equation,

$$Q^2\xi^{(0)''} - Qx^2\xi^{(0)} + D_s\xi^{(0)} = -Qx\Psi^{(0)} = -QxB_{II}^{(0)}. \tag{5.235}$$

With the present ordering of $x \sim Q^{1/4}$ and $D_s \sim Q^{3/2}$, each term on the LHS has the same order in Q. The solution of (5.235) that does not diverge exponentially for $x \to \pm\infty$ is odd, and approaches $\xi^{(0)} \to B_{II}^{(0)}/x$ in this limit. This is consistent with $\Psi^{(0)} \simeq B_{II}^{(0)}$.

Next, we examine the higher-order approximation for Ψ. Equation (5.217) gives

$$\Psi^{(1)''} = Q(\Psi^{(0)} - x\xi^{(0)}), \tag{5.236}$$

where $\Psi^{(1)}$ is smaller than $\Psi^{(0)}$ by a factor of $Q^{3/2}$ in the limit of small Q. For large x, $\Psi^0 \simeq x\xi^{(0)}$ and the RHS disappears. The corrections to (5.234) are

$$\begin{cases}\Psi^{(1)} \simeq B_{IIL}^{(1)} + A_{IIL}^{(1)}x & (x \to -\infty) \\ \Psi^{(1)} \simeq B_{IIR}^{(1)} + A_{IIR}^{(1)}x & (x \to +\infty).\end{cases} \tag{5.237}$$

Thus

$$\begin{cases} \Psi^{(1)'}|_{x\to-\infty} = A^{(1)}_{IIL} \\ \Psi^{(1)'}|_{x\to+\infty} = A^{(1)}_{IIR}. \end{cases} \tag{5.238}$$

We note that (5.236) can be integrated to yield

$$\Psi^{(1)'}\Big|_{x=-X}^{x=+X} = Q\int_{-X}^{X}(\Psi^{(0)} - x\xi^{(0)})dx \simeq Q^{5/4}B^{(0)}_{II}, \tag{5.239}$$

where $\Psi^{(0)} \simeq B^{(0)}_{II} \gg x\xi^{(0)}$ for $-X < x < X$ and X scales with Q as $X \sim Q^{1/4}$. It follows that $A^{(0)}_{II}$ in (5.234) will be of order $Q^{5/4}$. Thus, if $L_R/a \sim Q^{5/4}$, the RHS of (5.231) and (5.232) will be of order unity. With this ordering,

$$(r - r_s) \sim L_R Q^{1/4} \sim L_R(L_R/a)^{1/5} \sim \eta^{2/5} \tag{5.240}$$

is found and the growth rate is scaled as

$$\gamma_g \propto \eta^{3/5}. \tag{5.241}$$

This growth rate denotes that the modes considered in this section grow much more slowly than the unstable ideal modes, when the resistivity is small.

Now (5.235) and (5.239) can be evaluated explicitly. Since the LHS of (5.235) has a Hermite operator, we expand

$$\xi^{(0)} = \exp(-x^2/(2Q^{1/2}))\sum_{\ell=0}^{\infty} C_\ell H_\ell(\sqrt{2}x/Q^{1/4}), \tag{5.242}$$

where H_ℓ is a normalized Hermite polynomial. Use of (5.242) in the LHS of (5.235) yields

$$\begin{aligned} &Q^2\xi^{(0)''} - Qx^2\xi^{(0)} + D_s\xi^{(0)} \\ &= Q^{3/2}\exp(-x^2/(2Q^{1/2}))\sum_{\ell=0}^{\infty}\left(\frac{D_s}{Q^{3/2}} - 2\ell - 1\right)C_\ell H_\ell. \end{aligned} \tag{5.243}$$

Furthermore, it can be shown that

$$z = \sqrt{2}\exp(-z^2/2)\sum_{\ell=0}^{\infty}\frac{H_{2\ell+1}(z)}{4^\ell\Gamma(\ell+1)}, \tag{5.244}$$

where Γ is a Gamma function and $z = \sqrt{2}x/Q^{1/4}$. Equating the coefficient of H_ℓ in (5.235) to zero yields equations for the C_ℓ, and hence a formal solution for $\xi^{(0)}$. This solution can be inserted into (5.239), and $(\Psi^{(1)'}(X) - \Psi^{(1)'}(-X))$ is given in terms of an infinite series. Here we note the orthogonality condition for normalized Hermite polynomials:

$$\int_{-\infty}^{\infty} H_m(z)H_\ell(z)\exp(-z^2/2)dz = \delta_{m\ell}. \tag{5.245}$$

By using (5.243) and (5.244) in (5.235), we obtain

$$C_\ell = -\frac{1}{\sqrt{2}}\frac{Q^{5/4}B_{II}^{(0)}}{D_s-(2\ell+1)Q^{3/2}}\int_{-\infty}^{\infty} zH_\ell(z)\exp(-z^2/4)dz. \tag{5.246}$$

Use of (5.236) yields

$$\begin{aligned}
\delta &\equiv \frac{1}{B_{II}^{(0)}}[\Psi^{(1)'}(X)-\Psi^{(1)'}(-X)] \\
&= \frac{Q}{B_{II}^{(0)}}\int_{-X}^{X}(\Psi^{(0)}-x\xi^{(0)})dx \\
&= \frac{Q}{\sqrt{2}}^{5/4}\int_{-Z}^{Z}dz - \frac{Q^{3/2}}{2B_{II}^{(0)}}\sum_{\ell=0}^{\infty}C_\ell\int_{-Z}^{Z}zH_\ell(z)\exp(-z^2/4)dz \\
&= \frac{Q}{\sqrt{2}}^{5/4}\sum_{\ell=0}^{\infty}\Bigg[\left(\int_{-\infty}^{\infty}H_\ell(z)\exp(-z^2/4)dz\right)^2 \\
&\quad + \frac{1/2}{D_s/Q^{3/2}-(2\ell+1)}\left(\int_{-\infty}^{\infty}zH_\ell(z)\exp(-z^2/4)dz\right)^2\Bigg],
\end{aligned} \tag{5.247}$$

where $Z=\sqrt{2}X/Q^{1/4}$ and $\pm Z$ is replaced by $\pm\infty$ in the last expression of (5.247). Also, the Hermite polynomial expansion

$$1=\sum_{\ell=0}^{\infty}H_\ell(z)\exp(-z^2/4)\int_{-\infty}^{\infty}H_\ell(z)\exp(-z^2/4)dz \tag{5.248}$$

has been used to obtain the last equality. The integrals in the last expression are given by

$$\int_{-\infty}^{\infty}H_\ell(z)\exp(-z^2/4)dz = \begin{cases} 2^{3/4}\left\{\dfrac{\Gamma(\frac{\ell}{2}+\frac{1}{2})}{\Gamma(\frac{\ell}{2}+1)}\right\}^{1/2} & (\text{even } \ell) \\ 0 & (\text{odd } \ell)\end{cases} \tag{5.249}$$

and

$$\int_{-\infty}^{\infty} zH_\ell(z)\exp(-z^2/4)dz = \begin{cases} 0 & \text{(even } \ell) \\ 2^{9/4}\left\{\dfrac{\Gamma(\frac{\ell}{2}+1)}{\Gamma(\frac{\ell}{2}+\frac{1}{2})}\right\}^{1/2} & \text{(odd } \ell). \end{cases} \tag{5.250}$$

Substituting (5.249) with $\ell = 2m$ and (5.250) with $\ell = 2m+1$ into (5.247) yields

$$\delta = 2Q^{5/4}\sum_{m=0}^{\infty}\frac{\Gamma(m+\frac{1}{2})}{m!}\,\frac{\frac{1}{4}(D_s/Q^{2/3}-1)}{\frac{1}{4}(D_s/Q^{3/2}-3)-m}. \tag{5.251}$$

By noting the relation

$$\sum_{m=0}^{\infty}\frac{\Gamma(m+\frac{1}{2})}{m!}\,\frac{1}{m+\lambda} = \pi\,\frac{\Gamma(\lambda)}{\Gamma(\frac{1}{2}+\lambda)}, \tag{5.252}$$

we obtain

$$\delta = 2\pi Q^{5/4}\,\frac{\Gamma[\frac{1}{4}(3-D_s/Q^{3/2})]}{\Gamma[\frac{1}{4}(1-D_s/Q^{3/2})]}. \tag{5.253}$$

This δ is the only quantity needed from the inner layer to evaluate the dispersion relation.

From (5.238) we also obtain

$$\delta = \lim_{x\to\infty}\left(\frac{\Psi^{(1)'}(x)}{\Psi^{(0)}(x)} - \frac{\Psi^{(1)'}(-x)}{\Psi^{(0)}(-x)}\right) = \frac{A_{IIR}}{B_{IIR}} - \frac{A_{IIL}}{B_{IIL}} \tag{5.254}$$

for $B_{IIR} \simeq B_{II}^{(0)}$ and $B_{IIL} \simeq B_{II}^{(0)}$. According to (5.231) and (5.232), the individual terms on the RHS of (5.254) must be matched to the outside regions. Thus we obtain the dispersion relation,

$$\Delta = \Delta', \tag{5.255}$$

where

$$\Delta = \left(\frac{a}{L_R}\right)\delta = 2\pi\left(\frac{a}{L_R}\right)Q^{5/4}\,\frac{\Gamma[\frac{1}{4}(3-D_s/Q^{3/2})]}{\Gamma[\frac{1}{4}(1-D_s/Q^{3/2})]} \tag{5.256}$$

and

$$\Delta' = \frac{A_{III}}{B_{III}} - \frac{A_I}{B_I}, \tag{5.257}$$

since $s \simeq 0$. It is known that $\Gamma(z)$ becomes infinite at the infinite sequence of poles for $z < 0$, and $\Gamma(z) \sim \sqrt{2\pi} e^{-z} z^{z-(1/2)}$ asymptotically for $|z| \to \infty$. Thus Δ depends on the parameters of the inner region, and Δ' depends on the outer region. The dispersion relation (5.255) gives the growth rate Q. We now analyze (5.255) to obtain the dependence of A on D_s and Δ'. When $D_s > 0$, the Gamma functions have an infinite sequence of poles which accumulates at $Q = 0$. Thus Δ changes from $-\infty$ to $+\infty$ between two successive poles, and it passes through all values many times. Thus, for a given Δ', there is an infinite sequence of Q satisfying the dispersion relation. When Δ' vanishes or the denominator including the Gamma function diverges, these growth rates are given by

$$(1 - D_s/Q^{3/2})/4 = -\ell$$

or

$$Q^{3/2} = \frac{D_s}{4\ell + 1}, \tag{5.258}$$

which is the same as the dispersion relation (5.225) in the case of $s = 0$. The growth rate has the dependence of $\gamma_g \propto \eta^{1/3}$. This means that when $D_s > 0$, there is always an instability. When $D_s \lesssim 0$, Δ is always positive or Δ diverges as $Q^{5/4}$ for large Q, and vanishes as $(-D_s Q)^{1/2}$ for small Q, which is obtained from $\Gamma(az + b)/\Gamma(az + b') \sim (az)^{b-b'}$ for large z, where a, b, and b' are constant. Thus, there are no unstable solutions for $\Delta' < 0$. This means that there is an instability only when $\Delta' > 0$. The two kinds of instability, for $D_s > 0$ and $D_s \lesssim 0$, have quite different characters. The first kind with $D_s > 0$ and $\Delta' = 0$ only depends on the parameters within the resistive layer. This instability, which is similar to the ideal interchange mode, is called the resistive interchange mode.

It should be noted that D_s given by (5.179) is strongly affected by torodicity. The Mercier criterion is obtained by multiplying a factor $(1 - q^2)$ to the second term in (5.180) in the large aspect ratio tokamak with a circular cross-section. Here B_z must be replaced with a toroidal field B_0. Thus, in the toroidal case for $q > 1$ there are no unstable interchange modes. In the Heliotron E, the toroidal effect induces a Shafranov shift which produces a magnetic well in the central region with $\iota \sim 0.5$ or $q \sim 2$. This stabilizes both the ideal and resistive interchange modes; however, there is no such factor $(1 - q^2)$ as in tokamaks.

In the unstable modes with $D_s \lesssim 0$, matching conditions from the outer region play a crucial role. Thus these modes appear to be driven by forces in the outer regions. The role of the inner resistive layer is to allow motions that would be excluded in the ideal MHD theory. Such instabilities are called tearing modes. Marginal stability corresponds to $\Delta' = 0$. The growth rate of the tearing mode has the dependence given by (5.241), which is obtained from $\Delta' \propto (a/L_R)Q^{5/4}$. It is noted that the pressure γP_0 is assumed to be very small in this section. In toroidal plasmas, the finite-pressure effect may relax the tearing mode criterion $\Delta' > 0$ (unstable) somewhat.

The resistive interchange mode becomes unstable even for low m numbers, such as $m = 1, 2$, and 3. For heliotrons the three-field reduced MHD equations (3.200), (3.212), and (3.221) (or (3.231), (3.232), and (3.233)) in chapter 3 are suitable for the study of the linear stability and nonlinear behavior of the resistive interchange mode. In order to study the linear stability, we linearize (3.231), (3.232), and (3.233) in the normalized form as

$$\frac{\partial}{\partial t}\nabla_\perp^2 u_1 = -\nabla_{\parallel 0}\nabla_\perp^2 A_1 + \mathbf{\nabla} J_0 \times \mathbf{\nabla} A_1 \cdot \hat{\zeta} + \mathbf{\nabla} P_1 \times \mathbf{\nabla}\Omega \cdot \hat{\zeta}, \tag{5.259}$$

$$\frac{\partial A_1}{\partial t} = \nabla_{\parallel 0} u_1 + \eta \nabla_\perp^2 A_1, \tag{5.260}$$

$$\frac{\partial P_1}{\partial t} = -\mathbf{\nabla} P_0 \times \mathbf{\nabla} u_1 \cdot \hat{\zeta}, \tag{5.261}$$

where the subscripts 0 and 1 refer to the equilibrium and perturbed quantities, respectively. In the toroidal coordinates (ρ, θ, ζ),

$$\nabla_{\parallel 0} = \frac{\partial}{\partial z} + \mathbf{\nabla}\Psi_0 \times \hat{\zeta} \cdot \mathbf{\nabla}, \tag{5.262}$$

where Ψ_0 is a solution of the equilibrium solution of (4.41) (see (3.198)). The unit vector in the z direction is denoted by $\hat{\zeta}$. For heliotrons the second term on the RHS of (5.259) in negligible, since recent interest is in currentless MHD equilibria. When the currentless MHD equilibrium is obtained by solving the Grad–Shafranov type equation (4.44), the linear stability can be examined by solving (5.259)–(5.261) as an initial value problem after an adequate small perturbation is given. The RESORM and FAR codes are based on such an approach. When the perturbation grows exponentially, it corresponds to the unstable resistive interchange modes. In Fig. 5.6 the growth rates obtained by the RESORM code are plotted as a function of the magnetic Reynolds number $S \equiv 1/\eta$ in the normalized form for the $n = 1$ mode in the case of the Heliotron DR, where n is a toroidal mode number. The MHD equilibrium is the same as that for Fig. 5.3. The ideal interchange modes becomes unstable at β_0 (central beta value) $\simeq 0.7\%$. In Fig. 5.6 it is shown that the resistive interchange mode with $n = 1$ destabilized at the $\iota = 1$ surface in the Heliotron DR follows the growth rate scaling $\gamma \propto \eta^{1/3}$ given by (5.258) or $\gamma \propto S^{-1/3}$, when the ideal interchange modes are stable. When the Mercier criterion is violated, the growth rate scaling of resistive instability deviates from $\gamma \propto S^{-1/3}$ in the large S regime.

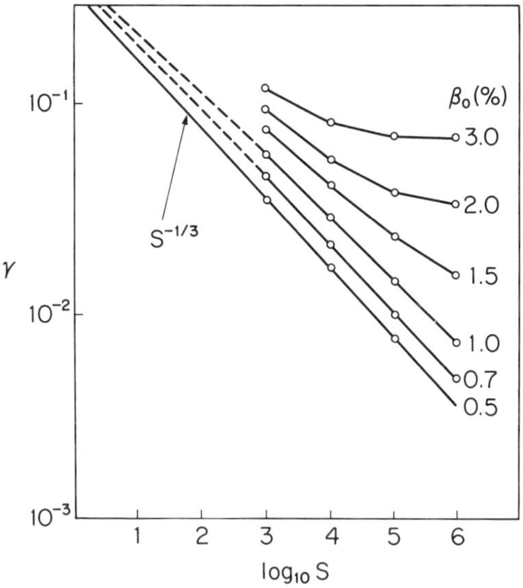

Fig. 5.6 The dependence of the growth rate of the $n = 1$ mode on the magnetic Reynolds number S for various β_0 values in the Heliotron DR. The pressure profile is assumed to be $P = P_0(1 - \rho^2)^2$, where $\rho =$ constant denotes a magnetic surface.

An example of eigenfunctions $u_1(\rho)$ and $A_1(\rho)$ is shown in Fig. 5.7. Here $\beta_0 = 2\%$ and $S = 10^5$, which exceeds the Mercier beta limit. The numbers in this figure denote poloidal mode numbers. It is seen that $(m, n) = (1, 1)$ is the dominant mode and is localized at the $\iota = 1$ resonant surface. Toroidal coupling produces the side band modes with $(m, n) = (2, 1)$ and $(m, n) = (0, 1)$. Even the lowest mode with $(m, n) = (1, 1)$ has a fairly localized mode structure in the neighborhood of the resonant surface.

5.6 THE ROLE OF LOCALIZED MODE STABILITY CRITERIA (SUYDAM CRITERION AND MERCIER CRITERION)

Stability against ideal interchange modes is the crucial problem in heliotrons, since they usually have a magnetic hill or an unfavorable averaged curvature region in the plasma edge. When resonant surfaces of low-n interchange modes exist inside the magnetic hill region, there appears to be a maximum beta value for the stability, called the beta limit, where n is a toroidal mode number. For example, the beta limit of the Heliotron E is about $\langle\beta\rangle \simeq 2\%$, which depends on the pressure profile and is determined by the low-n pressure-driven modes with $n \leq 4$. Here $\langle\beta\rangle$ is a volume-averaged beta value.

On the other hand, high-n mode stability is studied by using the Suydam criterion for cylindrical plasmas or the Mercier criterion for toroidal plasmas. Recently, numerical codes for studying the ideal MHD stability of heliotrons

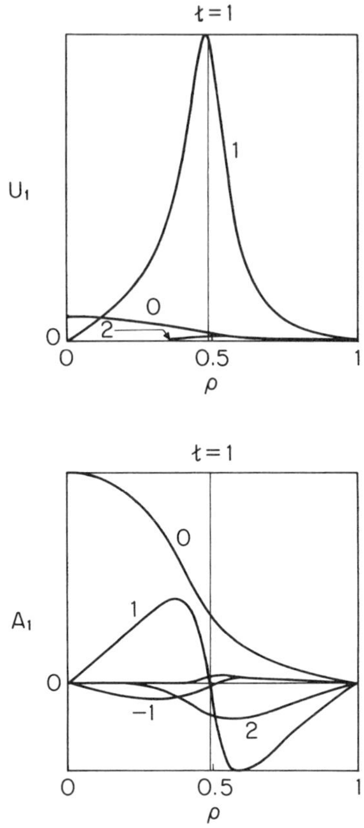

Fig. 5.7 The eigenfunctions $u_1(\rho)$ and $A_1(\rho)$ of the $n = 1$ mode at $\beta_0 = 2\%$ and $S = 10^5$. The numbers denote poloidal mode numbers. The pressure profile is assumed to be $P = P_0(1 - \rho^2)^2$, where $\rho = 0$ is the magnetic axis and $\rho = 1$ is the outermost magnetic surface.

have been developed to a practically usable level to analyze experimental results and to design a new stellarator and heliotron. High-n interchange mode stability is usually examined in the numerical codes solving three-dimensional MHD equilibria, such as the BETA and VMEC codes with the Mercier criterion. There are several numerical codes developed for low-n interchange mode stability, such as the STEP code, the FAR code, and the HERATO code. In these codes the averaged MHD equilibrium is solved first, and then the ideal MHD stability is examined to determine the stability beta limit. Recently, the TARPSHICORE and CAS3D codes have been developed to study the ideal MHD pressure-driven modes for the three-dimensional MHD equilibria obtained by the VMEC code. For heliotrons, the agreement between these two approaches is sufficiently good. The advantage of the former approach using the average MHD equilibrium is the numerical accuracy, since the stability calculation can be performed for a single assigned toroidal mode number, which is the same as the stability calculation for tokamaks. The Mercier criterion and the low-n unstable modes in the (β_0, Ψ) plane are shown in Figure 5.8 for a heliotron with $M = 14$. The pressure profile used here is also $P = P_0(1 - \Psi)^2$, and the vertical and horizontal axes denote the central beta value at the magnetic axis β_0 and the flux function Ψ corresponding to the radial coordinates, respectively. The critical beta values determined by the Mercier limit, defined as $D_I = 1/4 - D_M$, with the VMEC and STEP codes are in good agreement. The dotted lines in the figure show the contours of $D_I = -0.2$ and $D_I = -0.4$. It is noted that $D_I < 0$ $(D_I > 0)$ corresponds to instability (stability). The lines for $\iota = 3/4$ and $\iota = 4/5$ show how the resonant surfaces move in the currentless MHD equilibrium as β_0 increases. The heavy parts show that the low-n modes with $n = 3$ and $n = 4$ resonant at the surface of $\iota = 3/4$ and $\iota = 4/5$, respectively, are unstable. It should be noted that the lower ends of the heavy lines showing the marginal state correspond to $D_I \simeq -0.2$ rather than to $D_I = 0$. Although it seems that the low-n mode is more stable than the high-n mode, we consider that the difference is attributed to limitations in the numerical code using finite size meshes, because the low-n interchange modes also has the property of being highly localized around the resonant surface as beta approaches the critical value. This problem is always

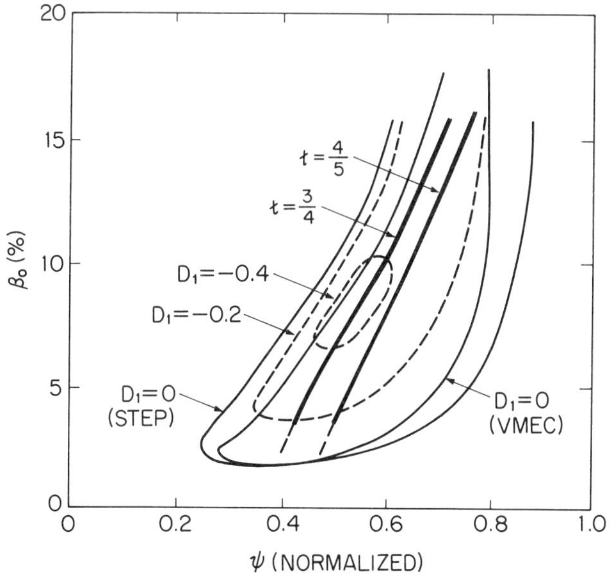

Fig. 5.8 The stability diagram in the β_0 versus Ψ representation. The Mercier boundaries from the VMEC and STEP codes are drawn as light curves. The positions of the resonant surfaces within the Mercier boundary from the STEP code are drawn as dashed curves, with superimposed heavy curves representing the range of β of the unstable low-n modes. The values of the Mercier criterion $D_I < 0$ (for instability) from the STEP code are drawn as dashed curves.

evident when finding the critical beta for the low-n ideal interchange modes in heliotrons.

In order to understand the relation between the Mercier criterion and the low-n interchange instability in the toroidal plasma, it is useful to study the relation between the Suydam criterion and a low-m interchange mode in a cylindrical plasma, where m is a poloidal mode number. Here the equilibrium plasma current is assumed to be zero, since the interchange mode is destabilized by the pressure gradient.

From the linearized reduced MHD equations for stellarators, (5.259), (5.260), and (5.261), the linear eigenmode equation including the growth rate is obtained by neglecting the equilibrium plasma current J_0 and the resistivity η, in the cylindrical limit

$$\gamma_g^2\left(\frac{d^2}{dr^2}+\frac{1}{r}\frac{d}{dr}-\frac{m^2}{r^2}\right)u_1 = k_\parallel\left(\frac{d^2}{dr^2}+\frac{1}{r}\frac{d}{dr}-\frac{m^2}{r^2}\right)(k_\parallel u_1)-\frac{m^2}{r^2}\frac{dP_0}{dr}\frac{d\Omega}{dr}u_1, \tag{5.263}$$

where the Fourier modes $\{u_1(\mathbf{r}),\ P_1(\mathbf{r}),\ A_1(\mathbf{r})\} = \{u_1(r),\ P_1(r),\ A_1(r)\}\exp(\gamma_g t + im\theta - inz/R)$ are assumed for a currentless MHD equilibrium described by pressure P_0 and flux function Ψ_0. Here $k_\parallel = m\iota - n$, the rotational transform is $\iota = -(1/r)(d\Psi_0/dr)$, and $2\pi R$ is a periodic length in the z-direction. The term including $(dP_0/dr)(d\Omega/dr)$ is the driving term for interchange modes for $d\Omega/dr > 0$. Equation (5.263) is obtained by substituting $\gamma_g P_1 = -(im/r)(dP_0/dr)u_1$ and $\gamma_g A_1 = ik_\parallel u_1$ into (5.259).

Here our concern is in the case in which the Suydam criterion is violated so that the system is unstable to interchange modes. Since the interchange mode is localized in the neighborhood of the resonant surface $r = r_s$, the eigenmode equation (5.263) is rewritten as

$$(\gamma_g^2 + k_\parallel'^2 x^2)\frac{d^2u_1}{dx^2} + 2k_\parallel'^2 x\frac{du_1}{dx} - k_\theta^2(P_0'\Omega' + k_\parallel'^2 x^2)u_1 - \gamma_g^2 k_\theta^2 u_1 = 0, \tag{5.264}$$

where $k_\theta = m/r_s$, $P_0' = (dP_0/dr)|_{r=r_s}$, and $\Omega' = (d\Omega/dr)|_{r=r_s}$. It is noted that $k_\parallel = k_\parallel' x$ and $|(1/r)(du_1/dr)|$ is neglected compared to $|d^2u_1/dr^2|$. By using $x = (\gamma_g/k_\parallel')\xi$ in (5.264), we obtain

$$(\xi^2+1)\frac{d^2u_1}{d\xi^2} + 2\xi\frac{du_1}{d\xi} + \lambda u_1 - \mu^2(\xi^2+1)u_1 = 0, \tag{5.265}$$

where

$$\lambda = -\frac{k_\theta^2}{k_\parallel'^2}P_0'\Omega' = -\frac{P_0'\Omega'}{(r_s\iota')^2}, \tag{5.266}$$

$$\mu = \frac{k_\theta\gamma_g}{k_\parallel'} = \frac{\gamma_g}{(r_s\iota')}. \tag{5.267}$$

Here (5.265) is supplemented by boundary conditions whereby the perturbation vanishes at infinity:

$$\lim_{\xi \to \pm\infty} u_1 = 0. \tag{5.268}$$

It is noted that (5.265) can be transformed into the form of the Schrödinger equation with the changes of variables

$$\begin{cases} \xi = \sinh(z) \\ u_1 = (\cosh(z))^{-1/2}\psi(z). \end{cases} \tag{5.269}$$

The resultant equation is shown as

$$\frac{d^2\psi}{dz^2} + [E - U(z)]\psi = 0, \tag{5.270}$$

with the potential

$$U(z) = \mu^2 \cosh^2(z) + \frac{1}{4\cosh^2(z)} \tag{5.271}$$

and the energy

$$E = \lambda - \tfrac{1}{4}. \tag{5.272}$$

Thus the linear stability problem of ideal interchange modes in the cylindrical plasma is reduced to finding energy levels in the potential well (5.271) with the boundary conditions

$$\lim_{z \to \pm\infty} \psi = 0. \tag{5.273}$$

One remarkable feature of (5.270) is that the energy (5.272) is expressed through the equilibrium parameter λ, while the shape of the potential well depends only on the growth rate of perturbation μ.

As the general properties of the Schrödinger equation are well-known, we can use them to estimate the growth rates of ideal interchange modes. It should be noted that only a positive energy level may appear in the potential well (5.271), since this well exists for $\mu^2 > 0$ only and satisfies $U(z) > 0$. The necessary condition for the existence of an energy level in the potential (5.271) gives a lower limit to the quantity λ,

$$\lambda - 1/4 > 0, \tag{5.274}$$

which implies localized instability. Inequality (5.274) corresponds to the Suydam criterion for interchange instabilities in heliotrons (see (5.180)). The inequality (5.274) is written as

$$\frac{1}{4}\left(\frac{\iota'}{\iota}\right)^2 + \frac{1}{\iota^2 r^2}\frac{dP_0}{dr}\frac{d\Omega}{dr} < 0, \tag{5.275}$$

with normalized variables. In Fig. 5.9 are shown growth rates of interchange modes with $(m, n) = (1, 1), (2, 2)$, and $(3, 3)$ as a function of the central beta value β_0. Here the pressure profile is assumed to be $P = P_0(1 - (r/a)^2)^2$ and the rotational transform is assumed to be $\iota(r) = 0.51 + 1.69(r/a)^{2.5}$ as an approximate expression for the Heliotron E. The growth rates and eigenfunctions are obtained by solving the eigenvalue equation (5.263) with a shooting method by imposing the boundary conditions $u_1(a) = 0$ and $u_1 \propto r^m$ at $r \simeq 0$. The condition $u_1(a) = 0$ corresponds to the fixed boundary or to the existence of the conducting wall at $r = a$. The condition $u_1 \propto r^m$ is obtained by noting that (5.263) reduces to $u_1'' + u_1'/r - (m^2/r^2)u_1 \simeq 0$ at $r \simeq 0$. In Fig. 5.9 the growth rate of the Suydam mode is plotted by using the expression given by Kulsrud (1963) under the assumption of $n = 1$. It is seen that the growth rates become extremly small when β_0 becomes close to the Suydam limit, $\beta_0 = 1.62\%$. In

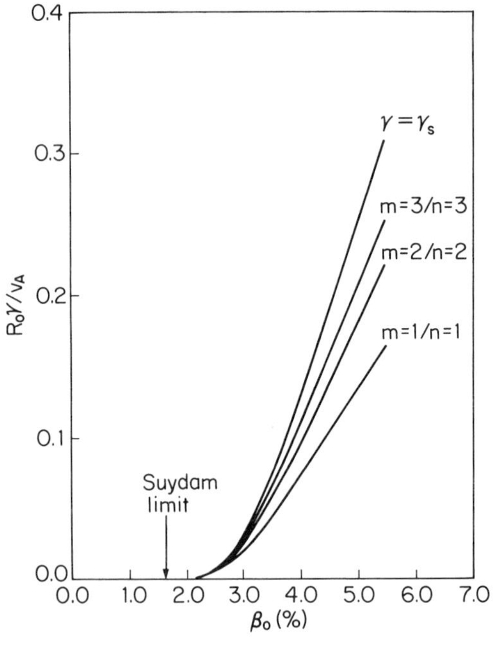

Fig. 5.9 The growth rates of the $(m, n) = (1, 1)$, $(m, n) = (2, 2)$, and $(m, n) = (3, 3)$ modes versus the central beta value β_0. The upper curve shows the analytic growth rates of the Suydam mode.

Fig. 5.10 are shown radial profiles of the eigenfunction $u_1(r)$ for the $(m, n) = (1, 1)$ mode at $\beta_0 = 4\%$ and $\beta_0 = 2.5\%$. It can be clearly seen that the eigenfunction becomes localized significantly when β_0 becomes close to the Suydam limit given by $\lambda = 1/4$.

In the usual beta limit analysis, the critical value has been obtained by linearly extrapolating the growth rates calculated numerically in the higher-beta region to the zero growth rate. The resultant critical beta value is not accurate, since the beta dependence of the growth rate obeys an exponential relationship, as shown by Kulsrud (1963). This point is different from the usual assumption $\gamma_g^2 \propto (\beta_0 - \beta_{c0})$, where β_{c0} is the critical beta value at $\gamma_g^2 = 0$. The gap between the Suydam limit and the critical beta determined by global mode stability is practically inevitable. However, it is interpreted that the interchange mode is marginally unstable in the gap region, becausee the growth rate is extremely small. This interpretation is also applicable to the beta limit calculation in the toroidal geometry, as shown in Fig. 5.8.

In addition to the shear and the magnetic well, there is a stabilizing effect on the interchange mode, which is derived from finiteness of the ion Larmor radius in the circulating motion around the magnetic line force. In magneto-

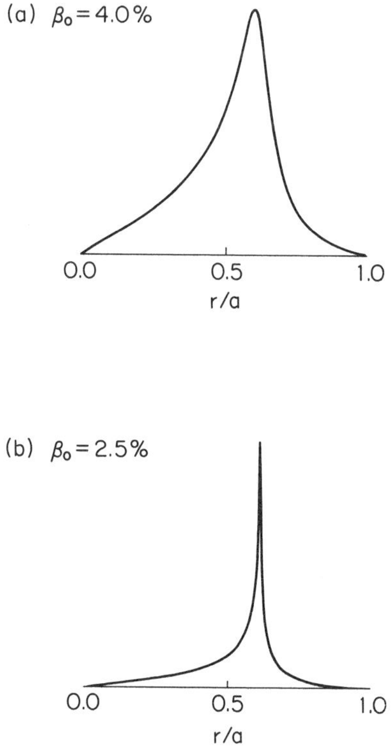

Fig. 5.10 The radial profiles of eigenfunction u_1 for $(m, n) = (1, 1)$.

hydrodynamics, we treat the small Larmor radius limit for both ions and electrons. The MHD equations can be modified to include the finite Larmor radius (FLR) effect.

Here we show that a new term including the FLR effect appears in the vorticity equation of the three-field reduced MHD model. The FLR effect can be introduced through the gyroviscosity given by Braginskii (1965),

$$\overleftrightarrow{\Pi}_{gi} = \frac{P_i}{4\omega_{ci}} \{\hat{\mathbf{b}} \times \overleftrightarrow{W} \cdot (\overleftrightarrow{I} + \hat{\mathbf{b}}\hat{\mathbf{b}}) + (I + \hat{\mathbf{b}}\hat{\mathbf{b}}) \cdot \overleftrightarrow{W} \times \hat{\mathbf{b}}\}, \tag{5.276}$$

where $\omega_{ci} = eB/m_i$ is the ion cyclotron frequency and $\overleftrightarrow{W}$ denotes the rate of strain tensor, defined by

$$W_{\alpha\beta} = \frac{\partial v_\alpha}{\partial x_\beta} + \frac{\partial v_\beta}{\partial x_\alpha} - \frac{2}{3}\,\delta_{\alpha\beta} \nabla \cdot \mathbf{v}, \tag{5.277}$$

and $\hat{\mathbf{b}} = \mathbf{B}/B$. By using the perpendicular velocity

$$\mathbf{v}_\perp = \frac{\hat{\mathbf{z}}}{B_0} \times \nabla\left(u + \frac{\tilde{P}_i}{n_0 e}\right) \tag{5.278}$$

and $|\mathbf{v}_\parallel| \ll |\mathbf{v}_\perp|$ in (5.276), the gyroviscosity is approximated as

$$\overleftrightarrow{\Pi}_{gi} = \frac{P_i}{2\omega_{ci}} [\nabla_\perp^2 F(\overleftrightarrow{I} - \hat{\mathbf{z}}\hat{\mathbf{z}}) - 2\nabla_\perp \nabla_\perp F], \tag{5.279}$$

where

$$F = \frac{1}{B_0}\left(u + \frac{\tilde{P}_i}{n_0 e}\right). \tag{5.280}$$

Here it is noted that $n = n_0 + \tilde{n}$ and $P_i = P_{i0} + \tilde{P}_i$ are assumed, and n_0 and P_{i0} are the volume-averaged density and ion pressure, respectively. B_0 is a constant longitudinal field in the z direction. It is considered that the gyroviscosity may be included by changing $nm_i d\mathbf{v}/dt$ into $nm_i d\mathbf{v}/dt + \nabla \cdot \overleftrightarrow{\Pi}_{gi}$, and divergence of (5.279) gives

$$\nabla \cdot \overleftrightarrow{\Pi}_{gi} = \nabla_\perp\left(\frac{P_i}{2\omega_{ci}} \nabla_\perp^2 F\right) - \nabla_\perp \cdot \left(\frac{P_i}{\omega_{ci}} \nabla_\perp \nabla_\perp F\right). \tag{5.281}$$

By noting that

$$\frac{\mathbf{B}}{B^2} \cdot \nabla \times \nabla \cdot \overleftrightarrow{\Pi}_{gi} = -\frac{m_i}{eB_0^2} \nabla_\perp \cdot (\hat{\mathbf{z}} \times \nabla P_i \cdot \nabla_\perp \nabla_\perp F), \tag{5.282}$$

the LHS of (3.212) becomes

$$n_0 m_i \left(\frac{\partial}{\partial t} + \hat{\mathbf{z}} \times \nabla F \cdot \nabla \right) \nabla_\perp^2 F - \frac{1}{\omega_{ci}} \nabla_\perp \cdot (\hat{\mathbf{z}} \times \nabla P_i \cdot \nabla_\perp \nabla_\perp F). \tag{5.283}$$

Next $n_0 m_i \frac{\partial}{\partial t} \nabla_\perp^2 \tilde{P}_i$ in (5.283) is replaced by $-\nabla_\perp^2 [\nabla u \times \hat{\mathbf{z}} \cdot \nabla P_i]/\omega_{ci}$ with the use of (3.221). The resultant expression becomes

$$n_0 m_i \left(\frac{\partial}{\partial t} + \hat{\mathbf{z}} \times \nabla u \cdot \nabla \right) \nabla_\perp^2 u + \frac{1}{\omega_{ci}} \nabla_\perp \cdot (\hat{\mathbf{z}} \times \nabla P_i \cdot \nabla_\perp \nabla_\perp u), \tag{5.284}$$

which gives $\alpha \nabla_\perp \cdot (\hat{\mathbf{z}} \times \nabla P \cdot \nabla_\perp \nabla_\perp u)$ in the normalized form (see (3.230)), where $\alpha = c/(\omega_{pi} a)$, $\omega_{pi} = (n_0 e^2 / m_i \varepsilon_0)^{1/2}$ is the ion plasma oscillation and $\nabla P \simeq \nabla P_i$ is assumed here. The relation $c^2 = 1/(\varepsilon_0 \mu_0)$ has been used, where c is the light velocity. By including this FLR correction introduced by gyroviscosity in the linearized vorticity equation in (5.259), we find that

$$\begin{aligned} &\frac{\partial}{\partial t} \nabla_\perp^2 u_1 + \alpha \nabla_\perp \cdot (\hat{\mathbf{z}} \times \nabla P_0 \cdot \nabla_\perp \nabla_\perp u_1) \\ &= -\nabla_{\parallel 0} \nabla_\perp^2 A_1 + \nabla J_0 \times \nabla A_1 \cdot \hat{\mathbf{z}} + \nabla P_1 \times \nabla \Omega \cdot \hat{\mathbf{z}}, \end{aligned} \tag{5.285}$$

where only the equilibrium part of ∇P_i equal to ∇P_0 is retained in the second term in the RHS. By eliminating P_1 and A_1 using (5.260) and (5.261) and assuming that $\eta \to 0$, we obtain

$$\frac{\partial^2}{\partial t^2} \nabla_\perp^2 u_1 + 2\alpha L \frac{\partial u_1}{\partial t} - M u_1 = 0, \tag{5.286}$$

where the linear operators L and M are defined by

$$\begin{aligned} M u_1 \equiv & -\nabla_{\parallel 0} \nabla_\perp^2 \nabla_{\parallel 0} u_1 + \nabla J_0 \times \nabla (\nabla_{\parallel 0} u_1) \cdot \hat{\mathbf{z}} \\ & + \nabla \Omega \times \nabla (\nabla P_0 \times \nabla u_1 \cdot \hat{\mathbf{z}}) \cdot \hat{\mathbf{z}} \end{aligned} \tag{5.287}$$

$$L u_1 \equiv \tfrac{1}{2} \nabla_\perp \cdot (\hat{z} \times \nabla P_0 \cdot \nabla_\perp \nabla_\perp u_1). \tag{5.288}$$

By assuming that $u_1(\mathbf{r}, t) = u_1(\mathbf{r}) e^{-i\omega t}$, (5.286) can be written as an eigenvalue equation,

$$\omega^2 \nabla_\perp^2 u_1 + 2i\omega\alpha L u_1 + M u_1 = 0, \tag{5.289}$$

where ω is the eigenvalue, and the eigenfunction u_1 is generally a complex function. Multiplying (5.289) by u_1^* and integrating in the whole plasma region yields

$$K\omega^2 - 2\alpha V \omega - W = 0, \tag{5.290}$$

where

$$\begin{cases} K = \frac{1}{2}\int |\nabla_\perp u_1|^2 d\mathbf{r}, \\ V = \frac{i}{2}\int u_1^* L u_1 d\mathbf{r}, \\ W = \frac{1}{2}\int u_1^* M u_1 d\mathbf{r}. \end{cases} \tag{5.291}$$

Here K is always positive. Since the operators M and L have the properties $\int vMud\mathbf{r} = \int uMvd\mathbf{r}$ and $\int vLud\mathbf{r} = -\int uLvd\mathbf{r}$ for any choice of functions u and v satisfying the boundary conditions, V and W are real. W is essentially the same as the energy integral shown by δW_F in (5.81). From (5.290) the eigenvalue is given by

$$\omega = \alpha \frac{V}{K} \pm \sqrt{\alpha^2 \left(\frac{V}{K}\right)^2 + \frac{W}{K}}. \tag{5.292}$$

This result suggests that, even if $W < 0$, the system becomes stable if

$$\alpha^2 \left(\frac{V}{K}\right)^2 + \frac{W}{K} > 0. \tag{5.293}$$

This is considered to be the FLR stabilization of the interchange mode.

The physical mechanism is as follows. When the beta value approaches the critical betal, the radial eigenfunction of the interchange mode becomes highly localized, as shown in Fig. 5.10. When this localization becomes comparable to the ion Larmor radius, the ion experiences the perturbation averaged over the ion orbit instead of the local one, which reduces the phase difference between the potential perturbation u_1 and the pressure perturbation P_1, and the stabilizing tendency appears.

Recently, poloidal shear flows driven by the radial electric field on various instabilities were intensively investigated from the point of view of confinement improvement, since a rapid increase in poloidal shear flow has been observed in the initial phase of transition from the L (low) mode of confinement to the H (high) mode in several tokamaks. The improvements in confinement in the stellarator and heliotron will be briefly discussed in chapter 8. Usually, the confinement time of the H mode is about twice as good as that of the L mode. The transition usually starts with a decrease of H_α or D_α of light emission, and an accompanying increase in density and temperature occurs, of the order of several tens of milliseconds in the present-day tokamaks.

In this section we will study the effect of the poloidal shear flow on the ideal interchange mode in the cylindrical plasma model for Heliotron E. This becomes possible by retaining the poloidal flow $v_E = du_0/dr$ in the normalized three-field reduced MHD equations (3.230), (3.231), and (3.232). The poloidal shear flow appears in the eigenvalue equation for the interchange mode (5.263) by retaining $u_0(r)$:

$$(\omega-\omega_E)^2\left(\frac{1}{r}\frac{d}{dr}r\frac{d}{dr}-k_\theta^2\right)u_1-(\omega-\omega_E)k_\theta(\nabla_\perp^2 u_0)'u_1$$
$$=k_\parallel(\omega-\omega_E)\left(\frac{1}{r}\frac{d}{dr}r\frac{d}{dr}-k_\theta^2\right)\left(\frac{k_\parallel u_1}{\omega-\omega_E}\right)+k_\theta^2 P_0'\Omega u_1, \qquad (5.294)$$

where $k_\theta=m/r$ and γ_g is replaced with $-i\omega$. The perturbations A_1 and P_1 are given by $A_1=k_\parallel u_1/(\omega-\omega_E)$ (see (5.260)) and $P_1=k_\theta P_0' u_1/(\omega-\omega_E)$ (see (5.261)), respectively. In (5.294), ω_E is assumed

$$\omega_E\equiv\frac{k_\theta v_E}{\varepsilon V_A}=\frac{m}{r}\frac{du_0}{dr}=\omega_{E0}\tanh\left(\frac{r-r_s}{L_E}\right), \qquad (5.295)$$

where r_s is the radial position of the resonant surface $\iota(r_s)=n/m$, and L_E corresponds to the characteristic width of the shear flow. Here $u_0(r)$ corresponds to an electric potential of a radial electric field. The flow profile described by $\tanh(x)$ is often used in the slab model.

We can follow a similar analysis as in the use of the Suydam criterion to obtain a stability criterion in the presence of poloidal shear flow. By using $x=r-r_s$, $k_\parallel=k_\parallel' x$, $k_\parallel'=m\iota'|_{r=r_s}$, and $\omega-\omega_E=-\omega_E'|_{r=r_s}x$, (5.294) can be written as

$$\frac{d^2}{dx^2}\hat{u}_1+\left\{\frac{k_{\theta 0}^2 P_0'\Omega'}{[(\omega_E')^2-(k_\parallel')^2]x^2}-k_{\theta 0}^2\right\}\hat{u}_1=0, \qquad (5.296)$$

where $k_{\theta 0}=m/r_s$. The necessary stability criterion becomes

$$-\frac{k_{\theta 0}^2 P_0'\Omega'}{(k_\parallel')^2[1-(\omega_E'/k_\parallel')^2]}<\frac{1}{4}. \qquad (5.297)$$

When the poloidal shear flow is negligibly small or $\omega_E'=0$, the stability becomes the Suydam criterion exactly or $\lambda<\frac{1}{4}$ (see (5.274)). Another point is that the poloidal mode number m disppears in the inequality (5.297) even for $\omega_E'\neq 0$, since both ω_E' and $k_\parallel'$ are proportional to m. It should be noted that poloidal shear flow affects the local stability criterion through $(\omega_E')^2$, which means that the destabilizing effect on the ideal interchange mode is independent of the sign of the radial electric field for the normal pressure profile with $P_0'<0$. The local stability diagram modified by the poloidal shear flow given by (5.297) for the $(m,n)=(1,1)$ mode in the Heliotron E model is shown in Fig. 5.11. Here the shaded region denotes stability. For example, poloidal shear flow with $\omega_E'=1$ reduces the stability beta limit from $\beta_0=1.62\%$ to $\beta_0=1.21\%$.

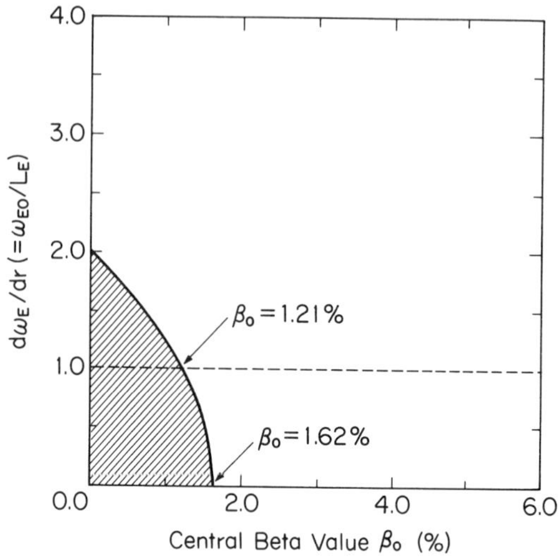

Fig. 5.11 The Suydam criterion modified by the poloidal shear flow for the Heliotron E model with a pressure profile $P = P_0(1 - (r/a)^2)^2$. The shaded region is stable to the interchange mode with $(m, n) = (1, 1)$ destabilized at the $\iota = 1$ surface.

5.7 PFIRSCH–SCHLÜTER CURRENT DRIVEN MAGNETIC ISLANDS IN STELLARATORS

The existence of MHD equilibrium in three-dimensional magnetic confinement systems such as stellarators and heliotrons is a problem of considerable subtlety, as explained in chapter 4. It is known that even vacuum magnetic fields in stellarators and heliotrons can possibly destroy magnetic surfaces without compensation for symmetry breaking by a judicious choice of external helical coils. This has been done in practice according to theoretical guidance. Once the vacuum surfaces are produced, it is necessary to consider the effect of finite plasma pressure. We need to study the formation of magnetic islands due to Pfirsch–Schlüter currents caused by plasma pressure gradients. If neighboring magnetic islands overlap, spatial stochasticity occurs and magnetic surfaces are destroyed completely, which leads to a loss of MHD equilibrium. As a simple model for the equilibrium near a magnetic island, we assume that the magnetic surfaces are labeled by a radial-like variable Φ in the limit that the island width is sufficiently small compared to equilibrium length scales. Here Φ represents the toroidal flux through a magnetic surface. The equilibrium magnetic field with the magnetic island is shown by

$$\mathbf{B} = \nabla\Phi \times \nabla\alpha + \nabla\zeta \times \nabla\Psi \tag{5.298}$$

in the magnetic coordinates (Φ, θ, ζ), where $\alpha = \theta - \iota_s\zeta$ is the helical resonant angle near the resonant surface satisfying $\iota(\Phi_s) = \iota_s = n_0/m_0$, θ is the poloidal angle, ζ is the toroidal variable, and Φ_s denotes Φ at the resonant surface. It is noted that the first term of (5.298) denotes the toroidal magnetic field. The helical flux function Ψ is given by

$$\Psi = \int (\iota - \iota_s)dx - \Psi_s \cos(m_0\alpha), \tag{5.299}$$

where $x = \Phi - \Phi_s$. The coordinate system defined by (Φ, α, ζ) will be used to describe the magnetic configuration in this section. The symmetry breaking magnetic field responsible for the magnetic island is given by $\Psi_s \cos(m_0\alpha)$, where we assume that Ψ_s varies weakly with Φ near the resonant surface which corresponds to the constant Ψ approximation in the tearing mode theory explained in section 5.5. Since Ψ in the expression (5.299) includes the toroidal angle ζ only through α in the coordinates (Φ, α, ζ), $\mathbf{B} \cdot \nabla\Psi = 0$ is obtained. When $\iota(\Phi)$ is expanded around the resonant surface of ι_s as $\iota(\Phi) = \iota_s + \iota_s' x$, an approximate analytic form for the magnetic surfaces near the resonant surface is given by

$$\Psi = \frac{\iota_s' x^2}{2} - \Psi_s \cos(m_0\alpha), \tag{5.300}$$

where ι_s' is evaluated at $\Phi = \Phi_s$. This representation of the helical flux function describes a magnetic island at $\Phi = \Phi_s$ with a half-width given by

$$w = 2(|\Psi_s/\iota_s'|)^{1/2}. \tag{5.301}$$

Here we will take Ψ_s to have the same sign as ι_s' so that the O-point of the magnetic island is located at $\alpha = 0$, and the X-point of the magnetic island is located at $m_0\alpha = \pm\pi$ (see Fig. 2.14 and (2.149)).

The variation of the magnetic field strength is quantified by the Jacobian

$$J = (\nabla\Phi \times \nabla\alpha \cdot \nabla\zeta)^{-1} \tag{5.302}$$

in the magnetic coordinates (Φ, α, ζ). For a three-dimensional equilibrium magnetic configuration, the Jacobian is written as

$$J = \sum_{m,n} J_{mn}(\Phi) \exp(im\theta - in\zeta), \tag{5.303}$$

using a Fourier series. The Jacobian is also related to the structure of the equilibrium magnetic field. In particular, the specific volume $V' \equiv dV/d\Phi$ is described using the component J_{00} in (5.303) as

$$V' = \int_0^{2\pi} \frac{d\theta}{2\pi} \int_0^{2\pi} \frac{d\zeta}{2\pi} J = J_{00}(\Phi). \tag{5.304}$$

The derivative of the specific volume corresponds to the normal curvature of the averaged magnetic field, where $V'' > 0$ (< 0) indicates a magnetic hill (well) and determines the stability of low-beta plasmas against resistive interchange modes, since shear stabilization is not expected for resistive modes. The specific volume of a flux tube closed on the resonant surface $\iota_s = n_0/m_0$ is given by

$$\oint \frac{d\ell}{B} = \oint \frac{d\zeta}{B_\zeta} = \oint \frac{d\zeta}{(\nabla\Phi \times \nabla\alpha)\cdot\nabla\zeta} = \bar{J} = \sum_\ell J_{\ell m_0, \ell n_0} \exp(i\ell m_0 \alpha). \tag{5.305}$$

For simplicity we assume that the resonant value of the Jacobian $\bar{J}$ is given by the first two terms in the Fourier expansion,

$$\bar{J} = J_{00}(\Phi) + J_{m_0 n_0} \cos(m_0\alpha), \tag{5.306}$$

where the reality condition has been employed.

The parallel current is determined by solving the MHD equilibrium near the magnetic island. From the quasi-neutrality condition $\nabla \cdot \mathbf{J} = 0$, which is equivalent to $\mu_0 \mathbf{J} = \nabla \times \mathbf{B}$,

$$\mathbf{B} \cdot \nabla\sigma = -\nabla \cdot \mathbf{J}_\perp, \tag{5.307}$$

where $\sigma = \mathbf{J}\cdot\mathbf{B}/B^2$. The force blance equation $\mathbf{J}\times\mathbf{B} = \nabla P$ gives $\mathbf{J}_\perp = \mathbf{B}\times\nabla P/B^2$. When this is approximated by $\mathbf{J}_\perp \simeq \nabla\zeta \times \nabla P/B_\zeta$ and substituted into the RHS of (5.307),

$$\nabla\zeta \cdot \nabla\Psi \times \nabla\sigma = -\nabla\zeta \cdot \nabla P \times \nabla\bar{J} \tag{5.308}$$

is obtained after averaging over the angle ζ. It is assumed that $\nabla\Phi \times \nabla\alpha$ in (5.298) corresponds to the constant toroidal field. Thus σ depends on Φ and α. Since $P = P(\Psi)$, (5.308) may have the solution $\sigma = -P'(\Psi)\bar{J} + h(\Psi)$, where $h(\Psi)$ is an undetermined function of Ψ. From Ohm's law for the equilibrium state,

$$-\mathbf{B}\cdot\nabla\phi = \eta \mathbf{J}\cdot\mathbf{B} \tag{5.309}$$

is obtained by taking the component along the magnetic field line, where ϕ is the electrostatic potential and η is the resistivity. By averaging (5.309) over a magnetic surface at $\Phi \simeq \Phi_s$, the LHS is annihilated and a constraint on the parallel current will be

$$\langle\sigma\rangle = 0, \tag{5.310}$$

where the magnetic surface average in the vicinity of the resonant surface with (5.300) is defined by

$$\langle Q \rangle = \frac{\oint Q\left(\frac{\partial \Psi}{\partial x}\right)^{-1} d\alpha}{\oint \left(\frac{\partial \Psi}{\partial x}\right)^{-1} d\alpha}. \tag{5.311}$$

It is noted that magnetic surface averaged quantities satisfy $\mathbf{B} \cdot \nabla \langle Q \rangle = 0$. Using the constraint (5.310), the parallel current profile is given by

$$\begin{aligned} \sigma &= P'(\Psi)[\langle \bar{J} \rangle - \bar{J}] \\ &\simeq P'(\Psi)V''[\langle x \rangle - x], \end{aligned} \tag{5.312}$$

where we have used (5.306) and expanded J_{00} as

$$J_{00} = V' + V''x \tag{5.313}$$

by noting (5.304). It is noted here that the second term of (5.306) is assumed to be negligible. The current proportional to V'' in (5.312) is interpreted as the parallel current arising from a resistive interchange perturbation for $V'' > 0$. It should be noted that $V'' > 0$ is an instability condition that is valid only for low-beta plasmas, and the stability criterion for resistive interchange modes is given by Glasser, Greene, and Johnson (1975) for finite-beta plasmas.

The local parallel current gives rise to a discontinuity in the magnetic field as the resonant surface is crossed. In tearing mode theory this discontinuity is measured by the parameter Δ'. Here we evalute Δ' from the local parallel current given by (5.312). The magnetic field produced by the parallel current is evaluated with a helical flux function Ψ through $\mathbf{B}_\perp = \nabla\zeta \times \nabla\Psi(\Phi, \alpha)$, where Ψ is a solution of

$$\nabla_\perp^2 \Psi = B_0 \sigma. \tag{5.314}$$

in the neighborhood of the resonant surface. Then the resonant symmetry breaking magnetic field is given by $\partial\Psi/\partial\alpha \neq 0$. The parameter Δ' is defined at the resonant surface by the relation

$$\Delta' = \left[\frac{\partial \Psi(\Phi_s^+)}{\partial x} - \frac{\partial \Psi(\Phi_s^-)}{\partial x}\right] / \Psi(\Phi_s), \tag{5.315}$$

under the assumption that the radial variation is largest, where $\Phi_s^\pm$ denote $\Phi_s^+ = \Phi_s + \delta\Phi$ and $\Phi_s^- = \Phi_s - \delta\Phi$, respectively, with an infinitesimal quantity of $\delta\Phi$ and $\Psi(\Phi_s) = \Psi_s$ in (5.300). It should be noted that Δ' given by (5.315) is evaluated by components of $\partial\Psi/\partial\alpha \neq 0$. From (5.314), $\Delta'\Psi_s$ is obtained by integrating the angle variable α after $\cos(m_0\alpha)$ is multipled, and then the radial variable $(-\infty < x < \infty$ except $x \simeq 0)$ with the boundary conditions

$\lim_{x\to\pm\infty}\sigma = 0$. The RHS of (5.314) can be evaluated with the parallel current (5.312) as

$$\Delta'\Psi_s = B_0\int_{-\infty}^{\infty}dx\int_0^{2\pi}\frac{d\alpha}{2\pi}\cos(m_0\alpha)\sigma$$
$$= B_0P'V''\int_{-\infty}^{\infty}dx\int_0^{2\pi}\frac{d\alpha}{2\pi}\cos(m_0\alpha)[\langle x\rangle - x]. \tag{5.316}$$

We note that x is obtained from (5.300) and approximated for small Ψ_s as

$$x \simeq \pm\left(\frac{2}{\iota_s'}\Psi\right)^{1/2}\left(1+\frac{\Psi_s}{2\Psi}\cos(m_0\alpha)\right). \tag{5.317}$$

Substituting this expression into (5.316) yields

$$\Delta' \simeq -\frac{B_0P'V''}{2\iota_s'}\int_{\Psi_s}^{\Psi_c}\frac{d\Psi}{\Psi}. \tag{5.318}$$

Here $\langle x\rangle \simeq x$ is assumed for $\Psi \gtrsim \Psi_c$. It is noted that (5.318) is a rough approximation for $\Psi \gtrsim \Psi_s$. More accurate evaluations of the RHS in (5.318) are given by Cary and Kotschenreuther (1985), and Hegna and Bhattacharjee (1989). However, the result that the RHS of (5.318) is positive for $P' < 0$, $V'' > 0$, and $\iota_s' > 0$ does not change, since the integral with respect to Ψ is positive. Here $\Delta' > 0$ means that the magnetic island can exist in the equilibrium state, which is the same as for the magnetic island due to the tearing mode.

This result suggests that a three-dimensional equilibrium can be viewed as a two-dimensional equilibrium perturbed with saturated magnetic islands driven by resistive interchange modes in the case of $V'' > 0$. The restriction that the pressure is a function of $\oint d\ell/B$ is not applicable at the resonant surfaces for three-dimensional stellarator equilibria when the existence of well-defined magnetic surfaces is not imposed. Consequently, the distinction between symmetry breaking magnetic perturbations in the saturated state of MHD instabilities and three-dimensional equilibrium magnetic fields may disappear. In heliotrons with a $V'' > 0$ region, magnetic islands appear due to resistive interchange modes in finite-pressure plasmas. From this point of view, the equilibrium beta limit can be defined at the beta value at which island overlapping to destroy magnetic surfaces occurs.

5.8 THE PRESSURE-DRIVEN SAWTOOTH IN THE HELIOTRON E

In tokamak experiments an oscillatory behavior in the form of a sawtooth is often seen in the soft X-ray emission of the central region (see Fig. 5.14, showing a current-driven sawtooth in the Heliotron E). A plausible explanation for this sawtooth oscillation is as follows. First, the central electron temperature rises because of ohmic heating. The resulting increase in electric conductivity leads to an increase in the current density on the magnetic axis and, as a consequence, the safety factor $q(0)$ decreases below unity. The plasma is then vulnerable to an $m = 1$ internal kink instability which produces a magnetic island due to the finite resistivity. The subsequent evolution of the magnetic island induces transport of particles and energy from the central region to the outer edge region. Finally, the plasma relaxes to an axisymmetric state with $q(0) > 1$, and the whole process is then repeated. Almost identical sawtooth oscillations with phase inversion around the $\iota = \iota_h + \iota_J = 1$ surface are found in the ohmic-heating low-beta Heliotron E plasma, as shown in Fig. 5.14, which will be discussed in the next section. Here ι_h is a rotational transform produced by the external helical magnetic field and ι_J is a rotational transform due to the net toroidal plasma current.

In stellarator and heliotron experiments, two approaches for currentless plasma confinement are pursued. Currentless high-density plasmas have been successfully obtained in the WVII-A stellarator (Max-Planck Institute, Germany) by reducing the ohmic-heating current during neutral-beam injection (NBI). On the other hand, the Heliotron E obtained currentless plasmas by injecting the neutral beam into electron-cyclotron-resonance-heated (ECRH) plasmas with a gyrotron frequency of 53 GHz at $B_0 = 19$ kGauss. The Heliotron E also produced high-beta currentless plasmas via reduction of the magnetic field strength at the magnetic axis to 9.4 kGauss with second-harmonic heating for target plasma production. Recently, these ECRH plasma production become standard in the stellarator and heliotron research worldwide.

In the Heliotron E high-beta experiments, two types of discharge were observed, which depended on the gas puffing conditions during NBI. The target plasma with $\bar{n}_e \simeq 1 \times 10^{13}\,\mathrm{cm}^{-3}$ was first produced by ECRH at $B_0 = 9.4$ kGauss, where $\bar{n}_e$ is an average density. The density was increased from $1 \times 10^{13}\,\mathrm{cm}^{-3}$ during about 30 msec and an NBI of about 1 MW was started. During this NBI, accompanied by intense gas puffing, the gyrotron was turned off and a density increased up to $7 - 8 \times 10^{13}\,\mathrm{cm}^{-3}$ was accomplished during about 60 msec. At this stage, the second NBI of 1.0–1.5 MW was turned on for about 40 msec; then the electron and ion temperatures were increased to $T_e(0) \simeq T_i(0) \simeq 300 - 400$ eV, which gave high-beta plasmas with $\beta(0) \leq 3.6\%$.

When the gas puffing is weakened during the second NBI, a sawtooth with precursor fluctuations is seen in the soft X-ray signal from the central chord for $\beta(0) \geq 1\%$, as shown in the lower part of Fig. 5.12. The other soft X-ray signal comes from the half-radius region, which shows an inverted sawtooth behavior. The upper part of Fig. 5.12 shows a similar inverted sawtooth behavior for the line densities along both the half-radius (#1) and the edge region (#0). The soft X-ray signal of the central chord typically decreases by 20–30% in 1.0–1.5 msec. The estimated mode number of the precursor oscillation is $(m, n) = (1, 1)$. Density and magnetic fluctuations (the latter one is not shown here) are also observed, which correlate with the soft X-ray fluctuations. At the time of the sawtooth crash, an electron energy flow toward the edge region was detected with a bolometer.

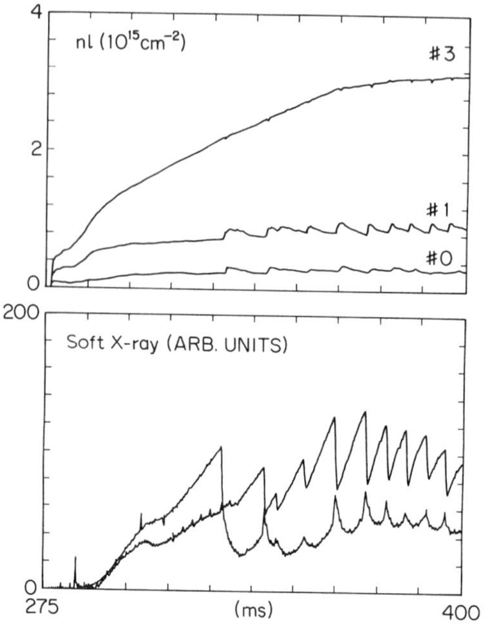

Fig. 5.12 The behavior of density (upper figure) and soft X-ray emissions (lower figure) in the case of sawtooth-like oscillation in a Heliotron E high-beta plasma. #0, #1, and #3 are the chords for density measurements with a FIR interferometer at the edge chord, the half-radius chord and the central chord, respectively. The upper trace in the lower figure shows soft X-ray emission from the central region and the lower trace from the half-radius region.

When intense gas puffing is continued during the NBI, the fluctuations in the soft X-ray signals become weak or even disappear at a fairly high beta value. The main difference between the unstable and stable (or marginal) cases is the pressure profile. The unstable discharges have peaked pressure profiles, while broader profiles are found in the weakly unstable or stable discharges. This tendency is consistent with results of linear analyses of ideal MHD stability for the low-n pressure-driven modes in the Heliotron E.

In this section we discuss the nonlinear evolution of the $(m, n) = (1, 1)$ pressure-driven mode in a resistive plasma in the limit of the straight cylindrical approximation of the Heliotron E. To study the nonlinear evolution, we solve the reduced MHD equations (3.230), (3.231), and (3.232) by adding several dissipative terms,

$$\frac{\partial}{\partial t} \nabla_\perp^2 u = [u, \nabla_\perp^2 u] + [\nabla_\perp^2 A, \Psi] + \frac{\partial}{\partial z} \nabla_\perp^2 A + [\Omega, P] + \bar{\mu} \nabla_\perp^4 u, \tag{5.319}$$

$$\frac{\partial A}{\partial t} = [u, \Psi] + \frac{\partial u}{\partial z} + \bar{\eta} \nabla_\perp^2 A, \tag{5.320}$$

$$\frac{\partial P}{\partial t} = [u, P] = \bar{\chi}\nabla_\perp^2 P, \tag{5.321}$$

where the viscosity $\bar{\mu}$ and the thermal conductivity $\bar{\chi}$ are added in (5.319) and (5.321), respectively, to obtain a saturated state. Here z denotes a coordinate in the longitudinal direction and the term including the toroidal effect is neglected in $[\Omega, P]$ in (5.319), since the $(m, n) = (1, 1)$ resistive interchange mode destabilized by the magnetic hill exists even in the straight cylindrical approximation of the Heliotron E. The resistivity included in Ohm's law induces reconnection of magnetic field lines during the growth of the resistive interchange mode.

In the high-beta Heliotron experiments, typical plasma parameters are $T_e(0) \simeq T_i(0) \simeq 300\,\text{eV}$, $n_e(0) \simeq 10^{14}\,\text{cm}^{-3}$, and $B_0 = 9.4\,\text{kGauss}$, which give a magnetic Reynolds number of $S = 1/\bar{\eta} \simeq 2.5 \times 10^5$ from an expression for hydrogen plasmas:

$$S = 2.6 \times 10^5 \frac{[a(\text{cm})][B(\text{Gauss})][T(\text{eV})]^{3/2}}{[n(\text{cm}^{-3})]^{1/2} \ln \Lambda}. \tag{5.322}$$

Here the poloidal magnetic field estimated with $R/a \simeq 10$ and $\iota_h(a) \simeq 2$ is substituted in (5.322) and $\ln \Lambda \simeq 20$ is used. When Z_{eff} (the effective ion charge) is increased up to 3 with impurity ions in hydrogen plasmas, S ($\propto 1/Z_{eff}$) may be about 10^5. The viscosity in the vorticity equation (5.319) is appropriately given to keep the numerical stability of the finite difference equations.

The numerical scheme to solve (5.319), (5.320), and (5.321) as an initial and boundary value problem is as follows. The three functions u, A, and P are Fourier expanded with respect to the variable $\zeta = m\theta - nz/R$,

$$\begin{cases} u = \sum i u_\ell e^{i\ell\zeta} \\ A = \sum A_\ell e^{i\ell\zeta} \\ P = \sum P_\ell e^{i\ell\zeta}, \end{cases} \tag{5.323}$$

where $2\pi R$ is the periodic length of the cylindrical plasma. These expansions are correct for the single helicity case in which only harmonics $(\ell m, \ell n)$ of the fundamental mode (m, n) are considered. The phase differences amongst u, A, and P imposed by the MHD equations are taken into account in (5.323). By substituting (5.323) into (5.319), (5.320), and (5.321), and eliminating the phase factor $e^{iN\zeta}$, we obtain

$$\begin{aligned} \frac{\partial V_N}{\partial t} = &- \sum_{\ell_1+\ell_2=N} \frac{\ell_2 m}{r}\left[\left(\frac{\partial u_{\ell_1}}{\partial r} V_{\ell_2} - \frac{\partial V_{\ell_1}}{\partial r} u_{\ell_2}\right) + \left(\frac{\partial J_{\ell_1}}{\partial r} A_{\ell_2} - \frac{\partial A_{\ell_1}}{\partial r} J_{\ell_2}\right)\right] \\ &+ \frac{Nm}{r} J_N \frac{d\Psi_h}{dr} + \frac{Nm}{r} P_N \frac{d\Omega}{dr} + \frac{Nn}{R} J_N \\ &+ \bar{\mu}\left(\frac{1}{r}\frac{d}{dr} r \frac{d}{dr} - \frac{N^2 m^2}{r^2}\right) V_N, \end{aligned} \tag{5.324}$$

$$\frac{\partial A_N}{\partial t} = -\sum_{\ell_1+\ell_2=N} \frac{\ell_2 m}{r}\left(\frac{\partial u_{\ell_1}}{\partial r} A_{\ell_2} - \frac{\partial A_{\ell_1}}{\partial r} u_{\ell_2}\right) - \frac{Nm}{r} u_N \frac{d\Psi_h}{dr} + \frac{Nn}{R} u_N - \bar{\eta} J_N, \tag{5.325}$$

$$\frac{\partial P_N}{\partial t} = -\sum_{\ell_1+\ell_2=N} \frac{\ell_2 m}{r}\left(\frac{\partial u_{\ell_1}}{\partial r} P_{\ell_2} - \frac{\partial P_{\ell_1}}{\partial r} u_{\ell_2}\right) + \bar{\chi}\left(\frac{1}{r}\frac{d}{dr} r \frac{d}{dr} - \frac{N^2 m^2}{r^2}\right) P_N, \tag{5.326}$$

where the vorticity $V = \nabla_\perp^2 u$ and the plasma current $J = -\nabla_\perp^2 A$ are shown as

$$V_N = \left(\frac{1}{r}\frac{d}{dr} r \frac{d}{dr} - \frac{N^2 m^2}{r^2}\right) u_N, \tag{5.327}$$

$$J_n = -\left(\frac{1}{r}\frac{d}{dr} r \frac{d}{dr} - \frac{N^2 m^2}{r^2}\right) A_N. \tag{5.328}$$

To develop a numerical code, the derivative with respect to r can be approximated by finite-difference forms: $df/dr = (f_{j+1} - f_{j-1})/(2\Delta r)$ and $d^2 f/dr^2 = (f_{j+1} - 2f_j + f_{j-1})/(\Delta r)^2$, where Δr is a grid size and j denotes the grid point at $r = (j-1)\Delta r$. The total grid number is $J_{\max}$ and the total number of grid points is $J_{\max} + 1$.

There are several methods of time advancement. For the time evolution equation in the following form,

$$\frac{\partial f}{\partial t} = F(f), \tag{5.329}$$

a predictor–corrector method is given by

$$\begin{cases} \dfrac{f^{i+1/2} - f^i}{(\Delta t/2)} = F(f^i) \\ \dfrac{f^{i+1} - f^i}{\Delta t} = F(f^{i+1/2}), \end{cases} \tag{5.330}$$

or an implicit method is given by

$$\frac{f^{i+1} - f^i}{\Delta t} = F(f^{i+1}). \tag{5.331}$$

here $F(f)$ denotes a functional of f and Δt denotes a time difference between the ith and $(i+1)$th steps. Numerical stability is ensured in these two numerical methods. Since $F(f)$ includes the first and second derivative of f, f^{i+1} in (5.331) is obtained by solving an algebraic equations given in a matrix form for f_j^{i+1} ($j = 1$ to $j = L$). The quantity J_N is found from A_N by the numerical derivatives through (5.328). The quantity u_N is found from V_N by solving (5.327). It is formally shown as

$$A_j u_{j+1} - B_j u_j + C_j u_{j-1} = D_j, \tag{5.332}$$

by applying the finite-difference forms to the derivatives in (5.327). This type of equation can be solved by assuming a recurrence formula, $u_{j+1} = E_j u_j + F_j$; and we find relations for calculating (E_{j-1}, F_{j-1}) from (E_j, F_j):

$$\begin{cases} E_{j-1} = -\dfrac{C_j}{A_j E_j - B_j} \\ F_{j-1} = -\dfrac{A_j}{A_j E_j - B_j} F_j + \dfrac{D_j}{A_j E_j - B_j}. \end{cases} \tag{5.333}$$

The boundary conditions for solving (5.324), (5.325), and (5.326) are given by

$$A_N(0) = J_N(0) = u_N(0) = V_N(0) = P_N(0) = 0 \quad (\text{for } N \neq 0) \tag{5.334}$$

at the magnetic axis and $dA_0/dr|_{r=0} = 0$, $dJ_0/dr|_{r=0} = 0$, and $dP_0/dr|_{r=0} = 0$. There is no $N = 0$ component for u and V, since the equilibrium state has no poloidal plasma flow here. Also, for the choice of (5.323), the nonlinear coupling terms in (5.324) cannot produce a u_0 component. At the plasma surface determined by the conducting wall, the boundary conditions are

$$A_N(1) = J_N(1) = u_N(1) = V_N(1) = P_N(1) = 0 \quad (\text{for } N \neq 0). \tag{5.335}$$

It is noted that $dA_0/dr|_{r=1} = -B_\theta|_{r=1}$, $J_0(1) = 0$, and $P_0(1) = 0$. It is noted that $B_\theta|_{r=1}$ is finite only for current-carrying plasmas. From $u_L(1) = 0$, $E_{J_{max}} = F_{J_{max}} = 0$ are imposed on the relations (5.333), and (E_j, F_j) with $j = 1, 2, \cdots, J_{max} - 1$ are obtained by using (5.333) iteratively.

For the Heliotron E model the rotational transform profile is chosen as $\iota_h(r) = 0.51 + 1.69(r/a)^{2.5}$, which is an approximation for ι_h obtained by the line-tracing method. The pressure profile is chosen as $P = P_0(1 - (r/a)^2)^2$, where P_0 is the central pressure directly related to β_0. In the straight approximation, for a configuration with finite pressure and no net plasma current, the RHS of (5.319) vanishes with $u = 0$ and $A = 0$, since Ω becomes a function of r only. This state corresponds to MHD equilibrium in the lowest order of the stellarator expansion. Linear stability of the ideal interchange mode with $(m, n) = (1, 1)$ will give the beta limit $\beta(0) = 1.62\%$ as shown in Fig. 5.9 by the Suydam criterion. However, the numerical solution using the finite difference approximation shows exponential growth only for $\beta(0) \gtrsim 2\%$, since the growth rate becomes extremely small for $\beta(0) < 2\%$. When the resistivity is included in Ohm's law (5.320), resistive interchange modes become unstable even for $\beta(0) < 1.62\%$ and the growth rate of the resistive interchange mode with $(m, n) = (1, 1)$ becomes sustantial for $\beta(0) < 2\%$. To demonstrate that the nonlinear resistive interchange mode with $(m, n) = (1, 1)$ is a probable candidate that can explain the pressure-driven sawtooth, the parameters are chosen as $\beta(0) = 2.7\%$, $S = 5 \times 10^3$, $\bar{\chi}_\perp = 4 \times 10^{-4}$, and $\bar{\mu} = 6 \times 10^{-6}$. The resistivity profile is assumed to be $\eta(r) \propto [(1 - (r/a)^2) + 0.01]^{-1}$. An initial perturbation with $(m, n) = (1, 1)$ is given adequately. In the nonlinear calculation of the interchange mode as the initial value problem, single helicity is assumed and ten harmonics are included. The time evolution of the pressure profile is shown in Fig. 5.13. Here the time is normalized with the poloidal Alfvén time, which

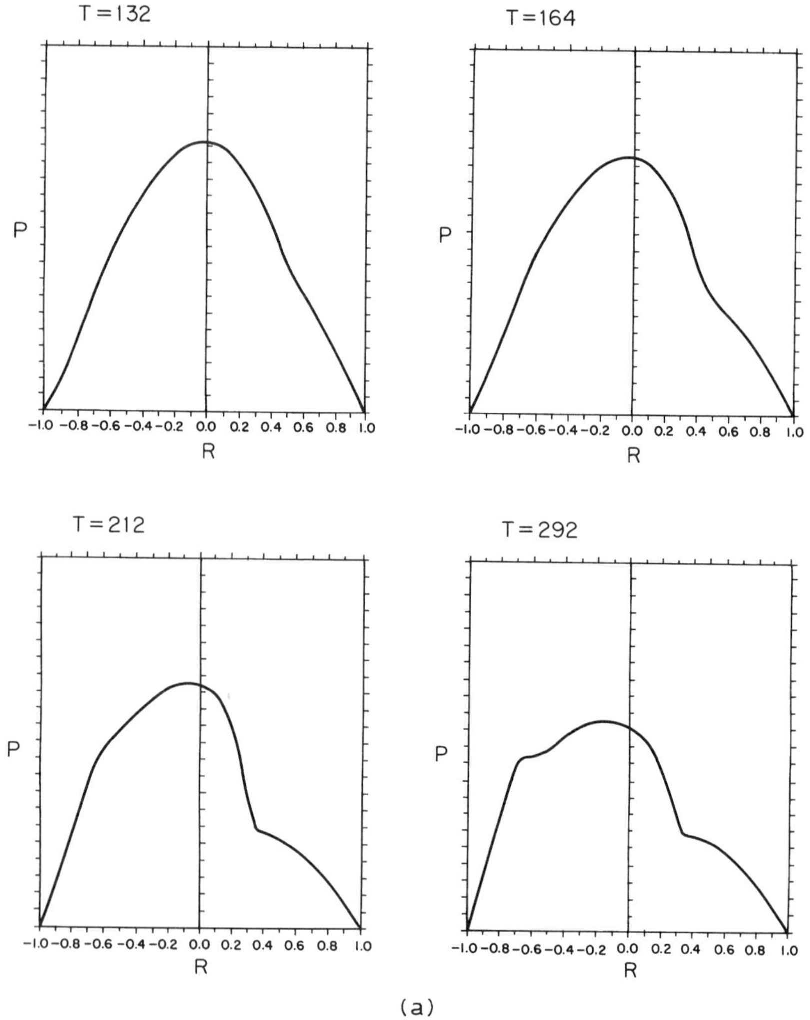

Fig. 5.13 (a) The time evolution of the pressure profiles along the horizontal diameters in (b).

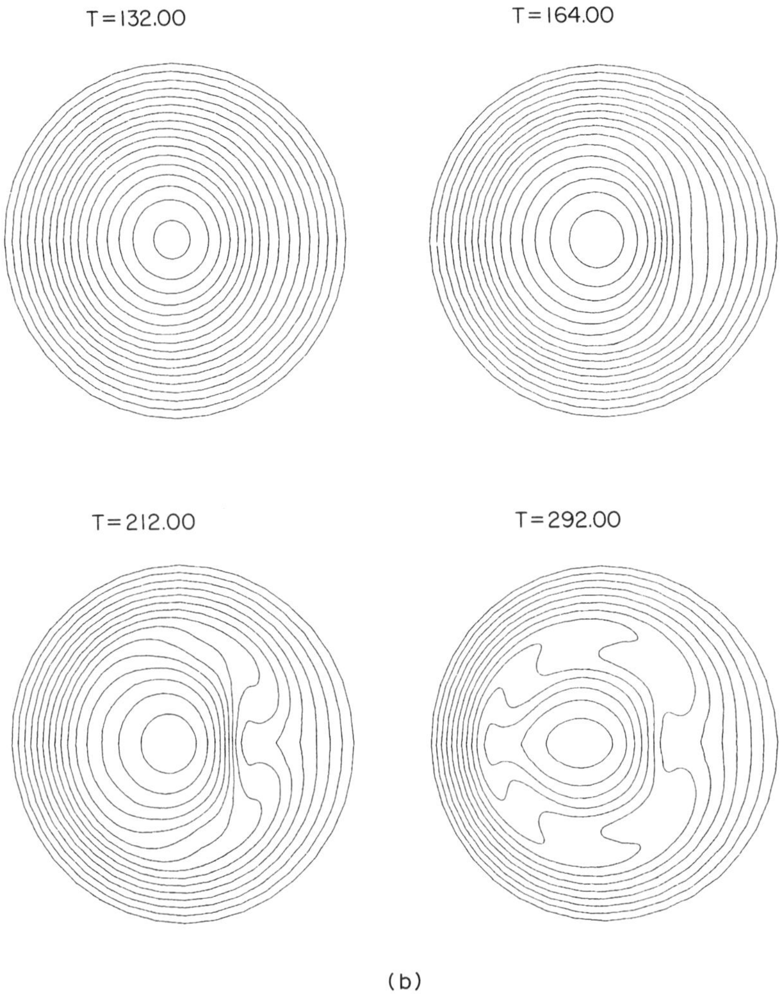

Fig. 5.13 (*Continued*) (b) The time evolution of pressure contours for the pressure gradient driven modes for $S = 5 \times 10^3$ and $\beta_0 = 2.7\%$. The initial pressure profile is assumed to be $P = P_0(1 - (r/a)^2)$.

is roughly $0.5\,\mu$sec for the typical Heliotron E high-beta plasma with $n(0) \simeq 10^{14}\,\text{cm}^{-3}$ at $B_0 = 9.4\,$kGauss under $\iota_h(a) \simeq 2$ and $R/a = 10$. Figure 5.13(a) shows the pressure profiles along a diameter through $\theta = 0$ and $\theta = \pi$. The pressure profile deforms because of the growth of the $(m, n) = (1, 1)$ interchange mode and thermal diffusion. The figure for $T = 132$ corresponds to the growing stage. At $T = 164$, reconnection of the magnetic field lines occurs. $T = 292$ corresponds to the saturation phase, and the beta value in the central region decreases to $\beta(0) \simeq 1.7\%$ from the initial value of $\beta(0) = 2.7\%$. After $T = 292$, the beta value still decreases gradually through the thermal diffusion. The corresponding contour plots of the pressure are shown in Fig. 5.13(b). The reconnection is triggered by the increase of the local parallel current density in the neighborhood of the $\iota_h = 1$ resonant surface for the resistive plasma. This local increase in the current density is induced by an accumulation of poloidal magnetic flux at the resonant surface due to the convective plasma motion driven by the pressure-driven interchange mode. The maximum local current density is about $65\,\mathrm{A/cm^2}$ at $T \simeq 164$ and $250\,\mathrm{A/cm^2}$ at $T \simeq 292$ for the example shown in Fig. 5.13.

Thus the nonlinear evolution of the pressure-driven resistive interchange mode with $(m, n) = (1, 1)$ seems a probable candidate that can explain the sawtooth crash observed in the currentless plasma of the Heliotron E.

5.9 THE CURRENT-DRIVEN SAWTOOTH IN THE HELIOTRON E

Since the Heliotron E has a rotational transform that increases toward the plasma surface with $\iota_h(0) \simeq 0.5$ and $\iota_h(a) \simeq 2.2$, two resonant surfaces satisfying $\iota = \iota_h + \iota_J = 1$ appear for current-carrying plasmas with highly peaked profiles. When a plasma current I_p is induced to increase the rotational transform due to the external helical magnetic field (additive case), sawtooth oscillation appears at $I_p \simeq 17\,\mathrm{kA}$. This current produces $\iota_J(a) \simeq 0.1$ in the Heliotron E in the case of $B_0 = 17\,\mathrm{kGauss}$. On the other hand, there is no observation of a sawtooth when I_p is induced to decrease the external rotational transform (subtractive case). In Fig. 5.14, the increase in I_p is related to

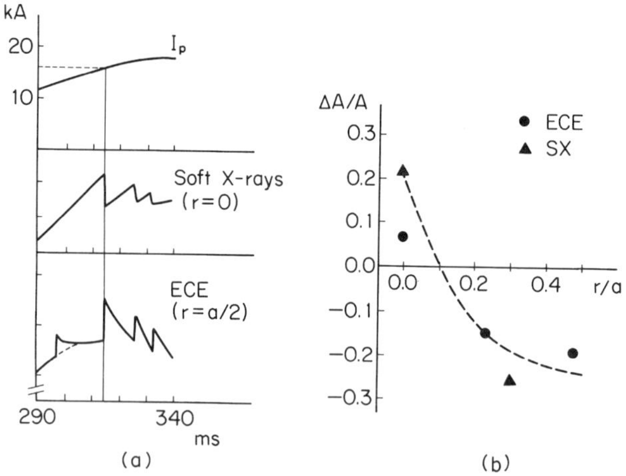

Fig. 5.14 (a) The time evolution of the plasma current I_p, soft X-ray emission at the central region $r \simeq 0$, and the second harmonic of electron cyclotron emission at the half-radius, $r \simeq a/2$. (b) Relative sawtooth amplitudes measured from electron cyclotron emission and soft X-ray emission are shown as a function of the radius. The sawtooth inversion radius is roughly $0.1(r/a)$.

the increase in the electron temperature by NBI heating at constant-voltage operation. The sawtooth phase inversion radius of soft X-ray and electron cyclotron emission seems to be $0.1r/a$, which may be interpreted as the position of the $\iota = 1$ (or $q = 1$) resonant surface. The overall characteristics of the current-driven sawtooth as shown in Fig. 5.14 are similar to those found in tokamak plasmas.

When the $\iota = \iota_h + \iota_J = 1$ resonant surface exists in a current-carrying stellarator, the internal kink mode with $(m, n) = (1, 1)$ becomes unstable under the ideal MHD approximation. It is known that a skin-like current appears in the neighborhood of the resonant surface due to convective motion driven by the internal kink mode under the frozen-in condition. When the plasma resistivity is finite, reconnection of magnetic field lines occurs in the nonlinear stage and a magnetic island is formed. This process is similar to the nonlinear evolution of the unstable interchange mode in the resistive currentless plasma discussed in section 5.8. The reconnection process continues until the hot plasma core is squeezed outside the inner $\iota = 1$ surface. This process coincides with a flattening of the ι-profile. It is noted that, when ι_J decreases rapidly in the radial direction for a peaked current profile, the inner $\iota = \iota_h + \iota_J = 1$ surface is near the magnetic axis and the outer $\iota = \iota_h + \iota_J = 1$ surface is near the surface, satisfying $\iota_h = 1$ in the Heliotron E.

Since the ohmic heating works effectively in the core plasma, the central electron temperature may increase again after the disappearance of the unstable $\iota > 1$ (or $q < 1$) region near the magnetic axis. By the associated increase of current density near the magnetic axis, the central ι-value will increase and the $\iota > 1$ region will also appear again. This physical picture of the current-drive sawtooth was described by Kadomtsev (1975).

The mode structure of the internal kink mode with $(m, n) = (1, 1)$ is distinguished from other modes by the fact that $m = 1$ implies a radial displacement the poloidal variation of which is $\cos(\theta + \theta_0)$, corresponding to a simple displacement of each flux surface, where θ_0 is a phase angle determining the direction of the displacement. The significance of this fact becomes apparent from the cylindrical approximation for the potential energy δW for internal modes obtained from (5.81),

$$\delta W = \frac{\mu_0 \delta W_F}{2\pi^2 R_0} = \int_0^a (f\xi'^2 + g\xi^2)dr, \tag{5.336}$$

where

$$f = \frac{r^3 F^2}{m^2 + k^2 r^2} \simeq r^3 \frac{B_z^2}{R_0^2}\left(\frac{1}{q} - \frac{n}{m}\right)^2 + O\left(r^3\left(\frac{nr}{R_0}\right)^2\right) \tag{5.337}$$

and

$$g = \frac{(m^2-1)rF^2}{m^2+k^2r^2} + \frac{k^2r^2}{m^2+k^2r^2}\left(2\mu_0 P' + rF^2 + \frac{2}{r}\frac{k^2r^2B_z^2 - m^2B_\theta^2}{m^2+k^2r^2}\right)$$
$$\simeq r\frac{B_z^2}{r_0^2}(m^2-1)\left(\frac{1}{q}-\frac{n}{m}\right)^2 + O\left(r\left(\frac{nr}{R_0}\right)^2\right). \tag{5.338}$$

Here ξ is the amplitude of the radial displacement, and $k = -n/R_0$ is used. Also, $nr/R_0 \ll 1$. It is seen that for $m = 1$, a rigid displacement, with $\xi =$ constant, reduces the integrand to its minimum value or to zero in the lowest order of (nr/R_0). We note that the boundary condition of the internal modes requires $\xi(a) = 0$ to minimize the surface term of the potential energy (see (5.81)). Thus the constant displacement ξ becomes zero. However, when there is a $q = 1$ resonant surface, the displacement $\xi = \xi_0 =$ constant within the $q = 1$ surface and $\xi = 0$ outside the $q = 1$ surface becomes possible. The inevitable variation of ξ between these two regions is taken up at the $q = 1$ surface. On this surface, the perturbation has the same helicity as the magnetic field or the parallel wave number vanishes: $k_\parallel = (m/r)(B_\theta/B) - (n/R_0)(B_z/B) = 0$. Consequently, the potential energy change can be made arbitrarily small by localization of $d\xi/dr$. From the expression (5.336), the potential energy change in the neighborhood of the $q = 1$ surface is proportional to

$$r_s^3\frac{B_z^2}{R_0^2}\int_{-\delta}^{0}\left(\frac{\xi}{\delta}\right)^2(q'x)^2dx, \tag{5.339}$$

where $x = r - r_s$, $\delta \ll 1$ and r_s denotes the radius of the $q = 1$ surface. The integral of (5.339) is $O(\delta)$ and becomes zero as $\delta \to 0$. Therefore, the minimum value of δW is zero in the leading order of (nr/R_0). Thus the mode is marginally stable to this order and the stability depends upon the sign of the $O((nr/R_0)^2)$ term in (5.336),

$$\delta W = \xi_0^2\int_0^{r_s} g_1 dr, \tag{5.340}$$

where g_1 is the value given by (5.338) for $m = 1$, correct to order $(kr)^2$ or $(nr/R_0)^2$,

$$g_1 = k^2r(2\mu_0 rP' + 3k^2r^2B_z^2 + 2krB_zB_\theta - B_\theta^2). \tag{5.341}$$

For any monotonically decreasing profile of pressure $P(r)$ and longitudinal current $J_z(r)$, $g_1 < 0$ for $kr \ll 1$ in $0 < r < r_s$; hence, $\delta W < 0$ and the $m = 1$ internal kink mode becomes unstable.

As an example, δW can be evaluated for the simple profiles

$$P(r) = \begin{cases} P_0(1-(r/r_0)^2)+P_1 & \text{for} \quad r < r_0 \\ P_1 & \text{for} \quad r > r_0 \end{cases}$$

$$J_z(r) = \begin{cases} J_{z0} & \text{for} \quad r < r_0 \\ 0 & \text{for} \quad r > r_0, \end{cases}$$

where r_0 is a radius less than r_s and P_1 = constant is much less than P_0. These profiles are somewhat unrealistic, but useful for an analytic estimation. For this cylindrical equilibrium model, we obtain

$$\delta W = -\tfrac{1}{2}k^2 r_0^2 \xi_0^2 (B_\theta(r_0))^2 [\beta_\theta + \tfrac{1}{2}(1-q_s) - \tfrac{1}{2}\ln q_s], \tag{5.342}$$

where $\beta_\theta = 2\mu_0 P_0/(B_\theta(r_0))^2$, and $q(r_0) = q_s = r_0^2/r_s^2$. Since $q_s < 1$, it is easily seen that $\delta W < 0$. It is noted that $k < 0$ is considered for the third term in the RHS of (5.341).

The calculation of the internal kink mode with $(m, n) = (1, 1)$ for tokamak plasmas is quite complicated. The result is that, for the large aspect ratio plasma with a circular cross-section, stability is obtained if $\beta_p \lesssim 0.3$. Here β_p is defined by

$$\beta_p = \frac{2\mu_0 R_0^2}{r_s^2 B_0^2} \int_0^{r_s} \left(-\frac{dP}{dr}\right) r^2 dr. \tag{5.343}$$

In heliotrons the function g appearing in the potential energy for the $(m, n) = (1, 1)$ internal mode is shown by

$$g = -k^2 B_0^2 r (3r\iota_h' + r^2 \iota_h'')(\iota_J + \iota_h - 1) \tag{5.344}$$

under the stellarator expansion for low-beta plasmas. The potential energy of the heliotron can be obtained from W in (5.291) by using $\xi = (m/r\omega)u_1$. An interesting point is that g does not ever become zero in the lowest order of (r/R_0) for the $m = 1$ mode in high-shear stellarators such as heliotrons. When both ι_h' and ι_h'' are positive, g makes a negative contribution to δW for $\iota_J + \iota_h > 1$ or $q < 1$ in the case of constant displacement. Therefore, the internal kink mode with $(m, n) = (1, 1)$ seems more unstable in heliotrons compared to tokamaks.

In the Heliotron E case, although there are two $\iota = 1$ resonant surfaces for the toroidal plasma current with a peaked profile, there are no unstable modes localized between the two resonant surfaces or localized between the outer $\iota = 1$ surface and the plasma surface. In the former case $\iota_J + \iota_h < 1$ or $q > 1$ eliminates such an instability, and in the latter case the rigid displacement ξ = constant cannot satisfy the boundary condition for the internal mode $\xi(a) = 0$.

According to the nonlinear calculation based on the reduced MHD equations for heliotrons, (5.319) and (5.320) under the assumption of zero pressure, $P(r) = 0$, and a finite toroidal current with a peak profile $J_z(r)$, it is shown that the observed current-driven sawtooth in the Heliotron E can be explained by the linearly unstable $(m, n) = (1, 1)$ internal kink mode and subsequent reconnection of the magnetic field lines in the nonlinear regime. This result seems consistent with the Kadomtsev model for the tokamak sawtooth in resistive plasmas with $S \lesssim 10^6$–10^7. The plasma parameters of the Heliotron E in the current-carrying NBI plasmas are $T_e(0) \simeq T_i(0) \simeq 600\,\mathrm{eV}$ and $n_e(0) = 4 \times 10^{13}\,\mathrm{cm}^{-3}$ for $B_0 = 18\,\mathrm{kGauss}$, which gives $S \simeq 2.3 \times 10^6$ by considering $\iota_h(a) \simeq 2.2$ and $R/a = 10$.

BIBLIOGRAPHY

Batemann, G. (1978). *MHD instabilities*. MIT Press, Cambridge.

Bernstein, I. B., Frieman, E. A., Kruskal, M. D., and Kulsrud, R. M. (1958). An energy principle for hydromagnetic stability problems. *Proc. R. Soc.*, **A244**, 17.

Braginskii, S. I. (1965). Transport processes in a plasma, in *Reviews of plasma physics*. Consultants Bureau, New York, vol. 1, p. 205.

Bussac, M. N., Pellat, R., and Edery, D. (1975). Internal kink modes in toroidal plasmas with circular cross sections. *Phys. Rev. Letters*, **35**, 1638.

Cary, J. R., and Kotshenreuther, M. (1985). Pressure induced islands in three-dimensional toroidal plasma. *Phys. Fluids*, **28**, 1392.

Connor, J. W., Hastie, R. J., and Taylor, J. B. (1978). Shear, periodicity, and plasma ballooning modes. *Phys. Rev. Letters*, **40**, 396.

Friedberg, J. P. (1987). *Ideal magnetohydrodynamics*. Plenum, New York.

Glasser, A. H., Greene, J. M., and Johnson, J. L. (1975). Resistive instabilities in general toroidal plasma configurations. *Phys. Fluids*, **18**, 875.

Greene, J. M. (1976). *Introduction to Resistive Instabilities, LRP 114/76*. Report of Centre de Recherches en Physique des Plasmas, Ecole Polytechnique Federale de Lansanne.

Hegna, C. C., and Bhattacharjee, A. (1989). Magnetic island formation in three-dimensional plasma equilibria. *Phys. Fluids*, **B1**, 392.

Kadomtsev, B. B. (1966). Hydromagnetic stability of a plasma, in *Reviews of plasma physics*. Consultants Bureau, New York, vol. 2, p. 153.

Kadomtsev, B. B. (1975). Disruptive instability in tokamaks. *Sov. J. Plasma Phys.*, **1**, 389.

Kulsrud, R. M. (1963). The interchange instability in the stellarator. *Phys. Fluids*, **6**, 904.

Laval, G., Mercier, C., and Pellat, R. (1965). Necessity of the energy principles for magnetostatic stability. *Nucl. Fusion*, **5**, 156.

Newcomb, W. A. (1958). Hydromagnetic stability of a diffuse linear pinch. *Ann. Phys.*, **3**, 347.

Pustovitov, V. D., and Shafranov, V. D. (1990). Equilibrium and stability of plasma in stellarators, in *Reviews of plasma physics*. Consultants Bureau, New York, vol. 15, p. 163.

Snydam, B. R. (1958). Stability of a linear pinch. *IAEA Geneva Conf.*, **31**, 157.

6

THE PARTICLE ORBIT IN HELIOTRONS

6.1 INTRODUCTION

In order to realize a continuously working fusion reactor, high-energy charged particles produced by fusion reactions must be confined to heat a background fuel plasma. In particular, the confinement of alpha particles generated by fusion reactions between deuterons and tritons is a crucial problem for a first-generation fusion reactor. After the alpha particles of 3.5 MeV slow down up to the fuel plasma temperature of several 10 keV, they become alpha ash. However, it is not necessary to confine all alpha particles. The confined alpha ash is harmful, since fuel ions are diluted by the increase of thermalized alpha particles. It is considered that the alpha particle density should be kept below 10% of the total ions in a fusion reactor using deuterium and tritium fuels.

In present-day experiments, high-energy ions are produced in tokamaks and stellarators by NBI (Neutral Beam Injection) with an energy of 30–100 keV or by the ICRF (Ion Cyclotron Range of Frequency) electromagnetic wave with a frequency of 20–50 MHz. In order to heat hydrogen or deuteron plasmas with NBI or ICRF, the confinement property of high-energy ions is directly related to the heating efficiency. In axisymmetric tokamaks there are two classes of particle orbit: one is passing or untrapped particles and the other is trapped particles. The particle trapped in a local mirror field produced by an inhomogeneity in the magnetic field due to the toroidal effect appears on the outside of the torus and moves around the torus slowly. When the guiding center orbit of the trapped particle is projected on the poloidal plane, it shows a banana shape. Due to this projected orbit shape, trapped particles in tokamaks are called banana particles.

In heliotrons the magnetic field has two types of ripples. One type consists of helical ripples, characterized by the pole number $\ell = 2$ and the pitch number M. The other type is the toroidal ripple, which is similar to that in tokamaks. If we plot $|\mathbf{B}(\psi, \theta, \phi)|$ as a function of ϕ, both a fast and a slow sinusoidal variation are seen. The fast variation corresponds to the helical ripple and the slow one to the toroidal ripple. When a particle is trapped in the helical ripple, it is called a *localized particle*, as shown in Fig. 6.1. On the other hand, the particle trapped in the toroidal ripple is called a *blocked particle*. Even in heliotrons, a particle that is not trapped in both the helical and toroidal ripples is called a

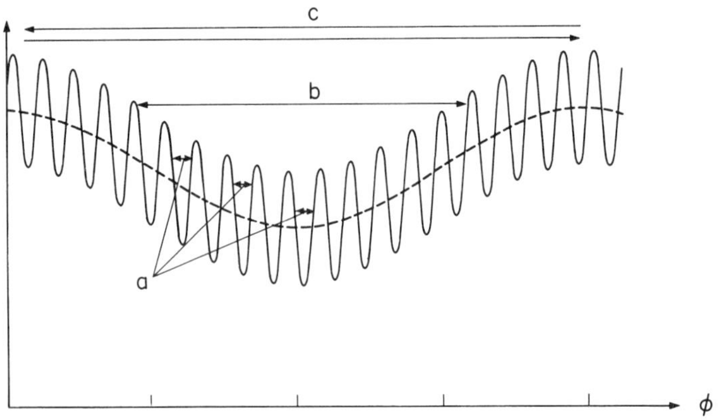

Fig. 6.1 Magnetic field ripples and three classes of particle orbit in a heliotron. a, Localized particle; b, blocked particle; c, passing particle.

passing particle. One significant characteristic of the particle orbit in heliotrons is that the transition between the localized particle orbit and the blocked particle orbit is possible without Coulomb collisons.

For the design of a stellarator, the first question is what magnetic field spectrum is suitable for the high-energy particle confinement. The next question is whether or not the finite beta effect on the magnetic field degrades particle confinement. For example, from the point of view of high-energy particle confinement, the configuration with a several centimeters inward shift of the magnetic axis by the vertical magnetic field is best in the Heliotron E at zero beta. In other heliotron devices there is also a favorable effect on particle confinement when the magnetic axis is shifted inward. However, particle confinement degrades due to the finite-beta effect in heliotrons. This result can be understood since the magnetic axis moves outward due to the finite-beta effect. In stellarators the ambipolar radial electric field is effective for confining trapped particles, since the poloidal motion due to $\mathbf{E} \times \mathbf{B}$ drift reduces the radial excursion of the trapped particle in the helical ripple. A mechanism to produce the ambipolar raidal electric field will be discussed in chapter 7.

To study the particle orbits in stellarators, Boozer introduced useful magnetic coordinates (Boozer coordinates), and formulated the guiding center drift equations by following the Hamiltonian dynamics formualtion. One significant advantage is that guiding center drift orbits are obtained only from knowledge of $|\mathbf{B}(\psi, \theta, \phi)|$ and that gradients of magnetic field vector are not necessary, as in the ordinary guiding center equations shown in section 3.6. Boozer's guiding center drift equations are given in section 6.2, and are applicable to finite-beta plasmas.

In order to analyze particle orbits in stellarators, the concept of the drift surface is useful, particularly for trapped particle orbits, and is defined by using the second adiabatic constant, $J = \oint v_{\parallel} ds$, where s shows the coordinate along the magnetic line of force. On the basis on both J conservaion and conservation of angular momentum, the averaged equation of motion over the bounce motion in the helical ripple can be derived. We explain these in section 6.3. Collisonless transitions between the localized particles and blocked particles is also discussed.

In section 6.4, we show the improvement of high-energy particle orbit confinement by using the so-called σ-optimization. Here σ is a parameter to control the position of the helical ripple, which is related to the $\ell = 1$ and $\ell = 3$ helical components or to the inward shift of the vacuum magnetic axis ($\sigma > 0$). It is noted that $\sigma > 0$ is necessary for trapped particle confinement. An example of σ-optimization from the point of view of neoclassical transport is shown in chapter 7. The behavior of particle orbits in the heliotron is directly related to the neoclassical transport, as explained in chapter 7. The neoclassical transport is developed based on the solution of the drift-kinetic equation. One approach is to solve the linearized drift-kinetic equation with the Coulomb collision term directly and to calculate the particle and heat fluxes. The other approach is to use the moment equations of the drift-kinetic equation with the Coulomb collision term. Flux–friction relations are thus derived. In addition to these approaches, Monte Carlo simulation is considered to be a useful method for investigating the neoclassical transport, particularly for three-dimensional magnetic configurations such as stellarators and heliotrons. When the Monte Carlo calculation of the Coulomb collision is added to the particle orbit calculation and particle motions of the order of several thousands are followed for several collision times, the global particle confinement time may be evaluated from the loss rate. When all particles start in the neighborhood of a particular magnetic surface, the diffusion coefficient is also evaluated using the deviation of the particle position from the original surface. We discuss Monte Carlo simulation in stellarators in section 6.5. This approach is also useful in the study of the efficiency of NBI or ICRF heating, as discussed in chapter 8.

6.2 DRIFT EQUATIONS OF MOTION IN BOOZER COORDINATES

When the magnetic moment μ is conserved, the particle orbit can be efficiently followed using the drift Hamiltonian defined by

$$H(\theta, P_\theta, \phi, P_\phi) = \tfrac{1}{2} m v_\parallel^2 + \mu B + e\Phi. \tag{6.1}$$

Here $v_\parallel = \mathbf{v} \cdot \mathbf{B}/B$, B is the magnetic field strength, m is the particle mass, e is the charge of the particle, and Φ is the electrostatic potential. The canonical angle variables are the poloidal angle θ and the toroidal angle ϕ. Their conjugate momenta are given by

$$P_\theta = m \frac{I v_\parallel}{B} + e\psi, \tag{6.2}$$

$$P_\phi = m\,\frac{gv_\|}{B} - e\chi, \tag{6.3}$$

where ψ is a toroidal flux function and χ is a poloidal flux function. These are obtain from $\mathbf{P} = m\mathbf{v} + e\mathbf{A}$ by noting that $\mathbf{B} = \nabla \times \mathbf{A}$ and $\mu_0 \mathbf{J} = \nabla \times \mathbf{B}$, where $\mathbf{A}$ is a vector potential and ψ and χ are equal to rA_θ and A_ϕ. Here r denotes a minor radius in the toroidal coordinates. The functions $g(\psi)$ and $I(\psi)$ are the toroidal field function and the toroidal current inside a toroidal magnetic surface. Toroidal magnetic surfaces can be labeled by the parameter ψ, denoting the toroidal flux inside a surface; and χ, denoting the poloidal flux outside a toroidal magnetic surface, can be equally used to label the surfaces. If the poloidal flux χ is considered as the independent coordinate, then the toroidal flux ψ has the form $\psi = \psi(\chi)$. Here we consider the choice $\chi(\psi)$; then the magnetic field strength and the electrostatic potential are $B(\psi, \theta, \phi)$ and $\Phi(\psi)$. It is noted that only the radial electric field is allowed at the equilibrium state with the scalar pressure. The coordinates (ψ, θ, ϕ) are magnetic coordinates, where a magnetic field line is seen as a straight line on the θ–ϕ plane.

Equations for particle drift motion are given from the drift Hamiltonian (6.1):

$$d\theta/dt = \partial H/\partial P_\theta, \tag{6.4}$$

$$d\phi/dt = \partial H/\partial P_\phi, \tag{6.5}$$

$$dP_\theta/dt = -\partial H/\partial \theta, \tag{6.6}$$

$$dP_\phi/dt = -\partial H/\partial \phi. \tag{6.7}$$

It is noted that these are the canonical equations of motion for describing particle drift motion.

In order to show that particle drift motion satisfies the Hamilton equation, we introduce appropriate magnetic coordinates. Since these coordinates can be constructed for arbitrary magnetic fields, we will consider magnetic fields in finite-beta plasmas of the stellarator and the heliotron. However, before deriving drift equations for finite-beta plasmas, we first consider a vacuum magnetic configuration for simplicity. It is written in two ways,

$$\mathbf{B} = \nabla\psi \times \nabla\theta_0, \tag{6.8}$$

$$\mathbf{B} = \nabla\chi, \tag{6.9}$$

where $\theta_0 = \theta - \iota\phi$, and $\iota(\psi)$ is the rotational transform. In (6.9), $\chi = g\phi + I\theta$, with $g = RB_\phi$ and $I = rB_\theta$ constants in the curl-free case. The magnetic coordinates (ψ, χ, θ_0) are chosen here for writing drift equations.

We note that the guiding center drift velocity is usually given by

$$\mathbf{v} = v_\| \frac{\mathbf{B}}{B} + \frac{m}{eB}\left(v_\|^2 + \tfrac{1}{2}v_\perp^2\right)\frac{\mathbf{B} \times \nabla B}{B^2} + \frac{\mathbf{B} \times \nabla \Phi}{B^2}. \tag{6.10}$$

Then the drift orbits in the (ψ, χ, θ_0) coordinates are

$$d\psi/dt = \mathbf{v} \cdot \nabla\psi, \tag{6.11}$$

$$d\chi/dt = \mathbf{v} \cdot \nabla\chi, \tag{6.12}$$

$$d\theta_0/dt = \mathbf{v} \cdot \nabla\theta_0. \tag{6.13}$$

To evaluate the terms in (6.10), we note that $\mathbf{B} \times \nabla\Phi = (\nabla\chi) \times (\partial\Phi/\partial\psi)/\nabla\psi$, $\rho_\| = v_\|(m/eB)$, and $v_\|\mathbf{B}/B = (e\rho_\|/m)\nabla\psi \times \nabla\theta_0$. Then we obtain

$$\frac{d\psi}{dt} = -\left(\frac{\mu}{e} + \frac{eB}{m}\rho_\|^2\right)\frac{\partial B}{\partial\theta_0}, \tag{6.14}$$

$$\frac{d\chi}{dt} = \frac{eB^2}{m}\rho_\|, \tag{6.15}$$

$$\frac{d\theta_0}{dt} = \frac{\partial\Phi}{\partial\psi} + \left(\frac{\mu}{e} + \frac{eB}{m}\rho_\|^2\right)\frac{\partial B}{\partial\psi}. \tag{6.16}$$

Here we interpret the drift Hamiltonian in (6.1) multiplied by the electric charge e as $\tilde{H}(\psi, \chi, \theta_0, \rho_\|)$. It is then easy to see that (6.14), (6.15), and (6.16) are equivalent to

$$d\psi/dt = -\partial\tilde{H}/\partial\theta_0, \tag{6.17}$$

$$d\chi/dt = \partial\tilde{H}/\partial\rho_\|, \tag{6.18}$$

$$d\theta_0/dt = \partial\tilde{H}/\partial\psi. \tag{6.19}$$

To obtain an equation for $\rho_\|$, we consider that the energy conservation implies $dH/dt = 0$, which gives

$$\frac{d\rho_\|}{dt} = -\frac{\partial\tilde{H}}{\partial\chi} = -\left(\frac{\mu}{e} + \frac{eB}{m}\rho_\|^2\right)\frac{\partial B}{\partial\chi}. \tag{6.20}$$

For actual use of the drift equations (6.14), (6.15), (6.16), and (6.20), concrete expressions for $\Phi(\psi)$ and $B(\psi, \chi, \theta_0)$ are required. One example for the electrostatic potential in scalar pressure equilibrium is

$$\Phi(\psi) = c_\phi(E_k/e)(1 - \psi/\psi_a)^2, \tag{6.21}$$

where E_k is the kinetic energy and c_ϕ is a dimensionless constant. The radial electric field given by this potential vanishes at the magnetic axis (since $\nabla\psi$ vanishes there) and at the plasma edge $\psi = \psi_a$.

The dependence of the magnetic field strength B on ψ, χ, and θ_0 depends, of course, on the type of stellarator or heliotron device. The relations between (ψ, χ, θ_0) and (ψ, θ, ϕ) imply, in the curl-free case with $\iota = 1/q$,

$$\begin{cases} \theta = (\iota\chi + g\theta_0)/(g + \iota I) = \theta_0 + \chi/gq, \\ \phi = (\chi - I\theta_0)/(g + \iota I) = \chi/g, \end{cases} \tag{6.22}$$

where we have ignored terms including I. The toroidal flux is approximated as

$$\psi \simeq \tfrac{1}{2} B_0 r^2, \tag{6.23}$$

where B_0 is the toroidal field at the magnetic axis. The toroidal magnetic field strength $B \simeq B_0[1 - \epsilon_a(r/a)\cos\theta]$, for example, is approximately

$$B \simeq B_0[1 - \epsilon_a(\psi/\psi_a)^{1/2}\cos(\chi/gq + \theta_0)], \tag{6.24}$$

where ϵ_a is the inverse aspect ratio at the plasma edge. The model magnetic field of a stellarator can be given by

$$B(r, \theta, \phi) = B_0\left[1 - \epsilon_a \frac{r}{a}\cos\theta - \delta_a\left(\frac{r}{a}\right)^{\ell}\sin(\ell\theta - m\phi)\right] \tag{6.25}$$

where δ_a is the helical ripple. Substituting the expressions for θ and ϕ in (6.22), one obtains the stellarator field in the magnetic coordinates (ψ, χ, θ_0) in dimensionless units:

$$\begin{aligned} B =& 1 - \epsilon_a(2\psi/\psi_a)^{1/2}\cos\left(\frac{\chi}{qg} + \theta_0\right) - \delta_a(2\psi/\psi_a)^{\ell/2}\times \\ &\times \sin\left((\ell - mq)\frac{\chi}{qg} + \ell\theta_0\right). \end{aligned} \tag{6.26}$$

To evaluate particle drifts we consider the stellarator with the magnetic field strength (6.26). The variation of field strength along a field line is given by varying χ while holding θ_0 and ψ constant. One can see that the variation of field strength along the field line consists of a slow oscillation due to the toroidal effect by the second term of (6.26) and a rapid oscillation due to the helical magnetic field by the third term of (6.26), as shown in Fig. 6.1. The toroidal and helical variations of the field strength allow two types of trapped particles, toroidally and helically trapped particles. An individual particle can switch from toroidally trapped to helically trapped and vice versa as a result of the variation in depth and width of the helical ripples along its collisionless drift orbit. The fact that particles can switch from toroidally to helically trapped implies that the longitudinal invariant $J = \oint mv_\parallel ds$ is not conserved over the entire drift motion. However, the longitudinal invariant is generally conserved for particles in the either helically or toroidally trapped state, and the quantity J can be used to calculate the transition from the helically to the toroidally trapped state.

For a magnetic field shown by $B(\psi, \theta, \phi)$ and an electric potential $\Phi(\psi)$ in a finite-beta plasma, (6.4) and (6.5) are written as

$$\frac{d\theta}{dt} = \left[\left(\mu + \frac{e^2\rho_\parallel^2 B}{m}\right)\frac{\partial B}{\partial\psi} + e\frac{\partial\Phi}{\partial\psi}\right]\frac{\partial\psi}{\partial P_\theta} + \frac{e^2B^2}{m}\rho_\parallel\frac{\partial\rho_\parallel}{\partial P_\theta}, \tag{6.27}$$

$$\frac{d\phi}{dt} = \left[\left(\mu + \frac{e^2 \rho_\parallel^2 B}{m} \right) \frac{\partial B}{\partial \psi} + e \frac{\partial \Phi}{\partial \psi} \right] \frac{\partial \psi}{\partial P_\phi} + \frac{e^2 B^2}{m} \rho_\parallel \frac{\partial \rho_\parallel}{\partial P_\phi}. \tag{6.28}$$

Also, the angular momenta (6.2) and (6.3) give

$$\frac{\partial P_\theta}{\partial t} = eI \frac{\partial \rho_\parallel}{\partial t} + e\rho_\parallel \frac{dI}{d\psi} \frac{\partial \psi}{\partial t} + e \frac{\partial \psi}{\partial t}, \tag{6.29}$$

$$\frac{\partial P_\phi}{\partial t} = eg \frac{\partial \rho_\parallel}{\partial t} + e\rho_\parallel \frac{dg}{d\psi} \frac{\partial \psi}{\partial t} - e \frac{d\chi}{d\psi} \frac{\partial \psi}{\partial t}. \tag{6.30}$$

From these equations

$$\frac{\partial \psi}{\partial t} = \left(\frac{\partial P_\theta}{\partial t} g - \frac{\partial P_\phi}{\partial t} I \right) / \Gamma \tag{6.31}$$

$$\frac{\partial \rho_\parallel}{\partial t} = -\left[\left(\rho_\parallel \frac{dg}{d\psi} - \iota \right) \frac{\partial P_\theta}{\partial t} - \left(\rho_\parallel \frac{dI}{d\psi} + 1 \right) \frac{\partial P_\phi}{\partial t} \right] / \Gamma \tag{6.32}$$

are obtained, where

$$\Gamma = -eI \left(\rho_\parallel \frac{dg}{d\psi} - \iota \right) + eg \left(\rho_\parallel \frac{dI}{d\psi} + 1 \right), \tag{6.33}$$

and $\iota = d\chi/d\psi$ has been used. The time derivatives of P_θ and P_ϕ in (6.31) and (6.32) are given from (6.6) and (6.7):

$$\frac{\partial P_\theta}{\partial t} = -\left(\mu + \frac{e^2 \rho_\parallel^2}{m} B \right) \frac{\partial B}{\partial \theta}, \tag{6.34}$$

$$\frac{\partial P_\phi}{\partial t} = -\left(\mu + \frac{e^2 \rho_\parallel^2}{m} B \right) \frac{\partial B}{\partial \phi}. \tag{6.35}$$

The derivatives of ψ and $\rho_\parallel$ with respect to the momenta P_θ and P_ϕ are obtained from (6.2) and (6.3):

$$1 = e \left(\rho_\parallel \frac{dI}{d\psi} + 1 \right) \frac{\partial \psi}{\partial P_\theta} + eI \frac{\partial \rho_\parallel}{\partial P_\theta}, \tag{6.36}$$

$$0 = e \left(\rho_\parallel \frac{dI}{d\psi} + 1 \right) \frac{\partial \psi}{\partial P_\phi} + eI \frac{\partial \rho_\parallel}{\partial P_\phi}, \tag{6.37}$$

$$0 = e \left(\rho_\parallel \frac{dg}{d\psi} - \iota \right) \frac{\partial \psi}{\partial P_\theta} + eg \frac{\partial \rho_\parallel}{\partial P_\theta}, \tag{6.38}$$

$$1 = e \left(\rho_\parallel \frac{dg}{d\psi} - \iota \right) \frac{\partial \psi}{\partial P_\phi} + eg \frac{\partial \rho_\parallel}{\partial P_\phi}, \tag{6.39}$$

They are

$$\frac{\partial\psi}{\partial P_\theta} = \frac{g}{\Gamma}, \tag{6.40}$$

$$\frac{\partial\rho_\parallel}{\partial P_\theta} = -\left(\rho_\parallel \frac{dg}{d\psi} - \iota\right)/\Gamma, \tag{6.41}$$

$$\frac{\partial\psi}{\partial P_\phi} = -\frac{I}{\Gamma}, \tag{6.42}$$

$$\frac{\partial\rho_\parallel}{\partial P_\phi} = \left(\rho_\parallel \frac{dI}{d\Psi} + 1\right)/\Gamma. \tag{6.43}$$

To carry out the collisionless drift orbit integration with (6.27), (6.28), (6.31), and (6.32), an eighth-order multistep generalization of the Runge–Kutta method, shown in (2.163), can be used to maintain numerical accuracy. It is noted that the change in energy per time step is a sensitive measure of the accuracy of the numerical integration. Thus particle orbit calculations require large computation times.

6.3 *J* INVARIANCE, TRAPPING, AND DETRAPPING

Here we again consider the Hamiltonian for charged particle motion in electromagnetic fields, as

$$H(\mathbf{q}, \mathbf{p}, t) = \frac{1}{2m}\left[\mathbf{p} - \frac{e}{\delta}\,\mathbf{A}(\mathbf{q}, \delta t)\right]^2 + e\Phi(\mathbf{q}, \delta t), \tag{6.44}$$

where $\mathbf{q}$ and $\mathbf{p}$ are the canonical spatial variable and the canonical momentum, and Φ and $\mathbf{A}$ are the electrostatic potential and the vector potential. Here δ is a small parameter and in (6.44) it is assumed that $\mathbf{A}$ and Φ change slowly; these electromagnetic fields are also assumed to be dominant in determining the charged particle motion. The Lagrangian L is also given by

$$\begin{aligned} L &= \mathbf{p}\cdot\dot{\mathbf{q}} - H \\ &= \mathbf{p}\cdot\dot{\mathbf{q}} - \frac{1}{2m}\left[\mathbf{p} - \frac{e}{\delta}\,\mathbf{A}(\mathbf{q}, \delta t)\right]^2 - e\Phi(\mathbf{q}, \delta t). \end{aligned} \tag{6.45}$$

Since the Lagrangian is invariant with respect to transformation of variables, the particle position $\mathbf{x}$ and the particle velocity $\mathbf{v}$ are introduced through the following relations:

$$\begin{cases} \mathbf{x} = \mathbf{q} \\ m\mathbf{v} = \mathbf{p} - \dfrac{e}{\delta}\,\mathbf{A}(\mathbf{q}, \delta t). \end{cases} \tag{6.46}$$

After these transformations,

$$L = \left[m\mathbf{v} + \frac{e}{\delta}\,\mathbf{A}(\mathbf{x}, \delta t)\right] \cdot \dot{\mathbf{x}} - \left[\frac{m}{2}\,v^2 + e\Phi(\mathbf{x}, \delta t)\right] \tag{6.47}$$

is obtained.

In order to obtain the Lagrangian for guiding center motion, an averaging over the fast gyration around the magnetic field line is applied. The velocity vector is written as $\mathbf{v} = v_{\|}\hat{\mathbf{b}} + v_{\perp}\hat{\mathbf{c}}$. Here $\hat{\mathbf{c}}$ is a unit vector perpendicular to the magnetic field line and includes rapid Larmor oscillations. Also, $\mathbf{x} = \mathbf{X} + \boldsymbol{\rho}$, where $\mathbf{X}$ denotes a position vector of the guiding center and $\boldsymbol{\rho}$ is a radial vector describing the Larmor radius. Usually, the Larmor radius is much shorter than the system size, and $\boldsymbol{\rho}$ belongs to $O(\delta)$ quantities, although the direction of $\boldsymbol{\rho}$ changes rapidly. By averaging the oscillatory components and keeping $O(1)$ quantities,

$$L = \left[\frac{e}{\delta}\,\mathbf{A}(\mathbf{X}, \delta t) + mv_{\|}\hat{\mathbf{b}}\right] \cdot \dot{\mathbf{X}} - [\tfrac{1}{2}mv_{\|}^2 + \tfrac{1}{2}mv_{\perp}^2 + e\Phi(\mathbf{X}, \delta t)] \tag{6.48}$$

is obtained. With the magnetic moment μ and $mv_{\|} = \pm mu = \pm\sqrt{2m}(E - \mu B - e\Phi)^{1/2}$, we obtain

$$L = \left[\frac{e}{\delta}\mathbf{A} + mv_{\|}\hat{\mathbf{b}}\right] \cdot \dot{\mathbf{X}} - [\tfrac{1}{2}mv_{\|}^2 + \mu B + e\Phi], \tag{6.49}$$

as the Lagrangian of the guiding center motion.

Here was used the magnetic coordinates (ψ, θ, ϕ) to write the Lagrangian (6.49) explicitly:

$$\begin{aligned} L = &\pm mub_{\psi}\dot{\psi} + \left(\pm mub_{\theta} + \frac{e}{\delta}\,\psi\right)\dot{\theta} \\ &+ (\pm mub_{\phi} + eA_{\phi})\dot{\phi} - E, \end{aligned} \tag{6.50}$$

where $\mathbf{A} = \psi\nabla\theta + \delta A_{\phi}\nabla\phi$ and $E = (1/2)mu^2 + \mu B + e\Phi$. The ordering in $\mathbf{A}$ corresponds to the fact that the rotational transform over one pitch length is sufficiently small. By replacing t with ϕ, or by the transformations $(\psi(t), \theta(t), \phi(t), E(t)) \to (\psi(\phi), \theta(\phi), t(\phi), E(\phi))$,

$$L^* = \pm mub_{\psi}\,\frac{d\psi}{d\phi} + \left(\pm mub_{\theta} + \frac{e}{\delta}\,\psi\right)\frac{d\theta}{d\phi} + (\pm mub_{\phi} + eA_{\phi}) - E\,\frac{dt}{d\phi}. \tag{6.51}$$

The guiding center equations of motion are given by the Euler–Lagrange equation for the stationary state of L^*,

$$\frac{d}{d\phi}\left(\frac{\partial L^*}{\partial\left(\dfrac{d\xi}{d\phi}\right)}\right) - \frac{\partial L^*}{\partial\xi} = 0, \tag{6.52}$$

where $\xi = \psi, \theta, t, E$. For $\xi = \psi$,

$$\frac{d}{d\phi}\left(\frac{\partial L^*}{\partial(d\psi/d\phi)}\right) = \frac{\partial}{\partial\psi}(\pm mub_\psi)\frac{d\psi}{d\phi} + \frac{\partial}{\partial\theta}(\pm mub_\psi)\frac{d\theta}{d\phi} + \frac{\partial}{\partial\phi}(\pm mub_\psi). \quad (6.53)$$

Here it is noted that mub_ψ is dependent on t. On the other hand,

$$\begin{aligned}\frac{\partial L^*}{\partial\psi} = &\frac{\partial}{\partial\psi}(\pm mub_\psi)\frac{d\psi}{d\phi} + \frac{\partial}{\partial\psi}(\pm mub_\theta)\frac{d\theta}{d\phi} + \frac{e}{\delta}\frac{d\theta}{d\phi}\\ &+ \frac{\partial}{\partial\psi}(\pm mub_\phi) + e\frac{dA_\phi}{d\psi},\end{aligned} \quad (6.54)$$

where $\iota = -dA_\phi/d\psi$. It is noted that A_ϕ depends on ψ only and $\iota \simeq O(\delta)$. From (6.53) and (6.54),

$$\frac{d\theta}{d\phi} = \frac{\delta}{D}\left[\iota + \frac{\partial}{\partial\phi}(\pm mub_\psi/e) - \frac{\partial}{\partial\psi}(\pm mub_\phi/e)\right], \quad (6.55)$$

where

$$D = 1 + \delta\left[\frac{\partial}{\partial\psi}(\pm mub_\theta/e) - \frac{\partial}{\partial\theta}(\pm mub_\psi/e)\right]. \quad (6.56)$$

Similarly, from $\xi = \theta$, we obtain

$$\frac{d\psi}{d\phi} = \frac{\delta}{D}\left[\frac{\partial}{\partial\theta}(\pm mub_\phi/e) - \frac{\partial}{\partial\phi}(\pm mub_\theta/e)\right] \quad (6.57)$$

from (6.52). Also, for $\xi = t$,

$$dE/d\phi = 0, \quad (6.58)$$

and, for $\xi = E$,

$$\frac{dt}{d\phi} = \frac{\partial}{\partial E}(\pm mu)\left(b_\phi + b_\psi\frac{d\psi}{d\phi} + b_\theta\frac{d\theta}{d\phi}\right) \quad (6.59)$$

are obtained from (6.52). By considering that $d\psi/d\phi = 0$ and $d\theta/d\phi = 0$ in the lowest order of δ, (6.59) becomes

$$\frac{dt}{d\phi} = \frac{\partial}{\partial E}(\pm mu)b_\phi. \quad (6.60)$$

These results are consistent with the assumption that the rotational transform over one pitch length is sufficiently small:

$$\iota/N \ll 1, \quad (6.61)$$

where N is a pitch number. Thus particles move along magnetic lines of force in the lowest order of δ. We also note that (6.58) means that the energy of the particle is conserved.

Now we discuss the adiabatic invariant J, which satisfies

$$dJ/d\phi = 0. \tag{6.62}$$

We introduce the notations J_r for trapped particles in helical ripples and $J_+(J_-)$ for untrapped particles with a positive (negative) parallel velocity. First, we expand $J_\alpha(\alpha = r \text{ or } +, -)$ with respect to the small parameter δ as

$$J_\alpha = J_{\alpha 0} + \delta J_{\alpha 1} + \cdots. \tag{6.63}$$

From (6.62),

$$dJ_{\alpha 0}/d\phi = 0, \tag{6.64}$$

in the zeroth order, which means that $J_{\alpha 0}$ is independent of ϕ. From (6.62) the first-order quantity $J_{\alpha 1}$ satisfies

$$\begin{aligned} \delta \frac{dJ_{\alpha 1}}{d\phi} &= -\frac{\partial J_{\alpha 0}}{d\psi}\frac{d\psi}{d\phi} - \frac{\partial J_{\alpha 0}}{\partial\theta}\frac{d\theta}{d\phi} \\ &= -\delta \frac{\partial J_{\alpha 0}}{\partial\psi}\left[\frac{\partial}{\partial\theta}(\pm mub_\phi/e) - \frac{\partial}{\partial\phi}(\pm mub_\theta/e)\right] \\ &\quad - \delta \frac{\partial J_{\alpha 0}}{\partial\theta}\left[\iota + \frac{\partial}{\partial\phi}(\pm mub_\psi/e) - \frac{\partial}{\partial\psi}(\pm mub_\phi/e)\right]. \end{aligned} \tag{6.65}$$

For untrapped particles we obtain $J_{\pm 1}$ by integrating (6.65) with respect to ϕ over one pitch length,

$$\{I_\pm, J_{\pm 0}\} \equiv \frac{\partial I_\pm}{\partial\theta}\frac{\partial J_{\pm 0}}{\partial\psi} - \frac{\partial I_\pm}{\partial\psi}\frac{\partial J_{\pm 0}}{\partial\theta} = 0, \tag{6.66}$$

under the condition (6.64), where

$$I_\pm \equiv \pm\frac{2\pi A_\phi}{N} + \int_0^{2\pi/N}\frac{mub_\phi}{e}\,d\phi. \tag{6.67}$$

We note that $J_{\pm 0}$ with no explicit ϕ dependence, and that u, b_ψ, and b_ϕ are periodic with respect to the pitch length. Since $\psi =$ constant and $\theta =$ constant in the lowest-order particle orbit, the integration with respect to ϕ has been carried out at fixed ψ and θ. From the relation (6.66), $J_{\pm 0}$ can be chosen as $J_{\pm 0} = eI_\pm$. By noting that magnetic field is shown as $\mathbf{B} = B_\psi\nabla\psi + B_\theta(\psi)\nabla\theta + B_\phi(\psi)\nabla\phi$ in the Boozer coordinates (ψ, θ, ϕ), $J_{\pm 0}$ is explicitly written as

$$J_{\pm 0} = \pm \frac{2\pi e A_\phi}{N} + B_\phi \int_0^{2\pi/N} \frac{mu}{B}\, d\phi. \tag{6.68}$$

To obtain J_{r0} we perform integration over one bounce motion between (ϕ_-, ϕ_+) at fixed ψ and θ, and the LHS of (6.65) gives

$$\oint \frac{\partial J_{r1}}{\partial \phi}\, d\phi = 0. \tag{6.69}$$

By noting that $u = 0$ at $\phi = \phi_+$ and $\phi = \phi_-$, and that $\partial J_{r0}/\partial\theta$ does not depend on ϕ explicitly,

$$\{I_r, J_{r0}\} \equiv \frac{\partial I_r}{\partial \theta}\frac{\partial J_{r0}}{\partial \psi} - \frac{\partial I_r}{\partial \psi}\frac{\partial J_{r0}}{\partial \theta} = 0, \tag{6.70}$$

where

$$I_r \equiv 2\int_{\phi_-}^{\phi_+} \frac{mub_\phi}{e}\, d\phi. \tag{6.71}$$

The relation (6.70) gives us

$$J_{r0} \equiv eI_r = 2B_\phi \int_{\phi_-}^{\phi_+} \frac{mu}{B}\, d\phi \tag{6.72}$$

in the same way as $J_{\pm 0}$ in (6.68). The obtained $J_{\pm 0}$ and J_{r0} are adiabatic invariant for particle motions in the stellarator or heliotron. Since $J_{\pm 0}$ and J_{r0} are functions of ψ and θ, $J_{\alpha 0} =$ constant describes the drift surface in the poloidal plane.

In the case of tokamaks, the Lagrangian L does not depend on ϕ and

$$\frac{d}{dt}\left(\frac{\partial L}{\partial \dot{\phi}}\right) - \frac{\partial L}{\partial \phi} = 0, \tag{6.73}$$

where $\dot{\phi} = d\phi/dt$. Thus

$$\frac{\partial L}{\partial \dot{\phi}} = \pm mub_\phi + eA_\phi = \text{constant} \tag{6.74}$$

is obtained from (6.50). On the other hand, for the axisymmetric case, (6.68) becomes

$$J_{\pm 0} = \pm\frac{2\pi}{N}(\pm mub_\phi + eA_\phi), \tag{6.75}$$

which corresponds to $\partial L/\partial\dot{\phi}$ in (6.74). We also note that the conjugate momentum of ϕ is given by $P_\phi = \partial L/\partial\dot{\phi}$, which yields

$$J = \frac{1}{2\pi}\oint P_\phi d\phi = \frac{1}{2\pi}\oint(\pm mub_\phi + eA_\phi)d\phi. \tag{6.76}$$

This results shows that $J_{\pm 0}$ is adiabatic invariant, since $J_{\pm 0}$ is equal to J with a constant factor in the case of an axisymmetric torus.

It is known that the adiabatic constant J_0 ($\alpha = +, -, r$) gives an averaged drift surface on the poloidal plane (ψ, θ). This drift surface is the same as that described by the averaged drift equations of motion. To show this, we first calculate periods T_α ($\alpha = +, -, r$) of the guiding center motion in the helical ripple,

$$\begin{cases} T_\pm = \displaystyle\int_0^{\pm 2\pi/N} \frac{dt}{d\phi}\, d\phi \\ T_r = \displaystyle\oint \frac{dt}{d\phi}\, d\phi. \end{cases} \tag{6.77}$$

Since

$$\frac{dt}{d\phi} = \pm\frac{\partial(mu)}{\partial E} b_\phi \tag{6.78}$$

in the lowest order of δ (see (6.59)), we obtain

$$\begin{cases} T_\pm = \dfrac{\partial}{\partial E}\displaystyle\int_0^{2\pi/N} mub_\phi d\phi = \dfrac{\partial J_{\pm 0}}{\partial E} \\ T_r = 2\displaystyle\int_{\phi_-}^{\phi_+} \frac{\partial(mu)}{\partial E} b_\phi d\phi = \dfrac{\partial J_{r0}}{\partial E}, \end{cases} \tag{6.79}$$

where the integration is carried out at fixed ψ and θ. Variations $\Delta\psi$ and $\Delta\theta$ are given by

$$\begin{cases} \Delta\psi = \displaystyle\int \frac{d\psi}{d\phi}\, d\phi \\ \Delta\theta = \displaystyle\int \frac{d\theta}{d\phi}\, d\phi, \end{cases} \tag{6.80}$$

where

$$\begin{cases} \dfrac{d\psi}{d\phi} = \dfrac{\partial}{\partial\theta}\left(\pm\dfrac{mub_\phi}{e}\right) - \dfrac{\partial}{\partial\phi}\left(\pm\dfrac{mub_\theta}{e}\right) \\ \dfrac{d\theta}{d\phi} = \iota + \dfrac{\partial}{\partial\phi}\left(\pm\dfrac{mub_\psi}{e}\right) - \dfrac{\partial}{\partial\psi}\left(\pm\dfrac{mub_\psi}{e}\right) \end{cases} \tag{6.81}$$

in the first order of δ, and the integration in (6.80) is also carried out at fixed ψ and θ. Thus we obtain

$$\begin{cases} \Delta\psi_\alpha = \dfrac{1}{e}\dfrac{\partial J_{\alpha 0}}{\partial\theta} \\ \Delta\theta_\alpha = -\dfrac{1}{e}\dfrac{\partial J_{\alpha 0}}{\partial\psi}. \end{cases} \tag{6.82}$$

From (6.79) and (6.82), the averaged drift equations of motion are given as

$$\begin{cases} \dfrac{d\psi_\alpha}{dt} = \dfrac{1}{e}\dfrac{\partial J_{\alpha 0}}{\partial\theta} \Big/ \dfrac{\partial J_{\alpha 0}}{\partial E} \\ \dfrac{d\theta_\alpha}{dt} = -\dfrac{1}{e}\dfrac{\partial J_{\alpha 0}}{\partial\psi} \Big/ \dfrac{\partial J_{\alpha 0}}{\partial E}. \end{cases} \tag{6.83}$$

By solving (6.83), we obtain a drift orbit in the poloidal plane which is equal to that given by $J_{\alpha 0} = \text{constant}$. These equations are simpler than the guiding center drift equations shown in section 6.2 and make it easier to understand particle orbits in the stellarator or heliotron.

Hereafter, we discuss particle orbits in the model magnetic field described as

$$B = B_0(\psi, \theta) - B_1(\psi, \theta)\cos(\ell\theta - N\phi), \tag{6.84}$$

For a simple case, $B_0(\psi, \theta) = \bar{B}_0(1 - \varepsilon_t(\psi)\cos\theta)$ and $B_1(\psi, \theta) = \bar{B}_0\varepsilon_h(\psi)$. The electrostatic potential for the ambipolar electric field is denoted by $\Phi(\psi)$. When a particle satisfies the inequality

$$E - E_X < 0 < E - \mu B_0 + \mu B_1 - e\Phi, \tag{6.85}$$

it becomes a trapped particle in a helical ripple, where $E_X = \mu B_0 + \mu B_1 + e\Phi$. On the other hand, untrapped particles satisfy $0 < E - E_X$. The inequality (6.85) can be written as $0 < \eta < 1$, where

$$\eta = \frac{E - \mu B_0 + \mu B_1 - e\Phi}{2\mu B_1}. \tag{6.86}$$

The boundary between the untrapped state and the trapped state is $\eta = 1$, and $\eta > 1$ is valid for untrapped particles. We note that $\eta = 1$ corresponds to $E = E_X$. Particles cannot enter the region with $\eta < 0$.

By introducing a parameter

$$\tilde{b} = \frac{2B_1}{B_0 - B_1}, \tag{6.87}$$

J_{r0} and $J_{\pm 0}$ are shown as

$$J_{r0} = \frac{8B_\phi}{N}\sqrt{\frac{m\mu}{B_1}}[(1 + \tilde{b}\eta)\Pi(\tilde{b}\eta, \eta) - K(\eta)] \tag{6.88}$$

and

$$J_{\pm 0} = \pm \frac{2\pi e A_\phi}{N} + \frac{4B_\phi}{N}\sqrt{\frac{m\mu}{B_1\eta}}\left[(1+\tilde{b}\eta)\Pi\left(\tilde{b}, \frac{1}{\eta}\right) - K\left(\frac{1}{\eta}\right)\right]. \tag{6.89}$$

Here an integral,

$$\begin{aligned}
&\int_0^{2\pi/N} \frac{mu(\psi, \theta, \phi, E)}{B(\psi, \theta, \phi)}\, d\phi \\
&= \sqrt{2m}\int_0^{2\pi/N} \frac{\sqrt{2\mu B_1\eta}\{1-(1/\eta)\sin^2((\ell\theta - N\phi)/2)\}^{1/2}}{(B_0 - B_1)\{1+\tilde{b}\sin^2((\ell\theta - N\phi)/2)\}}\, d\phi \\
&= \frac{8\sqrt{m\mu B_1}}{N(B_0 - B_1)}\int_0^{\pi/2} \frac{\sqrt{\eta}}{\tilde{b}\eta}\left[\frac{1+\tilde{b}\eta}{(1+\tilde{b}\sin^2 Z)\sqrt{1-(1/\eta)\sin^2 Z}}\right. \\
&\quad \left. - \frac{1}{\sqrt{1-(1/\eta)\sin^2 Z}}\right] dZ,
\end{aligned} \tag{6.90}$$

has been used to obtain $J_{\pm 0}$ shown by (6.89), where $Z = -(\ell\theta - N\phi)/2$ and $\int_{-\ell\theta/2}^{-\ell\theta/2+\pi} dZ = 2\int_0^{\pi/2} dZ$ are substituted here. Also, an integral

$$\begin{aligned}
&\int_{\phi_-}^{\phi_+} \frac{\{\eta - \sin^2((\ell\theta - N\phi)/2)\}^{1/2}}{1+\tilde{b}\sin^2((\ell\theta - N\phi)/2)}\, d\phi \\
&= \frac{2}{N}\int_{Z-}^{Z+} \frac{\sqrt{\eta - \sin^2 Z}}{1+\tilde{b}\sin^2 Z}\, dZ \\
&= \int_0^{\pi/2} \frac{1}{\tilde{b}}\left[\frac{1+\tilde{b}\eta}{(1+\tilde{b}\eta\sin^2 X)\sqrt{1-\eta\sin^2 X}}\right. \\
&\quad \left. - \frac{1}{\sqrt{1-\eta\sin^2 X}}\right] dX
\end{aligned} \tag{6.91}$$

has been used to obtain J_{r0} shown by (6.88), where ϕ_+ and ϕ_- satisfy $\eta - \sin^2((\ell\theta - N\phi)/2) = 0$ and Z_+ and Z_- satisfy $\eta - \sin^2 Z = 0$. The variable X is introduced by $\sin Z = \sqrt{\eta}\sin X$. The complete elliptic integrals in (6.88) and (6.89) are denoted as

$$K(k) = \int_0^{i/2} \frac{d\theta}{\sqrt{1-k\sin^2\theta}}, \tag{6.92}$$

$$E(k) = \int_0^{\pi/2} \sqrt{1-k\sin^2\theta}\, d\theta, \tag{6.93}$$

$$\Pi(c, k) = \int_0^{\pi/2} \frac{d\theta}{(1+c\sin^2\theta)\sqrt{1-k\sin^2\theta}}. \tag{6.94}$$

There are relations between the derivatives of $K(k)$, $E(k)$, and $\Pi(c, k)$:

$$\frac{dK(k)}{dk} = \frac{E(k)}{2k(1-k)} - \frac{K(k)}{2k}, \tag{6.95}$$

$$\frac{dE(k)}{dk} = \frac{E(k) - K(k)}{2k} \tag{6.96}$$

$$\frac{\partial \Pi(c,k)}{\partial k} = -\frac{\Pi(c,k)}{2(c+k)} + \frac{E(k)}{2(c+k)(1-k)}, \tag{6.97}$$

$$\frac{\partial \Pi(c,k)}{\partial c} = \frac{k-c^2}{2c(1+c)(c+k)}\,\Pi(c,k) + \frac{E(k)}{2(1+c)(c+k)} - \frac{K(k)}{2c(1+c)}. \tag{6.98}$$

In order to derive (6.97), we have used

$$\Pi(c,k) = \int_0^1 \frac{dx}{(1+cx^2)\sqrt{(1-x^2)(1-kx^2)}} \tag{6.99}$$

via the relation $x = \sin\theta$. From this integral,

$$\frac{\partial \Pi(c,k)}{\partial c} = -\frac{1}{c}\,\Pi(c,k) + \frac{1}{c^3}\int_0^1 \frac{dx}{(x^2+1/c)^2\sqrt{(1-x^2)(1-kx^2)}} \tag{6.100}$$

is given. The integral in (6.100) is written as

$$I = \int_0^1 \frac{dx}{(x^2-\alpha^2)^2\sqrt{(1-x^2)(1-kx^2)}} \tag{6.101}$$

with $\alpha = i/\sqrt{c}$. We note the relation

$$\begin{aligned}\frac{1}{(x\pm\alpha)^2\sqrt{\varphi(x)}} &= \pm\frac{\varphi'(\alpha)}{2\varphi(\alpha)}\frac{1}{(x\pm\alpha)\sqrt{\varphi(x)}} - \frac{1}{\varphi(\alpha)}\sqrt{\frac{1-kx^2}{1-x^2}} \\ &\quad + \frac{1}{(1-\alpha^2)\sqrt{\varphi(x)}} - \frac{1}{\varphi(\alpha)}\frac{d}{dx}\left(\frac{\sqrt{\varphi(x)}}{x\pm\alpha}\right),\end{aligned} \tag{6.102}$$

where $\varphi(x) = (1-x^2)(1-kx^2)$, and

$$\begin{aligned}&\frac{1}{(x-\alpha)^2\sqrt{\varphi(x)}} + \frac{1}{(x+\alpha)^2\sqrt{\varphi(x)}} \\ &= -\frac{\alpha\varphi'(\alpha)}{\varphi(\alpha)}\frac{1}{(x^2-\alpha^2)\sqrt{\varphi(x)}} - \frac{2}{\varphi(\alpha)}\sqrt{\frac{1-kx^2}{1-x^2}} \\ &\quad + \frac{2}{(1-\alpha^2)\sqrt{\varphi(x)}} - \frac{1}{\varphi(\alpha)}\frac{d}{dx}\left(\frac{2x\sqrt{\varphi(x)}}{x^2-\alpha^2}\right).\end{aligned} \tag{6.103}$$

By noting that $1/(x^2-\alpha^2)^2 = (1/4\alpha^2)[1/(x-\alpha)^2 + 1/(x+\alpha)^2] - (1/2\alpha^2)(1/(x^2-\alpha^2))$ and substituting (6.103) into (6.101), we obtain

$$I = \frac{1}{4\alpha^2}\left[-\left(\frac{\alpha\varphi'(\alpha)}{\varphi(\alpha)}+2\right)\int_0^1 \frac{dx}{(x^2-^2)\sqrt{\varphi(x)}} - \frac{2}{\varphi(\alpha)}\,E(k) + \frac{2K(k)}{1-\alpha^2}\right]$$
$$= -\frac{c}{4}\left[-\left\{\frac{2c(c+ck+2k)}{(1+c)(c+k)} + 2c\right\}\Pi(c,k) - \frac{2c^2}{(1+c)(c+k)}\,E(k) + \frac{2c}{1+c}\,K(k)\right], \tag{6.104}$$

where

$$K(k) = \int_0^1 \frac{dx}{\sqrt{(1-x^2)(1-k^2x^2)}}, \tag{6.105}$$

$$E(k) = \int_0^1 \sqrt{\frac{1-k^2x^2}{1-x^2}}\,dx, \tag{6.106}$$

$\varphi(\alpha) = (1+c)(c+k)/c^2$, and $\alpha\varphi'(\alpha) = 2(c+ck+2k)/c^2$. Since J_{r0} and $J_{\pm 0}$ have been obtained for the model magnetic field (6.84), T_α, $\Delta\psi_\alpha$, and $\Delta\theta_\alpha$ can be calculated.

The period of particle motion for trapped particles T_r or for untrapped particles $T_\pm$ is given by

$$T_r = \frac{\partial J_{r0}}{\partial E} = \frac{1}{2\mu B_1}\frac{\partial J_{r0}}{\partial \eta} = \frac{4B_\phi}{NB_1}\sqrt{\frac{m}{\mu B_1}}\,\frac{\tilde{b}}{2}\,\Pi(\tilde{b}\eta, \eta), \tag{6.107}$$

$$T_\pm = \frac{\partial J_{\pm 0}}{\partial E} = \frac{1}{2\mu B_1}\frac{\partial J_{\pm 0}}{\partial \eta} = \frac{4B_\phi}{NB_1}\sqrt{\frac{m}{\mu B_1}}\,\frac{\tilde{b}}{2\sqrt{\eta}}\,\Pi(\tilde{b}, 1/\eta). \tag{6.108}$$

Here we have used the following relations obtained with (6.95)–(6.98):

$$\frac{\partial}{\partial\eta}[(1+\tilde{b}\eta)\Pi(\tilde{b}\eta,\eta) - K(\eta)]$$
$$= (1+\tilde{b}\eta)\frac{\partial}{\partial\eta}\Pi(\tilde{b}\eta,\eta) + \tilde{b}\Pi(\tilde{b}\eta,\eta) - \frac{\partial}{\partial\eta}K(\eta)$$
$$= (1-\tilde{b}\eta)\tilde{b}\frac{\partial}{\partial c}\Pi(c,\eta) + (1+\tilde{b}\eta)\frac{\partial}{\partial k}\Pi(c,k) + \tilde{b}\Pi(\tilde{b}\eta,\eta) - \frac{\partial}{\partial\eta}K(\eta)$$
$$= \frac{\tilde{b}}{2}\,\Pi(\tilde{b}\eta,\eta), \tag{6.109}$$

$$\frac{\partial}{\partial \eta}[\{(1+\tilde{b}\eta)\Pi(\tilde{b},1/\eta)-K(1/\eta)\}/\sqrt{\eta}]$$
$$=\{-(1+\tilde{b}\eta)\Pi(\tilde{b},1/\eta)+K(1/\eta)\}/2\eta\sqrt{\eta}$$
$$+[(1+\tilde{b}\eta)\frac{\partial}{\partial \eta}\Pi(\tilde{b},1/\eta)+\tilde{b}\Pi(\tilde{b},1/\eta)-\frac{\partial}{\partial \eta}K(1/\eta)]/\sqrt{\eta}$$
$$=\frac{\tilde{b}}{2\sqrt{\eta}}\Pi(\tilde{b},1/\eta). \tag{6.110}$$

To obtain $\Delta\psi_\alpha$ and $\Delta\theta_\alpha$, we need $\partial J_{\alpha 0}/\partial\theta$ and $\partial J_{\alpha 0}/\partial\psi$. When B_ϕ is constant,

$$\frac{\partial J_{r0}}{\partial \xi}=\frac{4B_\phi}{NB_1}\sqrt{\frac{m\mu}{B_1}}\left[-\frac{e}{\mu}\frac{\partial\Phi}{\partial\xi}\frac{\tilde{b}}{2}\Pi(\tilde{b}\eta,\eta)\right.$$
$$-\frac{\partial B_0}{\partial\xi}\left\{\frac{\tilde{b}(2+\tilde{b})(1+\tilde{b}\eta)}{2(1+\tilde{b})}\Pi(\tilde{b}\eta,\eta)+\frac{\tilde{b}E(\eta)}{2(1+\tilde{b})}-\frac{\tilde{b}}{2}K(\eta)\right\}$$
$$\left.+\frac{\partial B_1}{\partial\xi}\left\{\frac{\tilde{b}^2(1+\tilde{b}\eta)}{2(1+\tilde{b})}\Pi(\tilde{b}\eta,\eta)+\frac{\tilde{b}(2+\tilde{b})}{2(1+\tilde{b})}E(\eta)-\frac{\tilde{b}}{2}K(\eta)\right\}\right], \tag{6.111}$$

where ξ denotes ψ or θ, and $\partial\Phi/\partial\theta=0$ is assumed in this section. By introducing

$$F_\xi\equiv\frac{2B_\phi}{NB_1}\sqrt{\frac{m\mu}{B_1}}\left[-\frac{e}{\mu}\frac{\partial\Phi}{\partial\xi}\frac{\tilde{b}}{2\sqrt{\eta}}\Pi(\tilde{b},1/\eta)\right.$$
$$-\frac{\partial B_0}{\partial\xi}\left\{\frac{\tilde{b}(2+\tilde{b})(1+\tilde{b}\eta)}{2\sqrt{\eta}(1+\tilde{b})}\Pi(\tilde{b},1/\eta)+\frac{\tilde{b}^2\sqrt{\eta}}{2(1+\tilde{b})}E(1/\eta)\right.$$
$$\left.-\frac{\tilde{b}(1+\tilde{b}\eta)}{2(1+\tilde{b})\sqrt{\eta}}K(1/\eta)\right\}$$
$$+\frac{\partial B_1}{\partial\xi}\left\{\frac{\tilde{b}^2(1+\tilde{b}\eta)}{2\sqrt{\eta}(1+\tilde{b})}\Pi(\tilde{b},1/\eta)+\frac{\tilde{b}(1-2\eta-\tilde{b}\eta)}{2\sqrt{\eta}(1+\tilde{b})}K(1/\eta)\right.$$
$$\left.\left.+\frac{\tilde{b}(2+\tilde{b})\sqrt{\eta}}{2(1+\tilde{b})}E(1/\eta)\right\}\right], \tag{6.112}$$

we also obtain

$$\frac{\partial J_{\pm 0}}{\partial\theta}=F_\theta, \tag{6.113}$$
$$\frac{\partial J_{\pm 0}}{\partial\psi}=\pm\frac{2\pi e}{N}\iota+F_\psi, \tag{6.114}$$

by noting $-dA_\phi/d\psi=\iota$. The derivation of (6.111) is shown as follows:

$$\frac{\partial J_{r0}}{\partial \xi} = \frac{\partial}{\partial \xi}\left(\frac{8B_\phi}{N}\sqrt{\frac{m\mu}{B_1}}\right)[(1+\tilde{b}\eta)\Pi(\tilde{b}\eta,\eta) - K(\eta)]$$
$$+ \frac{8B_\phi}{N}\sqrt{\frac{m\mu}{B_1}}\left(\frac{\partial \eta}{\partial \xi}\frac{\partial}{\partial \eta} + \frac{\partial \tilde{b}}{\partial \xi}\frac{\partial}{\partial \tilde{b}}\right)[(1+\tilde{b}\eta)\,\Pi(\tilde{b}\eta,\eta) - K(\eta)], \tag{6.115}$$

where

$$\frac{\partial \eta}{\partial \xi} = -\frac{1}{2B_1}\frac{\partial B_0}{\partial \xi} - \frac{1}{2B_1}(2\eta - 1)\frac{\partial B_1}{\partial \xi} - \frac{e}{2\mu B_1}\frac{\partial \Phi}{\partial \xi}, \tag{6.116}$$
$$\frac{\partial \tilde{b}}{\partial \xi} = -\frac{1}{2B_1}\tilde{b}^2\frac{\partial B_0}{\partial \xi} + \frac{1}{2B_1}\tilde{b}(\tilde{b}+2)\frac{\partial B_1}{\partial \xi}. \tag{6.117}$$

Here we also follow (6.109) as

$$\frac{\partial}{\partial \tilde{b}}[(1+\tilde{b}\eta)\Pi(\tilde{b}\eta,\eta) - K(\eta)]$$
$$= \eta\Pi(\tilde{b}\eta,\eta) + (1+\tilde{b}\eta)\frac{\partial}{\partial \tilde{b}}\Pi(\tilde{b}\eta,\eta)$$
$$= \left\{\eta + \frac{1-\tilde{b}^2\eta}{2\tilde{b}(1+\tilde{b})}\right\}\Pi(\tilde{b}\eta,\eta) + \frac{E(\eta)}{2(1+\tilde{b})} - \frac{K(\eta)}{2\tilde{b}}, \tag{6.118}$$

and use

$$\frac{\partial}{\partial \xi}\left(\frac{8B_\phi}{N}\sqrt{\frac{m\mu}{B_1}}\right) = -\frac{4B_\phi}{NB_1}\sqrt{\frac{m\mu}{B_1}}\frac{\partial B_1}{\partial \xi}, \tag{6.119}$$

in the case of $B_\phi = \text{constant}$. In (6.115), by gathering the terms including $\partial\Phi/\partial\xi$, $\partial B_0/\partial\xi$, and $\partial B_1/\partial\xi$ separately, the expression (6.111) appears. The expression (6.112) is obtained from

$$F_\xi = -\frac{2B_\phi}{NB_1}\sqrt{\frac{m\mu}{B_1}}\frac{\partial B_1}{\partial \xi}[(1+\tilde{b}\eta)\Pi(\tilde{b},1/\eta) - K(1/\eta)]/\sqrt{\eta}$$
$$+ \frac{4B_\phi}{N}\sqrt{\frac{m\mu}{B_1}}\left(\frac{\partial \eta}{\partial \xi}\frac{\partial}{\partial \eta} + \frac{\partial \tilde{b}}{\partial \xi}\frac{\partial}{\partial \tilde{b}}\right)[(1+\tilde{b}\eta)\Pi(\tilde{b},1/\eta) - K(1/\eta)]/\sqrt{\eta}. \tag{6.120}$$

Then we use (6.110) and

$$\frac{\partial}{\partial \tilde{b}}[(1+\tilde{b}\eta)\Pi(\tilde{b},1/\eta) - K(1/\eta)]$$
$$= \eta\Pi(\tilde{b},1/\eta) + (1+\tilde{b}\eta)\frac{\partial}{\partial \tilde{b}}\Pi(\tilde{b},1/\eta)$$
$$= \left\{\eta + \frac{1-\tilde{b}^2\eta}{2\tilde{b}(1+\tilde{b})}\right\}\Pi(\tilde{b},1/\eta) + \frac{\eta}{2(1+\tilde{b})}E(1/\eta) - \frac{1+\tilde{b}\eta}{2\tilde{b}(1+\tilde{b})}K(1/\eta). \tag{6.121}$$

By gathering the terms including $\partial\Phi/\partial\xi$, $\partial B_0/\partial\xi$, and $\partial B_1/\partial\xi$ separately, the expression (6.112) appears. With T_r, $T_\pm$, $\partial J_{r0}/\partial\psi$, $\partial J_{r0}/\partial\theta$, $\partial J_{\pm 0}/\partial\psi$, and $\partial J_{\pm 0}/\partial\theta$, we can write the averaged drift equations explicitly.

By solving the averaged drift equations it is found that there exist particles slowing a transition from a trapped state (localized particle) to an untrapped state (blocked particle) or vice versa, as shown in Fig. 6.2, where P and Q denote the transition points, r represents the trapped state (localized particle) in the helical ripple, and $+$ and $-$ represent the untrapped state (blocked particle) with a positive and a negative parallel velocity, respectively. It is noted that P and Q are on the line of $\eta = 1$. We note that P and Q have different characteristics. Particles that arrive at P are trapped (localized) particles or untrapped (blocked) particles with a positive parallel velocity, and particles that start at P are only untrapped (blocked) particles with a negative parallel velocity. Thus trapped (localized) particles or untrapped (blocked) particles with a positive parallel velocity that arrive at P should show the transition to untrapped (blocked) particles with a negative parallel velocity. On the other hand, particles that arrive at Q are only untrapped (blocked) particles with a negative parallel velocity, and particles that start at Q are trapped (localized) particles or untrapped (blocked) particles with a positive parallel velocity. We need further information to determine the transition from untrapped (blocked) particles with a negative parallel velocity to trapped (localized) particles with a positive parallel velocity. It is natural to introduce a transition probability, since we have lost the definite information concerning the toroidal angle ϕ by averaging the particle orbit with respect to ϕ.

In order to obtain the transition probability we consider that the particle drift motion occurs in the magnetic surface. We assume that the particle orbit is described with straight line on the surface of $\psi = \text{constant}$, as shown in Fig. 6.3. Here we also assume that the electrostatic potential Φ is a function of ψ. The inclined straight lines show magnetic lines of force. The inclination angle

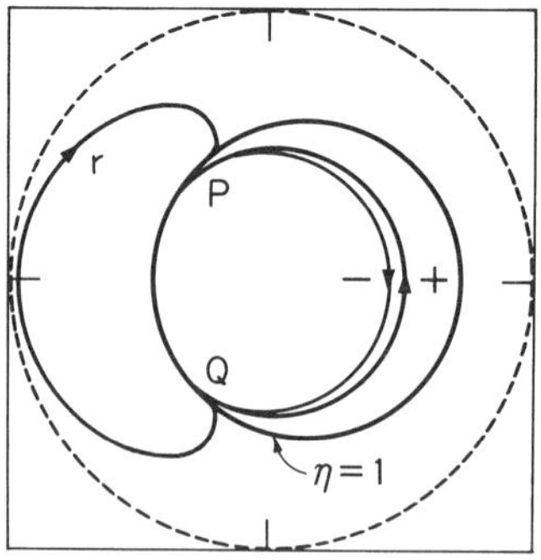

Fig. 6.2 Drift surfaces in a heliotron. $\pm$ and r classify particle states. P and Q denote transition points. The dotted circle shows the outermost magnetic surface.

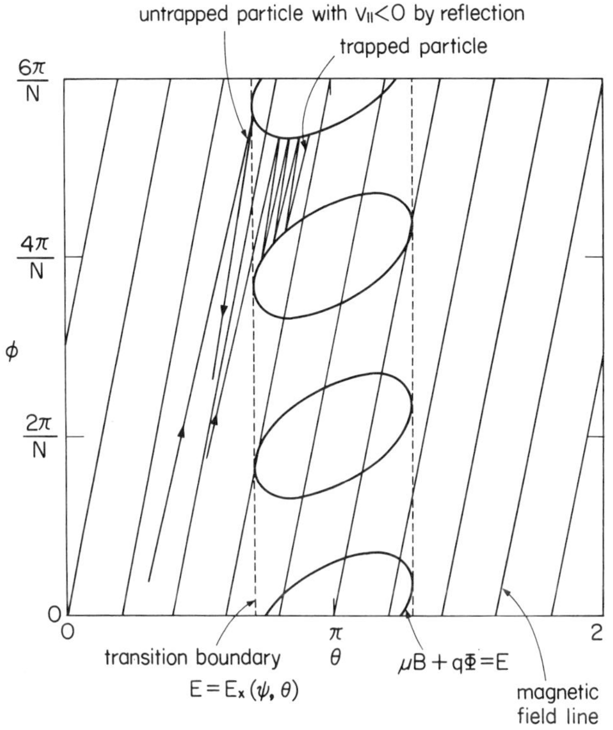

Fig. 6.3 Drift motion showing transition from the trapped state to untrapped state by crossing the transition boundary on a magnetic surface. The transition probability depends on the toroidal angle ϕ.

corresponds to a finite rotational transform. Elliptic shapes show lines satisfying

$$E = \mu B + e\Phi = \mu B_0(\psi, \theta) - \mu B_1(\psi, \theta)\cos(\ell\theta - M\phi) + e\Phi. \qquad (6.122)$$

Since the parallel drift motion along the magnetic line of force is given by $u = (2/m)^{1/2}(E - \mu B - e\Phi)^{1/2}$, u becomes zero on the elliptic lines, which means that reflection of particle motion occurs. The dotted lines correspond to

$$E = E_X(\psi, \theta) = \mu B_0(\psi, \theta) + \mu B_1(\psi, \theta) + e\Phi. \qquad (6.123)$$

The region between two dotted lines satisfies $\eta < 1$. The two lines with arrows describe particle motions in the present case. Since the particle motions include slow drift perpendicular to the magnetic lines of force, the orbits do not follow the magnetic lines of force. One particle is trapped and the other is not, which is considered to be dependent on the toroidal angle ϕ at the transition line $E = E_X(\psi, \theta)$. Since the averaged drift equations cannot give information concerning ϕ, we only give the probability of a particle being trapped at the reflection point.

We show a way of calculating the probability (see Fig. 6.4). The particle orbit A shows that the orbit goes through the boundary point P to arrive at the elliptic line C_1. On the other hand, the particle starting the position left of P cannot arrive at C_1. The particle orbit B shows that the orbit goes through the boundary point Q to arrive at C_1 and is trapped between C_1 and C_2. In other words, the particle starting at the position to the right of Q is reflected at C_1 and becomes trapped between C_1 and C_2. Thus we can say that particles that start from $\overline{PQ}$ are not trapped, while particles that start from $\overline{QR}$ are trapped. $\overline{QR}$ can be measured by a difference of poloidal angle $(\Delta\theta)_{QR}$ and $\overline{PR} = \overline{PQ} + \overline{QR}$ by $(\Delta\theta)_{PR}$. Then it is easily understood that the trapping probability is given by

$$P_r = (\Delta\theta)_{QR}/(\Delta\theta)_{PR}, \tag{6.124}$$

where particles are distributed uniformly with respect to θ.

In the more general problem, the angle θ is replaced with a parameter h, for which we can assume uniform particle distribution. We consider the case in which the particle is reflected at B in the magnetic field ripple, as shown in Fig. 6.5. Here h changes from A to B with Δh_1 during particle drift motion over the ripple just before the reflection (see Fig. 6.5(a)), and h changes from B to C with Δh_2 during particle drift motion over the ripple just after the reflection

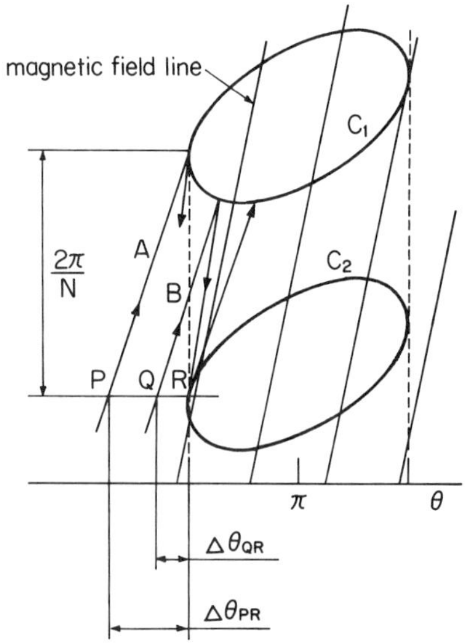

Fig. 6.4 A calculation of the transition probability with variation of the poloidal angle θ.

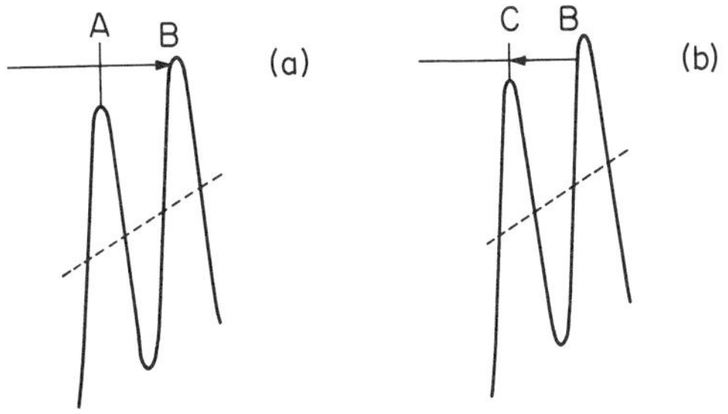

Fig. 6.5 Particle drift motion (a) just before and (b) just after reflection by the magnetic mirror at B.

(see Fig. 6.5(b)). In this case the transition probability that the particle reflected at B becomes trapped by the reflection at C is given by

$$P_r = \left| \frac{\Delta h_1 + \Delta h_2}{\Delta h_1} \right|. \tag{6.125}$$

We note that the sign of Δh_1 should be different from that of Δh_2, since $P_r < 1$.

Hereafter, we consider

$$h = E - E_X(\psi, \theta), \tag{6.126}$$

which denotes the difference between the particle energy and the transition boundary energy. For $h > 0$, trapped particles are characterized by $h < 0$ and untrapped particles by $h < 0$. In order to calculate the transition probability (6.125), we use Δh which is given by using a Taylor expansion at $h = 0$:

$$\Delta h \simeq -\frac{\partial E_X}{\partial \psi} \Delta\psi - \frac{\partial E_X}{\partial \theta} \Delta\theta. \tag{6.127}$$

Here we must know $\Delta\psi$ and $\Delta\theta$ near the transition boundary $h = 0$. It should be noted that the adiabatic invariants $J_{\pm 0}$ and J_{r0} change there, and they are written as

$$J_{\pm 0} = Y_{\pm 0} + (h/\omega)(1 + \ln |E_\pm/h|), \tag{6.128}$$

$$J_{r0} = Y_{r0} + (2h/\omega)(1 + \ln |E_r/h|), \tag{6.129}$$

where $Y_{\pm 0}$ and Y_{r0} denote $J_{\pm 0}$ and J_{r0} at $h = 0$, $E_\pm$ and E_r are of the order of E_X, and ω is a constant. Here $Y_{r0} = Y_{+0} + Y_{-0}$ and $E_r = (E_+ E_-)^{1/2}$ are obtained from $J_{r0} = J_{+0} + J_{-0}$ for $h \simeq 0$. Periods for trapped and untrapped particle orbits are obtained from $T_\alpha = \partial J_\alpha / \partial E$ as

$$T_\pm = \omega^{-1} \ln |E_\pm/h|, \tag{6.130}$$

$$T_r = 2\omega^{-1} \ln |E_r/h|. \tag{6.131}$$

First, we calculate $\Delta\psi_+$ and $\Delta\theta_+$ for an untrapped particle with $u > 0$. For the last path over the ripple,

$$\Delta\psi_+ = \int d\phi \frac{d\psi}{d\phi} = \int d\phi \left[\frac{\partial}{\partial\theta}(mub_\phi/e) - \frac{\partial}{\partial\phi}(mub_\theta/e)\right], \tag{6.132}$$

where the integral is taken under the conditions $\psi =$ constant and $\theta =$ constant at the transition boundary $h = 0$. By considering periodicity in the ϕ direction,

$$\Delta\psi_+ = \frac{1}{e}\frac{\partial}{\partial\theta}\int_0^{2\pi/N} d\phi(mub_\phi) = \frac{1}{e}\frac{\partial J_{+0}}{\partial\theta} \tag{6.133}$$

is obtained, which gives

$$\Delta\psi_+ = \frac{1}{e}\left(\frac{\partial Y_{+0}}{\partial\theta} - T_+ \frac{\partial E_X}{\partial\theta}\right) \tag{6.134}$$

with (6.128) and (6.130). By the same procedure,

$$\Delta\theta_+ = -\frac{1}{e}\left(\frac{\partial Y_{+0}}{\partial\psi} - T_+ \frac{\partial E_X}{\partial\psi}\right) \tag{6.135}$$

is obtained. Similarly, for an untrapped particle with $u < 0$,

$$\Delta\psi_- = \frac{1}{e}\left(\frac{\partial Y_{-0}}{\partial\theta} - T_- \frac{\partial E_X}{\partial\theta}\right), \tag{6.136}$$

$$\Delta\theta_- = -\frac{1}{e}\left(\frac{\partial Y_{-0}}{\partial\psi} - T_- \frac{\partial E_X}{\partial\psi}\right). \tag{6.137}$$

For a trapped particle, $\Delta\psi_r$ and $\Delta\theta_r$ denote variations over one bounce motion, which are given by

$$\Delta\psi_r = \Delta\psi_+ + \Delta\psi_- = \frac{1}{e}\left(\frac{\partial Y_{r0}}{\partial\theta} - T_r \frac{\partial E_X}{\partial\theta}\right), \tag{6.138}$$

$$\Delta\theta_r = \Delta\theta_+ + \Delta\theta_- = \frac{1}{e}\left(\frac{\partial Y_{r0}}{\partial\psi} - T_r \frac{\partial E_X}{\partial\psi}\right), \tag{6.139}$$

Again, we note that integrals are taken under $\psi =$ constant and $\theta =$ constant at the transition boundary $h = 0$, these being shown as $\psi_\times$ and $\theta_\times$. Thus all particles passing at $(\psi_\times, \theta_\times)$ on the poloidal plane have equal $\Delta\psi_\alpha$ and $\Delta\theta_\alpha$, independent of the toroidal angle ϕ.

Δh_α is obtained from (6.127):

$$\begin{aligned}\Delta h_\alpha &\simeq -\frac{\partial E_X}{\partial\psi}\,\Delta\psi_\alpha - \frac{\partial E_X}{\partial\theta}\,\Delta\theta_\alpha \\ &= \frac{1}{e}\left(\frac{\partial E_X}{\partial\theta}\frac{\partial Y_\alpha}{\partial\psi} - \frac{\partial E_X}{\partial\psi}\frac{\partial Y_\alpha}{\partial\theta}\right) \\ &= \frac{1}{e}\,\{E_X, Y_\alpha\}. \end{aligned} \tag{6.140}$$

This Δh_α is equal for all particles passing at the same transition boundary point. Since $Y_{r0} = Y_{+0} + Y_{-0}$, (6.140) gives

$$\Delta h_r = \Delta h_+ + \Delta h_-. \tag{6.141}$$

The transition probability is required for a point such as Q, shown in Fig. 6.2. Here we assume that untrapped particles with a negative parallel velocity enter the point Q. Since the particle with $h = E - E_X > 0$ in the untrapped state crosses the transition boundary $h = 0$, $\Delta h_- < 0$. On the contrary, for $\Delta h_- > 0$, the particle moves from the point Q as an untrapped particle with a positive parallel velocity. However, for trapped particles, the situation is reversed, and $\Delta h_r > 0$ corresponds to particle motion to the point Q and $\Delta h_r < 0$ to particle motion from the point Q.

The transition boundary point similar to the point Q satisfies the condition that only one of Δh_+, Δh_-, and $-\Delta h_r$ is negative. Since (6.141) becomes $\Delta h_+ + \Delta h_- - \Delta h_r = 0$, it is impossible for Δh_+, Δh_-, and $-\Delta h_r$ to have the same sign. The state with the negative sign corresponds to the majority state.

When the majority state is given, it is possible to calculate the transition probability to the other two states. First, we consider that the untrapped particle with a negative parallel velocity corresponds to the majority state. $h = (E - E_x)$ at the reflection point just after the particle crosses the transition boundary is denoted by h_0. Since this particle belongs to the trapped state at the reflection point, $h_0 < 0$. Also, since this particle belongs to the untrapped state before crossing the transition boundary, $h_0 - \Delta h_- > 0$. Thus the particle belonging to the untrapped state becomes reflected for

$$\Delta h_- < h_0 < 0, \tag{6.142}$$

after crossing the transition boundary $h = 0$. The same argument is applicable to the resultant trapped particle. This particle becomes untrapped with a positive parallel velocity by crossing the boundary $h = 0$ for $h_0 + \Delta h_+ > 0$, or

$$-\Delta h_+ < h_0 < 0. \tag{6.143}$$

On the other hand, this particle retains the trapped state for

$$\Delta h_- < h_0 < -\Delta h_+ = \Delta h_- - \Delta h_r. \tag{6.144}$$

It is noted that $\Delta h_- < 0$, $\Delta h_+ > 0$ and $-\Delta h_r > 0$ here. By admitting that particles are distributed uniformly for the range of $\Delta h_- < h_0 < 0$, the transition probability from the untrapped state with a negative parallel velocity to the trapped state is given as

$$P_r = \left|\frac{\Delta h_+ + \Delta h_-}{\Delta h_-}\right| = \left|\frac{\Delta h_r}{\Delta h_-}\right|, \tag{6.145}$$

which is applicable to the transition point Q in Fig. 6.2. This result may be generalized as

$$P_\beta = \left|\frac{\Delta h_\beta}{\Delta h_\alpha}\right|, \tag{6.146}$$

where the particle belongs to the majority state α and the transition from the state α to the state β occurs with probability P_β. It is remarked that, at the transition point P in Fig. 6.2, the trapped particle belongs to the majority state, and that the transition probability from the trapped particle to the untrapped particle with a negative parallel velocity is 1.

Finally, we show expressions $\Delta_{h\alpha}$ for the model magnetic field used to obtain $J_{\alpha 0}$. Since Y_r is the value of J_{r0} (see (6.88)) on the transition boundary shown by $\eta = 1$,

$$\begin{aligned} Y_r &= \frac{8B_\phi}{N}\sqrt{\frac{m\mu}{B_1}} \lim_{\eta\to 1-0} [(1+\tilde{b}\eta)\Pi(\tilde{b}\eta, \eta) - K(\eta)] \\ &= \frac{8B_\phi}{N}\sqrt{\frac{m\mu}{B_1}} \int_0^{\pi/2} \frac{\tilde{b}\cos\theta}{1+\tilde{b}\sin^2\theta}\, d\theta \\ &= \frac{8B_\phi}{N}\sqrt{\frac{m\mu}{B_1}} \sqrt{\tilde{b}}\tan^{-1}\sqrt{\tilde{b}}. \end{aligned} \tag{6.147}$$

The derivative of Y_r is given by

$$\begin{aligned} \frac{Y_r}{\partial\xi} = \frac{2B_\phi\sqrt{m\mu}}{NB_1^{3/2}}\tilde{b}\Bigg[&\left(\frac{2+\tilde{b}}{1+\tilde{b}} + \sqrt{\tilde{b}} + \tan^{-1}\sqrt{\tilde{b}}\right)\frac{\partial B_1}{\partial\xi} \\ &- \left(\frac{\tilde{b}}{1+\tilde{b}} + \sqrt{\tilde{b}}\tan^{-1}\sqrt{\tilde{b}}\right)\frac{\partial B_0}{\partial\xi}\Bigg], \end{aligned} \tag{6.148}$$

where ξ denotes ψ or θ, and we have used (6.117).

By taking the limit $\eta \to 1+0$ for $J_{\pm 0}$ shown by (6.89), we obtain

$$Y_{\pm} = \pm \frac{2\pi e A_\phi}{N} + \frac{1}{2} Y_r. \tag{6.149}$$

With $E_X = \mu B_0 + \mu B_1 + e\Phi$, (6.140) yields

$$\begin{aligned}
\Delta h_r &= \frac{1}{e}\{E_X, Y_r\} \\
&= \frac{1}{e}\left(\frac{\partial E_X}{\partial \theta}\frac{\partial Y_r}{\partial \psi} - \frac{\partial E_X}{\partial \psi}\frac{\partial Y_r}{\partial \theta}\right) \\
&= -\frac{2B_\phi \tilde{b}\sqrt{m\mu}}{NB_1^{3/2}}\left[\left(\frac{2+\tilde{b}}{1+\tilde{b}} + \sqrt{\tilde{b}}\tan^{-1}\sqrt{\tilde{b}}\right)\{B_1, \Phi\}\right. \\
&\quad - \left(\frac{\tilde{b}}{1+\tilde{b}} + \sqrt{\tilde{b}}\tan^{-1}\sqrt{\tilde{b}}\right)\{B_0, \Phi\} \\
&\quad \left. + \frac{2\mu}{e}(1+\sqrt{\tilde{b}}\tan^{-1}\sqrt{\tilde{b}})\{B_1, B_0\}\right],
\end{aligned} \tag{6.150}$$

and

$$\begin{aligned}
\Delta h_{\pm} &= \frac{1}{e}\{E_X, Y_{\pm}\} \\
&= \frac{1}{e}\left\{E_X, \pm\frac{2\pi e A_\phi}{N}\right\} + \frac{1}{2e}\{E_X, Y_r\} \\
&= \pm\frac{2\pi}{N}\{E_X, A_\phi\} + \frac{1}{2}\Delta h_r,
\end{aligned} \tag{6.151}$$

where we have assumed $A_\phi(\psi)$ and $\iota = -dA_\phi(\psi)/d\psi$.

For the case in which $\partial\Phi/\partial\theta = 0$, $\partial B_1/\partial\theta = 0$, and $\tilde{b} \ll 1$,

$$\begin{aligned}
\Delta h_r &\simeq \frac{2B_\phi \tilde{b}\sqrt{m\mu}}{NB_1^{3/2}}\frac{\partial B_0}{\partial \theta}\left(2\tilde{b}e\frac{\partial \Phi}{\partial \psi} + 2\mu\frac{\partial B_1}{\partial \psi}\right) \\
&\simeq \frac{16 B_\phi \varepsilon_t}{eN}\sqrt{\frac{m\mu\varepsilon_h}{\bar{B}_0}}\sin\theta\frac{\partial}{\partial \psi}\left(e\Phi + \frac{\mu\bar{B}_0}{2}\ln(\varepsilon_h)\right)
\end{aligned} \tag{6.152}$$

is given, where $1/\tilde{b} \simeq 1/(2\varepsilon_h)$ and $\partial B_1/\partial B_1/\partial\psi = \partial(\bar{B}_0\varepsilon_h)/\partial\psi$ have been used. For the model stellarator field with $\varepsilon_h = \delta_h\psi$, the derivative with respect to ψ in (6.152) becomes positive for $\Phi = 0$. Thus $\Delta h_r > 0$ for $\theta > 0$ and $\Delta h_r < 0$ for $\theta < 0$. This result means that the trapped particle enters P in Fig. 6.2, or in the upper half poloidal plane, and it moves out from Q in Fig. 6.2, or in the lower half poloidal plane.

6.4 THE CHARACTERISTICS OF TRAPPED PARTICLE CONFINEMENT

It is known that there are two types of particle in axisymmetric tokamaks; one type is called the untrapped or passing particle and the other is the trapped particle in the local magnetic mirrors produced by the toroidal curvature. In stellarator and heliotron devices, the particle orbits become more complicated due to the nonaxisymmetric toroidal magnetic configuration. In this section we survey the characteristics of particle orbits in stellarator and heliotron devices, with emphasis on the latter.

Here we consider the following model magnetic field in Boozer coordinates:

$$B = \bar{B}_0[1 - \varepsilon_t(\psi)\cos\theta - \varepsilon_h(\psi)\cos(2\theta - M\phi) - \varepsilon_1(\psi)\cos(\theta - M_\phi) - \varepsilon_3(\psi)\cos(3\theta - M\phi)], \tag{6.153}$$

where ε_t is called a toroidal ripple, ε_h is called a helical ripple of a heliotron device with the $\ell = 2$ helical coil, ε_1 is an $\ell = 1$ side band helical component, and ε_3 is an $\ell = 3$ side band helical component. The toroidal period number is denoted by M. In heliotron devices the side band components, ε_1 and ε_3, are produced by the toroidal effect through $\cos(2\theta - M\phi)\cos\theta = (\cos(3\theta - M\phi) + \cos(\theta - M\phi))/2$, where $\cos\theta$ arises from the toroidal magnetic field, $B = B_0/(1 + r\cos\theta/R) \simeq B_0(1 - r\cos\theta/R)$. It is possible to change the relative magnitudes of ε_h, ε_1, and ε_3 by applying the additional vertical magnetic field to change the magnetic axis position. In a vacuum heliotron configuration with $M = 12$, an average minor radius of 20 cm and a major radius of 2.1 m, with the line-tracing code given in Fig. 6.6, shows how the ε_1 and ε_3 side band helical components change when the magnetic axis position is changed

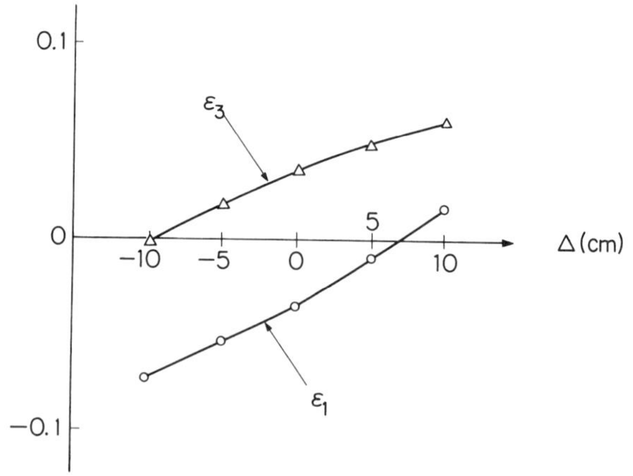

Fig. 6.6 The variation of the ε_1 and ε_3 side band helical components by the vertical magnetic field to change the magnetic axis position Δ in a vacuum configuration of $\ell = 2$ and $M = 12$, with $a = 20$ cm and $R = 2.1$ m.

from the standard position of $\Delta = 0$ by the vertical magnetic field. Here $\Delta > 0$ ($\Delta < 0$) corresponds to the outward (inward) shift of the magnetic axis. For the outward shift case, both $\varepsilon_1(< 0)$ and $\varepsilon_3(> 0)$ increase, and for the inward shift case, both $\varepsilon_1(< 0)$ and $\varepsilon_3(> 0)$ decrease. When the radial coordinate $\rho = r/a$ is introduced, $\varepsilon_t = \delta_t \rho$, $\varepsilon_h = \delta_h \rho^2$, $\varepsilon_1 = \delta_1 \rho$, and $\varepsilon_3 = \delta_3 \rho^3$are given approximately for the above model (6.153). It is noted that $\rho = 1$ corresponds to the outermost magnetic surface. It is interesting to compare (6.153) with the model magnetic field for the σ optimization of a stellarator or heliotron from the neoclassical transport point of view:

$$B(r, \theta, \phi) = B_0[1 - \varepsilon_t \cos\theta - \varepsilon_h^\ell \cos(\ell\theta - M\phi) + \tfrac{1}{2}\sigma(\varepsilon_h^{\ell-1} \cos((\ell - 1)\theta - M\phi) + \varepsilon_h^{\ell+1} \cos((\ell + 1)\theta - M\phi))], \quad (6.154)$$

where $\varepsilon_t = r/R_0$, $\varepsilon_h^\ell \propto r^\ell$, and $\varepsilon_h^{\ell\pm1} \propto r^{\ell\pm1}$. For a particular case, $\varepsilon_h^\ell(r_1) = \varepsilon_h^{\ell+1}(r_1) = \varepsilon_h^{\ell-1}(r_1)$ at $r = r_1$, (6.154) becomes

$$B(r, \theta, \phi) = B_0[1 - \varepsilon_t \cos\theta - \varepsilon_h^\ell \cos(\ell\theta - M\phi)(1 - \sigma\cos\theta)], \quad (6.155)$$

where the modulating envelope $(1 - \sigma\cos\theta)$ localizes the helical magnetic ripple to the interior (exterior) of the torus for $\sigma = 1$ (-1) (see Fig. 7.1). The $\sigma = 1$ configuration is the one with improved confinement from the point of view of the particle orbits and neoclassical transport; the $\sigma = -1$ configuration has poorer confinement than the standard one with $\sigma = 0$. This is understood from the poloidal projections of orbits showing transition from the passing particle to the helically trapped particle, as shown in Fig. 6.7, which are obtained by solving the drift equations (6.14)–(6.16) and (6.20) in Boozer coordinates.

From a comparison between (6.153) and (6.154), the decrease (increase) of both $\varepsilon_1(< 0)$ and $\varepsilon_3(> 0)$ corresponds to the positive (negative) σ case. Thus, the inward shift of the magnetic axis is favorable to the trapped particle confinement. However, when the magnetic axis is shifted inward by the vertical magnetic field in heliotron devices, the magnetic well always decreases and the magnetic hill always increases. This tendency degrades MHD stability against the pressure-driven modes or decreases the beta limit within the ideal MHD model. This result is the most serious contradiction in heliotron configurations. There have many trials to overcome this contradiction between the improved particle confinement and the high beta limit, such as the exploitation of quadrupole and hexapole fields to restore the magnetic well.

In order to study the deeply trapped particle orbit, it is useful to consider

$$B_{min}(\psi, \theta) = min_\phi |B(\psi, \theta, \phi)| \quad (6.156)$$

in Boozer coordinates (ψ, θ, ϕ), where $\theta = 0$ corresponds to the outer side of the toroidal plasma. When the initial velocity parallel to the magnetic field line is zero or $u = 0$ and the radial electric field is negligible, the contour of

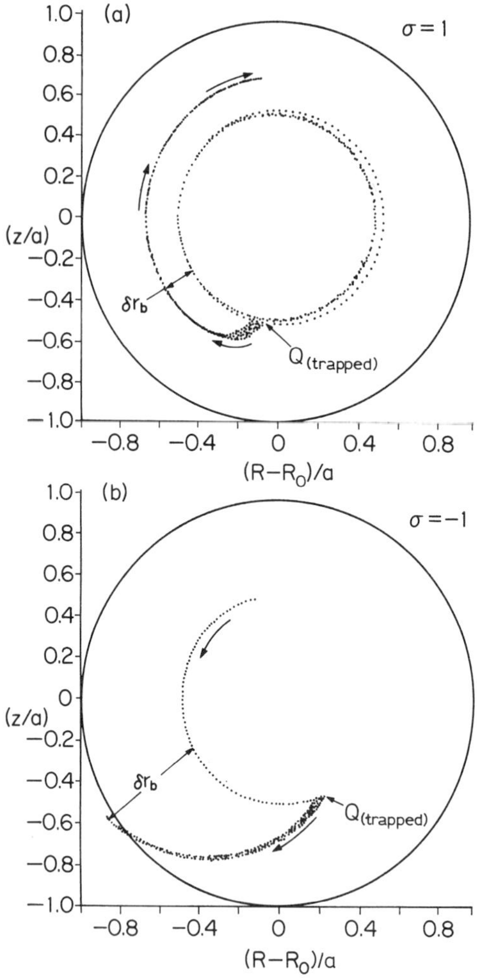

Fig. 6.7 Poloidal projections of orbits characteristic of those contributing to transport at low collosionality, for (a) the $\sigma = 1$ and (b) the $\sigma = -1$ configurations. For the $\sigma = -1$ case the particle moves straight out of the device, once helically trapped; however, for the $\sigma = 1$ case the particle makes small deviations from the magnetic surface, except for immediately after helical entrapment.

$B_{min}(\psi, \theta) =$ constant is equivalent to that of $J_{r0} \simeq 0$ (see (6.72)), since $u \simeq 0$ is equal to $E - \mu B_{min}(\psi, \theta) = 0$, and E and μ are constants of motion. By equating B_{min}/B_0 at $(\rho_*, 0)$ to B_{min}/B_0 at $(1, \pi)$, $\delta_h(1 - \rho_*) = \delta_t + \delta_1 + \delta_3(1 - \rho_* + \rho_*^2)$ is obtained. This equation means that the deeply trapped particle that started at $(\rho, 0)$ with $\rho < \rho_*$ can be confined. Using an approximation $1 - \rho_* + \rho_*^2 \sim 0(1)$ for $0 < \rho_* < 1$, we obtain

$$\rho_* \simeq 1 - (\delta_t + \delta_1 + \delta_3)/\delta_h. \qquad (6.157)$$

Thus the deeply trapped particle starting at $(\rho, 0)$ can be confined when ρ is smaller than in (6.157). It is noted that $\delta_h > \delta_t$ and $\delta_t + \delta_1 + \delta_3 > 0$ are also required here. In general, these conditions are satisfied in $\ell = 2$ heliotron configurations. This result means that *negative* δ_1 and δ_3 are favorable to confine the deeply trapped particles. This result is the same direction as to *shift the magnetic axis inward*, as shown in Fig. 6.6.

In heliotron devices, the deeply trapped particles are not always lost from the confinement region surrounded by the outermost magnetic surfaces. Sometimes the transition particles with alrge excursions are lost from the confinement region. By following many particle orbits with different energies E and pitches $v_\perp/v_\parallel$ initially, we can draw a figure showing the velocity space loss region, as shown in Fig. 6.8. The loss particle crosses the outermost magnetic surface during the orbit calculation with drift equations (6.14)–(6.16) and (6.20). Figure 6.8 shows that deeply trapped particles with $v_\parallel \simeq 0$ can be confined in the heliotron device.

It is not easy to eliminate the velocity space loss region completely in the region with $\rho > 0.5$. In the design of the LHD the condition that there should be no velocity space loss region for $\rho < 0.3$ was imposed.

One way to reduce the velocity space loss region is to exploit the radial electric field. From the guiding center drift velocity shown by (6.10), it is easily seen that the radial electric field produces the poloidal drift motion. When the poloidal drift velocity due to the radial electric field sufficiently exceeds the ∇B drift velocity, all trapped particles (both localized and blocked particles) are untrapped and the particle orbit confinement is improved significnatly for both positive and negative electric field cases compared to the zero or weak radial electric field case. However, it should be noted that cancellation of the poloidal drift due to ∇B drift with the $\mathbf{E} \times \mathbf{B}$ drift is possible in the poloidal direction,

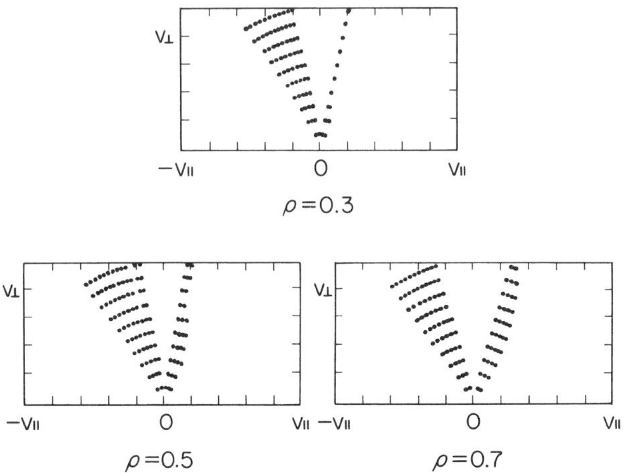

Fig. 6.8 Typical velocity space loss regions of the heliotron configuration for $\rho = 0.3$, $\rho = 0.5$, and $\rho = 0.7$. Here the angle variables are $\theta = 0$ and $\phi = 0$. The maximum particle energy is 100 keV and the magnetic field is 2 T.

although the vertical drift due to ∇B drift remains. For ions ($e > 0$), the $\nabla\Phi > 0$ case can produce the cancellation for a specific particle energy, since the average of the ∇B vector due to the helical ripple field over the trapped particle orbit is directed inward. On the contrary, for electrons ($e < 0$), the $\nabla\Phi < 0$ case can produce cancellation of the poloidal drift velocity.

Here we will show the effect of the radial electric field with the model stellarator field

$$B(x, y, \phi) = B_0\left[1 - \varepsilon_t\left(\frac{r}{a}\right)\cos\theta - \varepsilon_h\left(\frac{r}{a}\right)^2\cos(\ell\theta - M\phi)\right] \tag{6.158}$$

in the toroidal coordinates (r, θ, ϕ), where $x = r\cos\theta$ and $y = r\sin\theta$. We also assume that the magnetic surfaces averaged in the ϕ direction are circular (see Fig. 6.9) and described as

$$(x - \Delta)^2 + y^2 = \rho^2. \tag{6.159}$$

This averaging is the same as used in the stellarator expansion method in chapter 3. In (6.159), ρ is an average radius of the corresponding magnetic surface and Δ is its shift with respect to the center of the $\ell = 2$ helical coils in heliotron devices. We assume $\Delta =$ constant for simplicity. From J_{r0} of (6.72) for the deeply trapped particles, orbits in the (x, y) plane are described by the relation

$$B_{min}(x, y) + \frac{e}{\mu}\,\Phi(x, y) = \frac{E}{\mu}, \tag{6.160}$$

where the definition of B_{min} is the same as in (6.156).

First, we consider the $\Phi = 0$ case. By noting that

$$B_{min}(x, y) = B_0\left[1 - \varepsilon_t\left(\frac{x}{a}\right) - \varepsilon_h\left(\frac{x^2 + y^2}{a^2}\right)\right] \tag{6.161}$$

for the model stellarator field (6.158), the orbit given by (6.160) becomes circular in the (x, y) plane:

$$(x - x_{dtp})^2 + y^2 = \rho_{dtp}^2 \tag{6.162}$$

and

$$x_{dtp} = \frac{\varepsilon_t}{2\varepsilon_h}\,a. \tag{6.163}$$

For heliotron devices with $\varepsilon_t < \varepsilon_h$, $x_{dtp} < a$. For $x_{dtp} < \Delta$, the orbit passing the point $(x, y) = (\Delta - a, 0)$ corresponding to the innermost point of the averaged circular surface shown by (6.159) and (6.162) gives $\rho_{dtp} = x_{dtp} + a - \Delta$. This orbit also passes the outermost point $(x, y) = (2x_{dtp} + a - \Delta, 0)$. Thus deeply

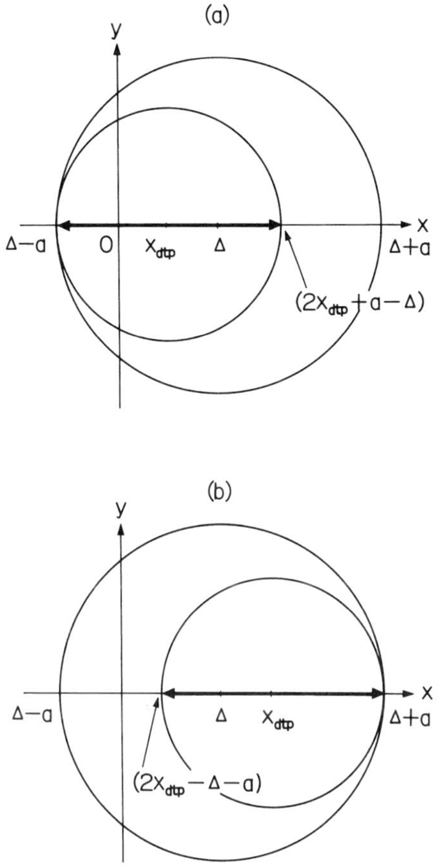

Fig. 6.9 For the deeply trapped particle starting at $(x, 0)$, (a) it is confined for $\Delta - a \leq x \leq 2x_{dtp} + a - \Delta$, when it crosses the inner region of the outermost circular magnetic surface, and (b) it is confined for $2x_{dtp} - \Delta - a \leq x \leq \Delta + a$, when it crosses the outer region of the outermost circular magnetic surface.

trapped particles starting from the point with $\Delta - a \leq x \leq 2x_{dtp} + a - \Delta$ and $y = 0$ can be confined completely (see Fig. 6.9(a)).

Next we include the radial electric field through the electric potential profile $\Phi(\rho) = \Phi_0(1 - (\rho/a)^2)$. From (6.159) and (6.160), the deeply trapped particle orbit also becomes circular for $\Phi_0 \neq 0$ and is given by (6.162) with

$$x_{dtp} = -[\varepsilon_t a - 2(e\Phi_0/\mu B_0)\Delta]/2(\varepsilon_h + e\Phi_0/\mu B_0) \tag{6.164}$$

and

$$\rho_{dtp} = \{x_{dtp}^2 + [(1 - E/\mu B_0)a^2 + (a^2 - \Delta^2)(e\Phi_0/\mu B_0)]/(\varepsilon_h + e\Phi_0/\mu B_0)\}^{1/2}. \tag{6.165}$$

We note here that (6.164) can also be written as

$$\frac{e\Phi_0}{\mu B_0} = \frac{\varepsilon_t a + 2\varepsilon_h x_{dtp}}{2(\Delta - x_{dtp})}. \tag{6.166}$$

We consider the deeply trapped particle starting at $(x, 0)$ here. When it crosses the inner region of the outermost averaged circular magnetic surface for the loss case or $\Delta - a \leq x_{dtp} \leq \Delta$ (see Fig. 6.9(a)), the confinement condition becomes

$$\Delta - a \leq x \leq 2x_{dtp} - \Delta + a \equiv x_R^c, \tag{6.167}$$

which is obtained from the similar way for the case of $\Phi_0 = 0$. From the RHS in (6.167),

$$\frac{\mu B_0}{e\Phi_0} = \frac{\Delta + a - x_R^c}{\varepsilon_t a + \varepsilon_h(x_R^c + \Delta - a)} \tag{6.168}$$

is obtained as the boundary of loss region.

Next, when the deeply trapped particle crosses the outer region of the outermost magnetic surface for the loss case, or $\Delta \leq x_{dtp} \leq \Delta + a$ (see Fig. 6.9(b)), the confinement condition becomes

$$x_L^c \equiv 2x_{dtp} - \Delta - a \leq x \leq \Delta + a. \tag{6.169}$$

From the LHS relation in (6.169),

$$\frac{\mu B_0}{e\Phi_0} = \frac{\Delta - a - x_L^c}{\varepsilon_t a + \varepsilon_h(x_L^c + \Delta + a)} \tag{6.170}$$

is obtained as the boundary of the loss region.

Using (6.168) and (6.170), we can draw the deeply trapped particle loss region schematically, as shown in Fig. 6.10. When the $\mathbf{E} \times \mathbf{B}$ poloidal drift cancels the ∇B drift, $\mu B_0/e\Phi_0 = -1/\varepsilon_h$ is satisfied. For this case, x_{dtp} of (6.164) becomes infinite, and this type of particle cannot be confined. The related loss region is shown in Fig. 6.10. For $x_R^c = \Delta$ and $x_L^c = \Delta$, (6.168) and (6.170) give $\mu B_0/e\Phi_0 = +\{2(\varepsilon_h/a)(\Delta + \varepsilon_t a/2\varepsilon_h) - \varepsilon_h\}^{-1}$ and $\mu B_0/e\Phi_0 = -\{2(\varepsilon_h/a)(\Delta + \varepsilon_t a/2\varepsilon_h) + \varepsilon_h\}^{-1}$, respectively. Also, for $x_R^c = \Delta + a$ and $x_L^c = \Delta + a$, (6.168) and (6.17) give $\mu B_0/e\Phi_0 = 0$ and $\mu B_0/e\Phi_0 = -2\{2(\varepsilon_h/a)(\Delta + \varepsilon_t a/2\varepsilon_h) + 2\varepsilon_h\}^{-1}$, respectively, and for $x_R^c = -a + \Delta$ and $x_L^c = -a + \Delta$, $\mu B_0/e\Phi_0 = 2\{2(\varepsilon_h/a)(\Delta + \varepsilon_t a/2\varepsilon_h) - 2\varepsilon_h\}^{-1}$ and $\mu B_0/e\Phi_0 = 0$, respectively. We also note that, for $x_R^c = -\Delta + (\varepsilon_h - \varepsilon_t)a/\varepsilon_h$, (1.168) gives $\mu B_0/e\Phi_0 \to \infty$. Then the shaded regions in Fig. 6.10 show the deeply trapped particle loss region in the $(x, \mu B_0/e\Phi_0)$ plane.

This result for the deeply trapped particle orbit can be extended for different electric potential profiles. One disadvantage in the case of a parabolic profile is

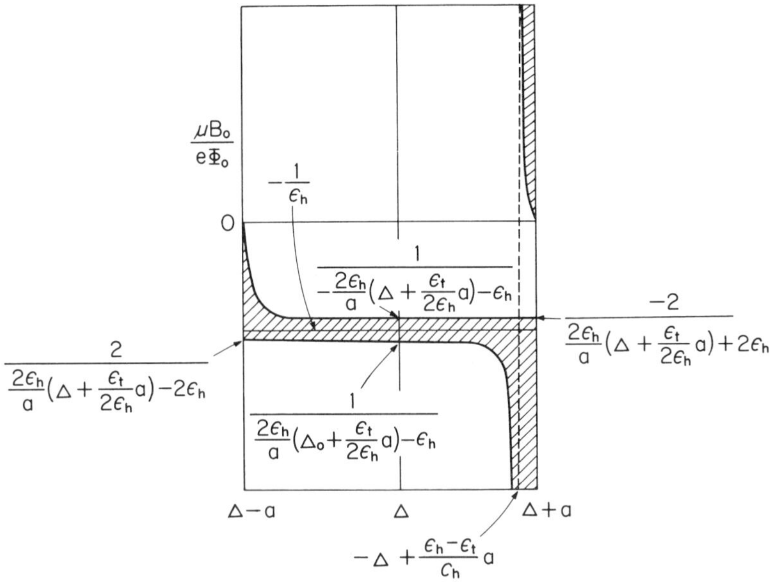

Fig. 6.10 The schematic loss region of deeply trapped particles, starting on the horizontal plane of the torus in the presence of a radial electric field with a parabolic potential profile. The shaded regions denote the loss region. $\Delta - a(\Delta + a)$ is the inside (outside) of the torus.

that the resonance at which the $\mathbf{E} \times \mathbf{B}$ poloidal drift is equal to the ∇B drift always exists for $\Delta - a \leq x \leq \Delta + a$ in Fig. 6.10.

Next, we discuss the effects of the magnetic perturbation on particle orbits. By considering energy conservation, $E = \mu B + mv_\parallel^2/2 + e\Phi = \text{constant}$,

$$\nabla E = \mu \nabla B + \nabla(\tfrac{1}{2} mv_\parallel^2) + e\nabla\Phi = 0 \tag{6.171}$$

is given. With (6.171), the guiding center drift shown by (6.10) becomes

$$\begin{aligned}\mathbf{v} &= v_\parallel \frac{\mathbf{B}}{B} + \frac{\mathbf{B}}{eB^2} \times \left(mv_\parallel^2 \frac{\nabla B}{B} - \nabla(\tfrac{1}{2} mv_\parallel^2) \right) \\ &= v_\parallel \frac{\mathbf{B}}{B} - \frac{v_\parallel}{B} \mathbf{B} \times \nabla \rho_\parallel,\end{aligned} \tag{6.172}$$

where $\rho_\parallel = v_\parallel m/eB$. By noting that $\nabla \times (\rho_\parallel \mathbf{B}) = \nabla\rho_\parallel \times \mathbf{B} - \rho_\parallel \nabla \times \mathbf{B}$,

$$\mathbf{v} = \frac{v_\parallel}{B} (\mathbf{B} \times \nabla \times (\rho_\parallel \mathbf{B})) \tag{6.173}$$

is obtained for a vacuum magnetic configuration satisfying $\nabla \times \mathbf{B} = 0$. When the magnetic perturbation is described by $\delta\mathbf{B} = \nabla \times (\alpha\mathbf{B})$, the drift velocity including $\delta\mathbf{B}$ can be written as

$$\mathbf{v} = \frac{v_{\parallel}}{B}(\mathbf{B} + \nabla \times (\rho_{\parallel} + \alpha)\mathbf{B}), \tag{6.174}$$

where α is a function of position. This result suggests that the drift equations including the magnetic perturbation can be derived by rewriting the Hamiltonian (6.1) as

$$H(\psi, \chi, \theta_0, \rho_c) = \frac{1}{2}\frac{e^2B^2}{m}(\rho_c - \alpha)^2 + \mu B + e\Phi, \tag{6.175}$$

where $\rho_c = \rho_{\parallel} + \alpha$. The most significant effect of $\delta\mathbf{B}$ on particle orbits appears when the magnetic surfaces are destroyed or the behavior of the magnetic field line becomes stochastic, as discussed in chapter 2. In this case passing particles that are almost following the magnetic field line escape easily from the central region of a toroidal plasma.

In stellarator and heliotron devices, magnetic islands and the stochastic magnetic region are generated by the finite-beta effect (see chapter 5). The drift equations (6.27), (6.28), (6.31), and (6.32) can be extended to include the magnetic perturbation $\delta\mathbf{B}$ in finite-beta plasmas. In heliotron configurations the magnetic axis shifts outward when the plasma pressure increases. This Shafranov shift usually degrades particle orbit confinement, as can be seen from Fig. 6.6, since Δ in Fig. 6.6 is interpreted as the Shafranov shift. It is considered that control of the vertical magnetic field is necessary to maintain good collisionless particle confinement for finite-beta plasmas in heliotron devices.

One way to improve the particle orbit confinement is to generate the magnetic field approximately shown as

$$B \simeq B_0[1 - \varepsilon_h(\psi)\cos(\ell\theta - M\phi)] \tag{6.176}$$

in Boozer coordinates (ψ, θ, ϕ). Since helical symmetry exists in (6.176), the particle confinement is better than in other stellarators.

The toridal effect shown by $B_0\varepsilon_t(\psi)\cos\theta$ can be suppressed if we apply strong pitch modulation to a winding law of the helical coil, as

$$\theta = (M/\ell)(\phi - \alpha\sin(M\phi/\ell)) + \pi, \tag{6.177}$$

where α is a pitch modulation parameter. However, in this approach a bumpy magnetic field appears, with $\ell = 0$ proportional to $B_0\varepsilon_0(\psi)\cos(M\phi)$. Nührenberg and Zille (1988) showed that a quasi-helically symmetric stellarator described by (6.176) is possible using the modular coil system. One disadvantage of a quasi-helically symmetric stellarator is that the bootstrap current which will be discussed in chapter 7 becomes large and changes the vacuum rotational transform substantially, which may produce a low-order resonant surface and destroy magnetic surfaces, since the quasi-helically symmetric stellarator usually has weak magnetic shear.

In order to suppress the bootstrap current in the quasi-helically symmetric stellarator, the bumpy magnetic field is introduced in the design of the W-7X Stellarator. In the W-7X, an interesting tendency for the diamagnetic effect due to finite pressure to improve trapped particle orbit confinement is realized.

6.5 THE MONTE CARLO METHOD FOR TRANSPORT PHENOMENA

Transport coefficients such as the diffusion coefficient and thermal diffusivity depend fundamentally on the number of symmetry directions of the plasma configuration. The enhancement of the diffusion coefficients in a tokamak or in an axisymmetric torus over those in a straight circular cylinder is well known, as Pfirsch–Schlüter diffusion, plateau diffusion, and banana diffusion. Actually, in the low-collisionality banana regime, any destruction of the two symmetry directions of the straight circular cylinder gives a similar enhancement. If the straight circular cylinder is distorted into an elliptic cylinder, a bumpy cylinder along its axis, a helically symmetric cylinder, or an axisymmetric torus, the low-collisionality transport is enhanced as in banana diffusion. As long as one symmetry direction remains, this enhancement can be calculated with methods developed for the neoclassical transport of tokamaks. In stellarators there is usually no absence of a symmetry direction, and the analytic techniques of a neoclassical transport become complicated even if good magnetic surfaces remain (see chapter 7). Since there is no symmetric direction, transport coefficients become much larger than those in tokamaks, particularly in the low-collisionality regime. Due to the sensitivity of the transport coefficients to asymmetry, reliable methods of evaluation are required. The analytic theory of transport in stellarators and heliotrons suffers from difficulties, since there are several types of magnetic ripples along the magnetic field lines. When both helical and toroidal ripples exist, the longitudinal adiabatic invariant $J = \oint m v_\parallel d\ell$ experiences jumps as particles suffer collisionless trapping and detrapping due to the change of trapping state.

The Monte Carlo method of evaluating transport coefficients avoids the serious problem of analytic theory, the need for a simple expression for the particle drift motions. In the Monte Carlo method for asymmetric devices, the particle drift orbits are obtained numerically. Coulomb scattering is included as an appropriate random variation of the constants of particle motion. Here we describe the Monte Carlo method, which offers several advantages. The use of the magnetic coordinates given by Boozer allows on trivially to make the ambipolar electrostatic potential constant on a magnetic surface. In addition, the magnetic field can be described by its magnitude alone.

In stellarator and heliotron devices, the dominant part of the collision operator is generally pitch angle scattering. The full collison operator changes both the energy and the pitch angle in the velocity space. Neoclassical transport is related to the particle trapping in the magnetic ripples along the field lines. Energy scattering does not change the trapping in the magnetic ripples, but ptich angle scattering does. However, when the transport is dominated by a resonance between the magnetic and electric drifts or the magnetic poloidal drift is exactly cancelled by the electric poloidal drift (see section 6.4), the energy scattering is more important. In this section the primary emphasis is on the conceptually simpler pitch angle scattering dominated case. The energy scattering or slowing down process also becomes an important ingredient in analysis of the heating process.

The diffusion equation is

$$\frac{\partial f}{\partial t} = \frac{1}{g(\psi)} \frac{\partial}{\partial \psi} g(\psi) D(E, \psi) \frac{\partial f}{\partial \psi} \tag{6.178}$$

for the distribution function $f(E, \psi, t)$, where $g(\psi)$ is defined by the volume element in real space, $dV = g(\psi) d\psi$. For a given D, the energy and particle transport coefficients can be evaluated by taking appropriate moments of the diffusion equation. Here we consider a large number of particles of given energy that cover a small region in ψ (radial coordinate) space. In the diffusion equation (6.178) this can be represented by f being a delta function in energy space and highly localized in ψ. By assuming that $\int f(E, \psi) g(\psi) d\psi = 1$, $\langle \psi \rangle = \int \psi f g(\psi) d\psi$. Then

$$\frac{d}{dt} \langle \psi \rangle = \int \psi \frac{\partial f}{\partial t} g(\psi) d\psi = \left\langle \frac{1}{g(\psi)} \frac{\partial}{\partial \psi} (g(\psi) D(E, \psi)) \right\rangle, \tag{6.179}$$

with two integrations by parts with respect to ψ. Similarly,

$$\frac{d}{dt} \langle \psi^2 \rangle = 2\langle D \rangle + 2\left\langle \frac{\psi}{g(\psi)} \frac{\partial}{\partial \psi} (g(\psi) D(E, \psi)) \right\rangle. \tag{6.180}$$

Since ψ is highly localized,

$$\left\langle \frac{\psi}{g(\psi)} \frac{\partial}{\partial \psi} (g(\psi) D(E, \psi)) \right\rangle \simeq \langle \psi \rangle \left\langle \frac{1}{g(\psi)} \frac{\partial}{\partial \psi} (g(\psi) D(E \psi)) \right\rangle. \tag{6.181}$$

Equations (6.179)–(6.181) give

$$\langle D \rangle = \frac{1}{2} \frac{d}{dt} (\langle \psi^2 \rangle - \langle \psi \rangle^2) = \frac{1}{2} \frac{d}{dt} \langle (\psi - \langle \psi \rangle)^2 \rangle. \tag{6.182}$$

Now let $\psi_j(t_0)$ be the radial position of the jth particle after a time t_0 with the condition that it was located at ψ_0 initially. This position $\psi_j(t_0)$ can be determined numerically by following the drift motion of a particle with an appropriate pitch angle scattering model after each time step. We define

$$D_j = (1/2t_0)(\psi_j - \psi_0)^2. \tag{6.183}$$

The estimate of D from J particles is given by

$$D = \frac{1}{J}\sum_{j=1}^{J} D_j. \tag{6.184}$$

There are several limitation on the Monte Carlo method to obtain the diffusion coefficient in stellarator and heliotron devices. First, consider the choice of t_0, the time estimating D_j in (6.183). If t_0 is chosen to be too long, it will not adequately represent the time derivative of $\langle(\psi - \langle\psi\rangle)^2\rangle$. This means that $|\psi_j - \psi_0|/\psi_0$ must be small. If t_0 is chosen to be too short, then $\psi_j(t_0)$ is controlled by deterministic motion. Therefore, $|\psi_j - \psi_0|$ must be large compared with the radial excursion of collisionless particle drift motion. The practical procedure for choosing t_0 is apparently to make it greater than or equal to the collision time. In stellarator and heliotron devices that can confine particles for many collision times, the radial excursion after one collision time corresponding to $|\psi_j(t_0) - \psi_0|$ is compared with the minor radius of the device. Consequently, a particular particle can be followed for several collision times with several values of D_j, given by

$$D_j = \frac{1}{2t_0}\,[\psi(t_j) - \psi(t_{j-1})]^2 \tag{6.185}$$

for $t_j = jt_0$.

The second limitation of the Monte Carlo method is the use of a Lorentz collision operator to change the pitch of particles, $v_\parallel/v$. Here the momentum is not conserved by the Lorentz collision operator. If a momentum-conserving pitch angle scattering operator is used, the diffusion coefficient D consists of the Lorentz collision operator part, D_L, and the momentum-conserving correction part, D_M. The Lorentz collision operator describes scattering of a test particle by a Maxwellian plasma following the Fokker–Planck collision operator. The momentum-conserving correction is related to a modification of the Maxwellian plasma. If there is axisymmetry in the torodial plasma, one can show that momentum conservation implies no particle transport from like-particle collisions, since D_M exactly cancels the like-particle collision contribution in D_L to the particle transport. However, in particle transport in the asymmetric torus, such as the stellarator and heliotron, like-particle collisions may contribute to particle transport and it is possible to assume $D_M \simeq 0$. Consequently, the Monte Carlo method discussed in this section is to evaluate D_L rather than D.

Let us consider how energy scattering can be included in the Monte Carlo method. As shown later, the inclusion of energy scattering as well as pitch angle scattering in the guiding center drift motion, presents a way to calculate $\psi_j(t_0)$. However, the diffusion coefficient derived from $\psi_j(t_0)$ is not $D(E, \psi)$ but the usual diffusion coefficient $D(T, \psi)$,

$$D(T, \psi) = \int_0^\infty D(E, \psi) f_M 4\pi v^2 dv, \tag{6.186}$$

where f_M is a Maxwellian of temperature T.

Now we explain the Lorentz pitch angle collision model and the energy scattering model in the Monte Carlo method. The Lorentz collision operator is shown as

$$\frac{\partial f}{\partial t} = \frac{\nu_d}{2} \frac{\partial}{\partial \lambda} (1 - \lambda^2) \frac{\partial f}{\partial \lambda}, \tag{6.187}$$

where $\lambda = v_{\|}/v$ and ν_d is the deflection collision frequency. Let

$$\langle \lambda \rangle \equiv \int_{-1}^{1} \lambda f d\lambda. \tag{6.188}$$

Then one can show that

$$\begin{cases} \dfrac{d\langle \lambda \rangle}{dt} = -\nu_d \langle \lambda \rangle \\ \dfrac{d\langle \lambda^2 \rangle}{dt} = \nu_d (1 - 3\langle \lambda^2 \rangle) \end{cases} \tag{6.189}$$

by integrating by parts. The square of the standard deviation of f in the pitch angle space, $\sigma^2 = \langle \lambda^2 \rangle - \langle \lambda \rangle^2$, broadens in time, with

$$d\sigma^2/dt = -\nu_d (1 - 3\langle \lambda^2 \rangle + 2\langle \lambda \rangle^2). \tag{6.190}$$

We assume that f is a delta function around $\lambda = \lambda_0$ initially, and that

$$\begin{cases} d\langle \lambda \rangle/dt = -\nu_d \lambda_0 \\ d\sigma^2/dt = \nu_d (1 - \lambda_0^2) \end{cases} \tag{6.191}$$

at $t = 0$. After a short time t_0, we expect f to be a Gaussian centered at $\lambda = \lambda_0(1 - \nu_d t_0)$ with standard deviation $[(1 - \lambda_0^2)\nu_d t_0]^{1/2}$.

Here we think the broadening of f as being due to a large number of steps in ptich of equal size, chosen randomly. The distribution function for obtaining m plus values in n trials with equal probability for plus and minus is the binomial distribution:

$$p(m) = \frac{1}{2^n} \frac{n!}{m!(n - m)!}. \tag{6.192}$$

Let j be the number of pluses minus the number of minuses, $j = m - (n - m)$. Then, for $n \gg 1$,

$$p(j) = (2\pi n)^{-1/2} \exp(-j^2/2n), \tag{6.193}$$

which means that standard deviation of j is $\sqrt{n}$. To reproduce the standard deviation σ, all λ steps must have a magnitude of $[(1-\lambda_0^2)\nu_d\tau]^{1/2}$, with τ the length of time between the steps. After n steps at the time $t = n\tau$, the standard deviation of the λ distribution becomes the step size times $\sqrt{n}$, as seen from (6.193): $\sigma = [(1-\lambda_0^2)\nu_d n\tau]^{1/2}$. Thus, if the pitch is changed from λ_0 to λ_n after a time step of τ with

$$\lambda_n = \lambda_0(1-\nu_d\tau) \pm [(1-\lambda_0^2)\nu_d\tau]^{1/2}, \tag{6.194}$$

the effects of the Lorentz collision operator will be reproduced for $\nu_d\tau \ll 1$. The $\pm$ sign means that the sign is to be chosen randomly, with equal probability for plus and minus.

The operator given by (6.194) has the property that if $|\lambda_0| < 1$, then $|\lambda_n| < 1$. It is physically obvious that the pitch must be less than unity. To show that $|\lambda_n| < 1$, let us assume $\nu_d\tau < 1$. For $|\lambda_0| \lesssim 1$, let $\lambda_0 = 1 - \delta$ with $\ll 1$. The largest λ_n can be

$$\lambda_n = 1 - (\delta + \nu_d\tau) + (2\nu_d\tau\delta)^{1/2}. \tag{6.195}$$

The maximum value of λ_n as δ is varied is given at $\delta = \nu_d\tau/2$, and

$$\lambda_n < 1 - \frac{\nu_d\tau}{2} < 1. \tag{6.196}$$

For $|\lambda_0| \ll 1$, $|\lambda_n| < 1$ is also valid.

In the Monte Carlo transport calculations, the time step of the integrations, τ, is chosen to be small enough for the orbit calculation to be energy conserving. This choice insures several time steps while a particle transits any region in which the magnetic field strength changes significantly.

The energy scattering is described by

$$\frac{\partial f}{\partial t} = \frac{1}{v^2}\frac{\partial}{\partial v}\left[\nu_E v^2\left(vf + \frac{T}{m}\frac{\partial f}{\partial v}\right)\right], \tag{6.197}$$

where ν_E is the energy scattering collision frequency. This energy scattering operator can be derived from the Fokker–Planck collision operator by assuming that particles scatter on a background Maxwellian of temperature T. A Monte Carlo equivalent energy scattering operator can be found by evaluating

$$\frac{d}{dt}\langle E\rangle = \frac{d}{dt}\int_0^\infty \left(\frac{1}{2}mv^2\right) f 4\pi v^2 dv \tag{6.198}$$

and

$$\frac{d}{dt}\langle E^2\rangle = \frac{d}{dt}\int_0^\infty \left(\frac{1}{2}mv^2\right)^2 f 4\pi v^2 dv. \tag{6.199}$$

The operator is given by considering the energy change from E_0 to E_n after the time step of τ,

$$E_n = E_0 - (2\nu_E\tau)\left[E_0 - \left(\frac{3}{2} + \frac{E}{\nu_E}\frac{d\nu_E}{dE}\right)T\right] \pm 2(TE_0\nu_E\tau)^{1/2}. \tag{6.200}$$

One can show that the energy is always positive, or if $E_0 > 0$, then $E_n > 0$.

Thus, particle and energy transport coefficients can be derived from the broadening of an ensemble of simulation particles around a single flux surface on which they are located. Monte Carlo simulation techniques in five-dimensional phase space have been used to analyze the transport properties in complex stellarator configurations and in ripple tokamaks, where the gyrophase of particle motion is averaged out by introducing the magnetic moment μ. They have also been applied to the study of plasma heating by high-energy neutral beam heating in tokamaks and stellarator and heliotron devices. An interesting application is the estimation of the bootstrap current in stellarators. An ensemble of monoenergetic simulation particles is pushed in time by drift orbit calculation with pitch angle scattering. Lost simulation particles are supplied until relaxation to a steady-state distribution function is obtained. With both ensemble and time averaging this numerical procedure leads to an averaged parallel current proportional to $\langle v_\parallel \rangle$ and to a particle density profile. This parallel current corresponds to the bootstrap current in stellarator and heliotron devices.

BIBLIOGRAPHY

Boozer, A. H. (1980). Guiding center drift equations. *Phys. Fluids*, **23**, 904.

Boozer, A. H., and Kuo-Petravic, G. (1981). Monte Carlo evaluation of transport coefficients. *Phys. Fluids*, **24**, 1999.

Cary, J. R., Hedrick, C. L., and Tolliver, J. S. (1988). Orbits in asymmetric toroidal magnetic fields. *Phys. Fluids*, **31**, 1586.

Dobrott, D., and Frieman, E. A. (1971). Magnetic and drift surfaces using a new stellarator expansion. *Phys. Fluids*, **14**, 349.

Fowler, R. H., Rome, J. A., and Lyon, J. F. (1985). Monte Carlo studies of transport in stellarators. *Phys. Fluids*, **28**, 338.

Gibson, A., and Taylor, J. B. (1967). Single particle motion in toroidal stellarator field. *Phys. Fluids*, **10**, 2653.

Kato, A., Nakamura, Y., and Wakatani, M. (1991). Trapped particle confinement studies in $\ell = 2$ torsatrons for additional helical coils, radial electric field and finite beta effect. *J. Phys. Soc. Japan*, **60**, 494.

Kuo-Petravic, G., Boozer, A. H., Rome, J. A., and Fowler, R. H. (1983). Numerical evaluation of magnetic coordinates for particle transport studies in asymmetric plasmas. *J. Comput. Phys.*, **51**, 261.

Littlejohn, R. G. (1981). Hamiltonian formulation of guiding center motion. *Phys. Fluids*, **24**, 1730.

Mynick, H. E., Chu, T. K., and Boozer, A. H. (1982). Class of model stellarator fields with enhanced confinement. *Phys. Rev. Letters*, **48**, 322.

Nührenberg, J., and Zille, R. (1988). Quasi-helically symmetric toroidal stellarator. *Phys. Letters*, **A129**, 113.

7

NEOCLASSICAL TRANSPORT IN THE STELLARATOR AND HELIOTRON

7.1 INTRODUCTION

It was first shown by Pfirsch and Schlüter (1962) that classical diffusion, $D_{c\ell}$, is enhanced with the factor $(1 + 2q^2)$ compared to the classical diffusion in a toroidal stellarator, where $q = 1/\iota\ (> 1)$ is a safety factor. This enhancement is related to the presence of the Pfirsch–Schlüter current required to maintain the MHD equilibrium of a toroidal plasma. When resistivity is finite, an electric field along magnetic field lines appears. Since the magnetic field has a poloidal component, the poloidal electric field is also finite. $\mathbf{E} \times \mathbf{B}$ drift due to this electric field produces a plasma flow across a magnetic surface, which is the origin of Pfirsch–Schlüter diffusion.

Galeev and Sagdeev (1968) showed that the classical diffusion in a tokamak is more enhanced, by a factor of $\varepsilon^{-3/2}q^2$, when, for rare-collisional tokamak plasmas, particles can be trapped in local mirrors along the magnetic field lines where $\varepsilon\ (< 1)$ is an inverse aspect ratio. The mirror field exists intrinsically in toroidal plasmas, since the magnetic field becomes weak in the outer region of the torus. The enhancement of collisional diffusion is related to trapped particles showing banana orbits which are derived from the shape of the orbit projected in the poloidal plane. The banana width $\rho_L q/\sqrt{\varepsilon}$ is much larger than the Larmor radius ρ_L, which leads to a larger correlation length in the diffusion process than the Larmor radius for classical diffusion. The correlation time becomes shorter, since the collision frequency for trapped particles ν/ε, is effectively enhanced. It is noted that the trapped particle fraction is about $\sqrt{\varepsilon}$. Thus the diffusion coefficient in the rare-collisional regime is estimated as $D_L = \sqrt{\varepsilon}(\nu/\varepsilon)(\rho_L q/\sqrt{\varepsilon})^2 = \varepsilon^{-3/2}q^2\nu\rho_L^2 = \varepsilon^{-3/2}q^2 D_{c\ell}$, which is often called banana diffusion.

Galeev and Sagdeev (1968) also showed that collisional diffusion becomes independent of the collision frequency for the collision frequency regime with $\varepsilon^{3/2}v_T/(qR) < v < v_T/(qR)$, which is called plateau diffusion. Here qR is the connection length of the magnetic field line. Neoclassical diffusion implies these three types of collisional diffusion: banana, plateau, and Pfirsch–Schlüter.

In neoclassical diffusion theory, the essential mechanism is the existence of a trapped particle with a velocity component perpendicular to the magnetic surfaces due to ∇B drift motion. From this point of view, neoclassical diffusion exists even in straight stellarators with perfect helical symmetry. Here, neoclassical transport is first introduced for a straight stellarator in section 7.2. We note the similarity between the tokamak with axisymmetry and the straight stellarator with helical symmetry.

An interesting phemonemon in the context of neoclassical transport theory is the appearance of the diffusion-driven current or bootstrap current, which is also related to trapped particles. We explain this briefly in section 7.2. We give a more complete theory of the bootstrap current and plasma flow in stellarators in section 7.5. Recently, the existence of the bootstrap current has been confirmed in several large tokamak experiments. The bootstrap current is considered to be an important ingredient for the tokamak fusion reactor, since it saves current drive that is inevitable for continuous operation of the tokamak. On the other hand, in stellarators and heliotrons, the bootstrap current, which usually increases the rotational transform, may have detrimental effects on plasma confinement. In low-shear stellarators, a weak shear is produced by the bootstrap current, which increases the possibility of the appearance of a low-order resonant surface, and magnetic islands might be generated. In high-shear heliotrons, the increase in the rotational transform reduces the Shafranov shift of the magnetic axis, which has a negative effect on MHD stability, since the magnetic well becomes narrow and shallow. Thus the next generation of stellarator devices, such as the W-7X Stellarator, are designed to reduce the bootstrap current significantly.

Although the neoclassical diffusion in straight stellarators is similar to that in tokamaks, there is a significant difference between the toroidal stellarator and the standard tokamak from the particle orbit point of view (see section 6.3). In a toroidal stellarator, two types of ripple exist in the magnitude of the magnetic field along the magnetic field line. One is known as helical ripples, which are generated by the helical coils in heliotrons and exist even in the straight stellarator, and the other is known as toroidal ripples, which are related to the toroidal curvature. Usually, the helical ripples have shorter periodic lengths than the toroidal ones. Since particles trapped in the helical ripples suffer toroidal drift motion in the toroidal system, the particle orbit does not close after one bounce motion in the helical ripple. The deviation is cumulative and produces a large excursion of the particle orbit from the original magnetic surface. When the collisional diffusion of the toroidal stellarator is analyzed by including the particles trapped in the helical ripples, the diffusion coefficient becomes proportional to $1/\nu$, where ν is a collision frequency. This dependence on a collision frequency does not appear in the tokamak neoclassical diffusion. Neoclassical diffusion with a $1/\nu$ dependence is discussed in section 7.3. Here, several types of helical ripples are included simultaneously in the calculation of neoclassical particle and heat fluxes.

In sections 7.2 and 7.3, neoclassical transport is studied on the basis of the drift-kinetic equation with a Coulomb collision term. There is another approach to the study of neoclassical transport, which uses the flux–friction relations derived from the moment equations of the Boltzmann equation or

Vlasov equation with a Coulomb collision term (see section 3.7). Neoclassical particle and heat fluxes are given in section 7.4 using the moment equation approach.

In several stellarator experiments it has been shown that the observed net plasma current is explained reasonably by the bootstrap current theory. The bootstrap current in present stellarators is less than 10–20 kA, and much less than the bootstrap current in large tokamaks, which is of the order of several hundreds of kA. However, the behavior of the observed plasma current in stellarators is consistent with the bootstrap currrent theory, as discussed in section 7.5.

In the moment equation approach to neoclassical transport in section 7.4, the parallel viscosity drives the neoclassical fluxes. In section 7.6 the neoclassical fluxes and the bootstrap current are calculated from the parallel viscosity.

With the neoclassical heat fluxes, we may formulate energy transport equations to evaluate the space and time evolution of electron and ion temperatures in stellarators and heliotrons. Since plasmas confined in nonaxisymmetric tori do not satisfy the ambipolar condition $\Gamma_e = \Gamma_i$ automatically, the radial electric field E_r to satisfy the ambipolar condition plays an important role in the transport process, where Γ_e and Γ_i are the electron and ion particle fluxes, respectively. The energy transport equations shown in section 7.7 include the radial electric field E_r explicitly. The term including E_r disappears consistently in the limit of the tokamak approximation.

7.2 NEOCLASSICAL TRANSPORT IN A STRAIGHT STELLARATOR

A simple model for a straight stellarator with helical symmetry is obtained from the magnetic flux function Ψ given by (2.78), which is applicable to low-beta stellarator plasmas. It is also assumed that the helix is loosely wound, or $\ell\alpha r \ll 1$ in (2.78), and that only ℓ component is present. For this case we find that

$$\Psi = -\frac{1}{2} B_0 \alpha r^2 + \frac{r B_H}{\ell} \cos(\ell\zeta), \tag{7.1}$$

$$\mathbf{B} = B_0 \hat{\zeta} + \hat{\zeta} \times \nabla\Psi = B_0 \mathbf{e}_z + \mathbf{B}_H, \tag{7.2}$$

$$\mathbf{B}_H = B_H(\mathbf{e}_r \sin(\ell\zeta) + \mathbf{e}_\theta \cos(\ell\zeta) - \mathbf{e}_z \alpha r \cos(\ell\zeta)), \tag{7.3}$$

where the magnitude of the helical field, B_H, is proportional to $r^{\ell-1}$, $\zeta = \theta - \alpha z$, and $\hat{\zeta} = (\alpha r \mathbf{e}_\theta + \mathbf{e}_z)/(1 + \alpha^2 r^2)$. From (7.1) it is clear that the surface of constant Ψ, or the magnetic surface, will be nearly cylindrical if $B_H/B_0 \ll \ell\alpha r/2$. In obtaining the diffusion coefficient analytically for the helically symmetric geometry, we will assume that $B_H/B_0 \ll \ell\alpha r/2 \ll 1$, which is the usual loosely wound helix limit. We will evaluate two quantities that are useful for radial particle transport study. These are the magnitude of magnetic field B, and the quantity $\mathbf{B}\cdot\nabla\zeta = \nabla\zeta\cdot\hat{\zeta}\times\nabla\psi = (1/r)(\partial\Psi/\partial r)$. If we define the small parameter ε by

$$\varepsilon = \alpha r B_H / B_0, \tag{7.4}$$

we obtain, in the loosely wound helix case,

$$\mathbf{B}\cdot\nabla\zeta \simeq -\alpha B_0\left(1 - \frac{\varepsilon}{\alpha^2 r^2}\cos(\ell\zeta)\right), \tag{7.5}$$

and

$$B \simeq B_z \simeq B_0(1 - \varepsilon\cos(\ell\zeta)). \tag{7.6}$$

In the axisymmetric case, the ratio of the poloidal field to the toroidal field is shown as $\Theta = r\mathbf{B}\cdot\nabla\theta/B$, where θ is the poloidal angle. The quantity that plays a similar role in the helically symmetric case is $\Theta_H = r|\mathbf{B}\cdot\nabla\zeta|/B \simeq \alpha r$.

The curvature vector $\boldsymbol{\kappa}$ of a magnetic field $\mathbf{B}$ is given by $\boldsymbol{\kappa} = (\mathbf{B}/B)\cdot\nabla(\mathbf{B}/B)$, which, for the curl-free field, can be written as $\boldsymbol{\kappa} = (\nabla_\perp B)/B$. By definition, the radius of curvature is equal to the inverse of the magnitude of $\boldsymbol{\kappa}$. With B given by (7.6), and with $\varepsilon \propto r^\ell$, we find the radius of curvature for the helically symmetric field to be $R_H = 1/|\boldsymbol{\kappa}| \simeq r/(\ell\varepsilon)$. The connection length, L_H, is defined by

$$L_H = \frac{1}{2\pi\ell}\int_0^{2\pi}\frac{B d\zeta}{|\mathbf{B}\cdot\nabla\zeta|} \simeq \frac{1}{\ell\alpha}, \tag{7.7}$$

where the expressions (7.5) and (7.6) have been used. Here, $2\pi L_H$ corresponds to the distance along the field line from a given point to the nearest identical point by helical symmetry. In analogy with the toroidal case, we now write

$$\Theta_H \equiv \varepsilon R_H / L_H \simeq \alpha r. \tag{7.8}$$

In order to evaluate neoclassical transport in a straight stellarator, we will first give an expression for the drift velocity in a curl-free field,

$$\mathbf{v}_D = \frac{v_\perp^2 + 2v_\parallel^2}{2\Omega B^2}(\mathbf{B}\times\nabla B), \tag{7.9}$$

with the cyclotron frequency $\Omega = eB/m$. In the long mean free path case, the transport is primarily determined by the $\sqrt{\varepsilon}$ fraction of particles which are trapped in the magnetic field. The typical drift time for such particles is $\tau \simeq L_H/v_\parallel \simeq L_H/\sqrt{\varepsilon}v_T$, where v_T is the thermal velocity. Since the drift velocity is shown as $v_D \simeq v_T\rho_L/R_H$ by using R_H and $\rho_L = v_T/\Omega$, the orbit size is given by

$$\Delta \simeq v_D\tau \simeq (\rho_L L_H)/(R_H\sqrt{\varepsilon}) \simeq \sqrt{\varepsilon}\rho_L/\Theta_H, \tag{7.10}$$

where ρ_L is the Larmor radius. This result for Δ is identical to the width of a banana orbit in a large aspect ratio, axisymmetric tokamak, if we replace Θ_H by Θ. Therefore, we expect the expression for neoclassical transport in a straight stellarator with a loosely wound helix to be essentially identical to that in the tokamak.

In the collisional limit with short mean free path compared to L_H, one also obtains identical formulas for the Pfirsch–Schlüter transport in a large aspect ratio tokamak and a straight stellarator with a loosely wound helix. However, the helical quantity, q_H, which is analogous to the safety factor, q, is much less than one, so that the Pfirsch–Schlüter correction in such a stellarator is of no great interest. To see this, we note that the typical effective drift time for a particle in the collisional regime is $\tau \simeq (L_H/\lambda)^2\tau_c$, with τ_c the collision time and $\lambda = v_T\tau_c$ the mean free path. By using $\Delta \simeq v_D\tau$ and $v_D \simeq v_T\rho_L/R_H$, the diffusion coefficient, $D \simeq \Delta^2/\tau$, becomes

$$D \simeq \rho_L^2 L_H^2/(R_H^2\tau_c) = \rho_L^2 q_H^2/\tau_c, \tag{7.11}$$

where we have defined q_H as $q_H \equiv L_H/R_H \simeq \varepsilon/\Theta_H$. This expression is identical to the tokamak result with q equal to ε/Θ. However, q_H is small compared to unity, since $q_H \simeq \varepsilon/\Theta_H \simeq B_H/B_0 \ll 1$ from (7.4) and (7.8).

In a torus, q is equal to the inverse of the rotational transform. However, in a straight stellarator, q_H is not so simply related to the rotational transform. By the rotational transform per period, ι_p, we mean the angle through which a field line rotates in the θ direction during one helical period, i.e., while $\ell\zeta$ increases by 2π. In order to calculate ι_p we need to know the radius of a magnetic surface as a function of ζ. Setting Ψ equal to a constant, $\Psi = -B_0\alpha r_0^2/2$, yields

$$r = r_0(1 + \delta\cos(\ell\zeta)), \quad \delta = B_H(r_0)/(B_0\ell\alpha r_0) = \varepsilon(r_0)/(\ell\alpha^2 r_0^2), \tag{7.12}$$

to the lowest order in ε. With the differrentials to be taken along the field line, we obtain

$$\frac{d\theta}{d(\ell\zeta)} = \frac{\mathbf{B}\cdot\nabla\theta}{\mathbf{B}\cdot\nabla(\ell\zeta)} = \frac{B_H\cos(\ell\zeta)}{-\ell\alpha r B_0(1 - \varepsilon\cos(\ell\zeta)/(\alpha^2 r^2))}$$
$$\simeq -\delta\cos(\ell\zeta)[1 + 2(\ell - 1)\delta\cos(\ell\zeta)], \tag{7.13}$$

where we have used (7.12) Averaging $-d\theta/d(\ell\zeta)$ over ζ and dividing it by 2π, we find immediately that

$$\iota_p \simeq (\ell - 1)\delta^2 = (\ell - 1)\varepsilon^2/(\ell^2(\alpha r_0)^4), \tag{7.14}$$

which is consistent with expression (4.49), since δ is proportional to $r^{\ell-2}$.

Now we will solve the drift-kinetic equation to obtain the particle and energy fluxes. We choose as the velocity space coordinates E, the total energy, μ, the magnetic moment, and $\sigma = v_\parallel/|v_\parallel|$, the sign of the velocity component in the **B** direction, where $E = (1/2)mv^2 + e\varphi$ and $\mu = mv_\perp^2/(2B)$. Here φ is an electric potential. The phase of the gyration motion around the magnetic line of force has been averaged out in the drift-kinetic equation (see section 3.6). The parallel velocity is shown as $v_\parallel = \sigma[2(E - \mu B - e\varphi)/m]^{1/2}$, which is a function of the real space coordinates r and ζ (through B and φ) as well as the velocity space coordinates. A particularly convenient expression for the guiding center drift velocity is given by

$$\mathbf{v} = \frac{v_\parallel}{B}\,[\mathbf{B} + \nabla \times (\rho_\parallel \mathbf{B})], \tag{7.15}$$

where $\rho_\parallel = v_\parallel/\Omega$ is the parallel velocity divided by the cyclotron frequency. The expression (7.15) is equivalent to (3.112) for $\mathbf{E} = 0$. It is noted that the component of $\mathbf{v}$ parallel to $\mathbf{B}$ differs from the conventional one through the presence of the second term in (7.15), which represents a correction, of first order in the Larmor radius, to the dominant first term. This difference is immaterial for the calculation presented here. It is noted, however, that the expression (7.15) is the correct one, since it leads to a particle-conserving drift-kinetic equation (see section 3.6).

If we insert expressions for $\mathbf{B}$ and $\nabla \times \mathbf{B}$ given by $g\hat{\zeta} + \hat{\zeta} \times \nabla\Psi$ and $\bar{H}\hat{\zeta} - \hat{\zeta} \times \nabla g$, respectively, into (7.15), where $\bar{H}$ is related to the plasma current in the $\hat{\zeta}$-direction, we obtain

$$\mathbf{v} = \frac{v_\parallel}{B}\,[\hat{\zeta}(g + \rho_\parallel \bar{H} + \nabla\rho_\parallel \cdot \nabla\Psi) + \hat{\zeta} \times \nabla(\Psi - \rho_\parallel g)], \tag{7.16}$$

since $\nabla\rho_\parallel \cdot \hat{\zeta} = 0$. For a guiding-center distribution function, $f(r, \zeta, v)$, by noting that $\hat{\zeta}$ is perpendicular to ∇r and $\nabla\zeta$, we arrive at the following time-independent drift-kinetic equation:

$$\mathbf{v} \cdot \nabla f = \frac{v_\parallel}{B}\,\hat{\zeta} \times \nabla\tilde{\Psi} \cdot \nabla f = C(f), \tag{7.17}$$

where $C(f)$ is a collision term, and $\tilde{\Psi} = \Psi - \rho_\parallel g$ is a constant of the motion in the absence of collisions. Here $v = |\mathbf{v}|$.

For a rare-collisional regime or a collision frequency that is low compared to the bounce frequency of trapped particles, we may write (7.17) in the following way:

$$f = f_V + f_C, \tag{7.18}$$

$$\mathbf{v} \cdot \nabla f_V = 0, \tag{7.19}$$

$$\mathbf{v} \cdot \nabla f_C = C(f_V). \tag{7.20}$$

It is clear that the collisionless equation (7.19) for f_V, $\hat{\zeta} \times \nabla\tilde{\Psi} \cdot \nabla f_V = 0$, is satisfied by an arbitrary function of $\tilde{\Psi}$. We will assume that f_V is close to a local Maxwellian, $f_M(E, \tilde{\Psi})$, of energy E, the density n and temprature T of which are functions of $\tilde{\Psi}$:

$$f_V = f_M(E, \tilde{\Psi}) + h(E, \mu, \sigma, \tilde{\Psi}). \tag{7.21}$$

Here, h, which represents the deviation from the Maxwellian, is assumed to be of first order in the Larmor radius. Since $\tilde{\Psi}$ differs from Ψ by a term of the order of the Larmor radius, it is convenient to expand (7.21) in the following way:

$$\begin{cases} f_V = f_M(E, \Psi) + f_1, \\ f_1 = -\dfrac{\partial f_M}{\partial \Psi} \rho_\| g + h(E, \mu, \sigma, \Psi). \end{cases} \tag{7.22}$$

Here it is noted that h depends on Ψ only spatially.

We now consider f_C. To the lowest order in the Larmor radius, (7.20) takes the form

$$\frac{v_\|}{B} \hat{\zeta} \times \nabla\Psi \cdot \nabla f_C = C(f_V). \tag{7.23}$$

If we assume f_C to be a function of Ψ and ζ, rather than of r and ζ, only the ζ-derivative of f_C exists and (7.23) becomes

$$\frac{\partial f_C}{\partial \zeta} = C(f_V)B/(v_\| \mathbf{B} \cdot \nabla\zeta), \tag{7.24}$$

with $\hat{\zeta} \cdot \nabla\zeta = 0$. Integrating this equation in ζ, keeping Ψ constant, and demanding that f_C be single-valued and periodic in ζ, we arrive at

$$0 = \oint d\zeta C(f_V)/(\rho_\| \mathbf{B} \cdot \nabla\zeta), \tag{7.25}$$

where the integral extends from $-\pi$ to π for untrapped particles, and over a full cycle of the bounce motion for trapped particles.

Equation (7.25) is valid for an arbitrary collision term. It can be simplified considerably if we consider a linearized collision operator for which $C(f_M) = 0$. For untrapped particles, (7.25) then reduces to

$$\langle C(h)/\rho_\| \rangle = g \frac{\partial f_M}{\partial \Psi} \left\langle \frac{C(v_\|)}{v_\|} \right\rangle, \tag{7.26}$$

where the bracket is defined as the following average over ζ:

$$\langle f \rangle = \frac{\displaystyle\int_{-\pi}^{\pi} d\zeta f / \mathbf{B} \cdot \nabla \zeta}{\displaystyle\int_{-\pi}^{\pi} d\zeta / \mathbf{B} \cdot \nabla \zeta}. \tag{7.27}$$

If we also assume that the collision operator is even in $v_\parallel$, then, for trapped particles, the contribution to (7.25) from the first term in f_1 shown in (7.22) vanishes. Thus, from the LHS of (7.26),

$$h = 0, \quad \mu \geq \mu_c, \tag{7.28}$$

is obtained, where $\mu_c = E/B_{max}$ is the value of μ at the boundary between trapped and untrapped particles, and B_{max} is the maximum value of B on a magnetic surface.

As an example of collision term that satisfies the above restrictions, we consider the Lorentz operator

$$C(f) = \nu \frac{m}{B} v_\parallel \frac{\partial}{\partial \mu}\left(v_\parallel \mu \frac{\partial f}{\partial \mu}\right), \tag{7.29}$$

where ν represents the electron–ion collision frequency $\nu = (Z^2 e^4 n_i)\ell n\Lambda / (4\pi\varepsilon_0^2 m^2 v_T^3)$ (see (3.58)) and m is the electron mass. With this collision term, we have $C(v_\parallel)/v_\parallel = -\nu$, and (7.26) for electrons becomes

$$\frac{\partial}{\partial \mu}\langle v_\parallel \rangle \mu \frac{\partial h}{\partial \mu} = -\frac{g}{e}\frac{\partial f_M}{\partial \Psi}. \tag{7.30}$$

It is easy to show that (7.30) is satisfied by

$$h = \frac{g}{e}\frac{\partial f_M}{\partial \Psi}\int_{\mu}^{\mu_c} \frac{d\mu'}{\langle v_\parallel(\mu')\rangle}, \quad 0 \leq \mu \leq \mu_c, \tag{7.31}$$

where the upper limit of integration has been chosen such that the two expressions for h, given by (7.28) and (7.31), agree for $\mu = \mu_c$. Thus, for the Lorentz collision model, the integral condition on f_V (7.25) is sufficient to determine f_V completely. It is given explicitly through (7.22), (7.28), and (7.31). It should be noted that every equation in this section applies equally well to helically symmetric and axisymmetric plasmas by changing the angle variable from ζ to θ.

We consider now collisional transport across magnetic surfaces. For a helically symmetric or toroidally symmetric plasma with nested magnetic surfaces, we calculate the particle flux in the $\nabla\Psi$ direction as

$$\Gamma = \int d^3v v_\Psi f = \sum_\sigma \int_0^\infty dE \int_0^{E/B} d\mu \frac{2\pi B}{m^2|v_\parallel|} v_\Psi f, \tag{7.32}$$

where the sum is over the sign of $v_\parallel$, and v_Ψ is the component of the drift velocity in the $\nabla\Psi$ direction. For simplicity, we will assume that there is no electric field, or that $\varphi = 0$. Taking $\rho_\parallel$ to be a function of Ψ and ζ and noting that g is a function of Ψ, we obtain

$$v_\Psi = \frac{v_\parallel}{B} g \frac{\partial \rho_\parallel}{\partial \zeta} \nabla\zeta \times \hat{\zeta} \cdot \frac{\nabla\Psi}{|\nabla\Psi|} \tag{7.33}$$

from (7.16), where $\hat{\zeta} \cdot \nabla\Psi = 0$.

Here our interest is in the particle flux averaged over the surface of constant Ψ. Since there is no need to average in the direction of symmetry, the surface average reduces to an average over ζ with the appropriate weighting factor,

$$\bar{\Gamma} = \int_{-\pi}^{\pi} \frac{d\zeta|\nabla\Psi|\Gamma}{\mathbf{B}\cdot\nabla\zeta} \Bigg/ \int_{-\pi}^{\pi} \frac{d\zeta|\nabla\Psi|}{\mathbf{B}\cdot\nabla\zeta}. \tag{7.34}$$

From (7.32), (7.33), and (7.34), we obtain

$$\bar{\Gamma} = \frac{2\pi g}{m^2} \int_{-\pi}^{\pi} d\zeta \sum_\sigma \sigma \int_0^\infty dE \int_0^{E/B} d\mu \frac{\partial \rho_\parallel}{\partial \zeta} f \Bigg/ \int_{-\pi}^{\pi} \frac{d\zeta|\nabla\Psi|}{\mathbf{B}\cdot\nabla\zeta}. \tag{7.35}$$

Here the integral over ζ is considered. In interchanging the integrals in ζ and μ, however, we must take account of the ζ dependence of E/B for the limit of integration in μ. By examining the area of integration in (μ, ζ) space,

$$\int_{-\pi}^{\pi} d\zeta \int_0^{E/B} d\mu = \int_0^{E/B_{min}} d\mu \int_{-\gamma}^{\gamma} d\zeta, \tag{7.36}$$

is given, where B_{min} is the minimum value of B on the magnetic surface, and $\pm\gamma$ is equal to $\pm\pi$ for untrapped particles, or equal to $\pm\zeta_T$ for trapped particles. Here $\pm\zeta_T$ are the values of ζ at the turning points. Integrating by parts in ζ yields

$$\int_{-\gamma}^{\gamma} d\zeta \frac{\partial \rho_\parallel}{\partial \zeta} f = -\int_{-\gamma}^{\gamma} d\zeta \rho_\parallel \frac{\partial f}{\partial \zeta}, \tag{7.37}$$

where we have noted that $\rho_\parallel$ vanishes at $\pm\zeta_T$ for trapped particles, and that $\rho_\parallel f$ takes the same value at the two limits, $\pm\pi$, for untrapped particles. It is noted that the f_V part of f does not contribute to this integral, since

$$\frac{\partial f_V}{\partial \zeta} = -\frac{\partial f_M}{\partial \Psi} g \frac{\partial \rho_\parallel}{\partial \zeta}, \tag{7.38}$$

and

$$\int_{-\gamma}^{\gamma} d\zeta \rho_\parallel \frac{\partial \rho_\parallel}{\partial \zeta} = 0. \tag{7.39}$$

The remaining part of the integral (7.37) is the contribution from f_C:

$$\int_{-\gamma}^{\gamma} d\zeta \frac{\partial \rho_\parallel}{\partial \zeta} f_C = -\int_{-\gamma}^{\gamma} d\zeta \rho_\parallel \frac{\partial f_C}{\partial \zeta}. \tag{7.40}$$

If we change the order of integration in ζ and μ back again, and substitute (7.24) into (7.40), the averaged particle flux (7.35) becomes

$$\bar{\Gamma} = -\frac{2\pi g}{em\langle|\nabla\Psi|\rangle} \sum_\sigma \sigma \int_0^\infty dE \left\langle \int_0^{E/B} d\mu C(f_V) \right\rangle, \tag{7.41}$$

where the bracket indicates the average over ζ defined by (7.27). This result is applicable to every collision operator in both helically symmetric and axially symmetric plasmas. No particular assumption has been made about the ζ dependence of the magnetic field or the shape of the surface of constant Ψ in straight stellarators.

From here on, we will use the Lorentz collision operator given by (7.29), in which case we find that

$$\int_0^{E/B} d\mu C(f_1) = -\nu \int_0^{E/B} \mu \frac{\partial f_1}{\partial \mu} = -\nu \int_0^{E/B} d\mu f_1, \tag{7.42}$$

where we have integrated by parts twice, and noted that μ vanishes at one limit and f_1 vanishes at the other. Furthermore, the function f_1, given by (7.22), (7.28), and (7.31), is linear in σ, the sign of $v_\parallel$, so that $\sum_\sigma \sigma f_1 = 2f_1^+$, where f_1^+ is equal to f_1 with $\sigma = +1$. Therefore, (7.41) becomes

$$\bar{\Gamma} = -\frac{4\pi\nu g}{em\langle|\nabla\Psi|\rangle} \int_0^\infty dE \left\langle \int_0^{E/B} d\mu f_1^+ \right\rangle. \tag{7.43}$$

This particle flux is considered as an electron flux, since the Lorentz collision term (7.29) is appropriate for electron–ion collisions. In toroidal plasmas the averaged radial particle flux must satisfy the ambipolar condition $\bar{\Gamma}_e = \bar{\Gamma}_i$, where $\bar{\Gamma}_e$ and $\bar{\Gamma}_i$ are the electron and ion particle flux, respectively. Usually, $|\bar{\Gamma}_i|$ is much larger than $|\bar{\Gamma}_e|$ in neoclassical diffusion, and a radial electric field is produced to reduce the ion particle flux and realize the ambipolar state. Thus the electron particle flux is conisdered to be the neoclassical one.

We note that the function h vanishes for $\mu \geq \mu_c$, where $\mu_c = E/B_{max}$. For the h part of f_1, therefore, the integral over μ in (7.43) may be taken from 0 to μ_c. With (7.22) and (7.31), we obtain

$$\left\langle \int_0^{E/B} d\mu f_1^+ \right\rangle = -\frac{g}{e}\frac{\partial f_M}{\partial \Psi}(S_1 - S_2), \tag{7.44}$$

$$S_1 = \left\langle \frac{m}{B}\int_0^{E/B} d\mu |v_\parallel| \right\rangle, \tag{7.45}$$

$$S_2 = \int_0^{\mu_c} d\mu \int_\mu^{\mu_c} \frac{d\mu'}{\langle |v_\parallel(\mu')|\rangle}.$$

The integral in S_1 can be completed as

$$S_1 = \tfrac{1}{3}(2E)^{3/2} m^{1/2} \langle B^{-2}\rangle, \tag{7.47}$$

and S_2, after integration by parts, becomes

$$S_2 = vintop_0^{\mu_c} \frac{d\mu\mu}{\langle |v_\parallel|\rangle}. \tag{7.48}$$

In S_2, before integration over μ, we must average $|v_\parallel|$ over ζ according to (7.27). This requires knowledge of the ζ dependence of B. In general, on a magnetic surface of constant Ψ, B varies periodically in ζ between B_{min} and B_{max}. If we define $\tilde{\varepsilon}$ by $\tilde{\varepsilon} = (B_{max} - B_{min})/2B_{min}$, we can always express the B field by

$$B = B_{min}(1 + 2\tilde{\varepsilon}\tau(\zeta)), \tag{7.49}$$

where B_{min} and $\tilde{\varepsilon}$ depend only on Ψ, and τ is a periodic function of ζ, with period of 2π, varying between 0 and 1. Then $|v_\parallel|$ may be written as

$$|v_\parallel| = [2(E - \mu B_{min})/m]^{1/2}(1 - p^2\tau)^{1/2}, \tag{7.50}$$

where

$$p = [2\tilde{\varepsilon}\mu B_{min}/(E - \mu B_{min})]^{1/2}. \tag{7.51}$$

If we change the variable of integration from μ to p, S_2 takes the form

$$S_2 = 2E^{3/2}m^{1/2}B_{min}^{-2}\tilde{\varepsilon}^{1/2}\int_0^1 \frac{p^3\,dp}{(p^2+2\tilde{\varepsilon})^{5/2}\langle(1-p^2\tau)^{1/2}\rangle}. \tag{7.52}$$

Combining (7.47) and (7.52) yields

$$S_1 - S_2 = 2Km^{1/2}E^{3/2}B_{min}^{-2}, \tag{7.53}$$

where K is the dimensionless, energy-independent factor,

$$K = \frac{\sqrt{2}}{3}\langle(1+2\tilde{\varepsilon}\tau)^{-2}\rangle - \sqrt{\tilde{\varepsilon}}\int_0^1 \frac{p^3\,dp}{(p^2+2\tilde{\varepsilon})^{5/2}\langle(1-p^2\tau)^{1/2}\rangle}. \tag{7.54}$$

With this expression for $S_1 - S_2$, and with

$$\frac{\partial f_M}{\partial \Psi} = f_M\left[\frac{1}{n}\frac{dn}{d\Psi} + \frac{1}{T}\frac{dT}{d\Psi}\left(\frac{E}{T}-\frac{3}{2}\right)\right], \tag{7.55}$$

the E dependence in (7.44) is determined, and the integral over E in (7.43) may be carried out. The result is

$$\bar{\Gamma} = -\frac{3K}{\sqrt{2}}\frac{\nu m g^2}{e^2 B_{min}^2\langle|\nabla\Psi|\rangle}\frac{dP}{d\Psi}, \tag{7.56}$$

where $P = nT$ is the pressure of the particle species under consideration. This expression for the particle flux across the magnetic surface is also valid for toroidally symmetric plasmas, when g, B_{min}, and Ψ are interpreted adequately. Aside from the symmetry, no other conditions have been imposed on Ψ or B. It is clear that the numerical factor K cannot be fully determined, unless the function $\tau(\zeta)$, defined by (7.49), is specified.

In practice, the parameter $\tilde{\varepsilon}$ is often quite small, $\tilde{\varepsilon} \ll 1$. In this case, we can expand (7.54) in a power series in $\sqrt{\tilde{\varepsilon}}$, with the result to the lowest order being

$$K \simeq \sqrt{\tilde{\varepsilon}}\left[1 - \int_0^1\left(\frac{1}{\langle(1-p^2\tau)^{1/2}\rangle} - 1\right)\frac{dp}{p^2}\right]. \tag{7.57}$$

It is noted here that the integral in (7.54) is divergent if $\tilde{\varepsilon}$ is completely neglected. The integral over p in (7.57) will be evaluated numerically. However, we can obtain a reasonable estimate by expanding the integrand in p^2,

$$K \simeq \sqrt{\tilde{\varepsilon}}[1 - \tfrac{1}{2}\langle\tau\rangle - \tfrac{1}{12}\langle\tau\rangle^2 - \tfrac{1}{24}\langle\tau^2\rangle]. \tag{7.58}$$

Furthermore, to the lowest order in ε, the average indicated by the bracket $\langle f \rangle$, defined in (7.27), reduces to $(2\pi)^{-1} \int_{-\pi}^{\pi} d\zeta f$.

As an important example, we now consider a helically symmetric plasma, for which the B field is given by $B \sim B_0[1 + 2\varepsilon \sin^2(\ell\zeta/2)]$. With τ equal to $\sin^2(\ell\zeta/2)$, we have

$$\langle (1 - p^2\tau)^{1/2} \rangle = \frac{2}{\pi} E(p), \tag{7.59}$$

where $E(p)$ is the complete elliptic integral. A numerical integration over p yields

$$K \simeq 0.69\sqrt{\varepsilon} \tag{7.60}$$

and, to the lowest order in $\sqrt{\varepsilon}$, we may rewrite the radial particle flux (7.56) as follows:

$$\bar{\Gamma} \simeq -1.46\sqrt{\varepsilon}\nu m g^2 \frac{dP}{dr} \bigg/ \left[e^2 B_0^2 \left(\frac{d\Psi}{dr}\right)^2 \right], \tag{7.61}$$

where we have ignored the slight ζ-dependence of Ψ. We note that, for an axisymmetric torus, $g^2/[B_0^2(d\Psi/dr)^2]$ is equal to B_θ^{-2} and ε is an inverse aspect ratio. The expression (7.61) then gives, essentially, the result first obtained by Galeev and Sagdeev (1968) for the neoclassical diffusion in the banana regime. On the other hand, for the helically symmetric plasma, $g^2/[B_0^2(d\Psi/dr)^2]$ is equal to $(B_0\alpha r)^{-2}$, where α is defined through $\zeta = \theta - \alpha z$, and ε is the parameter given by $\varepsilon = \alpha r B_H/B_0$.

Finally, we will end this section with a remark about the bootstrap current. It is clear from the form of the distribution function (see (7.22), (7.28), and (7.31)) that there exists an electron current in the direction of the B field,

$$J_\parallel = e \int d^3v v_\parallel f = \frac{4\pi e B}{m^2} \int_0^\infty dE \int_0^{E/B} d\mu f_1^+. \tag{7.62}$$

It follows from (7.43) that there is a simple relationship between the outward particle flux and the bootstrap current averaged on a magnetic surface:

$$\bar{\Gamma} = \frac{\nu m g}{e^2 \langle |\nabla\Psi| \rangle} \left\langle \frac{J_\parallel}{B} \right\rangle. \tag{7.63}$$

In calculating $\bar{\Gamma}$, we can determine $\langle J_\parallel/B \rangle$, and vice versa. Thus, by combining (7.56) and (7.63), we obtain

$$\left\langle \frac{J_\parallel}{B} \right\rangle = \frac{3K}{\sqrt{2}} \frac{g}{B_{min}^2} \frac{dP}{d\Psi}. \tag{7.64}$$

When the conditions to derive (7.61) are assumed, (7.64) is reduced to

$$J_{\parallel} \simeq 1.46 \frac{\sqrt{\varepsilon}}{B_0 \alpha r} \frac{dP}{dr} \tag{7.65}$$

for the helically symmetric case. With toroidal symmetry, $B_0 \alpha r$ in (7.65) should be replaced by $(-B_\theta)$. Thus the direction of the bootstrap current in the helically symmetric case is *opposite* to that in the toroidally symmetric case. The bootstrap current usually increases the rotational transform in tokamaks due to the ohmic current. However, in straight stellarators, it flows to decrease the external rotational transform.

7.3 NEOCLASSICAL TRANSPORT OF A TOROIDAL HELIOTRON WITH MULTIPLE-HELICITY MAGNETIC FIELDS IN THE LOW-COLLISIONALITY REGIME

Fourier expansion of the magnitude of the magnetic field in the magnetic coordinates shows that there are a number of poloidal and helical components in a heliotron device. For a general case, the magnetic field strength B is approximately described as

$$\frac{B}{B_0} = 1 + \epsilon_t \cos\theta + \epsilon_d \cos \ell\theta + \sum_{n=-\infty}^{\infty} \epsilon^{(n)} \cos(n\theta + \eta), \tag{7.66}$$

where ϵ_t, ϵ_d, and $\epsilon^{(n)}$ are the amplitudes of the corresponding harmonics. We note that $\epsilon^{(0)}$ describes the dominant helical modulation corresponding to ϵ_h. In (7.66) $\eta \equiv \ell\theta - M\phi$, where ℓ and M are the poloidal and toroidal mode numbers of the dominant helical component, and θ and ϕ are the poloidal and toroidal angles, respectively. It is noted that $\theta = 0$ corresponds to the innermost side of the torus here.

To calculate the second adiabatic invariant $\hat{J}$ as shown in section 6.3, we usually approximate the summation (7.66) with a single helicity of the dominant component. In this section we show that it can be extended to include other poloidal and helical harmonics. Without loss of generality in applying to heliotrons, we keep only $n = 0, \pm 1$, and ± 2 terms in (7.66). The scheme shown here will be applicable for an arbitrary number n. Considering the relation $\cos(\pm n\theta + \eta) = \cos n\theta \cos \eta \mp \sin n\theta \sin \eta$, (7.66) can be simplified to

$$\frac{B}{B_0} = 1 + \epsilon_t \cos\theta + \epsilon_d \cos \ell\theta + (C^2 + D^2)^{1/2} \times$$
$$\times \left(\frac{C}{(C^2 + D^2)^{1/2}} \cos\eta - \frac{D}{(C^2 + D^2)^{1/2}} \sin\eta \right), \tag{7.67}$$

where $C = \epsilon^{(0)} + (\epsilon^{(+1)} + \epsilon^{(-1)} \cos\theta + (\epsilon^{(+2)} + \epsilon^{(-2)}) \cos 2\theta$ and $D = (\epsilon^{(+1)} - \epsilon^{(-1)} \sin\theta + (\epsilon^{(+2)} - \epsilon^{(-2)}) \sin 2\theta$. If we define a phase angle χ as $(\epsilon^{(+2)} - \epsilon^{(-2)}) \sin 2\theta$. If we define a phase angle χ as

$$\begin{cases} \cos\chi = \dfrac{C}{(C^2 + D^2)^{1/2}} \\ \sin\chi = \dfrac{D}{(C^2 + D^2)^{1/2}}, \end{cases} \tag{7.68}$$

we obtain a simple model for a toroidal heliotron with multiple harmonics:

$$\frac{B}{B_0} = 1 + \epsilon_T + \epsilon_H \cos(\eta + \chi), \tag{7.69}$$

where $\epsilon_T = \epsilon_t \cos\theta + \epsilon_d \cos \ell\theta$ and $\epsilon_H = (C^2 + D^2)^{1/2}$. The magnetic field model given by (7.69) can be reduced to the model field proposed by Mynick, Chu, and Boozer (1982) by setting $\epsilon_d = 0$, $\epsilon^{(+1)} = \epsilon^{(-1)}$, and $\epsilon^{(+2)} = \epsilon^{(-2)} = 0$:

$$B/B_0 = 1 + \epsilon_t \cos\theta + \epsilon^{(0)}(1 - \sigma\cos\theta)\cos\eta, \tag{7.70}$$

where $\sigma = -(\epsilon^{(+1)} + \epsilon^{(-1)})/\epsilon^{(0)}$ is a parameter to control the helical ripple position in the poloidal plane. Usually, $\ell = 2$ and $M \gg 1$ for heliotrons. Equation (7.70) shows two types of oscillation of B along the magnetic field line; one is related to the rapidly oscillating helical ripple characterized by $\epsilon^{(0)}$ and the other to the slowly oscillating toroidal ripple by ϵ_t. In Fig. 7.1 is shown the variation of the helical ripple profile for $\sigma = 0, 1, -1$. When $\sigma = 1$ ($\sigma = -1$), the large helical ripples are localized on the inner side (outer side) of the torus and trapped particle confinement improves (is degraded) as discussed in section 6.4. It is noted that the bottom (top) of the helical ripples becomes constant for $\sigma = 1$ ($\sigma = -1$).

With the model field given by (7.69), we can calculate the second adiabatic invariant $\hat{J} \equiv \oint v_\| ds \simeq \oint v_\| R_0 d\phi$ for the helically trapped particles and obtain

$$\hat{J} = \frac{16R_0}{M}\left(\frac{\mu B_0 \epsilon_H}{m}\right)^{1/2} [E(k^2) - (1 - k^2)K(k^2)], \tag{7.71}$$

where R_0 is the major radius, μ is the magnetic moment, m is the mass of the particle, K and E are complete elliptic integrals of the first and second kind, respectively, and the pitch angle parameter is

$$k^2 = [W - \mu B_0(1 + \epsilon_T - \epsilon_H)]/(2\mu B_0 \epsilon_H), \tag{7.72}$$

where W is the energy of the particle. For the helically trapped particles, $0 \le k^2 \le 1$. To obtain (7.71), we have assumed that $M/\iota \gg \ell$. This assumption

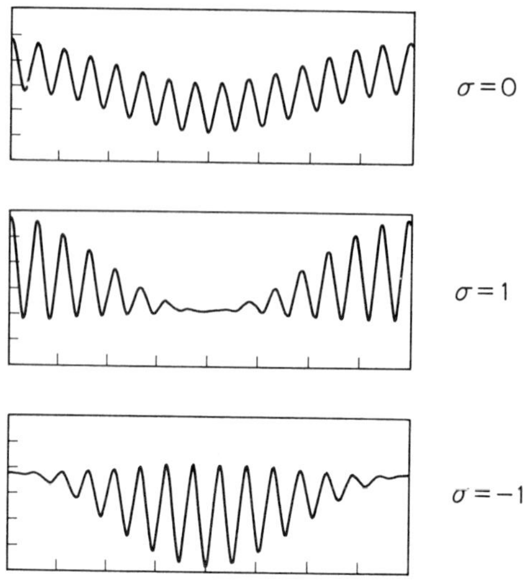

Fig. 7.1 Variation of the magnitude of the magnetic field along the magnetic line of force for $\sigma = 0$ (top), 1 (middle), and -1 (bottom). The left and right sides correspond to the innermost side of the torus and the middle point corresponds to the outermost side of the torus.

implies that within one helical period, the magnetic field line moves mainly in the ϕ direction due to the small rotational transform. Thus, we can treat θ as a parameter in the expression for $\hat{J}$. Note that $\hat{J}$ as given in (7.71) has the same form as for the conventional stellarator, except that now ϵ_T and ϵ_H are functions of θ.

The bounce-averaged drift velocities for the helically trapped particles, which are already shown in section 6.3 (see (6.83) and note that $\theta = 0$ corresponds to the outermost side of the torus there), are then obtained by noting that $\hat{J}(r, \theta) =$ constant describes a bounce-averaged drift surface:

$$r\dot{\theta} = \frac{\mu B_0}{m\Omega}\left[\frac{\partial \epsilon_H}{\partial r}\left(\frac{2E}{K} - 1\right) - \frac{\partial \epsilon_T}{\partial r}\right], \tag{7.73}$$

and

$$r\dot{r} = -\frac{\mu B_0}{m\Omega}\left[\frac{\partial \epsilon_H}{\partial \theta}\left(\frac{2E}{K} - 1\right) - \frac{\partial \epsilon_T}{\partial \theta}\right], \tag{7.74}$$

where Ω is the cyclotron frequency. Here we have used $\partial[E(k^2) - (1 - k^2) K(k^2)]/\partial k^2 = K(k^2)/2$. Equations (7.71), (7.72), (7.73), and (7.74) are the fundamental equations needed to calculate the transport coefficients in the low-collisionality regime. In this section, we calculate only the fluxes in the $1/\nu$ regime, which are fairly large sometimes, comparable to the anomalous fluxes that will be discussed in chapter 8.

Here we show the standard procedure for calculating particle and heat fluxes in the $1/\nu$ regime for a nonaxisymmetric system. The $1/\nu$ regime is characterized by the relation $\omega_E < \nu_{eff} < \omega_b$, where ν_{eff} denotes the effective collision frequency, $\nu_{eff} = \nu/\epsilon_H$. This expression is obtained by considering that Coulomb collisions are described by a diffusion process in the velocity space and that trapped particles in the helical ripple have a typical velocity of $\sqrt{\epsilon_H} v_T$, or $\nu_{eff} = \nu v_T^2/(\sqrt{\epsilon_H} v_T)^2$. It is noted that ν_{eff} corresponds to pitch angle scattering with an angle of about $\sqrt{\epsilon_H}$ instead of ν characterized by that of 90°. The bounce frequency of a particle trapped in the helical ripple is shown by ω_b and is estimated as $\omega_b \simeq \sqrt{\epsilon_H} v_T/(2\pi R_0/M)$. In the collisionless regime of heliotrons, the radial electric field is usually generated to realize the ambipolar diffusion. This electric field produces a poloidal rotation through $\mathbf{E} \times \mathbf{B}$ drift motion. The rotation frequency ω_E is defined as $\omega_E \equiv E_r/(B_0 r)$. For the effective collision frequency $\nu_{eff} < \omega_E$, the transport mechanism changes from the $1/\nu$ regime and the effect of radial electric field on the diffusion becomes significant. This diffusion coefficient is shown in the last part of this section.

Since we are mainly interested in the $1/\nu$ regime in which $\omega_E < \nu_{eff} < \omega_b$, we can employ the bounce-averaged drift-kinetic equation for the first-order distribution function f:

$$\dot{\theta}\,\frac{\partial f}{\partial \theta} + \dot{r}\,\frac{\partial F}{\partial r} = \langle C(f)\rangle. \tag{7.75}$$

Here $\langle C(f)\rangle \equiv \oint (C(f)/v_\parallel)ds/\oint(1/v_\parallel)ds$ is the bounce-averaged collision operator, and $\dot{\theta}$ and $\dot{r}$ are the bounce-averaged poloidal and radial drift velocities, respectively. F is an equilibrium distribution function. Also, s denotes the direction of the magnetic field line. Assuming first that $\nu \sim \dot{\theta} \gg \dot{r}/\Delta r$, where Δr is the typical scale length of the plasma, we have, to the lowest order of $(\dot{r}/\Delta r)/\nu$,

$$\dot{\theta}\,\frac{\partial f_0}{\partial \theta} = \langle C(f_0)\rangle, \tag{7.76}$$

where f is expanded as $f = f_0 + f_1 + \cdots$ by assuming that $(\dot{r}/\Delta r)/\nu \ll 1$. The appropriate solution of (7.76) is a Maxwellian $f_0 = f_M$ satisfying $C(f_M) = 0$, where f_M depends only on r, the minor radius. To the next order we have

$$\dot{\theta}\,\frac{\partial f_1}{\partial \theta} + \dot{r}\,\frac{\partial f_0}{\partial r} = \langle C(f_1)\rangle. \tag{7.77}$$

Since $\omega_E < \nu_{eff} < \omega_b$, we can neglect the $\dot{\theta}(\partial f_1/\partial\theta)$ term compared with the $\langle C(f_1)\rangle$ term. Thus, to determine f_1 we only need to solve

$$\dot{r}\,\frac{\partial f_0}{\partial r} = \langle C(f)\rangle, \tag{7.78}$$

or, explicitly,

$$\dot{r}\frac{\partial f_0}{\partial r} = \frac{(\nu/B)\partial/\partial\mu[\mu\hat{J}(\partial f_1/\partial\mu)]}{\partial\hat{J}/\partial W}, \tag{7.79}$$

where we note that $(m/2)\partial\hat{J}/\partial W = \oint(1/v_\parallel)ds \simeq \oint(1/v_\parallel)R_0 d\phi$. To obtain (7.79), we have employed the Lorentz collision operator (7.29). With the approximate relation $\partial/\partial\mu \simeq -(2\mu\epsilon_H)^{-1}\partial/\partial k^2$ under an assumption of $W \simeq \mu B_0$ for trapped particles, (7.79) can be integrated directly to obtain

$$\frac{\partial f_1}{\partial k^2} = \frac{\epsilon_H}{\nu}\frac{\mu B_0}{m\Omega r}\frac{\partial f_M}{\partial r} \times \frac{\int_0^{k^2} d(k^2)[(\partial\epsilon_T/\partial\theta)K - (\partial\epsilon_H/\partial\theta)(2E-K)]}{E-(1-k^2)K}. \tag{7.80}$$

Here the boundary condition for $\partial f_1/\partial k^2$ is regularity at $k^2 = 0$. We can integrate (7.80) to obtain f_1. We note that an explicit expression for f_1 is not needed to calculate particle and heat fluxes here.

The particle flux Γ given by averaging (7.32) over the magnetic surface is shown as

$$\Gamma = \frac{1}{(2\pi)^2}\int d\theta \int d\phi \int d^3v v_{dr} f_1, \tag{7.81}$$

where v_{dr} is the radial drift velocity. Here, for simplicity, we omit the bar corresponding to (7.35). Using the identity $d^3v = \sum 2\sigma\pi dW d\mu B/m^2|v_\parallel|$, where σ is the sign of $v_\parallel$, and integrating over $d\phi$ in $[0, 2\pi/M]$ first, we obtain

$$\Gamma = \frac{1}{2\pi}\int\frac{d\theta}{2\pi}\int\frac{4\pi}{m^2}dW d\mu\, B\frac{M}{R_0}\dot{r}\frac{m}{2}\frac{\partial\hat{J}}{\partial W}f_1. \tag{7.82}$$

With the approximate relation $d\mu \simeq -2\mu\epsilon_H dk^2$ and $\partial\hat{J}/\partial W = (4R_0/M)(1/m\mu B_0\epsilon_H)^{1/2}K$, and integrating by parts with respect to k^2, we finally obtain

$$\Gamma = -\frac{4}{\pi}\frac{1}{m^{7/2}\Omega^2 r^2}\int_0^\infty dW\frac{W^{5/2}}{\nu}\frac{\partial f_M}{\partial r} \times \left\{\int_0^{2\pi} d\theta\epsilon_H^{3/2}\left[G_1\left(\frac{\partial\epsilon_T}{\partial\theta}\right)^2 - 2G_2\frac{\partial\epsilon_T}{\partial\theta}\frac{\partial\epsilon_H}{\partial\theta} + G_3\left(\frac{\partial\epsilon_H}{\partial\theta}\right)^2\right]\right\}, \tag{7.83}$$

where

$$\begin{cases} G_1 = \int_0^1 dk^2 \dfrac{\{\int_0^{k^2} d(k^2)K\}^2}{E-(1-k^2)K} = \dfrac{16}{9}, \\ G_2 = \int_0^1 dk^2 \dfrac{\{\int_0^{k^2} d(k^2)K\}\{\int_0^{k^2} d(k^2)(2E-K)\}}{E-(1-k^2)K} = \dfrac{16}{15}, \\ G_3 = \int_0^1 dk^2 \dfrac{\{\int_0^{k^2} d(k^2)(2E-K)\}^2}{E-(1-k^2)K} = 0.684. \end{cases} \tag{7.84}$$

Obviously, (7.83) will reproduce the usual result for a standard model of stellarator and heliotron if $\partial\epsilon_H/\partial\theta = 0$ and $\partial\epsilon_T/\partial\theta = \epsilon_t \sin\theta$. Here, to obtain (7.83), we have assumed that $\mu B_0 \simeq W$. Similarly, for the heat flux Q,

$$Q = -\frac{4}{\pi}\frac{1}{m^{7/2}\Omega^2 r^2}\int_0^\infty dW \frac{W^{7/2}}{\nu}\frac{\partial f_M}{\partial r} G, \tag{7.85}$$

where G represents the function inside the braces of (7.83), which depends only on the geometric parameters ϵ_T and ϵ_H. We see that both particle and heat fluxes have the same geometric dependencies through the function G. Thus, under certain constraints, the particle and heat fluxes can be minimized simultaneously by the control of ϵ_T and ϵ_H.

As an example, we consider the following model for a heliotron with side band fields of $\ell = 1$ and $\ell = 3$:

$$B = \bar{B}_0[1 + \epsilon_t \cos\theta + \epsilon_h \cos(2\theta - M\phi) + \epsilon_1 \cos(\theta - M\phi) + \epsilon_3 \cos(3\theta - M\phi)]. \tag{7.86}$$

Both ϵ_T and ϵ_H are given by $\epsilon_T = \epsilon_t \cos\theta$ and $\epsilon_H = [\epsilon_h^2 + \epsilon_1^2 + \epsilon_2^3 + 2\epsilon_h(\epsilon_1 + \epsilon_3)\cos\theta + 2\epsilon_1\epsilon_3 \cos 2\theta]^{1/2}$, respectively. Here $\epsilon_t = 0.13$ and $\epsilon_h = 0.20$ are fixed. The minimum of the function G in this example is shown by the contour plot in the ϵ_1–ϵ_3 plane, as given in Fig. 7.2. The particle and heat fluxes at the minimum are one order of magnitude smaller than those of the conventional stellarator for the minimum point, $\epsilon_1 = \epsilon_3 = -0.1$, with the same ϵ_h and ϵ_t. We note that these constraints give rise to a model field that is very similar to the σ-optimized stellarator proposed by Mynick, Chu, and Boozer (1982).

If the spectrum of B under the constraint expressed by (7.66) for finite-beta equilibrium is given (see chapter 4), G functions can also be calculated. Usually, the effect of a finite beta degrades the neoclassical transport and it is considered that the plasma pressure has a stronger effect on neoclassical transport in stellarators and heliotrons than in tokamaks.

For the collisionless regime with $\nu_{eff} \lesssim \omega_E$ in the presence of a radial electric field E_r, the diffusion coefficient is estimated on the basis of the random walk model. The velocity of drift motion from the magnetic surface is

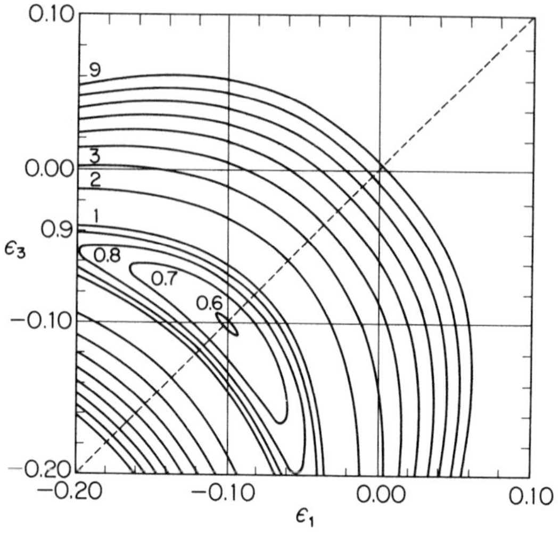

Fig. 7.2 The contours of function G for $\epsilon_t = 0.13$ and $\epsilon_h = 0.20$. The minimum point of about 0.6 appears at $\epsilon_1 = \epsilon_3 = -0.1$.

$v_\perp/\omega_E = (\epsilon_t T/eBr)/\omega_E$ and $\omega_E = E_r/(Br)$. Then the diffusion coefficient becomes

$$D = (\Delta v_\parallel/v_T)(v_\perp/\omega_E)^2 \nu_{eff}. \tag{7.87}$$

Here the effective collision frequency is considered as $\nu_{eff} \simeq \nu/(\Delta v_\parallel/v_T)^2$ from the boundary layer $\Delta v_\parallel$ in the velocity space, which exists at the boundary between the trapped particle region and the untrapped particle region. Then, for $\nu_{eff} \simeq \omega_E$ and $(\Delta v_\parallel/v_T)^2 \sim \nu/\omega_E$, the expression (7.87) gives

$$D = \epsilon_t^2 \left(\frac{\nu}{\omega_E}\right)^{1/2} \left(\frac{T}{eE_r r}\right)\left(\frac{T}{eB}\right). \tag{7.88}$$

It is noted that, for $\nu_{eff} \propto \nu$ or $\Delta v_\parallel$ independent of ν at the lower collision frequency, the diffusion coefficient $D \propto \nu/E_r^2$ may appear. The variation in the diffusion coefficient D with collision frequency ν is shown in Fig. 7.3, where $\epsilon_h > \epsilon_t$ has been assumed. Here D_h is the diffusion coefficient obtained from the particle flux given by (7.83) and D_E corresponds to that given by (7.88). D_E may become proportional to ν in the lower ν regime. The three diffusion coefficients, D_{PS}, D_P, and D_{GS}, are relevant to tokamak neoclassical diffusion: D_{PS} denotes Pfirsch–Schlüter diffusion, D_P plateau diffusion, and D_{GS} banana diffusion. The latter two, D_P and D_{GS}, were first given by Galeev and Sagdeev (1968).

For stellarators and heliotrons, the diffusion coefficients, D_h and D_E, become significant in the rare-collisional regime relevant to a fusion reactor.

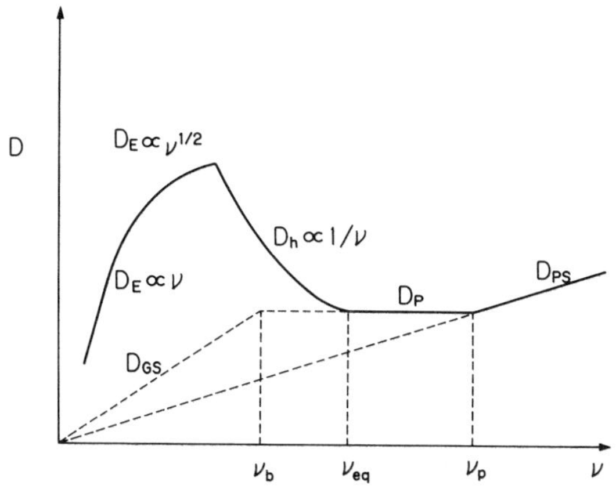

Fig. 7.3 The dependence of the diffusion coefficient on the collision frequency for heliotrons with $\epsilon_h > \epsilon_t$. Here $\nu_p = v_T/(qR)$, $\nu_b = \epsilon_t^{3/2}\nu_p$, and $\nu_{eq} = \epsilon_h^{3/2}\nu_p$.

It should be noted that the magnitude of D_h or D_E depends on the spectrum of the magnetic field and the radial electric field. Thus optimization of the magnetic configuration is crucial to the design of an efficient stellarator reactor.

7.4 FLUX–FRICTION RELATIONS

As in axisymmetric tokamaks, the relation between the cross-field particle flux and the dissipative forces can be derived from the stationary momentum balance equation for species a shown by (3.134):

$$n_a e_a(\mathbf{E} + \mathbf{u}_a \times \mathbf{B}) + \mathbf{F}_{a1} - \nabla P_a - \nabla \cdot \overset{\leftrightarrow}{\Pi}_a = 0. \tag{7.89}$$

The friction force and viscosity tensor are defined by (3.131) and (3.132). For the magnetic coordinates we use the Hamada coordinates (V, θ, ζ), where the magnetic field is given by

$$\mathbf{B} = \nabla V \times \nabla(\psi'\theta - \chi'\zeta) = \mathbf{B}_t + \mathbf{B}_p. \tag{7.90}$$

Here V is the volume enclosed by a magnetic surface, and θ and ζ are the poloidal and toroidal angles. Also, $\psi' = d\psi/dV = \mathbf{B} \cdot \nabla\zeta$ and $\chi' = d\chi/dV = \mathbf{B} \cdot \nabla\theta$ are toroidal and poloidal flux densities, and $\mathbf{B}_p = -\chi'\nabla V \times \nabla\zeta$ and $\mathbf{B}_t = \psi'\nabla V \times \nabla\theta$ are the poloidal and toroidal magnetic fields in the contravariant representation. The Jacobian $(\nabla V \times \nabla\theta \cdot \nabla\zeta)^{-1}$ is unity in the Hamada coordinates. Taking the perpendicular component of (7.89) by $\mathbf{B}\times$ (7.89), which is the vector cross product, and averaging it over a magnetic (or flux) surface $V =$ constant, we obtain the surface-averaged radial particle flux Γ_r^a of species a:

$$\begin{aligned}\langle\Gamma_r^a\rangle &= \langle n_a \mathbf{u}_{\perp a}\cdot\nabla V\rangle \\ &= \left\langle n_a \frac{\mathbf{E}\cdot\hat{\mathbf{n}}\times\nabla V}{B}\right\rangle + \left\langle\frac{\mathbf{F}_{a1}\cdot\hat{\mathbf{n}}\times\nabla V}{m_a\Omega_a}\right\rangle \\ &\quad - \left\langle\frac{\hat{\mathbf{n}}\times\nabla V\cdot\nabla\cdot\overleftrightarrow{P}_a}{m_a\Omega_a}\right\rangle,\end{aligned}\tag{7.91}$$

where $\mathbf{E} = -\nabla\varphi$, $\overleftrightarrow{P}_a = P_a\overleftrightarrow{I} + \overleftrightarrow{\Pi}_a$ is the pressure tensor, $\Omega_a = e_a B/m_a$ is the cyclotron frequency, and the angular brackets denote the surface averaging $\langle A\rangle \equiv \int\int d\theta d\zeta A/\int\int d\theta d\zeta$. The unit vector along the magnetic field line is denoted by $\hat{\mathbf{n}}$. In (7.91), the classical flux is driven by the perpendicular friction force $\mathbf{F}_{a1}\cdot\hat{\mathbf{n}}\times\nabla V$, and the neoclassical flux is driven by the viscosity (banana-plateau regime) and the pressure variation (Pfirsch–Schlüter regime) within the flux surface. It is noted that $\mathbf{E}$ does not contribute to $\langle\Gamma_r^a\rangle$ if φ is a function of V only.

To derive the flux–friction relations, we take the toroidal component of (7.89) by $\nabla V\times\nabla\theta\cdot$(7.89), which is the product between $\nabla V\times\nabla\theta$ and (7.89), and obtain the surface-averaged particle flux as

$$\langle\Gamma_r^a\rangle = \Gamma_{cl}^a + \Gamma_{bp}^a + \Gamma_{ps}^a + \Gamma_{na}^a,\tag{7.92}$$

where

$$\Gamma_{cl}^a = -\frac{1}{e_a\chi'\psi'}\langle\mathbf{B}_t\cdot\mathbf{F}_{\perp a1}\rangle,\tag{7.93}$$

$$\Gamma_{bp}^a = -\frac{1}{e_a\chi'}\frac{\langle I\rangle\langle BF_{\parallel a1}\rangle}{\langle B^2\rangle},\tag{7.94}$$

$$\Gamma_{ps}^a = -\frac{1}{e_a\chi'}\left\langle\frac{F_{\parallel a1}}{B}\left(I - \langle I\rangle\frac{B^2}{\langle B^2\rangle}\right)\right\rangle,\tag{7.95}$$

$$\Gamma_{na}^a = \frac{1}{e_a\chi'\psi'}\langle\mathbf{B}_t\cdot\nabla\cdot\overleftrightarrow{\Pi}_a\rangle,\tag{7.96}$$

and $I \equiv \mathbf{B}_t\cdot\mathbf{B}/\psi'$, and $\mathbf{F}_{a1} = \mathbf{F}_{\parallel a1}\hat{n} + \mathbf{F}_{\perp a1}$. It should be noted that we have neglected the inductive electric field completely. This approximation is not correct when the toroidal flux has a time variation. The classical (Γ_{cl}^a), banana-plateau (Γ_{bp}^a), and Pfirsch–Schlüter (Γ_{ps}^a) fluxes have the same meaning as in axisymmetric tokamaks, and are intrinsically ambipolar (or $\Gamma^e = \Gamma^i$). The nonaxisymmetric flux characteristic in the stellarator or heliotron Γ_{na}^a given by (7.96) is induced by the toroidal viscosity $\langle\mathbf{B}_t\cdot\nabla\cdot\overleftrightarrow{\Pi}_a\rangle$ and is nonambipolar. In the low-collisionality regime (usually this means that the collision frequency ν is less than the bounce frequency of the trapped particle in the helical ripple), the dominant contribution to $\langle\Gamma_r^a\rangle$ usually comes from Γ_{na}^a. From the surface-averaged parallel component of (7.89), we have

$$\langle BF_{\parallel a1}\rangle = \langle\mathbf{B}\cdot\nabla\cdot\overleftrightarrow{\Pi}_a\rangle,\tag{7.97}$$

since $\mathbf{E}$ and P are assumed to be functions of V only. Hence, we can write (7.94) as

$$\Gamma_{bp}^{a} = -\frac{1}{e_a\chi'}\frac{\langle I\rangle\langle \mathbf{B}\cdot\nabla\cdot\overleftrightarrow{\Pi}_a\rangle}{\langle B^2\rangle}. \tag{7.98}$$

As in axisymmetric tokamaks, the banana-plateau flux can be viewed as driven either by the parallel friction force or parallel viscosity.

To find the equivalence between (7.91) and (7.92), we need a geometric relationship between $\mathbf{B}_t$ and $\hat{\mathbf{n}}\times\nabla V$. Components of $\mathbf{B}_t$ along $\hat{\mathbf{n}}$ and $\hat{\mathbf{n}}\times\nabla V$ are given as $\mathbf{B}_t\cdot\mathbf{B}/B$ and $\mathbf{B}_t\cdot\hat{\mathbf{n}}\times\nabla V = -\chi'\hat{\mathbf{n}}\cdot\nabla\zeta = -\chi'\psi'/B$, and $\mathbf{B}_t$ can be separated as

$$\frac{\mathbf{B}_t}{\psi'} = \frac{I}{B}\hat{\mathbf{n}} - \chi'\frac{\hat{\mathbf{n}}\times\nabla V}{B}. \tag{7.99}$$

Here $|\nabla V|^2 = 1$ have been used. With (7.99) we can confirm that (7.91) and (7.92) are equivalent.

We have derived the flux–friction relations (7.92)–(7.96) by taking the toroidal component of (7.89). However, there is no particular reason why we have to do so. We can also take the poloidal component of (7.89) by $\nabla V\times\nabla\zeta\cdot$ (7.89), which is the product between $\nabla V\times\nabla\zeta$ and (7.89), and obtain

$$\langle\Gamma_r^a\rangle = \Gamma_{cl}^a + \Gamma_{bp}^a + \Gamma_{ps}^a + \Gamma_{na}^a, \tag{7.100}$$

where

$$\Gamma_{cl}^a = \frac{1}{e_a\chi'\psi'}\langle\mathbf{B}_p\cdot\mathbf{F}_{\perp a1}\rangle, \tag{7.101}$$

$$\Gamma_{bp}^a = \frac{1}{e_a\psi'}\frac{\langle I_p\rangle\langle BF_{\|a1}\rangle}{\langle B^2\rangle}, \tag{7.102}$$

$$\Gamma_{ps}^a = \frac{1}{e_a\psi'}\left\langle\frac{F_{\|a1}}{B}\left(I_p - \langle I_p\rangle\frac{B^2}{\langle B^2\rangle}\right)\right\rangle, \tag{7.103}$$

$$\Gamma_{na}^a = -\frac{1}{e_a\chi'\psi'}\langle\mathbf{B}_p\cdot\nabla\cdot\overleftrightarrow{\Pi}_a\rangle, \tag{7.104}$$

and $I_p = \mathbf{B}_p\cdot\mathbf{B}/\chi'$. Since $(I_p - \langle I_p\rangle B^2/\langle B^2\rangle)/\psi' = -(I - \langle I\rangle B^2/\langle B^2\rangle)/\chi'$, the Pfirsch–Schlüter fluxes defined by (7.95) and (7.103) are identical. Also, with this relation and (7.97), we can show that the sums of the banana-plateau and nonaxisymmetric fluxes $(\Gamma_{bp}^a + \Gamma_{na}^a)$ are identical in both (7.92)–(7.96) and (7.100)–(7.104). To transform from the expressions (7.94) and (7.96) to the expressions (7.102) and (7.104), we need only to add the flux $-\langle BF_{\|a1}\rangle/(e_a\chi'\psi')$ to (7.96) and subtract it from (7.94). Thus in general, to transform from one representation to the flux–friction relation to the other, we need only to add (or substract) a term proportional to the parallel friction $\langle BF_{\|a1}\rangle$ in Γ_{bp}^a and substract (or add) it in Γ_{na}^a. The physical reason why the representation of the flux–friction relation is not unique is that a nonaxisymmetric torus can be thought of as a perturbed state of an originally axisymmetric torus. For example, to obtain (7.92), we have implicity assumed that the nonaxisymmetric torus results from a *toroidally* symmetric torus with perturbations. On the other hand, to obtain (7.100), we have implicitly assumed that the original torus is *poloidally* symmetric.

So far, we have discussed the relationship between the particle flux and the dissipative force. A similar relationship can be obtained between the heat flux and the dissipative force for the steady state from the flux balance equation (3.135), which is accurate to the first order in the gyroradius expansion:

$$\frac{e_a}{m_a}[\mathbf{E}\cdot(\tfrac{5}{2}P_a\overleftrightarrow{I}+\overleftrightarrow{\Pi}_a)+\mathbf{Q}_a\times\mathbf{B}]+\mathbf{G}_a-\nabla\cdot\overleftrightarrow{r}_a=0. \tag{7.105}$$

Here the total energy flux $\mathbf{Q}_a$, the collisional rate of heat flux generation (or heat friction) $\mathbf{G}_a = T_a/m_a(5/2\mathbf{F}_{a1}+\mathbf{F}_{a2})$, and the energy-weighted stress tensor $\overleftrightarrow{r}_a$ for species a are given by (3.136), (3.137), and (3.138). Taking the toroidal component of (7.105) as $\nabla V\times\nabla\theta\cdot$ (7.105), which is the product between $\nabla V\times\nabla\theta$ and (7.105), we obtain the radial component of the surface-averaged heat flux:

$$\langle \mathbf{q}_r^a\cdot\nabla V\rangle \equiv \langle \mathbf{Q}_a\cdot\nabla V-\tfrac{5}{2}T_a\Gamma_r^a\rangle = \left\langle\frac{\hat{\mathbf{n}}\times\nabla\cdot\overleftrightarrow{r}_a}{\Omega_a}\cdot\nabla V\right\rangle$$
$$-\left\langle\frac{5}{2}\frac{T_a\hat{\mathbf{n}}\times\nabla\cdot\overleftrightarrow{\mathbf{P}}_a}{m_a\Omega_a}\cdot\nabla V\right\rangle+\left\langle\frac{T_a\mathbf{F}_{a2}\times\hat{\mathbf{n}}}{m_a\Omega_a}\cdot\nabla V\right\rangle \tag{7.106}$$

or

$$\left\langle\frac{\mathbf{q}_r^a\cdot\nabla V}{T_a}\right\rangle \equiv \left\langle\frac{\mathbf{q}_V^a}{T_a}\right\rangle = (q_{cl}^a+q_{bp}^a+q_{ps}^a+q_{na}^a)/T_a. \tag{7.107}$$

It is noted that $\langle \mathbf{Q}_a\cdot\nabla V\rangle$ is obtained by taking $\mathbf{B}\times$(7.105), which is the vector cross product, and averaging it over a magnetic surface. In (7.107) the classical q_{cl}^a, banana-plateau q_{bp}^a, Pfirsch–Schlüter q_{ps}^a, and nonaxisymmetric q_{na}^a heat fluxes are defined by

$$\frac{q_{cl}^a}{T_a} = -\frac{1}{e_a\chi'\psi'}\langle \mathbf{B}_t\cdot\mathbf{F}_{\perp a2}\rangle, \tag{7.108}$$

$$\frac{q_{bp}^a}{T_a} = -\frac{1}{e_a\chi'}\frac{\langle I\rangle\langle BF_{\|a2}\rangle}{\langle B^2\rangle}, \tag{7.109}$$

$$\frac{q^a_{ps}}{T_a} = -\frac{1}{e_a\chi'}\left\langle \frac{F_{\|a2}}{B}\left(I - \langle I\rangle \frac{B^2}{\langle B^2\rangle}\right)\right\rangle, \tag{7.110}$$

$$\frac{q^a_{na}}{T_a} = \frac{1}{e_a\chi'\psi'}\langle \mathbf{B}_t \cdot \nabla \cdot \overleftrightarrow{\Theta}_a\rangle, \tag{7.111}$$

$\mathbf{F}_{a2} = F_{\|a2}\hat{\mathbf{n}} + \mathbf{F}_{\perp a2}$ is the heat friction, and $\overleftrightarrow{\Theta}_a = [m_a(r_{\|a} - r_{\perp a})/T_a - 5(P_{\|a} - P_{\perp a})/2](\hat{\mathbf{n}}\hat{\mathbf{n}} - \overleftrightarrow{I}/3) = (\Theta_{\|a} - \Theta_{\perp a})(\hat{\mathbf{n}}\hat{\mathbf{n}} - \overleftrightarrow{I}/3)$ is the heat viscosity tensor. Here we have used the fact that the energy-weighted stress tensor $\overleftrightarrow{\mathrm{R}}_a$ has the Chew–Goldberger–Low form

$$\overleftrightarrow{\mathrm{r}}_a = r_a\overleftrightarrow{\mathrm{I}} + (r_{\|a} - r_{\perp a})(\hat{\mathbf{n}}\hat{\mathbf{n}} - \overleftrightarrow{I}/3) \tag{7.112}$$

as well as the pressure tensor

$$\overleftrightarrow{\mathrm{P}}_a = P_a\overleftrightarrow{I} + (P_{\|a} - P_{\perp a})(\hat{\mathbf{n}}\hat{\mathbf{n}} - \overleftrightarrow{I}/3). \tag{7.113}$$

From the surface-averaged parallel components of (7.105), we obtain the parallel heat flux balance equation

$$\langle BF_{\|a2}\rangle = \langle \mathbf{B} \cdot \nabla \cdot \overleftrightarrow{\Theta}_a\rangle. \tag{7.114}$$

Here we have used $\mathbf{E} \cdot \mathbf{B} = 0$, and (7.97) for evaluating $\langle BF_{\|a1}\rangle$. Thus, like the banana-plateau particle flux Γ^a_{bp}, the banana-plateau heat flux q^a_{bp} can be viewed as driven either by parallel heat flux friction $F_{\|a2}$ or parallel heat viscosity $\langle \mathbf{B} \cdot \nabla \cdot \overleftrightarrow{\Theta}_a\rangle$.

7.5 THE GEOMETRICAL FACTOR, $\langle G_{BS}\rangle_{1/\nu}$

The crucial point in neoclassical transport theory for nonsymmetric toroidal systems such as stellarators lies in evaluating the neoclassical fluxes that result from trapped particles in various magnetic ripples for the low-collisionality regime. In the fluid description of neoclassical transport as shown in section 7.4, the contribution from these particles becomes clear through the parallel viscous stress tensor.

In this section, analytic expressions for the parallel viscosity in the low-collisionality regime in arbitrary nonsymmetric toroidal plasmas will be derived without considering boundary layer effects in velocity space. They are expressed using the Fourier spectrum of the strength of the magnetic field $|\mathbf{B}|$. Here we concentrate on the geometric factor, denoted by $\langle G_{BS}\rangle_{1/\nu}$, which usually appears in discussions of the bootstrap current and neoclassical fluxes in stellarators and heliotrons.

When a distribution function $f_a(\mathbf{v})$ is given, the flow and heat fluxes for a species a are defined by

$$\mathbf{u}_a = \frac{1}{n_a}\int d\mathbf{v}\mathbf{v}f_a, \tag{7.115}$$

$$\mathbf{q}_a = \int d\mathbf{v} \, \frac{m_a}{2} \, |\mathbf{v} - \mathbf{u}_a|^2 (\mathbf{v} - \mathbf{u}_a) f_a. \tag{7.116}$$

Based on the transport ordering, $\rho_{La}/L \sim \omega/\Omega_a \gg 1$, the lowest-order flow and heat flux are written as

$$\mathbf{u}_{a1} = u_{\|a1} \hat{\mathbf{n}} + \mathbf{u}_{\perp a1}, \tag{7.117}$$

$$\mathbf{q}_{a1} = q_{\|a1} \hat{\mathbf{n}} + \mathbf{q}_{\perp a1}, \tag{7.118}$$

where the subscript 1 denotes the quantity of the order of ρ_{La}/L and Ω_a is the cyclotron frequency. Here ρ_{La} and L denotes the Larmor radius and characteristic length of the inhomogeneity in the radial direction, respectively. The transport ordering also requires $\partial/\partial t \sim \omega(\rho_{La}/L)^2$, where $\omega = v_{Ta}/L$ and v_{Ta} is the thermal velocity. The quantities in (7.117) and (7.118) have the following properties:

$$\begin{cases} \nabla \cdot \mathbf{u}_{a1} = 0, \\ \mathbf{u}_{\perp a1} = A_{a1}(V) \, \dfrac{\mathbf{B} \times \nabla V}{B^2}, \\ \mathbf{u}_{a1} \cdot \nabla V = 0, \end{cases} \tag{7.119}$$

$$\begin{cases} \nabla \cdot \mathbf{q}_{a1} = 0, \\ \mathbf{q}_{\perp a1} = \frac{5}{2} P_a A_{a2}(V) \, \dfrac{\mathbf{B} \times \nabla V}{B^2}, \\ \mathbf{q}_{a1} \cdot \nabla V = 0, \end{cases} \tag{7.120}$$

where

$$A_{a1}(V) = \frac{1}{e_a n_a} \frac{dP_a}{dV} + \frac{d\varphi}{dV} \tag{7.121}$$

and

$$A_{a2}(V) = \frac{1}{e_a} \frac{dT_a}{dV}. \tag{7.122}$$

Here P_a and T_a are the pressure and temperature, respectively, V is the volume covered by a magnetic surface, and ∇V is in the radial direction. The potential of the radial electric field is denoted by φ, which satisfies $e_a \varphi / T_a \lesssim 1$ for neoclassical transport. $A_{a1}(V)$ and $A_{a2}(V)$ are thermodynamic forces, which produce the particle and heat fluxes, the plasma flow, and the plasma current.

Let f_a be the distribution function of particle species a in a phase space $(V, \theta, \zeta, E, \mu, \delta)$, where (V, θ, ζ) are the Hamada coordinates and E, μ, and δ are the total particle energy, magnetic moment, and gyro-phase, respectively. The distribution function f_a is divided into the gyro-averaged part, $\bar{f}_a$, and

$\tilde{f}_a = f_a - \bar{f}_a$. Here $\tilde{f}_a$ depends on the gyro-phase δ. The lowest-order part of $\tilde{f}_a$, shown as $\tilde{f}_{a1}$, is given in terms of the gyroradius vector $\boldsymbol{\rho}_{La}$ and the local Maxwellian $f_{aM}(V, E)$:

$$\tilde{f}_{a1} = -\boldsymbol{\rho}_{La} \cdot \nabla f_{aM} = -\boldsymbol{\rho}_{La} \cdot \nabla V \frac{\partial f_{aM}}{\partial V}, \tag{7.123}$$

where $\boldsymbol{\rho}_{La} = (\hat{\mathbf{n}} \times \mathbf{v}_\perp)/\Omega_a$ and $\mathbf{v}_\perp = (2\mu B/m_a)^{1/2}(\hat{\mathbf{e}}_2 \cos\delta - \hat{\mathbf{e}}_3 \sin\delta)$. Here $(\hat{\mathbf{n}}, \hat{\mathbf{e}}_2, \hat{\mathbf{e}}_3)$ become an orthogonal right-handed coordinate system, and $\hat{\mathbf{n}} = \mathbf{B}/B$. From the definition of the local Maxwellian,

$$f_{aM}(V, E) = \frac{n_a}{\pi^{3/2} V_{Te}^3} \exp\left[-\frac{E - e_a\varphi}{T_a}\right], \tag{7.124}$$

$$\frac{\partial f_{aM}}{\partial V} = [A_{a1}(V)L_0^{(3/2)}(x_a) - A_{a2}(V)L_1^{(3/2)}(x_a)] \frac{e_a}{T_a} f_{aM} \tag{7.125}$$

are obtained, where $L_0^{(\alpha)}(x_a) = 1$ and $L_1^{(\alpha)}(x_a) = \alpha + 1 - x$ are the Laguerre polynomial functions with $x_a = (v/v_{Ta})^2$. Here it can be shown that

$$\mathbf{u}_{\perp a1} = \frac{1}{n_a} \int d\mathbf{v}\, \mathbf{v}_\perp L_0^{(3/2)}(x_a) \tilde{f}_{a1}, \tag{7.126}$$

$$\mathbf{q}_{\perp a1} = -T_a \int d\mathbf{v}\, \mathbf{v}_\perp L_1^{(3/2)}(x_a) \tilde{f}_{a1}. \tag{7.127}$$

Hence, the perpendicular flow $\mathbf{u}_{\perp a1}$ and heat flux $\mathbf{q}_{\perp a1}$ depend on the cyclotron motion, and the thermodynamic forces (7.121) and (7.122) are valid for the local Maxwellian.

Using $\nabla \cdot \mathbf{u}_{a1} = 0$ and $\nabla \cdot \mathbf{q}_{a1} = 0$ and noting that $\nabla \cdot (u_{\|a1}\mathbf{B}/B) = \mathbf{B} \cdot \nabla(u_{\|a1}/B)$, we find that

$$Bu_{\|a1} = -A_{a1}(V)\left\{g_2 - \frac{B^2}{\langle B^2 \rangle}\langle g_2 \rangle\right\} + \frac{B^2}{\langle B^2 \rangle}\langle Bu_{\|a1}\rangle, \tag{7.128}$$

$$Bq_{\|a1} = -\frac{5}{2}P_a A_{a2}(V)\left\{g_2 - \frac{B^2}{\langle B^2 \rangle}\langle g^2 \rangle\right\} + \frac{B^2}{\langle B^2 \rangle}\langle Bq_{\|a1}\rangle, \tag{7.129}$$

where the function g_2 is the solution of the equation

$$\mathbf{B} \cdot \nabla(g_2/B^2) = \mathbf{B} \times \nabla V \cdot \nabla(1/B^2), \tag{7.130}$$

and $g_2(B = B_{max}) = 0$. Here B_{max} is the maximum magnetic field along a magnetic field line. In (7.128) and (7.129), $\langle\ \rangle$ denotes the flux-surface average, which is taken to give an integration constant. To obtain (7.128), for example, $u_{\|a1}/B + A_{a1}(V)g_2/B^2 = h(V)$ is given, and then $h(V)$ is determined with the flux-surface average. It is noted that (7.128) shows a relation between the parallel flow and the diamagnetic flow.

The function g_2 is also useful in obtaining the parallel current $J_{\|}$ from $\nabla \cdot \mathbf{J} = 0$. For example, the Pfirsch–Schlüter current discussed in section 4.6 is given by

$$BJ_{\|ps} = -\sum_a e_a n_a A_{a1}(V)\left\{ g_2 - \frac{B^2}{\langle B^2 \rangle}\langle g_2 \rangle \right\}$$
$$= -\frac{dP}{dV}\left\{ g_2 - \frac{B^2}{\langle B^2 \rangle}\langle g_2 \rangle \right\}, \tag{7.131}$$

where the charge neutrality $\sum_a e_a n_a = 0$ and $\langle \mathbf{J}_{ps} \cdot \mathbf{B} \rangle = 0$ have been used.

Here we show the solvability conditions of (7.130), which belongs to the magnetic differential equation with a general form,

$$\mathbf{B} \cdot \nabla G = F, \tag{7.132}$$

where G and F are single-valued functions. The solvability conditions are given by

$$\int d\tau F = 0, \tag{7.133}$$

$$\oint \frac{d\ell}{B} F = 0 \quad \text{(on a rational surface)}, \tag{7.134}$$

where $d\tau$ and $d\ell$ are the volume element and the line element along a magnetic field line, respectively. In the case of (7.130), condition (7.133) is satisfied automatically. Also, (7.134) is satisfied on the rational surface, by noting that $\oint d\ell/B$ is a surface quantity.

To determine the flux-surface averaged parallel flow $\langle Bu_{\|a1} \rangle$ and the heat flux $\langle Bq_{\|a1} \rangle$ in (7.128) and (7.129), the flux-surface averaged parallel momentum and heat flux balance equations (7.97) and (7.114) are shown as

$$\begin{bmatrix} \langle \mathbf{B} \cdot \nabla \cdot \overleftrightarrow{\Pi}_a \rangle \\ -\langle \mathbf{B} \cdot \nabla \cdot \overleftrightarrow{\Theta}_a \rangle \end{bmatrix} = \begin{bmatrix} \langle \mathbf{B} \cdot \mathbf{F}_{a1} \rangle \\ -\langle \mathbf{B} \cdot \mathbf{F}_{a2} \rangle \end{bmatrix}. \tag{7.135}$$

It is noted that the inductive electric field has been neglected. The first-order parallel momentum and heat friction, $\langle \mathbf{B} \cdot \mathbf{F}_{a1} \rangle$ and $\langle \mathbf{B} \cdot \mathbf{F}_{a2} \rangle$, are expressed in terms of $\langle Bu_{\|a1} \rangle$ and $\langle Bq_{\|a1} \rangle$ of all particle species,

$$\begin{bmatrix} \langle \mathbf{B} \cdot \mathbf{F}_{a1} \rangle \\ -\langle \mathbf{B} \cdot \mathbf{F}_{a2} \rangle \end{bmatrix} = \sum_b \begin{bmatrix} \ell_{11}^{ab} & \ell_{12}^{ab} \\ \ell_{21}^{ab} & \ell_{22}^{ab} \end{bmatrix} \begin{bmatrix} \langle Bu_{\|b1} \rangle \\ -\dfrac{2}{5P_b}\langle Bq_{\|b1} \rangle \end{bmatrix}, \tag{7.136}$$

where ℓ_{ij}^{ab} $(i, j = 1, 2)$ are the friction coefficients, which are independent of the magnetic field $\mathbf{B}$. The expression (7.136) is valid for an arbitrary toroidal system and for all collisionality regimes.

When the first-order flux-surface averaged parallel viscosity and the heat viscosity are expressed in terms of $\langle Bu_{\|a1}\rangle$ and $\langle Bq_{\|a1}\rangle$, then we can solve (7.135) to obtain $\langle Bu_{\|a1}\rangle$ and $\langle Bq_{\|a1}\rangle$. To the first order of ρ_L/L, the viscosities in (7.135) are written as

$$\langle \mathbf{B}\cdot\nabla\cdot\overleftrightarrow{\Pi}_a\rangle = \langle (P_{\perp a} - P_{\|a})\hat{\mathbf{n}}\cdot\nabla B\rangle \simeq \left\langle \int d\mathbf{v} m_a P_2(\xi) v^2 L_0^{(3/2)}(x_a)\bar{f}_{a1}\hat{\mathbf{n}}\cdot\nabla B\right\rangle \tag{7.137}$$

and

$$-\langle \mathbf{B}\cdot\nabla\cdot\overleftrightarrow{\Theta}_a\rangle = -\langle (\Theta_{\perp a} - \Theta_{\|a})\hat{\mathbf{n}}\cdot\nabla B\rangle \simeq -\left\langle \int d\mathbf{v} m_a P_2(\xi) v^2 L_1^{(3/2)}(x_a)\bar{f}_{a1}\hat{\mathbf{n}}\cdot\nabla B\right\rangle, \tag{7.138}$$

where $\xi = v_\|/v$, and $P_2(\xi) = 3\xi^2/2 - 1/2$ is the second-order Legendre polynomial. Here $\bar{f}_{a1}$ is obtained by solving the first-order drift-kinetic equation in $(V, \theta, \zeta, E, \mu)$ space,

$$\frac{v_\|}{B}\mathbf{B}\cdot\nabla\bar{f}_{a1} + \mathbf{v}_{Da}\cdot\nabla V\frac{\partial f_{aM}}{\partial V} = C_a(\bar{f}_{a1}), \tag{7.139}$$

where the radial drift $\mathbf{v}_{Da}\cdot\nabla V$ is given by

$$\mathbf{v}_{Da}\cdot\nabla V = \frac{v_\|}{\Omega_a}\mathbf{B}\times\nabla V\cdot\nabla(v_\|/B). \tag{7.140}$$

The linearized drift-kinetic equation (7.139) is usually solved in the asymptotic limit of collisionality.

From (7.137) and (7.138) we see that the parallel viscosity and the heat viscosity are expressed in terms of the flow and the heat flux in the direction in which the magnetic field strength changes. In other words, when the flow and the heat flux in the direction of the magnetic field strength are constant, both viscosities become zero.

Now we consider the $1/\nu$ regime to show the expression of the geometrical factor $\langle G_{BS}\rangle_{1/\nu}$, since neoclassical transport in stellarators and heliotrons is typically characterized by this collisionality regime. For this regime, $\nu_{eff}/\omega_b \ll 1$ is used as the auxiliary expansion parameter, where ν_{eff} and ω_b are the effective collision frequency and the bounce frequency of trapped particles, respectively. First we consider the distribution function for the passing particles, denoted by $\bar{f}_{a1c}$. The zeroth-order equation of (7.139) with respect to ν_{eff}/ω_b becomes

$$\frac{v_\|}{B}\mathbf{B}\cdot\nabla\bar{f}_{a1c}^{(0)} + \mathbf{v}_{Da}\cdot\nabla V\frac{\partial f_{aM}}{\partial V} = 0. \tag{7.141}$$

According to (7.130), we introduce a function g_3 that satisfies

$$\mathbf{B} \cdot \nabla\left(\frac{v_\parallel}{B} g_3\right) = \mathbf{B} \times \nabla V \cdot \nabla\left(\frac{v_\parallel}{B}\right) \tag{7.142}$$

and $g_3(B = B_{max}) = 0$. With the function g_3 and (7.140), we can solve (7.141) with a solution

$$\bar{f}^{(0)}_{a1c} = -\frac{m_a}{e_a}\frac{v_\parallel}{B} g_3 \frac{\partial f_{aM}}{\partial V} + g_{a1c}(V, E, \mu), \tag{7.143}$$

where g_{a1c} is an homogeneous solution of (7.141). The solvability condition of (7.141) is given by

$$\oint \frac{d\ell}{v_\parallel} \mathbf{v}_{Da} \cdot \nabla V = 0, \tag{7.144}$$

for a rational magnetic surface on which $\partial f_{aM}/\partial V$ is constant. This condition means that particles do not deviate from a flux surface without collisions.

The unknown function g_{a1c} is determined by the flux-surface averaging, $\langle C_a(\bar{f}^{(0)}_{a1c})B/v_\parallel\rangle = 0$, of the first-order equation of (7.139):

$$\frac{v_\parallel}{B} \mathbf{B} \cdot \nabla \bar{f}^{(1)}_{a1c} = C_a(\bar{f}^{(0)}_{a1c}). \tag{7.145}$$

It is noted that RHS of (7.145) disappears due to the flux-surface averaging. A general form for the function g_{a1c} is shown as

$$\begin{aligned} g_{a1c}(V, K, \lambda) = & \frac{2H(1-\lambda)V_\parallel(V, K, \lambda)G(V, K)}{v_{Ta}^2} f_{aM} \\ & + \sigma \frac{H(1-\lambda)v}{v_{Ta}^2 B_{max}} \frac{T_a}{e_a} \int_\lambda^1 d\lambda \frac{\langle g_4\rangle}{\langle g_1\rangle} \frac{\partial f_{aM}}{\partial V} \end{aligned} \tag{7.146}$$

by substituting $\bar{f}^{(0)}_{a1c}$ given by (7.143) into the flux-surface average, where the variables (E, μ) are transformed to (K, λ) with $K = E - e_a\varphi$ and $\lambda = B_{max}\mu/(E - e_a\varphi)$, and $H(1-\lambda)$ is the Heaviside function such that $H(1-\lambda) = 0$ for $\lambda > 1$ (trapped particles) and $H(1-\lambda) = 1$ for $\lambda < 1$ (untrapped particles). In (K, λ) space, the parallel velocity is expressed by

$$v_\parallel = \sigma v g_1, \tag{7.147}$$

where $\sigma = \pm 1$, $K = m_a v^2/2$ and

$$g_1 = \sqrt{1 - \lambda \frac{B}{B_{max}}}. \tag{7.148}$$

The first term of (7.146) denotes the usual pitch angle or $v_\parallel$ dependence obtained for axisymmetric systems. It is characterized by $V_\parallel(V, K, \lambda)$, given by

$$V_{\parallel}(V, K, \lambda) = \sigma \frac{v}{2} \frac{\sqrt{\langle B^2 \rangle}}{B_{max}} \int_{\lambda}^{1} d\lambda \frac{1}{\langle g_1 \rangle}. \tag{7.149}$$

It is noted that $V_{\parallel}(V, K, \lambda) \to v_{\parallel}$ in a uniform magnetic field. Also, the average magnetic field is given by $\sqrt{\langle B^2 \rangle}$. The function g_4 is related to g_3 by

$$g_4 = -2B_{max} g_1 \frac{\partial}{\partial \lambda} \left(\frac{g_1}{B} g_3 \right), \tag{7.150}$$

which is also the solution of the magnetic differential equation

$$\mathbf{B} \cdot \nabla(g_4/g_1) = \mathbf{B} \times \nabla V \cdot \nabla(1/g_1) \tag{7.151}$$

with $g_4(B = B_{max}) = 0$. $V_{\parallel}(V, K, \lambda)$ and the second term of (7.146) come from the pitch angle scattering collision operator or the Lorentz operator (7.29). As seen in (7.150), the second term in (7.146) also orginates from the first term in (7.143). The unknown function $G(V, K)$ in (7.146) is determined by using the trapped particle disribution and the following Laguerre polynomial expansion:

$$G(V, K) = C_0(V) L_0^{(3/2)}(x_a) - C_1(V) L_1^{(3/2)}(x_a) + \cdots, \tag{7.152}$$

where $L_0^{(3/2)}(x_a) = 1$ and $L_1^{(3/2)}(x_a) = \frac{5}{2} - x_a^2$. The trapped particle distribution function $\bar{f}_{a1t}^{(0)}$, which is required to determine the coefficients $C_0(V)$ and $C_1(V)$, is an odd function with respect to σ. It is noted that $\bar{f}_{a1c}^{(0)}$ is also the odd function of $v_{\parallel}$ (see (7.143), (7.146), and (7.149)). Then, the trapped particle distribution function is given by

$$\bar{f}_{a1t}^{(0)} = \bar{F}_{a1t}, \tag{7.153}$$

where $\bar{F}_{a1t}$ has the same functional form as the first term in (7.143), although the velocity space concerned is different from that in (7.143). Here g_{a1t} corresponding to the second term in (7.143) is assumed zero, which implies neglect of the boundary layer effect. From the definitions

$$\frac{u_{\parallel a1}}{B} = \frac{1}{n_a} \left[\int_c d\mathbf{v} \frac{v_{\parallel}}{B} L_0^{(3/2)}(x_a) \bar{f}_{a1c}^{(0)} + \int_t d\mathbf{v} \frac{v_{\parallel}}{B} L_0^{(3/2)}(x_a) \bar{f}_{a1t}^{(0)} \right] \tag{7.154}$$

and

$$\frac{q_{\parallel a1}}{B} = -T_a \left[\int_c d\mathbf{v} \frac{v_{\parallel}}{B} L_1^{(3/2)}(x_a) \bar{f}_{a1c}^{(0)} + \int_t d\mathbf{v} \frac{v_{\parallel}}{B} L_1^{(3/2)}(x_a) \bar{f}_{a1t}^{(0)} \right], \tag{7.155}$$

we find

$$C_0(V) = \frac{\sqrt{\langle B^2 \rangle}}{f_c} \left\{ \frac{u_{\|a1}}{B} + A_{a1}(V) \left[\frac{g_2}{B^2} - \frac{3}{4B_{max}^2} \int_0^1 d\lambda \, \frac{\langle g_4 \rangle}{\langle g_1 \rangle} \lambda \right] \right\}, \tag{7.156}$$

$$C_1(V) = \frac{\sqrt{\langle B^2 \rangle}}{f_c} \left\{ \frac{2}{5P_a} \frac{q_{\|a1}}{B} + A_{a2}(V) \left[\frac{g_2}{B^2} - \frac{3}{4B_{max}^2} \int_0^1 d\lambda \, \frac{\langle g_4 \rangle}{\langle g_1 \rangle} \lambda \right] \right\}, \tag{7.157}$$

from (7.128) and (7.129), where f_c is the fraction of passing particles,

$$f_c = \frac{3}{4} \frac{\langle B^2 \rangle}{B_{max}^2} \int_0^1 d\lambda \, \frac{\lambda}{\langle g_1 \rangle} \tag{7.158}$$

and g_2, given by (7.130), is related to g_3 by

$$g_2 = \frac{3}{2B_{max}^2} \, B^2 \int_0^{B_{max}/B} d\lambda \, \frac{g_1}{B} \, g_3. \tag{7.159}$$

It is noted that in (K, λ) space (7.142) becomes

$$\mathbf{B} \cdot \nabla \left(\frac{g_1}{B} g_3 \right) = \mathbf{B} \times \nabla V \cdot \nabla \left(\frac{g_1}{B} \right) \tag{7.160}$$

and $g_3(B = B_{max}) = 0$. Substituting (7.128) and (7.129) into (7.156) and (7.157) and defining the coefficient dependent on the magnetic field configuration as

$$\langle G_{BS} \rangle_{1/\nu} = \frac{1}{f_t} \left\{ \langle g_2 \rangle - \frac{3 \langle B^2 \rangle}{4B_{max}^2} \int_0^1 d\lambda \, \frac{\langle g_4 \rangle}{\langle g_1 \rangle} \lambda \right\}, \tag{7.161}$$

we can construct the distribution function $\bar{f}_{a1}^{(0)}$ by combining the untrapped and trapped distribution functions, $\bar{f}_{a1c}^{(0)}$, given by (7.143), and $\bar{f}_{a1t}^{(0)}$, given by (7.153), in the following form:

$$\begin{aligned}
\bar{f}_{a1}^{(0)} = \frac{2}{v_{Ta}^2} f_{aM} \times \Bigg\{ & \Bigg[\frac{\langle B u_{\|a1} \rangle}{f_c \sqrt{\langle B^2 \rangle}} H(1-\lambda) V_\|(V, K, \lambda) \\
& + A_{a1}(V) \Bigg\{ \frac{f_t \langle G_{BS} \rangle_{1/\nu}}{f_c \sqrt{\langle B^2 \rangle}} H(1-\lambda) V_\|(V, K, \lambda) \\
& - \sigma v \left[\frac{g_1}{B} g_3 - \frac{H(1-\lambda)}{2B_{max}} \int_\lambda^1 d\lambda \, \frac{\langle g_4 \rangle}{\langle g_1 \rangle} \right] \Bigg\} \Bigg] L_0^{(3/2)}(x_a) \\
& - \Bigg[\frac{2}{5P_a} \frac{\langle B q_{\|a1} \rangle}{f_c \sqrt{\langle B^2 \rangle}} H(1-\lambda) V_\|(V, K, \lambda) \\
& + A_{a2}(V) \Bigg\{ \frac{f_t \langle G_{BS} \rangle_{1/\nu}}{f_c \sqrt{\langle B^2 \rangle}} H(1-\lambda) V_\|(V, K, \lambda) \\
& - \sigma v \left[\frac{g_1}{B} g_3 - \frac{H(1-\lambda)}{2B_{max}} \int_\lambda^1 d\lambda \, \frac{\langle g_4 \rangle}{\langle g_1 \rangle} \right] \Bigg\} \Bigg] L_1^{(3/2)}(x_a) \Bigg\},
\end{aligned} \tag{7.162}$$

where $f_t = 1 - f_c$ is the fraction of trapped particles. The coefficient $\langle G_{BS}\rangle_{1/\nu}$ is called the geometric factor, and was introduced by Shaing and Callen (1983) in the $1/\nu$ regime. Here we use the suffix *BS* to represent the *b*oot*s*trap current.

The parallel viscosities are calculated from $\bar{f}_{a1}^{(1)}$ satisfying the linearized drift kinetic equation

$$\frac{v_\parallel}{B}\mathbf{B}\cdot\nabla\bar{f}_{a1}^{(1)} = C_a(\bar{f}_{a1}^{(0)}). \tag{7.163}$$

Therefore, the parallel viscosities consist of the parallel flow $\langle Bu_{\parallel a1}\rangle$ and heat flux $\langle Bq_{\parallel a1}\rangle$, and the thermodynamic forces $A_{a1}(V)$ and $A_{a2}(V)$ multiplied by the coefficient dependent on the magnetic configuration. Note that the function g_4 in (7.161) and (7.162) is related to g_3 (see (7.150)) and that the function g_2 showing the divergence-free property of the first-order flow and heat flux is also related to g_3 (see (7.159)).

If the physical quantities have helical symmetry, i.e., $Q(V,\theta,\zeta) = Q(V,\phi = L\theta - M\zeta)$, then the function g_3 given by (7.142) becomes

$$g_3 = \frac{LJ + MI}{L\iota - M}\left[1 - \frac{B}{B_{max}}\frac{\sqrt{1-\lambda}}{g_1}\right]\frac{dV}{d\Phi}, \tag{7.164}$$

in the magnetic coordinates (Φ,θ,ζ), where $2\pi J(2\pi I)$ is the total poloidal (toroidal) current outside (inside) a flux surface, $\Phi = $ constant. Then, from (7.150), (7.159), and (7.161), we obtain

$$g_2 = \frac{LJ + MI}{L\iota - M}\left[1 - \left(\frac{B}{B_{max}}\right)^2\right]\frac{dV}{d\Phi}, \tag{7.165}$$

$$g_4 = \frac{LJ + MI}{L\iota - M}\left[1 - \frac{g_1}{\sqrt{1-\lambda}}\right]\frac{dV}{d\Phi}, \tag{7.166}$$

$$\langle G_{BS}\rangle_{1/\nu} = \frac{LJ + MI}{L\iota - M}\frac{dV}{d\Phi}. \tag{7.167}$$

It is noted that $\langle G_{BS}\rangle_{1/\nu}$ in (7.167) just means $\langle G_{BS}\rangle$ in a helically symmetric stellarator. Hence (7.162) also becomes

$$\begin{aligned}\bar{f}_{a1}^{(0)} = &-\frac{m_a}{e_a}\frac{v_\parallel}{B}\frac{LJ + MI}{L\iota - M}\frac{dV}{d\Phi}\frac{\partial f_{aM}}{\partial V}\\ &+\frac{2H(1-\lambda)V_\parallel(V,K,\lambda)}{v_{Ta}^2}\frac{\sqrt{\langle B^2\rangle}}{f_c}\left[u_{\phi a1}(V)L_0^{(3/2)}(x_a)\right.\\ &\left.-\frac{2}{5P_a}q_{\phi a1}(V)L_1^{(3/2)}(x_a)\right]f_{aM},\end{aligned} \tag{7.168}$$

where

$$u_{\phi a1}(V) = \frac{\mathbf{u}_{a1} \cdot \nabla\phi}{\mathbf{B} \cdot \nabla\phi}$$

and

$$q_{\phi a1}(V) = \frac{\mathbf{q}_{a1} \cdot \nabla\phi}{\mathbf{B} \cdot \nabla\phi}.$$

From the above derivations, we may show that the surface-averaged parallel viscosity and the heat viscosity have the following form:

$$\begin{bmatrix} \langle \mathbf{B} \cdot \nabla \cdot \overleftrightarrow{\Pi}_a \rangle \\ -\langle \mathbf{B} \cdot \nabla \cdot \overleftrightarrow{\Theta}_a \rangle \end{bmatrix} = \begin{bmatrix} \mu_{a1} & \mu_{a2} \\ \mu_{a2} & \mu_{a3} \end{bmatrix} \begin{bmatrix} \langle Bu_{\|a1} \rangle + \langle G_{BS} \rangle_{1/\nu} A_{a1}(V) \\ -\dfrac{2}{5P_a} \langle Bq_{\|a1} \rangle - \langle G_{BS} \rangle_{1/\nu} A_{a2}(V) \end{bmatrix}, \tag{7.169}$$

where $\mu_{aj}(j = 1, 2, 3)$ and $\langle G_{BS} \rangle_{1/\nu}$ are the viscosity coefficients and the geometric factor for particle species a. Here we note that μ_{aj} is related to the viscosity matrix K_{ij}^a $(i, j = 1, 2)$, $\mu_{a1} = K_{11}^a$, $\mu_{a2} = -K_{12}^a - 5K_{11}^a/2$, and $\mu_{a3} = K_{22}^a - 5K_{12}^a + 25K_{11}^a/4$. The coefficients K_{ij}^a are defined by $J_{\|}^a/e = K_{11}^a \bar{A}_{a1} + K_{12}^a \bar{A}_{a2}$, $q_{\|e}/T_e = K_{12}^e \bar{A}_{e1} - K_{22}^e \bar{A}_{e2}$, and $q_{\|i}/T_i = -K_{22}^i \bar{A}_{i2}$. The Onsager symmetry relation requires $K_{12}^a = K_{21}^a$. Here $\bar{A}_{a1} = \hat{\mathbf{n}} \cdot \nabla \ln P_a$ and $\bar{A}_{a2} = -\hat{\mathbf{n}} \cdot \nabla \ln T_a$ are introduced, where $\hat{\mathbf{n}} = \mathbf{B}/B$. By substituting (7.136) and (7.169) into (7.135), we can obtain the surface-averaged parallel flow and heat flux. These will be discussed in the next section. Here, in order to clarify the physics, we write another form of the surface-averaged parallel viscosity and heat viscosity. The poloidal and toroidal flow are given by

$$\langle \mathbf{u}_{a1} \cdot \nabla\theta \rangle = \frac{\iota}{J + \iota I} \langle Bu_{\|a1} \rangle + \frac{J}{J + \iota I} A_{a1}(V) \frac{dV}{d\Phi} \tag{7.170}$$

and

$$\langle \mathbf{u}_{a1} \cdot \nabla\zeta \rangle = \frac{1}{J + \iota I} \langle Bu_{\|a1} \rangle - \frac{I}{J + \iota I} A_{a1}(V) \frac{dV}{d\Phi}, \tag{7.171}$$

where

$$A_{a1}(V) \frac{dV}{d\Phi} = \langle \mathbf{u}_{a1} \cdot \nabla\theta \rangle - \iota \langle \mathbf{u}_{a1} \cdot \nabla\zeta \rangle \tag{7.172}$$

and

$$\langle Bu_{\|a1} \rangle = I \langle \mathbf{u}_{a1} \cdot \nabla\theta \rangle + J \langle \mathbf{u}_{a1} \cdot \nabla\zeta \rangle. \tag{7.173}$$

Similarly, we obtain

$$A_{a2}(V)\frac{dV}{d\Phi}=\frac{2}{5P_a}[\langle \mathbf{q}_{a1}\cdot\nabla\theta\rangle-\iota\langle \mathbf{q}_{a1}\cdot\nabla\zeta\rangle], \tag{7.174}$$

$$\langle Bq_{\|a1}\rangle=I\langle \mathbf{q}_{a1}\cdot\nabla\theta\rangle+J\langle \mathbf{q}_{a1}\cdot\nabla\zeta\rangle. \tag{7.175}$$

Substituting (7.172)–(7.175) into (7.169) yields

$$\begin{bmatrix}\langle \mathbf{B}\cdot\nabla\cdot\overleftrightarrow{\Pi}_a\rangle\\ -\langle \mathbf{B}\cdot\nabla\cdot\overleftrightarrow{\Theta}_a\rangle\end{bmatrix}=\left(I+\langle G_{BS}\rangle_{1/\nu}\right)\Big/\frac{dV}{d\Phi}\begin{bmatrix}\mu_{a1}&\mu_{a2}\\ \mu_{a2}&\mu_{a3}\end{bmatrix}\begin{bmatrix}\langle \mathbf{u}_{a1}\cdot\nabla\theta\rangle\\ -\dfrac{2}{5P_a}\langle \mathbf{q}_{a1}\cdot\nabla\theta\rangle\end{bmatrix}$$
$$+\left(J-\iota\langle G_{BS}\rangle_{1/\nu}\right)\Big/\frac{dV}{d\Phi}\begin{bmatrix}\mu_{a1}&\mu_{a2}\\ \mu_{a2}&\mu_{a3}\end{bmatrix}\begin{bmatrix}\langle \mathbf{u}_{a1}\cdot\nabla\zeta\rangle\\ \dfrac{2}{5P_a}\langle \mathbf{q}_{a1}\cdot\nabla\zeta\rangle\end{bmatrix}. \tag{7.176}$$

When helical symmetry is imposed, the geometric factor $\langle G_{BS}\rangle_{1/\nu}$ is given by (7.167), and the relation (7.176) becomes

$$\begin{bmatrix}\langle \mathbf{B}\cdot\nabla\cdot\overleftrightarrow{\Pi}_a\rangle\\ -\langle \mathbf{B}\cdot\nabla\cdot\overleftrightarrow{\Theta}_a\rangle\end{bmatrix}=\frac{J+\iota I}{L\iota-M}\begin{bmatrix}\mu_{a1}&\mu_{a2}\\ \mu_{a2}&\mu_{a3}\end{bmatrix}\begin{bmatrix}\langle \mathbf{u}_{a1}\cdot\nabla\phi\rangle\\ -\dfrac{2}{5P_a}\langle \mathbf{q}_{a1}\cdot\nabla\phi\rangle\end{bmatrix}. \tag{7.177}$$

It is noted that the toroidal symmetry corresponds to $M=0$ and the poloidal symmetry to $L=0$ in this relation. Thus, the role of the parallel viscosities is to damp the flow and heat flux in the direction in which the magnetic field strength varies. The damping is essentially determined by the function g_3 (see (7.164)). Since g_3 obtained from (7.142) is related to the radial drift motion of the guiding center given by (7.140), we can understand that the difference between the symmetric system and the nonsymmetric system is derived from the radial drift motion. It is also related to the collisionality regime. It is the geometric factor $\langle G_{BS}\rangle_{1/\nu}$ that indicates the degree of symmetry breaking in stellarators and heliotrons.

When the magnetic field, $B(B,\theta,\zeta)$, is given in Boozer coordinates or in Hamada coordinates, (7.160) for g_3 may be solved by using Fourier expansions withr espect to θ and ζ. When g_3 is obtained, g_2 and g_4 are calculated from (7.150) and (7.159), and finally $\langle G_{BS}\rangle_{1/\nu}$ is evaluated by using (7.161). An example of $\langle G_{BS}\rangle_{1/\nu}$ is shown in Fig. 7.4, where $(f_t/f_c)\langle G_{BS}\rangle_{1/\nu}$ is plotted as a function of the average radius, which will be useful in obtaining the bootstrap current, as shown in (7.220) in the next section. Here, the LHD configuration is taken as an example, and the magnetic axis position is controlled by using a vertical magnetic field. It can be seen that $\langle G_{BS}\rangle_{1/\nu}$ depends sensitively on the magnetic configuration. In the edge region $\langle G_{BS}\rangle_{1/\nu}$ becomes small or negative

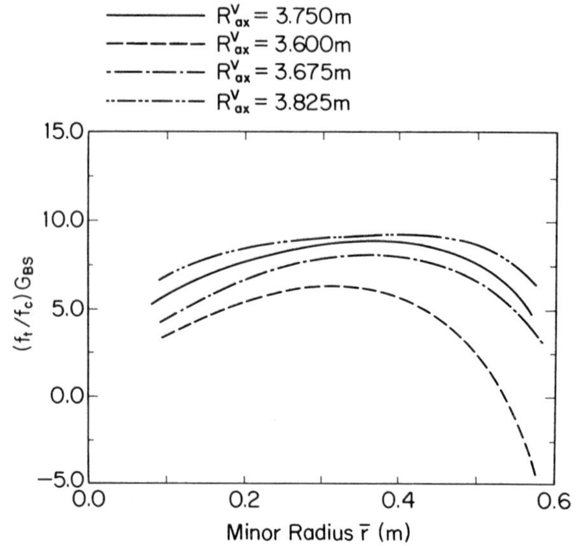

Fig. 7.4 The radial profile of $(f_t/f_c)\langle G_{BS}\rangle_{1/\nu}$ for various vacuum magnetic configurations of the LHD. $\bar{r}$(m) denotes the minor radius of the average flux surface. The solid line corresponds to the standard magnetic configuration with $R^v_{ax} = 3.75$ m, and the magnetic axis position R^v_{ax} is changed from 3.6 m to 3.825 m.

in heliotrons. This result suggests that neoclassical transport and the bootstrap current may be controlled by changing the magnetic field configuration using the vertical, quadrupole, and/or hexapole field. It is noted that a zero-beta vacuum configuration is assumed in Fig. 7.4.

Finite-beta equilibria are more important in estimating the bootstrap current with the use of $\langle G_{BS}\rangle_{1/\nu}$. In this case we must include the bootstrap current in the MHD equilibrium calculation consistently. One way in which to obtain finite-beta equilibrium with the bootstrap current is to rely on iterative calculations between $\mathbf{J}_{BS}$, which is proportional to $\langle G_{BS}\rangle_{1/\nu}$, and the MHD equilibrium with $\mathbf{J}_{BS}$. When the iteration converges, the obtained equilibrium will be a self-consistent finite-beta equilibrium in the presence of the bootstrap current.

7.6 PARALLEL VISCOSITY-DRIVEN FLUXES AND THE BOOTSTRAP CURRENT

To evaluate the neoclassical fluxes driven by the parallel viscosity, we need to determine the parallel particle and heat flows in terms of the radial gradients of the plasma density, the temperature, and the electrostatic potential. This can be achieved by considering the balance equations between the parallel frictional forces and the parallel viscosities,

$$\langle BF_{\|a1}\rangle = \langle \mathbf{B}\cdot\nabla\cdot\overset{\leftrightarrow}{\Pi}_a\rangle, \qquad (7.178)$$

$$\langle BF_{\|a2}\rangle = \langle \mathbf{B}\cdot\nabla\cdot\overleftrightarrow{\Theta}_a\rangle, \tag{7.179}$$

for all particle species a.

With $\overleftrightarrow{\Pi}_a = (P_{\|a} - P_{\perp a})(\hat{n}\hat{n} - \frac{1}{3}\overleftrightarrow{I})$ and $\overleftrightarrow{\Theta}_a = (\Theta_{\|a} - \Theta_{\perp a})(\hat{n}\hat{n} - \frac{1}{3}\overleftrightarrow{I})$ (see (7.112) and (7.113)), we can simplify the parallel viscosities as follows:

$$\langle \mathbf{B}\cdot\nabla\cdot\overleftrightarrow{\Pi}_a\rangle = \langle (P_{\perp a} - P_{\|a})\hat{n}\cdot\nabla B\rangle, \tag{7.180}$$

$$\langle \mathbf{B}\cdot\nabla\cdot\overleftrightarrow{\Theta}_a\rangle = \langle (\Theta_{\perp a} - \Theta_{\|a})\hat{n}\cdot\nabla B\rangle. \tag{7.181}$$

In principle, we can calculate the parallel viscosities from (7.180) and (7.181) as long as we know the distribution function f_a of particle species a. It is well known that in the low-collisionality regime, $\nu \ll \omega_b$ (the bounce frequency in helical ripples), the particle flux is dominated in heliotrons by the nonaxisymmetric flux, Γ^a_{na}. However, since the parallel viscosity $\langle \mathbf{B}\cdot\nabla\cdot\overleftrightarrow{\Pi}_a\rangle$ is proportional to the banana-plateau flux Γ^a_{bp} (see (7.102)), it is natural to consider that $\langle \mathbf{B}\cdot\nabla\cdot\overleftrightarrow{\Pi}_a\rangle$ in heliotrons has a similar property to that in tokamaks. Indeed, we can show that the solution of the bounce-averaged drift-kinetic equation f_a, which gives rise to the nonaxisymmetric flux Γ^a_{na} in the low-collisionality regime $\nu < \omega_b$ will not contribute to the parallel viscosity. To show this, we note that since f_a is a solution of the bounce-averaged drift-kinetic equation, f_a will not depend on the variable that measures the distance along the field line. For convenience, we define $\mathbf{B} = \nabla V \times \nabla\alpha$ with $\alpha = \psi'\theta - \chi'\zeta$ and choose ζ as the variable that measures the distance along the field line. Then, f_a will have the functional form $f_a = f_a(E, \mu, V, \alpha)$, where $E = \frac{1}{2}m_a v^2$ and $\mu = \frac{1}{2}m_a v_\perp^2/B$ are the total energy and magnetic moment of particle species a, respectively. Neglecting the flow velocity $\mathbf{u}_a$ compared with the thermal velocity $\mathbf{v}_{ta}$, we have

$$\langle \mathbf{B}\cdot\nabla\cdot\overleftrightarrow{\Pi}_a\rangle = \left\langle \hat{n}\cdot\nabla B\int d\mathbf{v}\, m_a\left(\frac{v_\perp^2}{2} - v_\|^2\right)f_a\right\rangle, \tag{7.182}$$

from (7.180). Changing variables from $d\mathbf{v}$ to $4\pi dE d\mu B/(m_a^2|v_\||)$ and expressing the flux surface average explicitly, we can write (7.182) as

$$\langle \mathbf{B}\cdot\nabla\cdot\overleftrightarrow{\Pi}_a\rangle = \oint d\zeta \int \frac{4\pi}{m_a}\, dE d\mu\left[\frac{1}{|v_\||}\frac{\partial B}{\partial\zeta}\left(\frac{v_\perp^2}{2} - v_\|^2\right)\right]f_a. \tag{7.183}$$

Here we have noted that the flux surface average corresponds to $\oint d\zeta/B$ of the integrand. The term inside the large square bracket can be written as a total derivative with respect to ζ:

$$\frac{1}{|v_\||}\frac{\partial B}{\partial\zeta}\left(\frac{v_\perp^2}{2} - v_\|^2\right) = -\frac{\partial}{\partial\zeta}(|v_\||B). \tag{7.184}$$

Since f_a is independent of ζ, we can carry out the ζ integration first and obtain $\langle \mathbf{B} \cdot \nabla \cdot \overleftrightarrow{\Pi}_a \rangle = 0$ from the periodic condition of ζ. This result is physically related to the fact that the bounce-averaged drift-kinetic equation is not sensitive to the Coulomb collisional effects in the parallel momentum balance, and thus will make no contribution to the parallel friction force and Γ^a_{bp}. Similarly, one can show that f_a will not contribute to the parallel heat viscosity $\langle \mathbf{B} \cdot \nabla \cdot \overleftrightarrow{\Theta}_a \rangle$.

Before we start calculation of the parallel viscosities with the drift-kinetic equation, we first calculate the parallel viscosity $\langle \mathbf{B} \cdot \nabla \cdot \overleftrightarrow{\Pi}_a \rangle$ from magnetohydrodynamic theory. For a collision frequency ν less than the cyclotron frequency Ω, the pressure anisotropy has been given as

$$P_\parallel - P_\perp = -3P\tau[\hat{\mathbf{n}} \cdot \nabla(\hat{\mathbf{n}} \cdot \mathbf{u}) - (\hat{\mathbf{n}} \cdot \nabla\hat{\mathbf{n}}) \cdot \mathbf{u}], \tag{7.185}$$

where P is the plasma pressure and τ is the ion–ion collision time. To obtain (7.185), we have noted the fact that $\nabla \cdot \mathbf{u} = 0$, and neglected the heat flow. The expression that the anistropy is proportional to τ is shown by solving the drift-kinetic equation. The second term inside the square brackets of (7.185) can be simplified using

$$(\hat{\mathbf{n}} \cdot \nabla\hat{\mathbf{n}}) \cdot \mathbf{u} = (\mathbf{u}_\perp \cdot \nabla B)/B \tag{7.186}$$

In Hamada coordinates, the first term inside the square brackets of (7.185) can be written as

$$\hat{\mathbf{n}} \cdot \nabla(\hat{\mathbf{n}} \cdot \mathbf{u}) = \frac{1}{B}\left(\frac{\partial}{\partial\theta}(u_\parallel \chi') + \frac{\partial}{\partial\zeta}(u_\parallel \psi')\right), \tag{7.187}$$

where $\chi' = \mathbf{B} \cdot \nabla\theta$ and $\psi' = \mathbf{B} \cdot \nabla\zeta$. Using the flow relationship described by $u_p = u_\parallel \hat{\mathbf{n}} \cdot \nabla\theta + \mathbf{u}_\perp \cdot \nabla\theta$ and $u_t = u_\parallel \hat{\mathbf{n}} \cdot \nabla\zeta + \mathbf{u}_\perp \cdot \nabla\zeta$, which are obtained from $\mathbf{u} = u_\parallel \hat{\mathbf{n}} + \mathbf{u}_\perp$, we find that

$$\begin{aligned}\hat{\mathbf{n}} \cdot \nabla(\hat{\mathbf{n}} \cdot \mathbf{u}) = &\frac{1}{B}\frac{\partial B}{\partial\theta}(u_p - \mathbf{u}_\perp \cdot \nabla\theta) + \frac{1}{B}\frac{\partial B}{\partial\zeta}(u_t - \mathbf{u}_\perp \cdot \nabla\zeta)\\ &- \left(\frac{\partial}{\partial\theta}(\mathbf{u}_\perp \cdot \nabla\theta) + \frac{\partial}{\partial\zeta}(\mathbf{u}_\perp \cdot \nabla\zeta)\right).\end{aligned} \tag{7.188}$$

Combining (7.186) and (7.188), we have an expression for the pressure anisotropy:

$$P_\parallel - P_\perp = -3P\tau\left[\frac{1}{B}\frac{\partial B}{\partial\theta}u_p + \frac{1}{B}\frac{\partial B}{\partial\zeta}, u_t - \left(\nabla \cdot \mathbf{u}_\perp + 2\frac{\mathbf{u}_\perp \cdot \nabla B}{B}\right)\right], \tag{7.189}$$

since $\mathbf{u}_\perp \cdot \nabla V = 0$. Recalling that $\mathbf{u}_\perp$ is just the diamagnetic flow and $\nabla \times \mathbf{B} \cdot \nabla V = 0$, we can show that the term in parentheses in (7.189) is

$$\nabla \cdot \mathbf{u}_\perp + 2(\mathbf{u}_\perp \cdot \nabla B)/B = (1/B^2)\nabla \cdot (\mathbf{u}_\perp B^2) = 0. \tag{7.190}$$

Thus, the parallel viscosity in the Pfirsch–Schlüter regime becomes

$$\langle \mathbf{B} \cdot \nabla \cdot \overleftrightarrow{\Pi} \rangle = 3P\tau \left\langle \hat{\mathbf{n}} \cdot \nabla B \left(\frac{1}{B} \frac{\partial B}{\partial \theta} u_p + \frac{1}{B} \frac{\partial B}{\partial \zeta} u_t \right) \right\rangle. \tag{7.191}$$

Equation (7.191) reduces to the axisymmetric tokamak result by neglecting the $\partial B/\partial \zeta$ term.

In the expectation that the result should have the same form as (7.191), we can also calculate the parallel viscosities from the linearized drift kinetic equation:

$$v_{\parallel} \hat{\mathbf{n}} \cdot \nabla f_{a1} + \mathbf{v}_{Da} \cdot \nabla V \frac{\partial f_{a0}}{\partial V} = C_a(f_{a1}), \tag{7.192}$$

where $C_a(f_{a1})$ is the linearized collision operator, and

$$\begin{aligned} \mathbf{v}_{da} \cdot \nabla V &= \frac{v_{\parallel}}{B} \nabla \cdot \left(\frac{v_{\parallel}}{\Omega_a} \mathbf{B} \times \nabla V \right) \\ &= \frac{v_{\parallel}}{B} \left[\frac{\partial}{\partial \theta} \left(\frac{v_{\parallel}}{\Omega_a} \mathbf{B} \times \nabla V \cdot \nabla \theta \right) + \frac{\partial}{\partial \zeta} \left(\frac{v_{\parallel}}{\Omega_a} \mathbf{B} \times \nabla V \cdot \nabla \zeta \right) \right], \end{aligned} \tag{7.193}$$

is the radial drift velocity. Here f_{a0} is the Maxwellian distribution. In (7.192) we have neglected the parallel electric field $E_{\parallel}$.

Equation (7.192) can be solved in the Pfirsch–Schlüter regime by an auxiliary expansion with the small parameter ω_t/ν, where $\omega_t = v_{\parallel}/L_{\parallel}$ is the typical transit frequency, with $L_{\parallel}$ being the typical parallel scale length. The zeroth-order equation is then

$$C_a(f_{a1}^{(0)}) = 0, \tag{7.194}$$

which has a solution corresponding to the Maxwellian distribution, $f_{a1}^{(0)} = n_a(m_a/2\pi T_a)^{3/2} \times \exp(-E/T_a)$, where $E = \frac{1}{2} m_a v^2$ is the kinetic energy of the particle, and both the density n_a and the temperature T_a depend on V. The first-order equation is

$$v_{\parallel} (\bar{A}_{1a} L_0^{(3/2)}(x_a^2) - \bar{A}_{2a} L_1^{(3/2)}(x_a^2)) f_{a1}^{(0)} = C_a(f_{a1}^{(1)}), \tag{7.195}$$

where $\bar{A}_{1a} = \hat{\mathbf{n}} \cdot \nabla \ln P_a$, and $\bar{A}_{2a} = \hat{\mathbf{n}} \cdot \nabla \ln T_a$. The solution $f_{a1}^{(1)}$ of (7.195) can be expressed in terms of Laguerre polynomials:

$$f_{a1}^{(1)} = 2 \frac{v_{\parallel}}{v_{Ta}^2} \left(u_{\parallel a} - \frac{2}{5} \frac{q_{\parallel a}}{P_a} L_1^{(3/2)}(x_a^2) + \cdots \right) f_{a1}^{(0)}. \tag{7.196}$$

The second-order equation obtained from (7.192) is

$$v_{\parallel}\hat{n}\cdot\nabla f_{a1}^{(1)} + \mathbf{v}_{Da}\cdot\nabla V\,\frac{\partial f_{a0}}{\partial V} = C_a(f_{a1}^{(2)}). \tag{7.197}$$

Using Hamada coordinates and (7.193), we can write (7.197) explicitly as

$$\frac{v_{\parallel}}{B}\chi'\frac{\partial f_{a1}^{(1)}}{\partial\theta} + \frac{v_{\parallel}}{B}\frac{\partial}{\partial\theta}\left(\frac{v_{\parallel}}{\Omega_a}\mathbf{B}\times\nabla V\cdot\nabla\theta\right)\frac{\partial f_{a0}}{\partial V} + \frac{v_{\parallel}}{B}\psi'\frac{\partial f_{a1}^{(1)}}{\partial\zeta}$$
$$+\frac{v_{\parallel}}{B}\frac{\partial}{\partial\zeta}\left(\frac{v_{\parallel}}{\Omega_a}\mathbf{B}\times\nabla V\cdot\nabla\zeta\right)\frac{\partial f_{a0}}{\partial V} = C_a(f_{a1}^{(2)}). \tag{7.198}$$

Substituting (7.196) into (7.198) and using u_{pa}, u_{ta} and q_{pa}, q_{ta}, we obtain

$$2x_a^2\frac{1}{B}\frac{\partial B}{\partial\theta}f_{a0}P_2(\xi)\left(u_{pa}L_0^{(3/2)}(x_a^2) - \frac{2}{5}\frac{L_1^{(3/2)}(x_a^2)}{P_a}q_{pa}\right)$$
$$+\,2x_a^2\frac{1}{B}\frac{\partial B}{\partial\zeta}f_{a0}P_2(\xi)\left(u_{ta}L_0^{(3/2)}(x_a^2) - \frac{2}{5}\frac{L_1^{(3/2)}(x_a^2)}{P_a}q_{ta}\right)$$
$$= C_a(f_{a1}^{(2)}), \tag{7.199}$$

where $\xi = v_{\parallel}/v$ and $x_a = v/v_{ta}$. The solution $f_{a1}^{(2)}$ of (7.199) can also be found by expanding $f_{a1}^{(2)}$ in terms of Laguerre polynomials $L_j^{(5/2)}(x_a^2)$, as

$$f_{a1}^{(2)} = \tfrac{2}{3}\,x_a^2P_2(\xi)\sum_j P_{aj}L_j^{(5/2)}(x_a^2)f_{a0}, \tag{7.200}$$

and

$$P_{aj} = \frac{5\int d\mathbf{v}x_a^2L_j^{(5/2)}(x_a^2)f_{a1}^{(2)}P_2(\xi)}{n_a\{x_a^2(L_j^{(5/2)}(x_a^2))^2\}}, \tag{7.201}$$

where $\{A(x_a^2)\} \equiv (8/3\sqrt{\pi})\int_0^\infty dx_a\exp(-x_a^2)x_a^4A(x_a^2)$. It is noted that $L_0^{(5/2)}(x_a^2) = 1$, $L_1^{(5/2)}(x_a^2) = 7/2 - x_a^2, \cdots$. From (7.137), (7.138), and (7.200) we have $(P_{\parallel a} - P_{\perp a}) = P_aP_{a0}$ and $\Theta_{\parallel a} - \Theta_{\perp a} = (P_{a0} - 7/2P_{a1})P_a$. Thus, to calculate parallel viscosities we only need to know P_{a0} and P_{a1}. Substituting (7.200) into (7.199) to obtain P_{a0} and P_{a1} and utilizing the expressions (7.180) and (7.181), we obtain

$$\langle\mathbf{B}\cdot\nabla\cdot\overleftrightarrow{\Pi}_a\rangle = 3P_a\tau_a\left[\left\langle\hat{\mathbf{n}}\cdot\nabla B\frac{1}{B}\frac{\partial B}{\partial\theta}\right\rangle\left(\mu_{a1}u_{pa} + \tfrac{2}{5}\,\mu_{a2}\frac{q_{pa}}{P_a}\right)\right.$$
$$\left.+\left\langle\hat{\mathbf{n}}\cdot\nabla B\frac{1}{B}\frac{\partial B}{\partial\zeta}\right\rangle\left(\mu_{a1}u_{ta} + \tfrac{2}{5}\,\mu_{a2}\frac{q_{ta}}{P_a}\right)\right], \tag{7.202}$$

$$\langle \mathbf{B}\cdot\nabla\cdot\overleftrightarrow{\Theta}_a\rangle = 3P_a\tau_a\left[\left\langle \hat{\mathbf{n}}\cdot\nabla B\,\frac{1}{B}\frac{\partial B}{\partial\theta}\right\rangle\left(\mu_{a2}u_{pa}+\tfrac{2}{5}\,\mu_{a3}\,\frac{q_{pa}}{P_a}\right)\right.$$
$$\left.+\left\langle \hat{\mathbf{n}}\cdot\nabla B\,\frac{1}{B}\frac{\partial B}{\partial\zeta}\right\rangle\left(\mu_{a2}u_{ta}+\tfrac{2}{5}\,\mu_{a3}\,\frac{q_{ta}}{P_a}\right)\right], \tag{7.203}$$

where τ_a is the self-collision time. Since the RHS of (7.199) is proportional to the collision frequency ν_a, $\tau_a = 1/\nu_a$ appears in (7.202) and (7.203). The definition of μ_{aj}, for $j = 1, 2, 3$, is given for a simple electron–ion plasma: $\mu_{i1} = 1.365$, $\mu_{i2} = 2.31$, $\mu_{i3} = 8.78$, and $\mu_{e1} = 0.733$, $\mu_{e2} = 1.51$, $\mu_{e3} = 6.06$. Note that if we neglect the heat flux q_{pa} and q_{ta}, (7.202) is the same as (7.191) to within a factor of order unity.

The parallel plasma flows can be calculated from the surface-averaged parallel momentum and heat flux balance equations:

$$\langle B(F_{\|a1} + n_a e_a E_\|)\rangle = \langle B\cdot\nabla\cdot\overleftrightarrow{\Pi}_a\rangle \tag{7.204}$$
$$\langle BF_{\|a2}\rangle = \langle \mathbf{B}\cdot\nabla\cdot\overleftrightarrow{\Theta}_a\rangle, \tag{7.205}$$

where the parallel electric field $E_\|$ has been retained here to calculate the parallel current explicitly (compare with (7.135)). The parallel friction forces for a simple electron–ion plasma are shown as

$$F_{\|e1} = -F_{\|i1} = \ell^e_{11}(u_{\|i} - u_{\|e}) + \tfrac{2}{5}\ell^e_{12}(q_{\|e}/P_e) \tag{7.206}$$
$$F_{\|e2} = -\ell^e_{12}(u_{\|i} - u_{\|e}) + \tfrac{2}{5}\ell^e_{22}(q_{\|e}/P_e), \tag{7.207}$$
$$F_{\|i2} = -\tfrac{2}{5}\ell^i_{22}(q_{\|i}/P_i), \tag{7.208}$$

where $\ell^e_{11} = n_e m_e \nu_{ei}$, $\ell^e_{12} = 1.5\ell^e_{11}$, $\ell^e_{22} = 4.66\ell^e_{11}$, and $\ell^i_{22} = \sqrt{2}n_i m_i \nu_{ii}$.

Since $|BF_{\|e1}| \gg |\langle\mathbf{B}\cdot\nabla\cdot\overleftrightarrow{\Pi}_e\rangle|$, $|BF_{\|i2}| \gg |\langle\mathbf{B}\cdot\nabla\cdot\overleftrightarrow{\Theta}_i\rangle|$ and $|BF_{\|e2}| \gg |\langle\mathbf{B}\cdot\nabla\cdot\overleftrightarrow{\Theta}_e\rangle|$ for a large aspect ratio torus, (7.204)–(7.208) give

$$u_{\|i} = u_{\|e} = u_\|, \tag{7.209}$$
$$q_{\|e} = q_{\|i} = 0, \tag{7.210}$$

to the lowest order in $|\langle\mathbf{B}\cdot\nabla\cdot\overleftrightarrow{\Pi}_e\rangle|/|BF_{\|e1}|$, $|\langle\mathbf{B}\cdot\nabla\overleftrightarrow{\Theta}_i\rangle|/|BF_{\|i2}|$, or $|\langle\mathbf{B}\cdot\nabla\cdot\overleftrightarrow{\Theta}_e\rangle|/|BF_{\|e2}|$. Due to conservation of momentum, $F_{\|e1} = -F_{\|i1}$, and quasineutrality, (7.204) gives

$$\sum_{i,e}\langle\mathbf{B}\cdot\nabla\cdot\overleftrightarrow{\Pi}_a\rangle = 0. \tag{7.211}$$

Since $|\langle\mathbf{B}\cdot\nabla\cdot\overleftrightarrow{\Pi}_i\rangle|/|\mathbf{B}\cdot\nabla\cdot\overleftrightarrow{\Pi}_e| \sim \sqrt{m_i/m_e} \gg 1$, the lowest-order approximation of (7.211) is

$$\langle\mathbf{B}\cdot\nabla\cdot\overleftrightarrow{\Pi}_i\rangle = 0. \tag{7.212}$$

Next, we consider the $1/\nu$ regime of heliotrons. From the expressions for the ion-parallel viscosity, (7.169) and (7.212), we obtain the parallel ion flow speed in the $1/\nu$ regime:

$$\langle B u_{\|i}\rangle = -\langle G_{BS}\rangle_{1/\nu}\,\frac{T_i}{e}\left(\frac{P_i'}{P_i}+\frac{e\varphi'}{T_i}+\frac{\mu_{2i}}{\mu_{1i}}\frac{T_i'}{T_i}\right). \tag{7.213}$$

The particle and heat fluxes driven by the parallel viscosities for species a are

$$\Gamma^a_{bp} = -\frac{1}{e_a\chi'}\,\frac{\langle I\rangle\langle \mathbf{B}\cdot\nabla\cdot\overleftrightarrow{\Pi}_a\rangle}{\langle B^2\rangle}, \tag{7.214}$$

$$\frac{q^a_{bp}}{T_a} = -\frac{1}{e_a\chi'}\,\frac{\langle I\rangle\langle \mathbf{B}\cdot\nabla\cdot\overleftrightarrow{\Theta}_a\rangle}{\langle B^2\rangle}, \tag{7.215}$$

where $I = \mathbf{B}_t\cdot\nabla/\psi'$ (see (7.98), (7.109), and (7.114)). Substituting (7.213) into the expressions of $\langle \mathbf{B}\cdot\nabla\cdot\overleftrightarrow{\Pi}_a\rangle$ and $\langle \mathbf{B}\cdot\nabla\cdot\overleftrightarrow{\Theta}_a\rangle$ given by (7.169), we obtain

$$\Gamma^e_{bp} \simeq \Gamma^i_{bp} = -\frac{\langle I\rangle}{e^2\chi'}\,\nu_e m_e\,\frac{f_t}{f_c}\,\mu_{e1}\,\frac{\langle G_{BS}\rangle_{1/\nu}}{\langle B^2\rangle}\left(P' + \frac{\mu_{i2}}{\mu_{i1}}\,nT_i' + \frac{\mu_{e2}}{\mu_{e1}}\,nT_e'\right) \tag{7.216}$$

$$\frac{q^e_{bp}}{T_e} = -\frac{\langle I\rangle}{e^2\chi'}\,\nu_e m_e\,\frac{f_t}{f_c}\,\mu_{e1}\,\frac{\langle G_{BS}\rangle_{1/\nu}}{\langle B^2\rangle}\left(\frac{\mu_{e2}}{\mu_{e1}}\,P' + \frac{\mu_{e2}}{\mu_{e1}}\frac{\mu_{i2}}{\mu_{i1}}\,nT_i' + \frac{\mu_{e3}}{\mu_{e1}}\,nT_e'\right), \tag{7.217}$$

$$\frac{q^i_{bp}}{T_i} = -\frac{\langle I\rangle}{e^2\chi'}\,\nu_i m_i\,\frac{f_t}{f_c}\,\frac{\langle G_{BS}\rangle_{1/\nu}}{\langle B^2\rangle}\left(\mu_{i3} - \frac{(\mu_{i2})^2}{\mu_{i1}}\right) nT_i', \tag{7.218}$$

where $P' = P_e' + P_i'$ and $n = n_e = n_i$ is the plasma density. Here we have used (7.209) and (7.210). It should be noted that μ_{aj} in (7.169) is replaced with $\nu_a n_a m_a (f_t/f_c)\mu_{aj}$. We can also calculate the bootstrap current from the definition

$$J_{BS} = \frac{\sigma_s}{ne}\,\frac{B}{\langle B^2\rangle}\left(\langle \mathbf{B}\cdot\nabla\cdot\overleftrightarrow{\Pi}_e\rangle + \frac{\ell^e_{12}}{\ell^e_{22}}\,\langle \mathbf{B}\cdot\nabla\cdot\overleftrightarrow{\Theta}_e\rangle\right), \tag{7.219}$$

which gives

$$\begin{aligned} J_{BS} = &-2\frac{f_t}{f_c}\,\frac{\langle G_{BS}\rangle_{1/\nu}}{\langle B^2\rangle}\,B\mu_{e1}\left[\left(1+\frac{\mu_{e2}}{\mu_{e1}}\frac{\ell^e_{12}}{\ell^e_{22}}\right)P'\right. \\ &\left.+\left(1+\frac{\mu_{e2}}{\mu_{e1}}\frac{\ell^e_{12}}{\ell^e_{22}}\right)\frac{\mu_{i2}}{\mu_{i1}}\,nT_i' + \left(\frac{\mu_{e2}}{\mu_{e1}} + \frac{\mu_{e3}}{\mu_{e1}}\frac{\ell^e_{12}}{\ell^e_{22}}\right)nT_e'\right], \end{aligned} \tag{7.220}$$

where $\sigma_s = 2.0ne^2\tau_e/m_e$ is the Spitzer's conductivity. It is noted that in these expressions the radial electric field given by $-\varphi'$ disappears completely.

The bootstrap current is estimated for the LHD configuration by using $(f_t/f_c)\langle G_{BS}\rangle_{1/\nu}$, as shown in Fig. 7.4 and (7.220) (see Fig. 7.5). Since the collisonality does not belong to the $1/\nu$ regime in the whole of the LHD plasma, a bootstrap current of about 100 kA for the average beta $\langle\beta\rangle \simeq 1\%$ may be overestimated. Here, an iterative method is applied to obtain finite-beta equilibra with the bootstrap current for various pressure profiles shown in Fig. 7.5. It is noted that the bootstrap current has a tendency to increase the rotational transform of the LHD, which may change the MHD equilibrium property. In particular, the Shafranov shift is reduced and the associated reduction of the magnetic well region degrades the ideal MHD stability against pressure-driven interchange modes. It is also noted that $\langle G_{BS}\rangle_{1\nu}$ is directly related to the magnetic configuration described by $\mathbf{B}(V, \theta, \zeta)$. This fact suggests that the bootstrap current can be controlled by changing the magnetic configuration adequately. One such example for aiming at reduction of the bootstrap current is the Wenderstein 7-X (W7-X) device.

The modification of the parallel conductivity due to the trapped particle effect can be calculated approximately by considering the flux and friction balance equations, (7.204) and (7.205). For simplicity, we consider an electron–ion plasma. Since our interest is in the parallel conductivity, the diamagnetic terms in the parallel viscosities $A_{a1}(V)$ and $A_{a2}(V)$ in (7.169) are neglected. The appropriate parallel viscosities are written as

$$\langle \mathbf{B} \cdot \nabla \cdot \overleftrightarrow{\Pi}_a \rangle = n_a m_a \nu_a \frac{f_t}{f_c} \left(\mu_{a1} \langle u_{\parallel a} B \rangle - 2\mu_{a2} \langle q_{\parallel a} B \rangle / 5P_a \right), \tag{7.221}$$

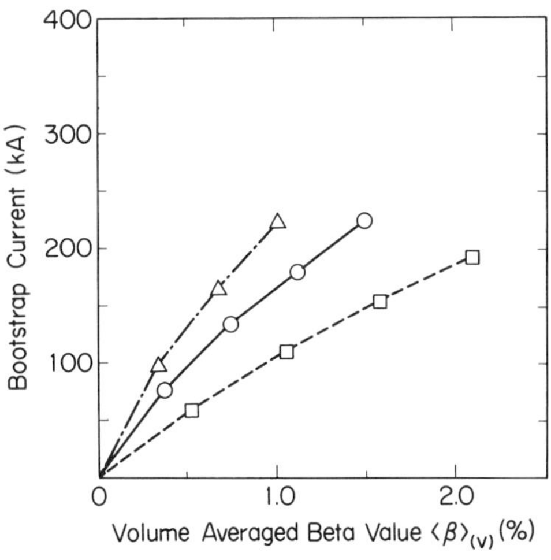

Fig. 7.5 The dependence of the total bootstrap current on the pressure profile. Circles correspond to $\langle\beta\rangle \sim \beta_0/3$, where $n = n_0(1-\psi)$ and $n_0 = 0.5 \times 10^{20}\,\mathrm{m}^{-3}$. Squares correspond to $\langle\beta\rangle \sim 0.45\beta_0$, where $n = n_0(1-\psi^3)$ and $n_0 = 0.65 \times 10^{20}\,\mathrm{m}^3$. Triangles correspond to $\langle\beta\rangle \sim \beta_0/4$, where $n = n_0(1-\psi)^2$ and $n_0 = 1.5 \times 10^{20}\,\mathrm{m}^{-3}$. The temperature profile is fixed at $T_i = T_e = T_0(1-\psi)$ and T_0 is varied according to $\langle\beta\rangle$. Here, β_0 is the central beta value and ψ is considered to be r^2 in the large aspect ratio limit.

$$\langle \mathbf{B} \cdot \nabla \cdot \overleftrightarrow{\Theta}_a \rangle = -n_a m_a \nu_a \frac{f_t}{f_c} (\mu_{a2} \langle u_{\parallel a} B \rangle - 2\mu_{a3} \langle q_{\parallel a} B \rangle / 5P_a). \tag{7.222}$$

In the large aspect ratio limit, we can again solve (7.204) and (7.205) by expanding them in terms of the small parallel parameter corresponding to $\langle \mathbf{B} \cdot \nabla \cdot \overleftrightarrow{\Pi}_e \rangle / |BF_{\parallel e1}| \sim \langle \mathbf{B} \cdot \nabla \cdot \overleftrightarrow{\Theta}_e \rangle / |BF_{\parallel e2}| \sim \langle \mathbf{B} \cdot \nabla \cdot \overleftrightarrow{\Theta}_i \rangle / |BF_{\parallel i2}| \ll 1$, and we obtain

$$\langle J_\parallel B \rangle_0 = \frac{n^2 e^2 \ell_{22}^e}{\ell_{11}^e \ell_{22}^e - (\ell_{12}^e)^2} \langle E_\parallel B \rangle \equiv \sigma_s \langle E_\parallel B \rangle, \tag{7.223}$$

in the lowest order, where $\sigma_s = n^2 e^2 \ell_{22}^e / [\ell_{11}^e \ell_{22}^e - (\ell_{12}^e)^2]$ is the Spitzer conductivity. To the next order, we have the modification due to trapped particles on $\langle J_\parallel B \rangle_0$,

$$\begin{aligned} \langle J_\parallel B \rangle_1 = & -\sigma_s \Bigg[\left(n m_e \nu_e \frac{f_t}{f_c} \frac{\mu_{e1}}{\ell_{11}^e} \right) \\ & \times \left(\frac{(\ell_{12}^e/\ell_{11}^e)(\mu_{e2}/\mu_{e1}) + \ell_{22}^e/\ell_{11}^e}{(\ell_{22}^e/\ell_{11}^e) - (\ell_{12}^e/\ell_{11}^e)^2} + \frac{\ell_{12}^e/\ell_{11}^e}{\ell_{22}^e/\ell_{11}^e} \right. \\ & \left. \times \frac{(\ell_{12}^e/\ell_{11}^e)(\mu_{e3}/\mu_{e1}) + (\ell_{22}^e/\ell_{11}^e)(\mu_{e2}/\mu_{e1})}{(\ell_{22}^e/\ell_{11}^e) - (\ell_{12}^e/\ell_{11}^e)^2} \right) \Bigg] \langle E_\parallel B \rangle. \end{aligned} \tag{7.224}$$

The terms in the brackets constitute the trapped particle effects on the parallel conductivity. For an electron–ion plasma, $\langle J_\parallel B \rangle_1 \simeq -\sigma_s (1.82\sqrt{\varepsilon}) \langle E_\parallel B \rangle$ is obtained for axisymmetric tokamaks.

In the large aspect ratio limit, we can approximate Hamada coordinates by cylindrical coordinates (r, θ, ζ). For symmetric tokamaks $\partial B / \partial \zeta = 0$ and $\langle G_{BS} \rangle = RrB_t/B_p$, where R is the major radius, r is the minor radius, and $B_t(B_p)$ is the toroidal (poloidal) magnetic field strength. This corresponds to (7.167) with $M = 0$. From (7.216) and (7.220), the ambipolar particle flux $(\Gamma_{bp}^e)_T$ and the bootstrap current $(J_{BS})_T$ are

$$(\Gamma_{bp}^e)_T = -\tfrac{1}{2} \nu_e \rho_{pe}^2 \frac{f_t}{f_c} \mu_{e1} (1 + T_i/T_e) \frac{dn}{dr} \tag{7.225}$$

and

$$(J_{BS})_T = -2 \frac{f_t}{f_c} \frac{\mu_{e1}}{B_p} \left[1 + \frac{\mu_{e2}}{\mu_{e1}} \frac{\ell_{12}^e}{\ell_{22}^e} \right] (T_e + T_i) \frac{dn}{dr}, \tag{7.226}$$

if the temperature gradients are neglected. Here ρ_{pe} is an electron poloidal Larmor radius, $\rho_{pe} = v_{Te}/(eB_p/m_e)$. These results are in agreement with the known results due to Hinton and Hazeltine (1976).

For a straight stellarator with $B = B_0[1 - \epsilon_h \cos(\ell\theta - m\zeta)]$, where ϵ_h is the helical modulation of the B field, $\langle G_{BS}\rangle \simeq -(\ell/mq) \times (rRB_t/B_p)$, where $mq \ll \ell$ is assumed. This corresponds to (7.167) with $L\iota \ll M$ and $I = 0$. The ambipolar particle flux $(\Gamma^e_{bp})_S$ and the bootstrap current $(J_{BS})_S$ are

$$(\Gamma^e_{bp})_S = -\tfrac{1}{2}\, \nu_e \rho_{pe}^2 \frac{f_t}{f_c}\, \mu_{e1}\left(1 + \frac{T_i}{T_e}\right)\frac{\ell}{mq}\frac{dn}{dr} \tag{7.227}$$

and

$$(J_{BS})_S = -2\frac{f_t}{f_c}\frac{\mu_{e1}}{B_p}\left[1 + \frac{\mu_{e2}}{\mu_{e1}}\frac{\ell^e_{12}}{\ell^e_{22}}\right]\left(-\frac{\ell}{mq}\right)(T_e + T_i)\frac{dn}{dr}. \tag{7.228}$$

If we set $\alpha = m/\ell$ and $q = (r/R_0)(B_0/B_p)$, where α is the helical pitch parameter for the helical angle variable $(\theta - \alpha z)$ in (r, θ, z) coordinates, the resulting $(\Gamma^e_{bp})_S$ and $(J_{BS})_S$ are again in agreement with the straight stellarator results, (7.61) and (7.65). We note that the bootstrap current in a straight stellarator is in the opposite direction to that in a tokamak.

7.7 ENERGY TRANSPORT EQUATIONS IN THE PRESENCE OF A RADIAL ELECTRIC FIELD

In order to study the evolution of electron and ion temperature profiles, usually energy transport equations are solved numerically for heliotrons and stellarators.

It is known that, in the energy transport equations, the term $\Gamma_r E_r$, which represents the energy exchange between the radial electric field E_r and the plasma through the radial particle flux Γ_r, does not appear explicitly for tokamaks. When the energy exchange is combined with the work done by both the friction force and the electric field on the toroidal plasma rotation, the energy transport equations can be expressed without the radial electric field. However, the energy exchange term with E_r appears explicitly in the energy transport equations for stellarators. In this section the reason why this term disappears in the energy transport equations for tokamaks, but appears in those for stellarators, will be discussed. Since it is common to compare the energy confinement properties of a stellarator with those of a similar tokamak, the differences in energy transport equations for these two systems should be clarified.

We first evaluate the energy exchange terms in the transport equations for nonaxisymmetric toroidal plasmas. To compare our results with the known tokamak results, we express the energy exchange quantities in terms of the poloidal flow, the pressure gradient, and the radial electric field. In this form, we find that, in addition to the poloidal flow and pressure gradient terms, there is a radial electric field term proportional to $\Gamma^a_{na}E_r$, where Γ^a_{na} is the nonaxisymmetric particle flux for species a. For tokamaks, $\Gamma^a_{na} = 0$ for both electrons and ions, and we can reproduce the usual tokamak results. However, for stellarators, we note that $\Gamma^a_{na} \neq 0$. When the energy exchange quantities are expressed in terms of the parallel flow, the pressure gradient, and the radial electric field, the radial electric field term is now proportional to $(\Gamma^a_{na} + \Gamma^a_{bp})E_r$, where Γ^a_{bp} is the banana-plateau flux for species a. The term that is proportional to the parallel flow is important for the cancellation of the $\Gamma^a_{bp}E_r$ term in the tokamak or the axisymmetric limit.

The energy transport equation for plasma species a is

$$\frac{\partial}{\partial t}\left(\tfrac{3}{2}P_a\right) + \nabla \cdot \mathbf{Q}_a = Q_a + \mathbf{u}_a \cdot (\mathbf{F}_{a1} + n_a e_a \mathbf{E}). \tag{7.229}$$

This equation is obtained from (3.130) and (3.136) by neglecting the energy source and the macroscopic flow energy. For simplicity, we discuss an electron–ion plasma. The energy transport equations for electrons and ions are

$$\frac{\partial}{\partial t}\left(\tfrac{3}{2}P_e\right) + \nabla \cdot \mathbf{Q}_e = -Q_i + \mathbf{J} \cdot \mathbf{E} - \mathbf{u}_i \cdot (\mathbf{F}_{i1} + n_i e_i \mathbf{E}), \tag{7.230}$$

$$\frac{\partial}{\partial t}\left(\tfrac{3}{2}P_i\right) + \nabla \cdot \mathbf{Q}_i = Q_i + \mathbf{u}_i \cdot (\mathbf{F}_{i1} + n_i e_i \mathbf{E}). \tag{7.231}$$

For a magnetically confined toroidal system, we can adopt a small gyroradius expansion to evaluate the energy exchange term $\mathbf{u}_i \cdot (\mathbf{F}_{i1} + n_i e_i \mathbf{E})$ in (7.230) and (7.231). To the accuracy of the second order in the gyroradius expansion, the surface-averaged quantity $\langle \mathbf{u}_i \cdot (\mathbf{F}_{i1} + n_i e_i \mathbf{E})\rangle$ is given by

$$\langle \mathbf{u}_i \cdot (\mathbf{F}_{i1} + n_i e_i \mathbf{E})\rangle = -e_i \langle n_i \mathbf{u}_{i2} \cdot \nabla V\rangle \frac{\partial \varphi}{\partial V} + \langle \mathbf{u}_{i1} \cdot (\mathbf{F}_{i1} + n_i e_i \mathbf{E})\rangle, \tag{7.232}$$

where $\varphi(V)$ is the electrostatic potential and the subscripts 1 and 2 in the ion flow velocity $\mathbf{u}_i$ denote the order of the gyroradius expansion. The Hamada coordinates (V, θ, ζ) are used here. The magnetic field $\mathbf{B}$ is $\mathbf{B} = \psi' \nabla V \times \nabla \theta - \chi' \nabla V \times \nabla \zeta$, with $\psi' = \mathbf{B} \cdot \nabla \zeta$ and $\chi' = \mathbf{B} \cdot \nabla \theta$. The first-order ion flow velocity $\mathbf{u}_{i1}$ is given by

$$\mathbf{u}_{i1} = (u_{\|i1}/B)\mathbf{B} + \mathbf{u}_{\perp i1}, \tag{7.233}$$

where $u_{\|i1}$ is the parallel component of $\mathbf{u}_{i1}$, and $\mathbf{u}_{\perp i1}$ includes both the diamagnetic flow and the $\mathbf{E} \times \mathbf{B}$ drift,

$$\mathbf{u}_{\perp i1} = \frac{\mathbf{B} \times \nabla V}{B^2} \frac{\partial \varphi}{\partial V} + \frac{\mathbf{B} \times \nabla V}{n_i m_i \Omega_i B} \frac{\partial P_i}{\partial V}. \tag{7.234}$$

With the vector identity $(B^2/\chi')\nabla V \times \nabla \theta = (\mathbf{B} \cdot \nabla V \times \nabla \theta)(\mathbf{B}/\chi') + \nabla V \times \mathbf{B}$, we can express $\mathbf{u}_{i1}$ as

$$\mathbf{u}_{i1} = \left(\frac{u_{pi}}{\chi'}\right)\mathbf{B} - \left(\frac{1}{B^2}\frac{\partial \varphi}{\partial V} + \frac{1}{n_i m_i \Omega_i B}\frac{\partial P_i}{\partial V}\right)\frac{B^2}{\chi'}(\nabla V \times \nabla \theta), \tag{7.235}$$

where $u_{pi} = \mathbf{u}_{i1} \cdot \nabla\theta$ is the poloidal ion flow velocity to be determined from the ion momentum balance equation. In deriving the vector identity, $|\nabla V|^2 = 1$ has been used. Also, the radial component of the lowest-order force balance equation $\mathbf{E} + \mathbf{u}_i \times \mathbf{B} - \nabla P_i/en_i = 0$, and $\mathbf{B} \simeq \mathbf{B}_t$ and $\mathbf{u}_{\|i} \simeq \mathbf{u}_{ti}$ have been used, where $\mathbf{B}_t$ and $\mathbf{u}_{ti}$ are the toroidal magnetic field and the toroidal flow velocity, respectively. Note that $\langle n_i \mathbf{u}_{i2} \cdot \nabla V \rangle = \Gamma^i$ is simply the ion radial particle flux and can be written as

$$\begin{aligned}\langle n_i \mathbf{u}_{i2} \cdot \nabla V \rangle = & -(1/e_i\chi')\langle \nabla V \times \nabla\theta \cdot (\mathbf{F}_{i1} + n_i e_i \mathbf{E})\rangle \\ & + (1/e_i\chi')\langle \nabla V \times \nabla\theta \cdot \nabla \cdot \overleftrightarrow{\Pi}_i \rangle. \end{aligned} \tag{7.236}$$

This is obtained from (7.89) by taking the scalar product with $(\nabla V \times \nabla\theta)$. Substituting (7.235) and (7.236) into (7.232) yields

$$\begin{aligned}\langle \mathbf{u}_i \cdot (\mathbf{F}_{i1} + n_i e_i \mathbf{E})\rangle = & (u_{pi}/\chi')\langle \mathbf{B} \cdot (\mathbf{F}_{i1} + n_i e_i \mathbf{E})\rangle \\ & - \frac{\partial P_i/\partial V}{n_i e_i \chi'} \langle \nabla V \times \nabla\theta \cdot (\mathbf{F}_{i1} + n_i e_i \mathbf{E})\rangle - e_i \Gamma^i_{na} \frac{\partial \varphi}{\partial V}, \end{aligned} \tag{7.237}$$

where $\Gamma^i_{na} = (1/e_i\chi')\langle \nabla V \times \nabla\theta \cdot \nabla \cdot \overleftrightarrow{\Pi}_i \rangle$ is the nonambipolar, nonaxisymmetric, ion particle flux. If we neglect the slow evolution of the magnetic flux surface and employ the flux–force relation, (7.237) can be written as

$$\langle \mathbf{u}_i \cdot (\mathbf{F}_{i1} + n_i e_i \mathbf{E})\rangle = -\frac{\langle B^2 \rangle}{\langle I \rangle} e_i \Gamma^i_{bp} u_{pi} + \frac{1}{n_i}(\Gamma^i_{bp} + \Gamma^i_{ps} + \Gamma^i_{c\ell}) \frac{\partial P_i}{\partial V} - e_i \Gamma^i_{na} \frac{\partial \varphi}{\partial V}, \tag{7.238}$$

where $\langle I \rangle = \langle \mathbf{B} \cdot \nabla V \times \nabla\theta \rangle$ and $\Gamma^i_{c\ell}$, Γ^i_{ps}, and Γ^i_{bp} are the ion classical, Pfirsch–Schlüter, and banana-plateau fluxes, respectively, which are already given by (7.93), (7.94), and (7.95).

We can transform (7.238) into the following form:

$$\langle \mathbf{u}_{i1} \cdot (\mathbf{F}_{i1} + n_i e_i \mathbf{E})\rangle = -\langle u_{\|i} B \rangle \frac{e_i \chi'}{\langle I \rangle} \Gamma^i_{bp} + \frac{1}{n_i}(\Gamma^i_{ps} + \Gamma^i_{c\ell}) \frac{\partial P_i}{\partial V} - e_i(\Gamma^i_{na} + \Gamma^i_{bp}) \frac{\partial \varphi}{\partial V}. \tag{7.239}$$

To obtain (7.239), we have expressed u_{pi} as

$$u_{pi} = \frac{\chi'}{\langle B^2 \rangle} \langle u_{\|i} B \rangle + \left(\frac{\partial \varphi}{\partial V} + \frac{1}{n_i e_i} \frac{\partial P_i}{\partial V} \right) \frac{\langle I \rangle}{\langle B^2 \rangle}, \tag{7.240}$$

which is given by the surface average of (7.235) after dot products with $\mathbf{B}$ are taken.

The consistent parallel flow $\langle u_{\parallel i}B\rangle$ and radial electric field $\partial\varphi/\partial V$ can be determined from the steady-state momentum balance equations within the flux surface,

$$\sum_a \langle \mathbf{B}\cdot\nabla\cdot\overleftrightarrow{\Pi}_a\rangle = 0, \tag{7.241}$$

as shown in section 7.6, and

$$\sum_a \Gamma^a_{na} = 0, \tag{7.242}$$

which is called the ambipolar condition. In the axisymmetric limit, (7.241) and (7.242) are not linearly independent, and only a relation between $\langle u_{\parallel i}B\rangle$ and $\partial\varphi/\partial V$ can be determined for given pressure and temperature gradients.

By imposing the ambipolar condition and the continuity equation in the stationary state, Kovrizhnykh (1989) discussed the sign of the radial electric field in the context of neoclassical transport for stellarators and heliotrons. The electron collisionality is assumed to be in the $1/\nu$ regime and the ion collisionality is assumed to be in the ν regime, where the transport coefficients are proportional to the collision frequency ν (see section 7.3). This transport appears in the rare-collisional regime less than the $1/\nu$ regime, where the super-banana orbits shown in chapter 6 contribute to the transport dominantly (see Fig. 7.3).

The nonaxisymmetric radial particle flux is approximately shown as

$$\Gamma^{na}_e = -D_e\left(\frac{n'_e}{n_e} + \eta\,\frac{E}{a}\right) \tag{7.243}$$

for electrons and

$$\Gamma^{na}_i = -\frac{D_i}{E^2+\delta^2}\left(\frac{n'_e}{n_e} - \frac{E}{a}\right) \tag{7.244}$$

for ions, where $E \equiv Z_i e a E_r/T_i$, $\eta \equiv T_i/T_e$, and δ is a small parameter. In (7.243) and (7.244), the particle flux proportional to the temperature gradient is neglected for simplicity. Also, the quasi-neutrality $n_e = Z_i n_i$ is assumed, where Z_i is a charge number of the ion. The time evolution of the density and radial electric field profiles is described using

$$\frac{\partial n_e}{\partial t} = -\frac{1}{r}\frac{\partial}{\partial r}\left[r(\Gamma_e - \Gamma^0_e)\right], \tag{7.245}$$

$$\epsilon_\perp \frac{\partial E}{\partial t} = -\frac{Z_i e^2 a}{T_i}\sum_{j=e,i} Z_j\Gamma_j, \tag{7.246}$$

where Γ_e^0 is the particle flux in the stationary state and $\epsilon_\perp$ is the perpendicular dielectric constant, given by

$$\epsilon_\perp = \left(1 + \frac{c^2}{V_A^2}\right)\epsilon_0, \tag{7.247}$$

where ϵ_0 is the dielectric constant of the vacuum, V_A is the Alfvén velocity, and c is the velocity of light. In the stationary state, an equation for determining E is given from (7.242),

$$E^2 - 2\hat{a}E - \hat{b} = 0, \tag{7.248}$$

where $\hat{a} = (1+\eta)D_i/(2a\Gamma_e^0)$ and $\hat{b} = D_i/D_e - \delta^2$. There are two solutions for E in (7.248), as shown in Fig. 7.6; however, only the smaller solution is stable with respect to the linear radial electric perturbation. In Fig. 7.6, if a perturbation to reduce the negative electric field is given for the negative E_r (left-hand side) solution satisfying (7.242), the ion flux becomes larger than the electron flux, which has a tendency to increase the negative electric field. If the opposite perturbation is given, the electron flux becomes larger than the ion flux and the negative electric field is reduced. These results mean that the negative E_r solution is linearly stable. By applying a similar argument, we may understand that the positive E_r (right-hand side) solution is unstable.

The physical mechanism to determine the radial electric field is as follows. If E_r is discontinuous in the radial direction, the particle flux also becomes discontinuous. However, the continuity equation (7.245) prevents such a discontinuous particle flux if the particle source term $(1/r)\partial(r\Gamma_e^0)/\partial r$ is continuous. This means that the continuous steady-state particle source requires that E_r is

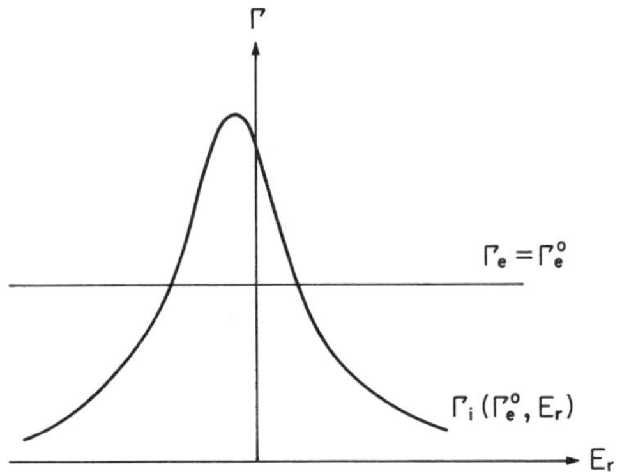

Fig. 7.6 Electron and ion particle fluxes as a function of E_r in the case of pure neoclassical diffusion in heliotrons. Here the fluxes are expressed using Γ_e^0 and E_r by eliminating n_e'/n_e.

continuous in the steady state. This also implies that it is difficult to obtain a large positive electric field from neoclassical transport. Even if the particle flux proportional to the temperature gradient is included, the above result does not change qualitatively.

In real plasmas confined in heliotrons, the particle flux includes the contribution driven by turbulence or by anomalous diffusion. Here we add the anomalous particle flux to Γ_e and Γ_i shown by (7.243) and (7.244), as

$$\Gamma_e = -D_e\left(\frac{n_e'}{n_e} + \eta\,\frac{E}{a}\right) - D^{an}\,\frac{n_e'}{n_e}, \tag{7.249}$$

$$\Gamma_i = -\frac{D_i}{E^2+\delta^2}\left(\frac{n_e'}{n_e} - \frac{E}{a}\right) - D^{an}\,\frac{n_e'}{n_e}, \tag{7.250}$$

by assuming that the anomalous diffusion coefficient is independent of the radial electric field. It is also assumed that the anomalous transport is intrinsically ambipolar. In the stationary state satisfying $\Gamma_e = \Gamma_e^0$ and $\Gamma_e = \Gamma_i$, an equation for determining E is given as

$$E^3 + \tilde{a}E^2 + \tilde{b}E + \tilde{c} = 0, \tag{7.251}$$

where $\tilde{a} = -\Gamma_e^0 a/(D^{an}\eta)$, $\tilde{b} = D_i/(D_e\eta) + (D_i/D^{an})(1/\eta + 1) + \delta^2$, and $\tilde{c} = a\Gamma_e^0(D_i/D_e - \delta^2)/(D^{an}\eta)$. It should be noted that the ambipolar condition and the continuity equation give a cubic algebraic equation instead of (7.248).

There is a case in which (7.251) has three roots for E, as shown in Fig. 7.7. Here Γ_i is shown after n_e'/n_e is eliminated using $\Gamma_e = \Gamma_e^0$. In this case the negative E_r (left-hand side) solution and the larger positive E_r (right-hand side) solution are stable with respect to the small radial electric field perturbation, while the smaller positive E_r (middle) solution is unstable. One interesting

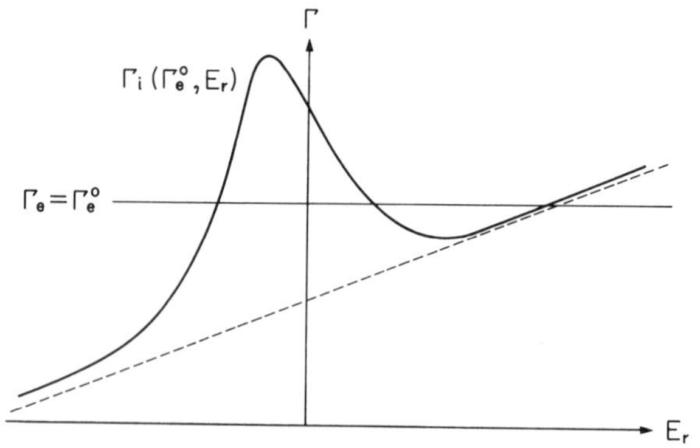

Fig. 7.7 Electron and ion particle fluxes as a function of E_r in the case in which both anomalous diffusion independent of E_r and neoclassical diffusion in heliotrons are included. Here the fluxes are expressed using Γ_e^0 and E_r by eliminating n_e'/n_e.

point is that (7.251) allows the existence of a large positive radial electric field. The reason why the intrinsically ambipolar anomalous particle flux affects the determination of E_r through (7.249) and (7.250) is that it changes the density profile in the stationary state.

The important consequence of the large radial electric field given by (7.251) is the significant reduction of the particle flux (see (7.250)). Since this result is not affected by including the particle flux proportional to the temperature gradient, it is also expected to reduce the energy flux by realizing the large positive electric field. In research into stellarators and heliotrons, the positive electric field is called the *electron root* and the negative electric field is called the *ion root*.

Finally, we discuss the poloidal flow damping due to the viscosity of the tokamak plasma in Hamada coordinates. The momentum evolution equation can be written as

$$nm_i \frac{\partial \mathbf{u}}{\partial t} = -\nabla(P_e + P_i) + \mathbf{J} \times \mathbf{B} + \nabla \cdot \overleftrightarrow{\Pi}. \tag{7.252}$$

For simplicity, we assume that the flow velocity is subsonic, so that we can neglect the convective nonlinearity or $(\mathbf{u} \cdot \nabla)\mathbf{u}$ term. We also neglect the electron contribution to the $nm_i \partial \mathbf{u}/\partial t$ and $\nabla \cdot \overleftrightarrow{\Pi}$ terms, because they are smaller than the ion contributions due to the factors of m_e/m_i and $\sqrt{m_e/m_i}$, respectively. For simplicity, the momentum source or sink associated with ion orbit loss is also neglected. The coordinates that we are interested in are the Hamada coordinates, in which the magnetic field is described by

$$\mathbf{B} = \psi' \nabla V \times \nabla \theta^H + \chi' \nabla \zeta^H \times \nabla V, \tag{7.253}$$

where $\psi' = \mathbf{B} \cdot \nabla \zeta^H$, $\chi' = \mathbf{B} \cdot \nabla \theta^H$, V is the volume enclosed by the flux surface, θ^H is the poloidal angle, and ζ^H is the toroidal angle. Both ψ' and χ' are functions of V only. The Jacobian is $\nabla V \times \nabla \theta^H . \nabla \zeta^H = 1$. In Hamada coordinates, an incompressible flow that satisfies the neoclassical approximations can be written as

$$\mathbf{u} = u_t \nabla V \times \nabla \theta^H + u_P \nabla \zeta^H \times \nabla V, \tag{7.254}$$

where $u_t = \mathbf{u} \cdot \nabla \zeta^H$ and $u_p = \mathbf{u} \cdot \nabla \theta^H$. Also, u_t and u_p are functions of V only; i.e., the flow velocity vector is a straight line in Hamada coordinates.

By taking the dot products of (7.252) with $\mathbf{B}_p^H = \chi' \nabla \zeta^H \times \nabla V$ and $\mathbf{B}_t^H = \psi' \nabla V \times \nabla \theta^H$ and surface averaging of the resultant equations, we obtain

$$nm_i \frac{\partial}{\partial t} \langle \mathbf{u} \cdot \mathbf{B}_p^H \rangle + \langle \mathbf{B}_p^H \cdot \nabla \cdot \overleftrightarrow{\Pi} \rangle = -\psi' \chi' \langle \mathbf{J} \cdot \nabla V \rangle, \tag{7.255}$$

$$nm_i \frac{\partial}{\partial t} \langle \mathbf{u} \cdot \mathbf{B}_t^H \rangle + \langle \mathbf{B}_t^H \cdot \nabla \cdot \overleftrightarrow{\Pi} \rangle = -\psi' \chi' \langle \mathbf{J} \cdot \nabla V \rangle. \tag{7.256}$$

In axisymmetric tokamaks, it is straightforward to show that $\langle \mathbf{B}_t^H \cdot \nabla \cdot \overleftrightarrow{\Pi} \rangle = 0$ in Hamada coordinates, since $\nabla \cdot \overleftrightarrow{\Pi}$ has no toroidal component. Then one concludes that

$$\langle \mathbf{B}_p^H \cdot \nabla \cdot \overleftrightarrow{\Pi} \rangle = \langle \mathbf{B} \cdot \nabla \cdot \overleftrightarrow{\Pi} \rangle. \tag{7.257}$$

To obtain an explicit expression for $\partial u_p/\partial t$, we adopt the large aspect ratio ordering so that ε, the inverse aspect ratio, is much less than unity; i.e., $\varepsilon \ll 1$. In this ordering the vectors $\mathbf{e}_\theta = \nabla \zeta^H \times \nabla V$ and $\mathbf{e}_\zeta = \nabla V \times \nabla \theta^H$ have the following approximate forms in (r, θ, ζ) coordinates:

$$\mathbf{e}_\theta = 2\pi r \left\{ (1 - \varepsilon \cos\theta)\hat{\theta} - \left[2\cos\theta + \frac{\varepsilon}{2}(1 - 2\cos^2\theta) \right] q\hat{\zeta} \right\}, \tag{7.258}$$

$$\mathbf{e}_\zeta = 2\pi R_0 (1 + \varepsilon \cos\theta)\hat{\zeta}, \tag{7.259}$$

where R_0 is the major radius, q is the safety factor, and $\varepsilon = r/R_0$. Note that $\mathbf{e}_\theta$ is nonorthogonal to $\mathbf{e}_\zeta$ and has an order of unity component in the $\hat{\zeta}$ direction.

By using (7.258) and (7.259) in (7.255) and (7.256), and eliminating the $\partial u_t/\partial t$ term, we find that

$$nm_i \chi' \left(\langle \mathbf{e}_\theta \cdot \mathbf{e}_\theta \rangle - \frac{\langle \mathbf{e}_\theta \cdot \mathbf{e}_\zeta \rangle^2}{\langle \mathbf{e}_\zeta \cdot \mathbf{e}_\zeta \rangle} \right) \frac{\partial u_P}{\partial t} + \chi' \left(\chi' \frac{\langle \mathbf{e}_\zeta \cdot \mathbf{e}_\theta \rangle}{\langle \mathbf{e}_\zeta \cdot \mathbf{e}_\zeta \rangle} + \psi' \right) \times$$

$$\times \langle \mathbf{J} \cdot \nabla V \rangle = -\langle \mathbf{B}_p^H \cdot \nabla \cdot \overleftrightarrow{\Pi} \rangle. \tag{7.260}$$

The leading-order contribution in the quantity $\langle \mathbf{e}_\theta \cdot \mathbf{e}_\theta \rangle - \langle \mathbf{e}_\theta \cdot \mathbf{e}_\zeta \rangle^2 / \langle \mathbf{e}_\zeta \cdot \mathbf{e}_\zeta \rangle$ is

$$\langle \mathbf{e}_\theta \cdot \mathbf{e}_\theta \rangle - \langle \mathbf{e}_\theta \cdot \mathbf{e}_\zeta \rangle^2 / \langle \mathbf{e}_\zeta \cdot \mathbf{e}_\zeta \rangle \simeq 4\pi^2 r^2 (1 + 2q^2). \tag{7.261}$$

By substituting (7.261) into (7.260) and recognizing that the $\langle \mathbf{J} \cdot \nabla V \rangle$ term showing an average radial current vanishes at the MHD equilibrium and is much smaller than the $\partial u_p/\partial t$ term, we have

$$nm_i \chi' 4\pi^2 r^2 (1 + 2q^2) \frac{\partial u_p}{\partial t} = -\langle \mathbf{B}_p^H \cdot \nabla \cdot \overleftrightarrow{\Pi} \rangle. \tag{7.262}$$

By noting that $\chi' = \mathbf{B} \cdot \nabla \theta^H = B_0/(2\pi q R_0)$ (see (4.182)), where B_0 is the magnetic field strength on the axis, we can write (7.262) as

$$nm_i \left(B_0^2 \frac{\varepsilon^2}{q^2} \right) (1 + 2q^2) \frac{\partial (u_p/\chi')}{\partial t} = -\langle \mathbf{B}_p^H \cdot \nabla \cdot \overleftrightarrow{\Pi} \rangle. \tag{7.263}$$

This equation shows that the poloidal flow u_p is damped by the poloidal viscosity given by $\langle \mathbf{B}_p^H \cdot \nabla \cdot \overleftrightarrow{\Pi} \rangle$ in tokamaks.

By taking the dot products of (7.252) with $\mathbf{B}$ and $\mathbf{B}_t^H$, we obtain an equation similar to (7.260), except that $\langle \mathbf{B}_p^H \cdot \nabla \cdot \overleftrightarrow{\Pi} \rangle$ is replaced by $\langle \mathbf{B} \cdot \nabla \cdot \overleftrightarrow{\Pi} \rangle$. However, because $\langle \mathbf{B}_p^H \cdot \nabla \cdot \overleftrightarrow{\Pi} \rangle = \langle \mathbf{B} \cdot \nabla \cdot \overleftrightarrow{\Pi} \rangle$, there is no difference in the physical meaning. Thus, in Hamada coordinates, the poloidal rotation is damped by either the poloidal viscosity or the parallel viscosity.

BIBLIOGRAPHY

Bickerton, R. J., Connor, J. W., and Taylor, J. B. (1971). Diffusion driven currents and bootstrap tokamak. *Nature, Physical Science*, **229**, 110.

Connor, J. W., and Hastie, R. J. (1974). Neoclassical diffusion in an $\ell = 3$ stellarator. *Phys. Fluids*, **17**, 114.

Coronado, M., and Wobig, H. (1986). Parallel and toroidal viscosity for nonaxisymmetric toroidal plasmas in the plateau regime. *Phys. Fluids*, **29**, 527.

Galeev, A. A., and Sagdeev, R. Z. (1968). Transport phenomena in a collisionless plasma in a toroidal magnetic system. *Sov. Phys.—JETP*, **26**, 233.

Hinton, F. L., and Hazeltine, R. D. (1973). Collision dominated plasma transport in toroidal confinement systems. *Phys. Fluids*, **16**, 1883.

Hinton, F. L., and Hazeltine, R. D. (1976). Theory of plasma transport. *Rev. Mod. Phys.*, **48**, 239.

Hinton, F. L., and Oberman, C. (1969). Electrical conductivity of plasma in a spatially inhomogeneous magnetic field. *Nucl. Fusion*, **9**, 319.

Kikuchi, M., and Azumi, M. (1995). Experimental evidence for the bootstrap current in a tokamak. *Plasma Phys. Contr. Fusion*, **37**, 1215.

Kovrizhnykh, L. M. (1984). Neoclassical theory of transport processes in toroidal magnetic confinement systems, with emphasis on non-axisymmetric configurations. *Nucl. Fusion*, **24**, 851.

Kovrizhnykh, L. M. (1989). Ambiguous solutions for the ambipolar electric field in stellarator. *Comments Plasma Phys. Contr. Fusion*, **8**, 85.

Morozov, A. I., and Solovév, L. S. (1966). The structure of magnetic fields, in *Reviews of plasma physics*. Consultants Bureau, New York, vol. 2, p. 1.

Mynick, H. E., Chu, T. K., and Boozer, A. H. (1982). Class of model stellarator fields with enhanced confinement. *Phys. Rev. Letters*, **48**, 322.

Nakajima, N., and Okamoto, M. (1992). Neoclassical flow, current, and rotation in general toroidal systems. *J. Phys. Soc. Japan*, **61**, 833.

Nührenberg, J., and Zille, R. (1988). Quasi-helically symmetric toroidal stellarators. *Phys. Letters*, **A129**, 113.

Pfirsch, D., and Schlüter, A. (1962). Der Einfluss der electrischen Leitfähigkeit auf das Gleichgewichtsverhalten von Plasmen niedrigen Drucks in Stellaratoren. *Report MPI/PA/7/62*. Max-Planck Institute.

Pytte, A., and Boozer, A. H. (1981). Neoclassical transport in helically symmetric plasmas. *Phys. Fluids*, **24**, 88.

Shaing, K. C., and Callen, J. D. (1983). Neoclassical flows and transport in nonaxisymmetric toroidal plasmas. *Phys. Fluids*, **26**, 3315.

Shaing, K. C., and Hoking, S. A. (1983). Neoclassical transport in a multiple-helicity torsatron in the low collisionality ($1/\nu$)-regime. *Phys. Fluids*, **26**, 2136.

Shaing, K. C., Hirshman, S. P., and Callen, J. D. (1986). Neoclassical transport fluxes in the plateau regime in nonaxisymmetric toroidal plasmas. *Phys. Fluids*, **26**, 521.

Shaing, K. C., Hirshman, S. P., and Tolliver, J. S. (1986). Parallel viscosity-driven neoclassical fluxes in the banana regime in nonsymmetric toroidal plasmas. *Phys. Fluids*, **29**, 2548.

Watanabe, K., Nakamura, Y., and Wakatani, M. (1991). Transport study for radial electric field and poloidal ion flow in Heliotron E. *J. Phys. Soc. Japan*, **60**, 884.

8

THE HEATING AND CONFINEMENT OF STELLARATOR AND HELIOTRON PLASMAS

8.1 INTRODUCTION

The Heliotron E was completed in 1980 to provide experimental proof of principles of the heliotron device. Table 8.1 shows the parameters of the heliotron devices built to investigate currentless toroidal plasma confinement in the 1980s; the Heliotron E (H-E), the Heliotron DR (H-DR), the ATF, and the CHS. In 1982, the H-E demonstrated that high-power ECRH (Electron Cyclotron Resonance Heating) using a gyrotron is the most suitable method for producing currentless plasmas. Recent development of the high-power and high-frequency gyrotron was remarkable and one gyrotron can produce an RF power output of about 500 kW at 140 GHz. The next target of research and development is a continuously working gyrotron. This is necessary for the next generation of experiments, such as the LHD and the W7-X. However, in a heliotron reactor, ECRH may be only required to produce an initial target plasma for NBI (Neutral Beam Injection) or ICRF (Ion Cyrotron Range of Frequency) heating, and pulsed operation of the gyrotron seems sufficient for ignition.

In the Heliotron E with a nominal magnetic field of 1.9 T, several gyrotrons with a frequency of 53 GHz and an output power of 200 kW were applied for the initial plasma production and subsequent ECR heating of currentless toroidal plasmas. In order to apply additional heating, such as NBI or ICRF

Table 8.1 Device Parameters of Heliotrons

	H–E	H–DR	ATF	CHS
Aspect ratio, $R/\bar{a}$	11	13	8	5
Major radius, R (m)	2.2	0.9	2.1	1.0
Minor radius, $\bar{a}$ (m)	0.2	0.07	0.27	0.2
Rotational transform				
$\iota(0)$	0.5	0.82	0.3	0.3
$\iota(a)$	2.5	2.2	1.0	1.0
Magnetic field, max. $B(T)$	1.9	0.6	2.0	2.0
Pitch number, M	19	15	12	8

heating, to the target plasma produced by ECRH, the plasma density was increased using gas-puffing. The ECRH-produced low-density plasmas were of $\bar{n}_e \sim 1 \times 10^{19}\,\mathrm{m}^{-3}$ $T_e(0) \sim 1\,\mathrm{keV}$, and $T_i(0) \sim 200\,\mathrm{eV}$, where $\bar{n}_e$ is an average density, and $T_e(0)\,(T_i(0))$ is a central electron (ion) temperature. On the other hand, ions were heated with NBI or ICRF, and $T_i(0) \gtrsim 1\,\mathrm{keV}$ was realized in the Heliotron E at $\bar{n}_e \sim 2.0$–$2.5 \times 10^{19}\,\mathrm{m}^{-3}$. The maximum NBI (ICRF) heating power was about 4 MW (2 MW) in the Heliotron E. The NBI heating also could produce high-beta plasmas in the high-density regime of $\bar{n}_e \gtrsim 6 \times 10^{19}\,\mathrm{m}^{-3}$ when the magnetic field was reduced to 0.95 T. The maximum average beta value in the Heliotron E was about 2%. The results of these heating experiments, including those in the Heliotron DR and the CHS, are shown in section 8.2.

By applying regression analysis to the experimental data for the Heliotron E, a scaling law of currentless plasma confinement was first presented. The obtained scaling is $\tau_E \propto \bar{n}_e^{\alpha 1} B^{\alpha 2} P_h^{\alpha 3}$, where $\alpha_1 \simeq 0.6$, $\alpha_2 \simeq 0.8$, and $\alpha_3 \simeq -0.5$. Here, $\bar{n}_e$ is an average density, B is a magnetic field strength at the magnetic axis, and P_h is a heating power. Recently, this scaling law was extended by including results from other stellarator and heliotron devices: it is then called LHD scaling, since it was obtained during the design study of the LHD. The scaling law of heliotron plasmas is discussed and compared to those for tokamak plasmas in section 8.3.

In the heliotron plasmas, particularly in the ATF and the CHS, the bootstrap current and plasma rotation were measured systematically. The ion temperature profile, and the poloidal and toroidal velocity profiles, were measured systematically with use of the charge exchange recombination spectroscopy in the CHS. These results are shown and compared with the theoretical expectations in section 8.4.

In the Heliotron E, pellet injection was successfully applied to increase the plasma density over a short period, with more peaked profiles than those by gas-puffing. Since there was no major disruption in the currentless heliotron plasmas, the rate of density increase could be made larger than that in tokamaks by using a larger-sized pellet or multiple-pellet injections. The results of pellet injection in the Heliotron E are explained in section 8.5.

Density fluctuations of heliotron plasmas were measured by using microwave scattering, CO_2-laser light scattering and the reciprocal Langmuir probe, and so on. In the edge plasma region, both the potential and density fluctuations were measured simultaneously using Langmuir probes. It was found that the density fluctuation, $\tilde{n}$, divided by the local background plasma density, $\bar{n}$, $\tilde{n}/\bar{n}$, increases from the central region to the edge region, and both the frequency spectrum and the wave number spectrum of the density fluctuations show some features of strong turbulence. Relations between the density and poloidal magnetic fluctuations were also investigated.

From theoretical studies of MHD instability, pressure-gradient driven turbulence may be considered as the origin of turbulence in heliotron devices, when the plasma beta exceeds $\beta(0) \gtrsim 1\%$, since the magnetic hill in the edge plasma region destabilizes both the ideal and resistive interchange modes. For lower-beta plasmas with $\beta(0) \lesssim 1\%$, the resistive interchange turbulence becomes a candidate for the anomalous transport.

In section 8.6, typical results of fluctuation measurements in heliotron devices are first explained. Then a theoretical model of pressure-gradient driven turbulence is given and the experimental properties of turbulence and anomalous transport in heliotron plasmas are compared with the pressure-gradient driven turbulence theory.

In section 8.7, theoretical approaches to evaluate the anomalous transport due to the pressure-gradient driven turbulence are explained; one is scale invariance and the other is mixing length theory. The former method gives the anomalous transport coefficient from the scale invariance constraint of the model equations for stellarator and heliotron devices. This approach is similar to dimensional analysis in fluid dynamics. The parameter dependence of the obtained transport coefficient is essentially the same as that given by mixing length theory, which is widely used to estimate the order of magnitude of the transport coefficent and the dependence of the transport coefficient on global parameters.

8.2 PLASMA HEATING IN HELIOTRON DEVICES

By applying various heating methods to Heliotron E plasmas, the maximum electron temperatures are shown in Fig. 8.1 as a function of the average density $\bar{n}_e$. In the low-density regime, $\bar{n}_e \lesssim 2 \times 10^{19}\,\mathrm{m}^{-3}$, high electron temperature plasmas with $T_e(0) \gtrsim 1\,\mathrm{keV}$ were obtained. When $\bar{n}_e$ increases, $T_e(0)$ decreases

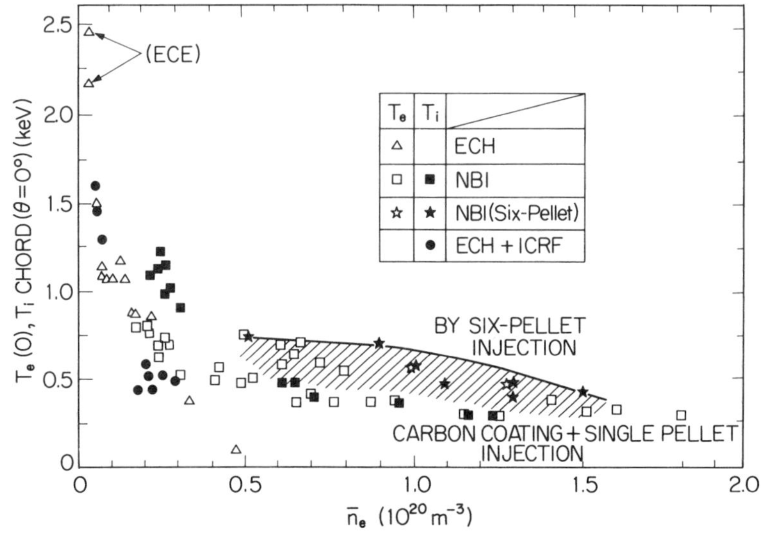

Fig. 8.1 The central electron temperature $T_e(0)$ by Thomson scattering and T_i along the central chord by a neutral particle energy analyzer versus the average density $\bar{n}_e$ in the Heliotron E currentless plasmas.

gradually. This is a general trend in plasma confinement experiments for a limited heating power. The maximum temperature is usually given by the transport process or the confinement time for the appropriate heating power (see section 8.3).

In the higher-density regime, $\bar{n}_e \gtrsim 2\text{–}3 \times 10^{19}\,\mathrm{m}^{-3}$, currentless plasmas were sustained by the neutral beam heating, since the electron cyclotron wave cannot propagate due to the cutoff, as shown later in this section. In the low-density NBI-heated plasma of $\bar{n}_e \simeq 2.5 \times 10^{19}\,\mathrm{m}^{-3}$, the central ion temperature, $T_i(0)$, increased up to about 1 keV and became $T_i(0) > T_e(0)$ with a heating power of about 3 MW and an injection energy of 26 keV. With a further increase in density, $T_i(0)$ decreases and $T_i(0) \simeq T_e(0) \sim 400\text{–}500\,\mathrm{eV}$ in the high-density regime of $\bar{n}_e \simeq 10^{20}\,\mathrm{m}^{-3}$. The maximum density is determined by the power balance, since there is no major disruption in heliotron devices. When the sum of the radiation loss and the transport loss may exceed the heating power, the plasma temperature decreases rapidly.

In order to study heating with electromagnetic waves, we must first analzye the wave propagation in the plasma column. Without appropriate wave propagation from the edge region to the central region, the heating efficiency is substantially degraded.

Ray tracing based on geometric optics is particularly useful for studying electromagnetic wave propagation in heliotron devices with three-dimensional magnetic field configurations. In geometric optics, the characteristic length of a spatial inhomogeneity, L, is assumed to be much larger than the wavelength:

$$L \gg 2\pi/|\mathbf{k}|, \tag{8.1}$$

where $\mathbf{k}$ is a wave number vector. The local dispersion relation in an inhomogeneous plasma is described as

$$G(\mathbf{k}, \omega, \mathbf{r}, t) = 0. \tag{8.2}$$

Here G is a smooth function with respect to the arguments. Along the ray trajectory, (8.2) can be written as

$$\delta G = \frac{\partial G}{\partial \mathbf{k}} \cdot \delta \mathbf{k} + \frac{\partial G}{\partial \omega} \delta \omega + \frac{\partial G}{\partial \mathbf{r}} \cdot \delta \mathbf{r} + \frac{\partial G}{\partial t} \delta t = 0. \tag{8.3}$$

For arbitrary $\delta \mathbf{r}$, δt, $\delta \mathbf{k}$, and $\delta \omega$, (8.3) is satisfied when

$$d\mathbf{r}/d\tau = \partial G/\partial \mathbf{k}, \tag{8.4}$$

$$d\mathbf{k}/d\tau = -\partial G/\partial \mathbf{r}, \tag{8.5}$$

$$dt/d\tau = -\partial G/\partial \omega, \tag{8.6}$$

$$d\omega/d\tau = \partial G/\partial t, \tag{8.7}$$

where τ is the parameter along the ray trajectory, which is defined by (8.6). From (8.4) and (8.6),

$$\frac{d\mathbf{r}}{dt} = \frac{d\mathbf{r}/d\tau}{dt/d\tau} = -\frac{\partial G/\partial \mathbf{k}}{\partial G/\partial \omega} = \frac{\partial \omega}{\partial \mathbf{k}} = \mathbf{v}_g \tag{8.8}$$

is obtained, where $\mathbf{v}_g$ is the group velocity. The spatial position of the ray is given by (8.2) and the variation of $\mathbf{k}$ along the ray trajectory is given by (8.5). When G is independent of t in the stationary plasma, (8.4) and (8.5) become the canonical form of the Hamiltonian dynamics.

According to the cold plasma wave theory, the plasma dispersion relation of an electromagnetic wave is shown as

$$\left| \overleftrightarrow{K} - n^2 \left(\overleftrightarrow{I} - \frac{\mathbf{nn}}{n^2} \right) \right| = 0, \tag{8.9}$$

where the refractive index vector is $\mathbf{n} = \mathbf{k}c/\omega$, $\mathbf{nn}$ is a tensor, and the dielectric tensor is

$$\overleftrightarrow{K} = \begin{pmatrix} S & -iD & 0 \\ iD & S & 0 \\ 0 & 0 & P \end{pmatrix}. \tag{8.10}$$

Here S, D, and P are Stix expressions:

$$S \equiv \tfrac{1}{2}(R + L), \tag{8.11}$$

$$D \equiv \tfrac{1}{2}(R - L), \tag{8.12}$$

$$R \equiv 1 - \sum_j \frac{\omega_{pj}^2}{\omega^2} \left(\frac{\omega}{\omega + \epsilon_j \omega_{cj}} \right), \tag{8.13}$$

$$L \equiv 1 - \sum_j \frac{\omega_{pj}^2}{\omega^2} \left(\frac{\omega}{\omega - \epsilon_j \omega_{cj}} \right), \tag{8.14}$$

$$P \equiv 1 - \sum_j \frac{\omega_{pj}^2}{\omega^2}, \tag{8.15}$$

where ω_{cj} and ω_{cj} are the cyclotron frequency and the plasma frequency of the jth species, respectively. Also, $\epsilon_i = 1$ and $\epsilon_e = -1$. For a high-frequency electromagnetic wave satisfying $\omega \gg \omega_{pi}$ and ω_{ci}, expressions (8.13)–(8.15) are approximated as $R = 1 - \alpha/(1 - \beta)$, $L = 1 - \alpha/(1 + \beta)$, and $P = 1 - \alpha$, where $\alpha = \omega_{pe}^2/\omega^2$ and $\beta = \omega_{ce}/\omega$. Then the dispersion relation (8.9) becomes

$$G = an_\perp^4 - bn_\perp^2 + c = 0, \tag{8.16}$$

where a, b, and c are given by

$$a \equiv S = 1 - \frac{\alpha}{1-\beta^2}, \tag{8.17}$$

$$\begin{aligned} b &\equiv RL + PS - (P+S)n_\parallel^2 \\ &= 2 - \alpha + \frac{2\alpha^2 - 3\alpha}{1-\beta^2} - \left(2 - \alpha - \frac{\alpha}{1-\beta^2}\right)n_\parallel^2, \end{aligned} \tag{8.18}$$

$$\begin{aligned} c &\equiv P(RL - 2Sn_\parallel^2 + n_\parallel^4) \\ &= (1-\alpha)\left\{1 + \frac{\alpha^2 - 2\alpha}{1-\beta^2} - 2\left(1 - \frac{\alpha}{1-\beta^2}\right)n_\parallel^2 + n_\parallel^4\right\}. \end{aligned} \tag{8.19}$$

Here $n_\parallel = k_\parallel c/\omega$, $n_\perp = k_\perp c/\omega$, and $n^2 - n_\parallel^2 + n_\perp^2$. For numerical calculations of the ray trajectory it is convenient to multiply (8.16) by $(1-\beta^2)$, and this is shown as

$$(1-\beta^2)G = An_\perp^4 - Bn_\perp^2 + C = 0, \tag{8.20}$$

where

$$A = 1 - \alpha - \beta^2, \tag{8.21}$$

$$B = 2(A - \alpha + \alpha^2) + \alpha\beta^2 - (2A + \alpha\beta^2)n_\parallel^2, \tag{8.22}$$

$$C = (1-\alpha)\{\alpha^2 - \alpha + A - 2An_\parallel^2 + (1-\beta^2)n_\parallel^4\}. \tag{8.23}$$

In the case of ECRH, since the wave absorption is very localized in the cyclotron resonance region, it is reasonable to use the cold plasma dispersion relation (8.20) in the ray equations (8.4)–(8.7).

In order to obtain energy absorbed by electrons, the dielectric tensor $\overleftrightarrow{K}$ including the wave–particle interaction for electrons is required. The Vlasov equation for electrons is written as

$$\frac{\partial f}{\partial t} + \mathbf{v}\cdot\frac{\partial f}{\partial \mathbf{r}} - \frac{e}{m}(\mathbf{E} + \mathbf{v}\times\mathbf{B})\cdot\frac{\partial f}{\partial \mathbf{v}} = 0. \tag{8.24}$$

The first-order perturbed distribution function f_1 is obtained by solving the linearized Vlasov equation

$$\left.\frac{df_1}{dt}\right|_0 = \frac{e}{m}(\mathbf{E}_1 + \mathbf{v}\times\mathbf{B}_1)\cdot\frac{\partial f_0}{\partial \mathbf{v}}, \tag{8.25}$$

where f_0 is the lowest equilibrium distribution function and $\mathbf{E}_1$ and $\mathbf{B}_1$ are the perturbed electric and magnetic fields, respectively. The LHS of (8.25) shows the variation of f_1 along the *unperturbed orbit* shown by $\mathbf{v}' = d\mathbf{r}'/dt'$ and $\mathbf{r}' = \mathbf{r}'(t')$. By integrating (8.25), f_1 is obtained as

$$f_1(\mathbf{r}, \mathbf{v}, t) = \frac{e}{m}\int_{-\infty}^{t}[\mathbf{E}_1(\mathbf{r}', t') + \mathbf{v}'\times\mathbf{B}_1(\mathbf{r}', t')]\cdot\frac{\partial f_0}{\partial \mathbf{v}'}\,dt'. \tag{8.26}$$

Here the mobility tensor is defined as

$$\langle \mathbf{v} \rangle = \int \mathbf{v} f_1 d\mathbf{v} = \frac{\overleftrightarrow{M}}{B_0} \mathbf{E}_1, \tag{8.27}$$

where B_0 is the lowest-order uniform magnetic field. We describe the perturbed fields $\mathbf{E}_1$ and $\mathbf{B}_1$ as

$$\begin{cases} \mathbf{E}_1 = \mathbf{E} \exp(i\mathbf{k} \cdot \mathbf{r}' - i\omega t') \\ \mathbf{B}_1 = \dfrac{\mathbf{k}}{\omega} \times \mathbf{E}_1. \end{cases} \tag{8.28}$$

Substituting (8.28) into (8.26) yields

$$f_1(\mathbf{r}, \mathbf{v}, t) = \frac{e}{m} \int_{-\infty}^{t} e^{i(\mathbf{k}\cdot\mathbf{r}' - \omega t')} \left(\mathbf{E} + \frac{\mathbf{E} \cdot \mathbf{v}'}{\omega} \mathbf{k} - \frac{\mathbf{v}' \cdot \mathbf{k}}{\omega} \mathbf{E} \right) \cdot \frac{\partial f_0(\mathbf{v}')}{\partial \mathbf{v}'} dt'. \tag{8.29}$$

The unperturbed orbit is given by solving the equation of motion

$$d\mathbf{v}'/dt' = -\omega_{ce}(\mathbf{v}' \times \hat{\mathbf{z}}), \tag{8.30}$$

where $\mathbf{B}_0 = B_0 \hat{\mathbf{z}}$ is assumed. With the conditions $\mathbf{r}' = \mathbf{r}$ and $\mathbf{v}' = \mathbf{v}$ at $t' = t$, the solution of (8.30) is given by

$$\begin{cases} v'_x = v_x \cos(\omega_{ce}\tau) + v_y \sin(\omega_{ce}\tau) \\ v'_y = -v_x \sin(\omega_{ce}\tau) + v_y \cos(\omega_{ce}\tau) \\ v'_z = v_z \end{cases} \tag{8.31}$$

and

$$\begin{cases} x' = -\dfrac{v_x}{\omega_{ce}} \sin(\omega_{ce}\tau) - \dfrac{v_y}{\omega_{ce}} (1 - \cos(\omega_{ce}\tau)) + x \\ y' = \dfrac{v_x}{\omega_{ce}} (1 - \cos(\omega_{ce}\tau)) - \dfrac{v_y}{\omega_{ce}} \sin(\omega_{ce}\tau) + y \\ z' = -v_z\tau + z, \end{cases} \tag{8.32}$$

where $\tau = t - t'$.

The lowest-order distribution function is given by

$$\frac{e}{m} (\mathbf{v} \times \mathbf{B}_0) \cdot \frac{\partial f_0}{\partial \mathbf{v}} = 0, \tag{8.33}$$

which gives $f_0 = f_0(v_\perp^2/2, v_z)$, where $v_\perp^2 = v_x^2 + v_y^2$. Thus we define $f_{0\perp}$ and f_{0z} in the following way: $\partial f_0/\partial v_x = v_x \partial f_0/\partial(v_\perp^2/2) \equiv v_x f_{0\perp}$ and $\partial f_0/\partial v_z \equiv f_{0z}$. Since $v_\perp^2$ and v_z are constants of motion, or $v_\perp^2 = v_\perp'^2$ and $v_z = v'_z$, f_0, $f_{0\perp}$, and f_{0z} are the same in both the Lagrange and Euler coordinates. By transforming (8.29) from the former (t') to the latter (τ) coordinates,

$$f_1(\mathbf{r}, \mathbf{v}, t) = \frac{e}{m} \int_0^{\infty} U\, e^V\, d\tau, \tag{8.34}$$

where

$$\begin{aligned} U = {} & E_x[v_x \cos(\omega_{ce}\tau) + v_y \sin(\omega_{ce}\tau)]\left[f_{0\perp} + \frac{k_z}{\omega}(f_{0z} - v_z f_{0\perp})\right] \\ & + E_y[-v_x \sin(\omega_{ce}\tau) + v_y \cos(\omega_{ce}\tau)]\left[f_{0\perp} + \frac{k_z}{\omega}(f_{0z} - v_z f_{0\perp})\right] \\ & + E_z\left[\frac{v_x}{\omega}\{-k_x \cos(\omega_{ce}\tau) + k_y \sin(\omega_{ce}\tau)\}\right. \\ & \left. + \frac{v_y}{\omega}\{-k_x \sin(\omega_{ce}\tau) - k_y \cos(\omega_{ce}\tau)\}\right](f_{0z} - v_z f_{0\perp}) + E_z f)_{0z}, \end{aligned} \tag{8.35}$$

and

$$\begin{aligned} V = {} & i\mathbf{k} \cdot \mathbf{r} - i\omega t + i\,\frac{v_x}{\omega_{ce}}[-k_x \sin(\omega_{ce}\tau) + k_y(1 - \cos(\omega_{ce}\tau))] \\ & + i\,\frac{v_y}{\omega_{ce}}[-k_y \sin(\omega_{ce}\tau) - k_x|(1 - \cos(\omega_{ce}\tau))] \\ & + iv_z(-k_z\tau) + i\omega\tau. \end{aligned} \tag{8.36}$$

In the following calculations, f_1 denotes the Fourier amplitude of $f_1(\mathbf{k}, \omega, \mathbf{v})$, by dropping $\exp(i\mathbf{k} \cdot \mathbf{r} - i\omega t)$.

In order to proceed with the calculations, we assume a Maxwellian dependence on $v_\perp^2/2$:

$$f_0(v_\perp^2/2, v_z) = \frac{m}{2\pi k_B T_\perp} h(v_z) \exp\left[-\frac{m(v_x^2 + v_y^2)}{2k_B T_\perp}\right]. \tag{8.37}$$

Here we define an average of an arbitrary function $g(\mathbf{v})$ with respect to v_x and v_y for the given perturbed distribution function (8.34), as

$$\langle g(\mathbf{v}) f_1 \rangle_\perp = \int_{-\infty}^{\infty} dv_x \int_{-\infty}^{\infty} dv_y g(\mathbf{v}) f_1. \tag{8.38}$$

Substituting (8.34) into (8.38) yields the zeroth- and first-order moments,

$$\langle f_1 \rangle_\perp = \frac{e}{m}\int_0^\infty d\tau e^\phi \left(\frac{i\alpha_1 a_1 k_B T_\perp}{m} + \frac{i\alpha_2 a_2 k_B T_\perp}{m} + \xi_z \right), \tag{8.39}$$

$$\langle v_x f_1 \rangle_\perp = \frac{e}{m}\int_0^\infty d\tau e^\phi \left[\alpha_1 \left(\frac{k_B T_\perp}{m} - \frac{a_1^2 k_B^2 T_\perp^2}{m^2} \right) - \frac{\alpha_2 a_1 a_2 k_B^2 T_\perp^2}{m^2} + \frac{i a_1 k_B T_\perp}{m} \xi_z \right], \tag{8.40}$$

$$\langle v_y f_1 \rangle_\perp = \frac{e}{m}\int_0^\infty d\tau e^\phi \left[-\frac{\alpha_1 a_1 a_2 k_B^2 T_\perp^2}{m^2} + \alpha_2 \left(\frac{k_B T_\perp}{m} - \frac{a_2^2 k_B^2 T_\perp^2}{m^2} \right) + \frac{i a_2 k_B T_\perp}{m} \xi_z \right], \tag{8.41}$$

where $\phi = -\lambda(1 - \cos(\omega_{ce}\tau)) - ik_z v_z \tau + i\omega\tau$ with $\lambda \equiv (k_x^2 + k_y^2) k_B T_\perp / m\omega_{ce}^2$ and

$$\begin{cases} a_1 \equiv \dfrac{1}{\omega_{ce}} [-k_x \sin(\omega_{ce}\tau) + k_y\{1 - \cos(\omega_{ce}\tau)\}] \\ a_2 \equiv \dfrac{1}{\omega_{ce}} [-k_y \sin(\omega_{ce}\tau) - k_x\{1 - \cos(\omega_{ce}\tau)\}], \end{cases} \tag{8.42}$$

$$\begin{cases} \alpha_1 \equiv \xi_x \cos(\omega_{ce}\tau) - \xi_y \sin(\omega_{ce}\tau) \\ \alpha_2 \equiv \xi_x \sin(\omega_{ce}\tau) + \xi_y \cos(\omega_{ce}\tau), \end{cases} \tag{8.43}$$

$$\begin{cases} \xi_z \equiv -\dfrac{mh(v_z)E_x}{k_B T_\perp} + \left(\dfrac{dh(v_z)}{dv_z} + \dfrac{mv_z h(v_z)}{k_B T_\perp} \right) \left(\dfrac{k_z E_x}{\omega} - \dfrac{k_x E_z}{\omega} \right) \\ \xi_y \equiv -\dfrac{mh(v_z)E_y}{k_B T_\perp} + \left(\dfrac{dh(v_z)}{dv_z} + \dfrac{mv_z h(v_z)}{k_B T_\perp} \right) \left(\dfrac{k_z E_y}{\omega} - \dfrac{k_y E_z}{\omega} \right) \\ \xi_z \equiv E_z \dfrac{dh(v_z)}{dv_z}. \end{cases} \tag{8.44}$$

It is noted that the parameter λ is related to the ratio of Larmor radius to perpendicular wavelength in the uniform magnetic field.

We introduce a coordinate transformation from the orthogonal coordinates to the local rotational coordinates as

$$\begin{pmatrix} q+ \\ q- \\ q_z \end{pmatrix} = \overleftrightarrow{R} \begin{pmatrix} q_x \\ q_y \\ q_z \end{pmatrix}, \quad \begin{pmatrix} q_z \\ q_y \\ q_z \end{pmatrix} = \overleftrightarrow{R}^{-1} \begin{pmatrix} q+ \\ q- \\ q_z \end{pmatrix}, \tag{8.45}$$

where $\mathbf{q} = (q_x, q_y, q_z)$ is an arbitrary vector and

$$\overleftrightarrow{R} = \begin{pmatrix} 1 & -i & 0 \\ 1 & i & 0 \\ 0 & 0 & 1 \end{pmatrix}, \quad \overleftrightarrow{R}^{-1} = \begin{pmatrix} \frac{1}{2} & \frac{1}{2} & 0 \\ -\dfrac{1}{2i} & \dfrac{1}{2i} & 0 \\ 0 & 0 & 1 \end{pmatrix}. \tag{8.46}$$

The transformation (8.45) is applied to **v**, **k**, and $\mathbf{E}_1$ here. Also, the equality

$$\exp[\lambda \cos(\omega_{ce}\tau)] = \sum_{n=-\infty}^{\infty} I_n(\lambda) \exp(in\omega_{ce}\tau) \tag{8.47}$$

is introduced, with the modified Bessel function $I_n(\lambda)$. With the coordinate transformation (8.45) and (8.47), the expressions (8.39)–(8.41) are written as

$$\begin{aligned}\langle f_1 \rangle_\perp &= \frac{e}{m} \exp(-\lambda) \sum_{n=-\infty}^{\infty} \int_0^\infty d\tau \exp i(n\omega_{ce}\tau - k_z v_z \tau + \omega\tau) \times \\ &\times \left[\frac{\xi_+ k_- k_B T_\perp}{2m\omega_{ce}} (I_{n+1} - I_n) + \frac{\xi_- k_+ k_B T_\perp}{2m\omega_{ce}} (I_n - I_{n-1}) + \xi_z I_n \right],\end{aligned} \tag{8.48}$$

$$\begin{aligned}\langle v_\pm f_1 \rangle_\perp &= \frac{e}{m} \exp(-\lambda) \sum_{n=-\infty}^{\infty} \int_0^\infty d\tau \exp i(n\omega_{ce}\tau - k_z v_z \tau + \omega\tau) \times \\ &\times \left[\frac{\xi_\pm k_B T_\perp}{m} I_{n\pm1} + \frac{\xi_\pm k_+ k_- k_B^2 T_\perp^2}{2\omega_{ce}^2 m^2} (I_n - 2I_{n\pm1} + I_{n\pm2}) \right. \\ &\left. + \frac{\xi_\mp k_\pm^2 k_B^2 T_\perp^2}{2\omega_{ce}^2 m^2} (I_{n+1} - 2I_n + I_{n-1}) \mp \frac{\xi_z k_\pm k_B T_\perp}{\omega_{ce} m} (I_n - I_{n\pm1}) \right],\end{aligned} \tag{8.49}$$

where $v_\pm = v_x \mp iv_y$, $k_\pm = k_x \mp ik_y$, and $\xi_\pm = \xi_x \mp i\xi_y$.

For the parallel velocity distribution function $h(v_z)$, the macroscopic parallel velocity V_z is included as a shifted Maxwellian:

$$h(v_z) = \left(\frac{m}{2\pi k_B T_\parallel} \right)^{1/2} \exp\left[-\frac{m(v_z - V_z)^2}{2k_B T_\parallel} \right]. \tag{8.50}$$

By introducing $Q(v_z)$ to represent $\xi_\pm$, ξ_z, $v_z\xi_\pm$, and $v_z\xi_z$, the integral

$$\langle Q(v_z) \rangle_n \equiv k_z \int_{-\infty}^{\infty} dv_z Q(v_z) \int_0^\infty d\tau \exp i(n\omega_{ce}\tau - k_z v_z \tau + \omega\tau) \tag{8.51}$$

is required to calculate (8.48) and (8.49). Performing the integration with respect to v_z yields

$$\langle \xi_\pm \rangle_n = 2\left(\frac{m}{2k_B T_\parallel} \right)^{3/2} \left[-\frac{T_\parallel E_\pm}{T_\perp} F_0 + \left(\frac{k_z E_\pm}{\omega} - \frac{k_\pm E_z}{\omega} \right) \left(V_z F_0 - F_1 + \frac{T_\parallel}{T_\perp} F_1 \right) \right], \tag{8.52}$$

$$\langle v_z\xi_\pm\rangle_n = 2\left(\frac{m}{2k_BT_\parallel}\right)^{3/2}\left[-\frac{T_\parallel E_\pm}{T_\perp}F_1 + \left(\frac{k_zE_\pm}{\omega} - \frac{k_\pm E_z}{\omega}\right)\left(V_zF_1 - F_2 + \frac{T_\parallel}{T_\perp}F_2\right)\right], \tag{8.53}$$

$$\langle \xi_z\rangle_n = 2\left(\frac{m}{2k_BT_\parallel}\right)^{3/2} E_z(-F_1 + V_zF_0), \tag{8.54}$$

$$\langle v_z\xi_z\rangle_n = 2\left(\frac{m}{2k_BT_\parallel}\right)^{3/2} E_z(-F_2 + V_zF_1), \tag{8.55}$$

where $E_\pm = E_x \mp iE_y$ and $k_\pm = k_x \mp ik_y$. In (8.52)–(8.55),

$$F_0 = \frac{\sqrt{\pi}k_z}{|k_z|}\exp(-\zeta_n^2) - \frac{i}{\sqrt{\pi}}P\int_{-\infty}^{\infty}\frac{e^{-z^2}}{z-\zeta_n}\,dz \equiv iZ(\zeta_n), \tag{8.56}$$

$$F_1 = -i\left(\frac{2k_BT_\parallel}{m}\right)^{1/2} + \left(\frac{\omega + n\omega_{ce}}{k_z}\right)F_0, \tag{8.57}$$

$$F_2 = -\frac{i(\omega + k_zV_z + n\omega_{ce})}{k_z}\left(\frac{2k_BT_\parallel}{m}\right)^{1/2} + \left(\frac{\omega + n\omega_{ce}}{k_z}\right)^2 F_0, \tag{8.58}$$

which are obtained from

$$F_p = \frac{k_z}{\sqrt{\pi}}\int_{-\infty}^{\infty} dv_z \int_0^{\infty} d\tau v_z^p \exp\left[-\frac{m(v_z - V_z)^2}{2k_BT_\parallel} + i\tau(n\omega_{ce} + \omega - k_zv_z)\right]. \tag{8.59}$$

for $I_m\omega > 0$, where

$$\zeta_n \equiv \frac{\omega - k_zV_z + n\omega_{ce}}{k_z}\left(\frac{m}{2k_BT_\parallel}\right)^{1/2} \tag{8.60}$$

and $Z(\zeta_n)$ is the plasma dispersion function in (8.56).

We can now calculate the mobility tensor (8.27), M_{+-z}, in the local rotational coordinates as shown by Stix (1962) using (8.48), (8.49), (8.52)–(8.55) and the following quantities:

$$\Theta \equiv -\frac{mh(v_z)}{k_BT_\perp} + \frac{k_z}{\omega}\left(\frac{dh(v_z)}{dv_z} + \frac{mv_zh(v_z)}{k_BT_\perp}\right), \tag{8.61}$$

$$\Phi \equiv -\frac{\omega_{ce}}{\omega}\left(\frac{dh(v_z)}{dv_z} + \frac{mv_zh(v_z)}{k_BT_\perp}\right), \tag{8.62}$$

$$\Psi \equiv \frac{dh(v_z)}{dv_z}. \tag{8.63}$$

Here we note that $\Phi = 0$ when an isotropic temperature and no parallel flow along the magnetic field are assumed for the Maxwellian. Then the mobility tensor in orthogonal coordinates can be obtained from M_{+-z} by using

$$M_{xyz} = \overleftrightarrow{R}^{-1} \overleftrightarrow{M}_{+-z} \overleftrightarrow{R}. \tag{8.64}$$

It can be shown by inspection that, for example, M_{zz} is proportional to $\sum_{n=-\infty}^{\infty} \langle v_z \Psi \rangle_n I_n$, from (8.27), (8.44), (8.48), (8.51), and (8.63). Then the components of the mobility tensor $\overleftrightarrow{M}_{xyz}$ are shown as

$$M_{xx} = \frac{\omega_{ce} e^{-\lambda} k_B T}{m k_z} \sum_{n=-\infty}^{\infty} \langle \Theta \rangle_n \frac{n^2}{\lambda} I_n, \tag{8.65}$$

$$M_{yy} = \frac{\omega_{ce} e^{-\lambda} k_B T}{m k_z} \sum_{n=-\infty}^{\infty} \langle \Theta \rangle_n \left(\frac{n^2}{\lambda} I_n + 2\lambda I_n - 2 \frac{dI_n}{d\lambda} \right), \tag{8.66}$$

$$M_{zz} = \frac{\omega_{ce} e^{-\lambda}}{k_z} \sum_{n=-\infty}^{\infty} \langle v_z \Psi \rangle_n I_n, \tag{8.67}$$

$$M_{xy} = -M_{yx} = \frac{\omega_{ce} e^{-\lambda} k_B T}{m k_z} \sum_{n=-\infty}^{\infty} i \langle \Theta \rangle_n n \left(I_n - \frac{dI_n}{d\lambda} \right), \tag{8.68}$$

$$M_{xz} = M_{zx} = -\frac{e^{-\lambda} k_B T}{m k_z} \sum_{n=-\infty}^{\infty} \langle \Psi \rangle_n \frac{n k_x}{\lambda} I_n, \tag{8.69}$$

$$M_{yz} = -M_{zy} = \frac{e^{-\lambda} k_B T}{m k_z} \sum_{n=-\infty}^{\infty} i \langle \Psi \rangle_n k_x \left(I_n - \frac{dI_n}{d\lambda} \right), \tag{8.70}$$

and $T = T_{\parallel} = T_{\perp}$ and $V_z = 0$ are assumed. It is also noted that $k_y = 0$ can be assumed in orthogonal coordinates. In the components of the mobility tensor $\overleftrightarrow{M}_{xyz}$, the following relations are obtained by using (8.56)–(8.58):

$$\langle \Theta \rangle_n = 2i \left(\frac{m}{2k_B T} \right)^{3/2} Z(\zeta_n), \tag{8.71}$$

$$\langle \Psi \rangle_n = \frac{2i}{k_z} \left(\frac{m}{2k_B T} \right)^{3/2} \left[k_z \left(\frac{2k_B T}{m} \right)^{1/2} + (\omega + n\omega_{ce}) Z(\zeta_n) \right], \tag{8.72}$$

$$\langle v_z \Psi \rangle_n = \frac{2i}{k_z} \left(\frac{m}{2k_B T} \right)^{3/2} (\omega + n\omega_{ce}) \left[k_z \left(\frac{2k_B T}{m} \right)^{1/2} + (\omega + n\omega_{ce}) Z(\zeta_n) \right], \tag{8.73}$$

where $Z(\zeta_n)$ is a plasma dispersion relation with an argument of $\zeta_n = ((\omega + n\omega_{ce})/k_z)(m/2\pi k_B T)^{1/2}$.

In the case of $\lambda \ll 1$ or a small Larmor radius limit, $\overleftrightarrow{M}$ is shown in the following way:

$$M_{xx} = \frac{\omega_{ce} k_B T}{2mk_z} (\langle\Theta\rangle_1 + \langle\Theta\rangle_{-1} - \lambda\langle\Theta\rangle_1 - \lambda\langle\Theta\rangle_{-1} + \lambda\langle\Theta\rangle_2 + \lambda\langle\Theta\rangle_{-2}), \tag{8.74}$$

$$M_{yy} = \frac{\omega_{ce} k_B T}{2mk_z} (\langle\Theta\rangle_1 + \langle\Theta\rangle_{-1} + 4\lambda\langle\Theta\rangle_0 - 3\lambda\langle\Theta\rangle_1 - 3\lambda\langle\Theta\rangle_{-1} + \lambda\langle\Theta\rangle_2 + \lambda\langle\Theta\rangle_{-2}), \tag{8.75}$$

$$M_{zz} = \frac{\omega_{ce}}{2k_z} (2\langle v_z\Psi\rangle_0 - 2\lambda\langle v_z\Psi\rangle_0 + \lambda\langle v_z\Psi\rangle_1 + \lambda\langle v_z\Psi\rangle_{-1}), \tag{8.76}$$

$$M_{xy} = -M_{yx} = \frac{i\omega_{ce} k_B T}{2mk_z} (\langle\Theta\rangle_1 - \langle\Theta\rangle_{-1} - 2\lambda\langle\Theta\rangle_1 + 2\lambda\langle\Theta\rangle_{-1} + \lambda\langle\Theta\rangle_2 - \lambda\langle\Theta\rangle_{-2}), \tag{8.77}$$

$$M_{xz} = -M_{zx} = -\frac{k_B T}{2mk_z} k_x(-\langle\Psi\rangle_1 + \langle\Psi\rangle_{-1}), \tag{8.78}$$

$$M_{yz} = -M_{zy} = \frac{ik_B T}{2mk_z} k_x(2\langle\Psi\rangle_0 - \langle\Psi\rangle_1 - \langle\Psi\rangle_{-1}). \tag{8.79}$$

When the mobility tensor $\overleftrightarrow{M}$ is given, it is easy to calculate the dielectric tensor $\overleftrightarrow{K}$ using

$$K = 1 + i\sum_j \frac{\omega_{pj}^2}{\omega_{cj}} \overleftrightarrow{M}^{(j)}. \tag{8.80}$$

In (8.80), both electrons and ions are included; however, in a high-density plasma with a Maxwellian velocity distribution function,

$$\begin{cases} \dfrac{c\omega_{pe}^2}{v_{Te}\omega_{ce}^2} \gg 1 \\ |\zeta_\ell| = \dfrac{\omega - \ell\omega_{ce}}{k_z v_{Te}} \gg 1 \ (\ell \neq 1) \\ \lambda \ll 1 \end{cases} \tag{8.81}$$

can be assumed for the electron cyclotron wave. Under the assumption of (8.81), the dielectric tensor (8.80) may be approximated as

$$K_{xx} = 1 - \frac{1}{2}\frac{\alpha}{1+\beta} + i\sigma(1-\lambda), \tag{8.82}$$

$$K_{yy} = 1 - \frac{1}{2}\frac{\alpha}{1+\beta} + i\sigma(1-3\lambda), \tag{8.83}$$

$$K_{zz} = 1 - \alpha + \alpha\zeta_0\lambda\zeta_1\{1 + \zeta_1 Z(\zeta_1)\}, \tag{8.84}$$

$$K_{xy} = -K_{yx} = -\frac{i}{2}\frac{\alpha}{1+\beta} + \sigma(1 - 2\lambda), \tag{8.85}$$

$$K_{xz} = K_{zx} = \frac{1}{2}\frac{\alpha}{\beta}\tan\theta\{1 + \zeta_1 Z(\zeta_1)\}, \tag{8.86}$$

$$K_{zy} = -K_{yz} = iK_{xz}, \tag{8.87}$$

where

$$\sigma = -\frac{i}{2}\,\alpha\zeta_0 Z(\zeta_1) \tag{8.88}$$

In (8.88), $Z(\zeta_1)$ is the plasma dispersion function defined by (8.56). Substituting (8.82)–(8.87) into the dispersion relation (8.9) yields

$$A_0 n^4 + B_0 n^2 + C_0 = 0 \tag{8.89}$$

in the coordinates with $\mathbf{k} = (k\sin\theta, 0, k\cos\theta)$ or $\mathbf{n} = (n\sin\theta, 0, n\cos\theta)$. In (8.89), A_0, B_0, and C_0 are given by

$$A_0 = K_{xx}\sin^2\theta + 2K_{xz}\sin\theta\cos\theta + K_{zz}\cos^2\theta, \tag{8.90}$$

$$\begin{aligned} B_0 = {} & -K_{xx}K_{zz} - (k_{xx}K_{zz} - K_{xz}^2)\cos^2\theta - (K_{xx}^2 + K_{xy}^2)\sin^2\theta \\ & + 2K_{xz}(iK_{xy} - K_{yy})\cos\theta\sin\theta + K_{xz}^2, \end{aligned} \tag{8.91}$$

$$C_0 = K_{zz}(K_{xx}^2 + K_{xy}^2) - 2K_{xx}K_{xz}^2 + 2iK_{xy}K_{xz}^2. \tag{8.92}$$

It is known that there are two polarizations in the electron cyclotron wave: one is the ordinary wave (or O-mode), with the polarized electric field parallel to the magnetic field line; and the other is the extraordinary wave (or X-mode), with the polarized electric field perpendicular to the magnetic field line.

Under the assumption of (8.81) and $\omega \simeq \omega_{ce}$, $\sigma \gg 1$ can be used in (8.82)–(8.87). Then the dispersion relation (8.89) is solved as

$$n_\pm^2 = \frac{1}{\sin^2\theta}\{1 + \tfrac{1}{2}\sin^2\theta - \alpha \pm [(1 - \tfrac{1}{2}\sin^2\theta - \alpha)^2 - (1-\alpha)(2-\alpha)\sin^2\theta]^{1/2}\}, \tag{8.93}$$

where θ is the angle between $\mathbf{B}_0 = B_0\hat{z}$ and the direction of electromagnetic wave propagation. In (8.93), $\alpha \simeq \omega_{pe}^2/\omega_{ce}^2$: the plus sign corresponds to the X-mode and the minus sign corresponds to the O-mode. Equation (8.93) is obtained in the following way. For $\lambda \simeq 0$, $\alpha = \omega_{pe}^2/\omega^2 \simeq \omega_{pc}^2/\omega_{ce}^2$, $\beta = \omega_{ce}/\omega \simeq 1$, and the dielectric tensor becomes $K_{xx} = K_{yy} = 1 - \alpha/4 + i\sigma$, $K_{zz} = 1 - \alpha$, $K_{xy} = -K_{yx} = \sigma - i\alpha/4$, $K_{xz} = K_{zx} = \alpha\tan\theta\{1 + \zeta_1 Z(\zeta_1)\}/2$, and $K_{yz} = -K_{zy} = iK_{xz}$, where $\zeta_1 = (\omega - \omega_{ce})/k_z v_{Te}$. It is noted that $\alpha\zeta_0\lambda\zeta_1 = (\omega_{pe}^2/2\omega_{ce}^2)((\omega - \omega_{ce})/\omega)\tan^2\theta \simeq 0$ for $\omega \simeq \omega_{ce}$ and θ not close to

$\pi/2$. For the electron cyclotron resonance, $\omega - \omega_{ce} = k_z v_z$ yields $\zeta_1 = v_z/v_{Te} \ll 1$, and then the dispersion function becomes $Z(\zeta_1) \simeq i\sqrt{\pi}k_z/\ |k_z|$ and (8.88) is reduced to $\sigma \simeq (\sqrt{\pi}/2)(\omega_{pe}^2/\omega^2)(\omega/|k_z|v_{Te}) \simeq (\sqrt{\pi}/2n_z)(c\omega_{pe}^2/v_{Te}\omega_{ce}^2) \gg 1$. This means that $\sigma \gg 1$ for high-density plasmas with $(c\omega_{pe}^2/v_{Te}\omega_{ce}^2) \gg 1$. Thus $K_{xx} \sim K_{xy} \sim K_{yy} \sim \sigma$ and $K_{xz} \sim K_{zx} \sim K_{zz} \sim 1$.

Here we consider that the dispersion equation (8.89) can be written as

$$i\sigma\Lambda_0 + \Lambda_1 + \Lambda_2\zeta_1 Z(\zeta_1) = 0, \tag{8.94}$$

where

$$\Lambda_0 = n^4 \sin^2\theta - n^2(2 + \sin^2\theta - 2\alpha) + (1-\alpha)(2-\alpha), \tag{8.95}$$

$$\begin{aligned}\Lambda_1 = {} & \left\{1 - \alpha\left(1 - \frac{n}{4}\sin^2\theta\right)\right\}n^4 + \left\{(1-\alpha)\left(\frac{\alpha}{2} - 1\right)\sin^2\theta\right. \\ & \left. + \tfrac{1}{4}\alpha^2\tan^2\theta(1+\cos^2\theta) - (1-\alpha)\left(1 - \frac{\alpha}{4}\right)(1+\cos^2\theta)\right\}n^2 \\ & + (1-\alpha)\left(1 - \frac{\alpha}{2}\right) + \tfrac{1}{4}\alpha^2(\alpha-2)\tan^2\theta,\end{aligned} \tag{8.96}$$

$$\begin{aligned}\Lambda_2 = {} & (\alpha\sin^2\theta)n^4 - \left\{\alpha\left(1 - \frac{\alpha}{2}\right)\sin^2\theta - \tfrac{1}{2}\alpha^2\tan^2\theta(1+\cos^2\theta)\right\}n^2 \\ & - \tfrac{1}{2}\alpha^2(2-\alpha)\tan^2\theta.\end{aligned} \tag{8.97}$$

For $\sigma \gg 1$, we can neglect the second and third terms in (8.94). $\Lambda_0 = 0$ gives (8.93).

The next-order quantities are obtained by substituting $n = n_\pm + \delta n_\pm + i\kappa_\pm$ in (8.94), where $n_\pm \gg \delta n_\pm$, $\kappa_\pm$. Near the cyclotron resonance with $\omega \simeq \omega_{ce}$ the third term in (8.94) is negligible. When $\delta n_\pm$ and $\kappa_\pm$ are real, $n^2 \simeq n_\pm^2 + 2i\kappa_\pm n_\pm$ and $n^4 \simeq n_\pm^4 + 4i\kappa_\pm n_\pm^3$. By considering that $n_\pm^2$ satisfies $\Lambda_0 = 0$, we obtain $\kappa_\pm$ from (8.94):

$$\begin{aligned}\kappa_\pm &= \frac{\Lambda_1}{2\sigma n_\pm(2\sin^2\theta n_\pm^2 - 2 + 2\alpha - \sin^2\theta)} \\ &= \frac{(v_{Te}/c)\cos\theta\Lambda_1}{\sqrt{\pi}\alpha(2\sin^2\theta n_\pm^2 - 2 + 2\alpha - \sin^2\theta)}\frac{1}{W(\zeta_1)},\end{aligned} \tag{8.98}$$

where

$$W(\zeta_1) = \frac{1}{i\sqrt{\pi}}Z(\zeta_1) = e^{-\zeta_1^2}\left(1 + \frac{2i}{\sqrt{\pi}}\int_0^{\zeta_1} e^{z^2}\,dz\right). \tag{8.99}$$

When the real part of $\kappa_\pm$ is shown as $\kappa_\pm^r$,

$$\kappa_\pm^r = \frac{(v_{Te}/c)\cos\theta\Lambda_1}{\sqrt{\pi}\alpha(2\sin^2\theta n_\pm^2 - 2 + 2\alpha - \sin^2\theta)}f(\zeta), \tag{8.100}$$

where

$$f(\zeta_1) = e^{\zeta_1^2} \Bigg/ \left\{ 1 + \frac{4}{\pi} \left(\int_0^{\zeta_1} e^{z^2} dz \right)^2 \right\}. \tag{8.101}$$

Since $\kappa_\pm^r$ denotes the imaginary part of the refractive index n, the damping coefficient is given by $\kappa_\pm^r / k_0$, where $k_0 = \omega/c$. In Fig. 8.2, the dependence of $\kappa_\pm^r / f(\zeta_1)$ on both θ and α is shown schematically. The continuous lines correspond to the X-mode and the dotted lines correspond to the O-mode. The larger α denotes the higher density. From Fig. 8.2 the X-mode is strongly damped in the low-density regime at small θ. On the other hand, the O-mode is strongly damped in the high-density regime at $\theta \simeq 90°$.

Here we apply the ray equations (8.4) and (8.5) to ECRH in heliotron devices. In the case of ECRH, since the wave absorption is fairly localized at the cyclotron resonance, it is reasonable to use the cold plasma dispersion relation for G in the ray equations. It is also noted that the cold plasma approximation is valid when the wave–particle interaction is negligible or $k_\perp V_{Te}/\omega_{ce} \ll 1$. However, this condition is not necessarily satisfied near the electron cyclotron resonance. It is expected that mode conversion from the electromagnetic (EM) wave to the electrostatic (ES) wave will occur due to the finite electron temperature effect. However, the ES wave is completely absorbed due to Landau damping in the neighborhood of the mode conversion region or the cyclotron resonance region. Thus, it is possible to estimate the absorbed EM wave energy without considering the mode conversion process, since the EM wave is also completely damped in the cyclotron resonance

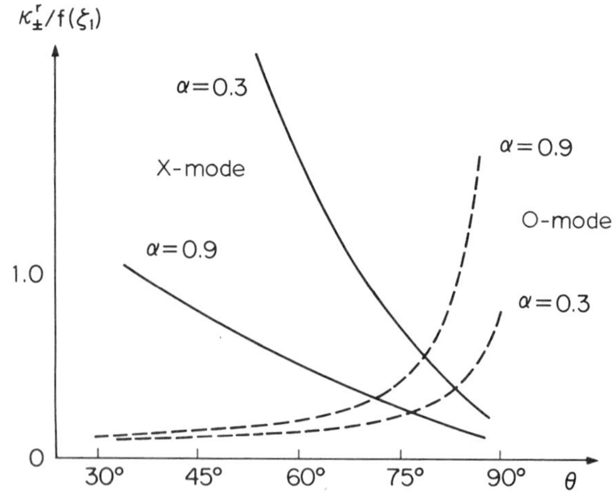

Fig. 8.2 The dependence of $k_\pm^r / f(\zeta_1)$ on the angle θ and the density parameter α. Continuous (dotted) lines correspond to the fundamental X-(O-)mode.

region. The power absorption due to both the cyclotron resonance and Landau damping is given by

$$P(\tau) = P_0 e^{-\Gamma}, \tag{8.102}$$

where P_0 is the power launched into the plasma column and Γ is the optical thickness obtained by

$$\Gamma_{\pm}(\tau) = 2\int_0^{\tau} \kappa_{\pm}^{r}\, d\ell, \tag{8.103}$$

where $\kappa_{\pm}^{r}$ is given by (8.100).

In order to demonstrate the ray trajectories of the electron cyclotron wave in the Heliotron E, we use a simple magnetic field model, $\mathbf{B} = (B_r, B_\theta, B_z)$, under the straight approximation with helical symmetry:

$$\begin{cases} B_r = 4B_0 \rho\rho_c K_2'(2\rho_c) \sin^2\left(\theta - \dfrac{2\pi}{L} z\right), \\ B_\theta = 4B_0 \dfrac{\rho\rho_c}{1+\rho^2} K_2'(2\rho_c) \cos 2\left(\theta - \dfrac{2\pi}{L} z\right), \\ B_z = B_0\left\{1 - 4\dfrac{\rho^2\rho_c}{1+\rho^2} K_2'(2\rho_c) \cos 2\left(\theta - \dfrac{2\pi}{L} z\right)\right\}, \end{cases} \tag{8.104}$$

where (r, θ, z) denote cylindrical coordinates. Here $\rho = 2\pi r/L$, $\rho_c = 2\pi a/L$, $a = 30\,\mathrm{cm}$, and $L = 2\pi R/9.5 = 145.5\,\mathrm{cm}$ for the Heliotron E. It is easily shown that $\mathbf{B}$ in (8.104) satisfies $\nabla \cdot \mathbf{B} = 0$ and $\mathbf{B} \cdot \nabla\psi = 0$ for the flux function $\psi = (B_0 L/4\pi)\{1 + 4\rho^2 \rho_c K_2'(2\rho_c) \cos 2(\theta - 2\pi z/L)\}$. $K_2'(2\rho_c)$ denotes the derivative of the modified Bessel function with respect to the argument. From expressions (8.104), $B = |\mathbf{B}|$ and ∇B can be calculated, these being required in the ray-tracing calculations. $|\mathbf{B}|$ contours and the outermost magnetic surface given by $\psi =$ constant in the poloidal cross-section with $z = 0$ are shown in Fig. 8.3.

To solve the ray equations (8.4) and (8.5) and calculate the optical thickness $\Gamma_{\pm}$, we need the local density, n_e, and temperature, T_e, in addition to the magnetic field. For simplicity, it is assumed that both n_e and T_e are parabolic,

$$\begin{cases} T_e(\psi) = (T_{e0} - T_{ew})\left(1 - \dfrac{\psi}{\psi_B}\right) + T_{ew}, \\ n_e(\psi) = (n_{e0} - n_{ew})\left(1 - \dfrac{\psi}{\psi_B}\right) + n_{ew}, \end{cases} \tag{8.105}$$

where ψ_B is the value of flux function ψ at the plasma boundary and the suffixes 0 and w indicate the values at the center and the boundary, respectively. The edge temperature $T_{ew} = 10\,\mathrm{eV}$ and the edge electron density,

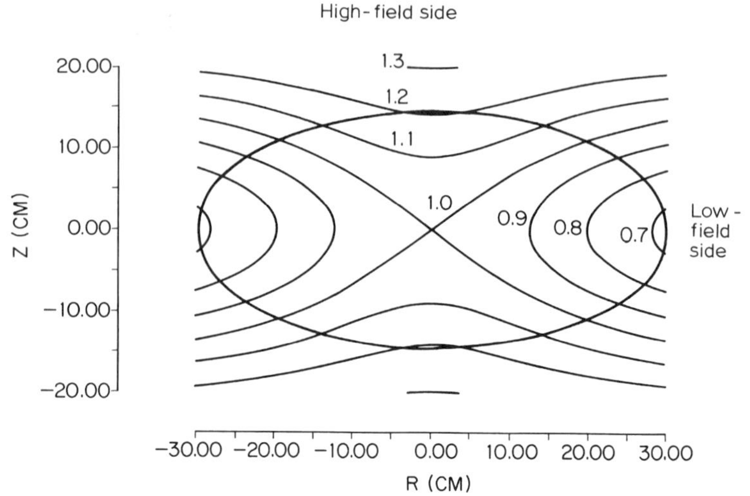

Fig. 8.3 $|\mathbf{B}|$ = constant contours and outermost magnetic surface of a straight Heliotron E model. The magnitude of the magnetic field is normalized with that at the magnetic axis.

$n_{ew} = 1 \times 10^{18}\,\mathrm{m}^{-3}$ are assumed here. Both T_{e0} and n_{e0} are varied as parameters.

The damping of the X-mode at the fundamental electron cyclotron resonance is larger than that of the O-mode, except for the case in which the angle between the propagation direction and the magnetic field is close to 90°, as shown in Fig. 8.2. However, when the X mode is launched from the low-field side (the LHS or RHS in Fig. 8.3), it does not contribute to the plasma heating, since it is reflected at the right-hand cutoff (see the R = constant line in Fig. 8.4) before it reaches the electron cyclotron resonance region. In the Heliotron E ECRH experiment, the electromagnetic waves are laucnhed from the low-field side, and the dominant contribution to the heating comes from the O-mode at the fundamental resonance. The ray trajectories of the O-mode within the outermost magnetic surface are shown in Fig. 8.4. The rays are not simply projected on the poloidal plane as in the axisymmetric tokamak case. When the ray moves in the z direction, the elliptic magnetic surface is rotated to maintain helical symmetry. The positions of the ray in the elliptic plasma cross-sections with different z positions are collected in this way in the horizontally elliptic plasma cross-section, as shown in Fig. 8.4. The plasma cutoff density satisfying $\alpha = 1$ or $\omega^2 = \omega_{pe}^2$ is $3.5 \times 10^{19}\,\mathrm{m}^{-3}$ for a gyrotron frequency of 53.2 GHz. As the plasma density increases, the rays are distorted considerably and cannot pass through the high-temperature region near the magnetic axis, where $\omega = \omega_{ce}$ is satisfied. Efficient heating may not be expected for plasmas with $n_{e0} \gtrsim 3 \times 10^{19}\,\mathrm{m}^{-3}$.

On the contrary, when n_{e0} is kept constant and T_{e0} is varied, the ray trajectories do not change, since the cold plasma dispersion relation is assumed in the

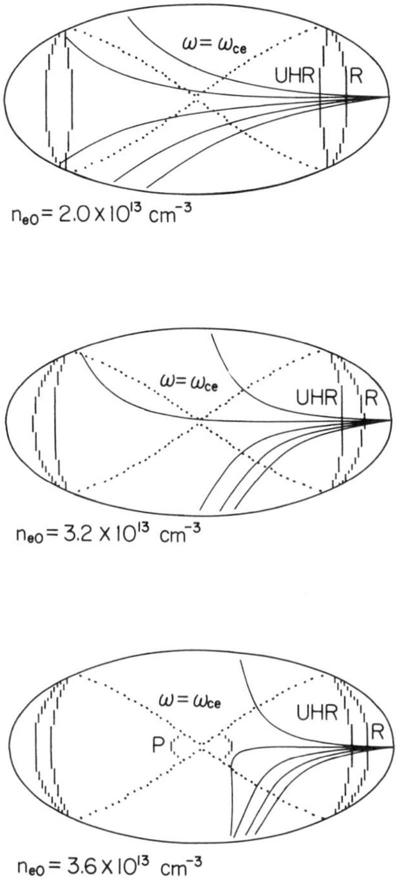

Fig. 8.4 Ray trajectories of the O-mode in the straight Heliotron E model for the case of $\omega = \omega_{ce}$ at the magnetic axis. The $R =$ constant surface denotes the right-hand cutoff, the $UHR =$ constant surface denotes the upper hybrid resonance, and the $P =$ constant surface corresponds to $\omega^2 = \omega_{pe}^2$.

ray tracing calculation. As T_e is increased, the total wave energy absorption inside the plasma column increases and the deposition profile of the absorbed wave energy becomes broad, since the $\omega \simeq \omega_{ce}$ region with a high T_{e0} spreads.

According to the theory of wave absorption, the X-mode is damped at the second harmonic resonance $\omega = 2\omega_{ce0}$, with an absorption coefficient comparable to that at the fundamental resonance, $\omega = \omega_{ce0}$, where ω_{ce0} is the electron cyclotron frequency at the magnetic axis. However, the O-mode has a much smaller absorption coefficient at $\omega = 2\omega_{ce0}$ than at $\omega = \omega_{ce0}$. The X-mode ray trajectories are shown in Fig. 8.5, when the magnetic field in Fig. 8.3 is decreased by a factor of 1/2. For the low-density regime less than the right-hand cutoff density, $n_{e0} = 1.75 \times 10^{19}\ \mathrm{m}^{-3}$, the X-mode reaches the second harmonic electron cyclotron resonance in the central region with an almost

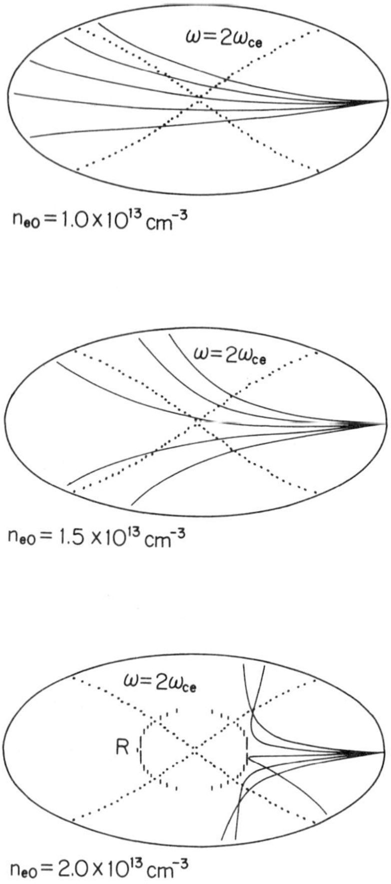

Fig. 8.5 Ray trajectories of the X-mode in the straight Heliotron E model for the case of $\omega = 2\omega_{ce}$ at the magnetic axis. The R = constant surface denotes the right-hand cutoff.

straight ray trajectory. When n_{e0} exceeds the right-hand cutoff density, rays are reflected before reaching the central region. For the absorption of the electromagnetic wave at $\omega = 2\omega_{ce0}$, the finite Larmor radius effect is essential, and an increase in T_{e0} due to electron heating improves the absorption rate significantly.

The variation of T_{e0} in the Heliotron E ECRH experiments when the magnetic field intensity was changed is shown in Fig. 8.6. The numerical results of wave energy absorption based on the straight Heliotron E model are shown in Fig. 8.7, which shows the ratio of the wave power deposited in the central region, $r \le a/3$, to the total radio frequency (RF) power. When ω_{ce0} is changed from $\omega_{ce0} = \omega_g$ to $2\omega_{ce0} = \omega_g$, or the cyclotron resonance shifts from the magnetic axis region, the deposition power in the region of $r \le a/3$ is decreased substantially, where ω_g is the gyrotron frequency. Figure 8.7 agrees with Fig.

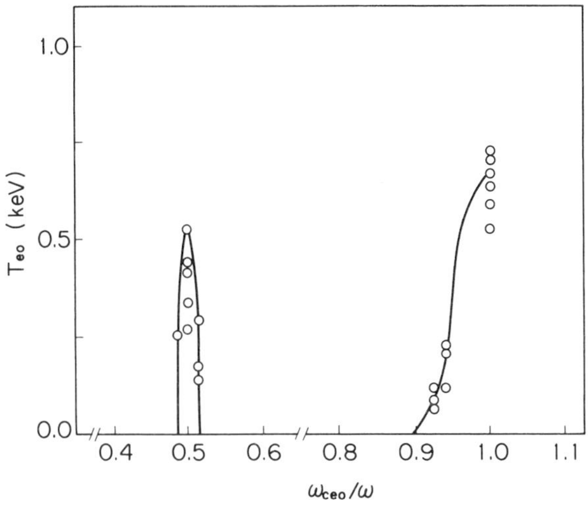

Fig. 8.6 The dependence of the central electron temperature T_{e0} on the magnetic field intensity in the ECRH experiment of the Heliotron E. The launched power from the gyrotron is assumed to be 160 kW. Here $\omega_{ce0}/\omega = 1$ and $\omega_{ce0}/\omega = 1/2$ corresponds to the fundamental harmonic ECRH and the second harmonic ECRH, respectively.

8.6 reasonably well, although T_{e0} is shown in Fig. 8.6 and the power deposition rate at $T_{e0} = 500\,\text{eV}$ in Fig. 8.7. This result suggests that the dominant wave energy absorption may come from the one-path ray trajectory, and that the wall reflection effect is weak. If we use a more realistic magnetic field model

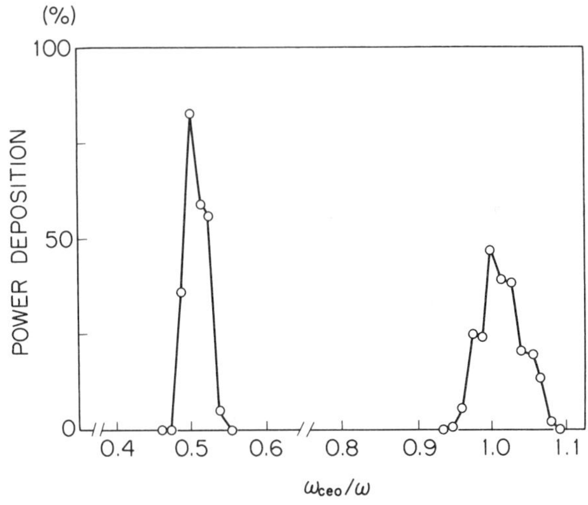

Fig. 8.7 The power deposition rate is the central region, $r/a \leq 1/3$, as a function of ω_{ce0}/ω, based on ray tracing calculations. T_{e0} is assumed to be 500 eV, and only one-path trajectories are taken into account in the absorption of wave energy.

and experimentally measured profiles of n_e and T_e in the ray tracing calculation, comparison between experiment and theory becomes more quantitative.

Next, we discuss ICRF heating results in Heliotron E experiments. High-power ICRF heating is first established as the additional heating method for a tokamak plasma. Two types of heating method are usually used in tokamaks; one is minority ion heating for plasmas including two ion species, which is based on ion–ion hybrid resonance defined by $n_{\|}^2 = S$, and the other is second ion cyclotron resonance heating. In tokamaks the ICRF waves heat both ions and electrons, and temperatures in excess of 10 keV have already been obtained in large tokamak experiments, such as JET and TFTR. An ICRF heating study was started in the C Stellarator, using Stix coils, in the mid-1960s. Also, in the mid-1970s, ICRF heating was successful in small- or medium-sized tokamaks such as the ST Tokamak, the ATC, and the TFR. In those days, the heating power level was several hundreds of kW. The present ICRF heating power exceeds 10 MW in large tokamaks. There has been significant progress in both antenna design for ICRF heating and suppression of the impurity level during ICRF heating.

The target plasma of ICRF heating is the currentless ECRH plasma produced using gyrotrons in the Heliotron E. For ICR frequencies, ω^2 is much less than ω_{pe}^2 and ω_{ce}^2, and therefore $|P| \gg |S| \sim |D|$. This corresponds to the fact that at these low frequencies the electron response along magnetic field lines is rapid enough to cancel perturbation of the parallel electric field, or $E_{\|} = 0$. A substantial simplification can be made in the wave propagation theory by setting $E_{\|} = 0$. If we then assume that the electric fields vary in space as $E_{\perp} = E_{\perp 0} \exp(ik_{\perp}x + ik_{\|}z)$, the dispersion relation reduces to

$$\begin{vmatrix} (s - n_{\|}^2) & iD \\ -iD & (S - n_{\|}^2 - n_{\perp}^2) \end{vmatrix} = 0. \tag{8.106}$$

The solution of (8.106) becomes

$$n_{\perp}^2 = \frac{(S - n_{\|}^2)^2 - D^2}{S - n_{\|}^2}, \tag{8.107}$$

which is the dispersion relation for a fast magnetosonic wave. The other slow magnetosonic wave branch with $n_{\perp}^2 \gg 1$ is given by including P in the dielectric tensor,

$$n_{\perp}^2 = \frac{P}{S}(S - n_{\|}^2), \tag{8.108}$$

which is obtained from the dispersion relation (8.16), and shown as

$$Sn_{\perp}^4 + [(S + P)(n_{\|}^2 - S) + D^2]n_{\perp}^2 + P[(S - n_{\|}^2)^2 - D^2] = 0, \tag{8.109}$$

by assuming $|P| \gg |S| \sim |D|$. This branch is always evanescent for $\omega > \omega_{ci}$, since $P < 0$ and $S < 0$. It is also noted that absorption is very weak for a pure plasma at the fundamental resonance $\omega = \omega_{ci}$ for the fast mode, because the polarization of the wave is to the right-hand side, or in the direction of electron gyration. An important modification of this situation takes place when more than one ion species is present. The contributions to the S term in the dielectric tensor from the two ion species can give opposite signs, or $\omega_{ce}^1 < \omega < \omega_{ci}^2$, and this leads to the reduction of S to the point at which a new resonance (an ion–ion hybrid resonance) occurs, where ω_{ci}^j is the cyclotron frequency of the jth ions. This resonant frequency is given by

$$\omega^2 = \omega_{ci}^1 \omega_{ci}^2 \frac{\bar{n}_1 \omega_{ci}^2 + \bar{n}_2 \omega_{ci}^1}{\bar{n}_1 \omega_{ci}^1 + \bar{n}_2 \omega_{ci}^2}, \tag{8.110}$$

where $\bar{n}_j = n_j/n_e$. Depending on the relative concentration of the two ion species, this may occur near to or far from the ion cyclotron frequency of one ion species. The former case is possible for dilute minority ions. The presence of the ion–ion hybrid resonance affects the polarization of the fast wave near the ion cyclotron resonance layer of the minority ions, and leads to enhanced damping.

In the region of conversion of the fast wave (FW) into the slow wave (SW) in a two ion species plasma, where S is close to $n_\parallel^2$, small perturbing factors can strongly affect the refractive index. Here we consider the influence of the thermal motion of ions, or the finiteness of the ion Larmor radius, on the conversion of FW into SW. To take the thermal motion of ions into account, we include

$$\delta\varepsilon = -\sum_j \frac{3k_\perp^2 v_{Tj}^2 \omega_{pj}^2}{(\omega^2 - \omega_{cj}^2)(\omega^2 - 4\omega_{cj}^2)} \tag{8.111}$$

in S, defined by (8.11). Then the dispersion relation (8.109) becomes

$$(S + \delta\varepsilon) n_\perp^4 + P(n_\parallel^2 - S - \delta\varepsilon) n_\perp^2 - PD^2 = 0 \tag{8.112}$$

for $|P| \gg |S| \sim |D|$. Neglecting the first term in (8.112) yields

$$n_\perp^2 = \frac{D^2}{n_\parallel^2 - S - \delta\varepsilon}. \tag{8.113}$$

By noting that $\delta\varepsilon$ is proportional to $n_\perp^2$, we obtain

$$n_\perp^2 = \frac{1}{2\beta_{eff}} [S - n_\parallel^2 \pm \sqrt{(S - n_\parallel^2)^2 + 4D^2 \beta_{eff}}], \tag{8.114}$$

where

$$\beta_{eff} = \frac{1}{c^2} \sum_j \frac{3 v_{Tj}^2 \omega_{pj}^2 \omega^2}{(\omega^2 - \omega_{cj}^2)(\omega^2 - 4\omega_{cj}^2)}. \tag{8.115}$$

If $\beta_{eff} < 0$ for the frequency (8.110), which is typical for a D (deuterium) plasma with a H (hydrogen) minority species, except for the case of highly different proton and deuteron temperatures, $T_H/T_D > \bar{n}_D/\bar{n}_H$, (8.114) determines the refractive indices of the two propagating waves (the fast wave (FW) and the slow wave (SW)). Far from the mode conversion point, we have

$$n_\perp^2 = \frac{D^2}{n_\parallel^2 - S} \tag{8.116}$$

for the minus sign (FW), and

$$n_\perp^2 = -\frac{n_\parallel^2 - S}{\beta_{eff}} \tag{8.117}$$

for the plus sign (SW), respectively. Receding further from the mode conversion point, the refractive index of the SW increases, since $k_\perp$ becomes of the order of $1/\rho_i$, where ρ_i is the ion Larmor radius. In this region, $k_\perp$ must be determined from the dispersion relation for ion cyclotron waves, $K_{xx} - n_\parallel^2 = 0$, where K_{xx} is given by (8.82) for ion species. When approaching the mode conversion point with $k_\perp \rho_i > 1$, the refractive index of the FW increases, while that of the SW decreases. At the mode conversion point, where $n_\parallel^2 - S = D\sqrt{-\beta_{eff}}$, both refractive indices coincide for

$$n_\perp^2 = \sqrt{-\frac{D^2}{\beta_{eff}}}. \tag{8.118}$$

With the cold plasma dispersion relation, the ion cyclotron resonance ($\omega = \omega_{CH} = 2\omega_{CD}$), the ion–ion hybrid resonance ($n_\parallel^2 = S$), and the right-hand cutoff ($n_\parallel^2 = R$) and left-hand cutoff ($n_\parallel^2 = L$) are shown in Fig. 8.8 for a hydrogen–deuterium plasma in the straight simple Heliotron E model, where ω_{CH} and ω_{CD} are the cyclotron frequencies of the proton and deuteron, respectively. The parallel wave number is assumed to be $k_\parallel = 8.64\,\mathrm{m}^{-1}$. The frequency of the ICRF wave is 26.7 MHz and the magnetic field at the magnetic axis is $B = 1.9\,\mathrm{T}$. The density ratio between the proton and hydrogen is $\bar{n}_H/(\bar{n}_H + \bar{n}_D) = 0.15$. By changing this ratio, the ion–ion hybrid resonance and the left-hand cutoff can be shifted from the ion cyclotron resonance. In Fig. 8.9 solutions of the local dispersion relation, including the hot plasma effect, along the line A–A' in Fig. 8.8 are shown. It is noted that $k_\parallel = 8.64\,\mathrm{m}^{-1}$ is fixed and the density and temperature profiles are assumed to be parabolic (see (8.105)) with $n_{e0} = 3 \times 10^{19}\,\mathrm{m}^{-3}$ and $T_{e0} = T_{i0} = 600\,\mathrm{eV}$. The fast waves and the ion Bernstein waves appear symmetrically along the line A–A'. In Fig. 8.9, A and A' correspond to the high-field side. Mode conversion from the fast wave to the ion Bernstein wave occurs. This situation is the same as that in tokamaks when the waves are propagated from the high-field side.

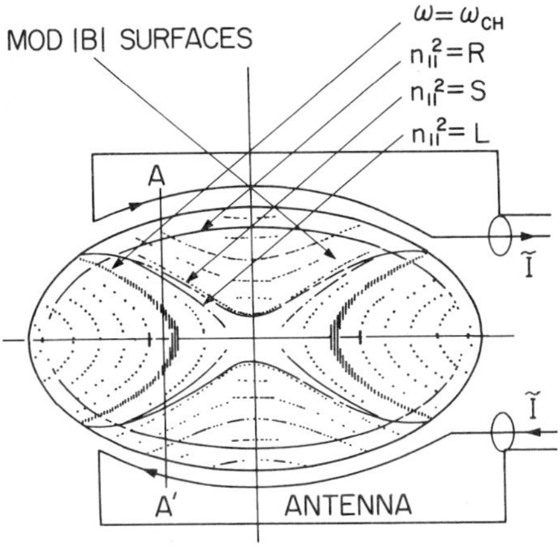

Fig. 8.8 The locations of $\omega = \omega_{CH} = 2\omega_{CD}$, $n_{\parallel}^2 = L$, $n_{\parallel}^2 = S$, and $n_{\parallel}^2 = R$ in the straight simple Heliotron E model. Antennas to excite ICRF waves are placed on both high-field sides of the heliotron plasma. Here $\omega = 26.7\,\text{MHz}$, $k_{\parallel} = 8.64\,\text{m}^{-1}$, and B at the magnetic axis is 1.9 T. Also, the density ratio between the proton and deuteron is $\bar{n}_H/(\bar{n}_H + \bar{n}_D) = 0.15$.

The ion–ion hybrid resonance is on the high-field side of the minority proton cyclotron resonance. An evanescent layer exists between $n_{\parallel}^2 = L$ surface and the $\omega = \omega_{CH}$ surface, as seen in Figs. 8.8 and 8.9. The ICRF heating from the high-field side requires wave penetration across this evanescent layer, the thickness of which depends on $k_{\parallel}$ and $k_{\perp}$. Then the wave propagates to the central

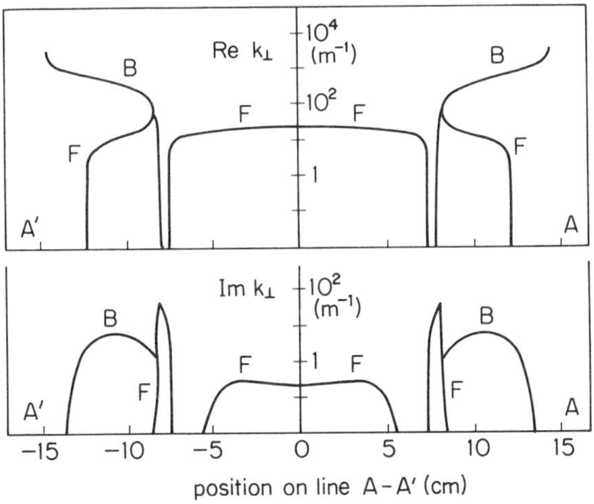

Fig. 8.9 Solutions of the local dispersion relation including the hot plasma effect along the line A–A′ in Fig. 8.8. F and B denote the fast and ion Bernstein waves, respectively. Here the n_e, T_e, and T_i profiles are assumed to be parabolic with $n_e(0) = 3 \times 10^{19}\,\text{m}^{-3}$ and $T_e(0) = T_i(0) = 600\,\text{eV}$.

region, where absorption, reflection, transmission, or mode conversion can take place. When we determine the wave number spectrum excited by the antenna, it is crucial to take into account both the coupling properties in the edge plasma and the absorption properties in the central region. The ICRF antenna for the heating of the Heliotron E currentless plasmas is shown in Fig. 8.10. Four antennas with Faraday shields were installed on the high-field side of an horizontally elliptic plasma. The length of the antenna was about 60 cm, and the width is 5 cm. The antennas faced toward the main plasma, the distance between the last closed magnetic surface and the antenna surface being 2.0–2.5 cm. The loading resistance of the antenna was 3–5 Ω for a typical Heliotron E plasma, which was comparable to that for tokamak ICRF-heated plasmas. The phase of the RF current between the top and bottom antenna pairs was 180°. Thus an ICRF wave with $m = 0$ (where m is the poloidal mode number) was dominantly excited.

In the ICRF experiments the electron temperature was measured with a Thomson scattering system and electron cyclotron emission (ECE) detectors, and the ion temperature was measured with a mass-resolved charge-exchange neutral particle analzyer (NPA) which gives a spectrum of a perpendicular particle energy.

The heating efficiencies corresponding to both ion and electron energy increases are plotted as functions of $\bar{n}_e$ in Fig. 8.11 for the fast (FW) and slow (SW) waves. Here the maximum RF power coupled to the plasma was 550 kW. The efficiency of minority heating by the fast wave (FW) increased with density. Data on second harmonic heating were obtained by reducing the magnetic field strength by a factor of two. This is the most effective heating scheme in the Heliotron E. There are no available data in the high-density regime because of the low cutoff density of target plasma produced by the second harmonic ($\omega = 2\omega_{ce}$) ECRH.

The slow wave (SW) heating experiments were performed with pure hydrogen plasmas; the energy increase disappeared at high densities ($\gtrsim 1.5\times$

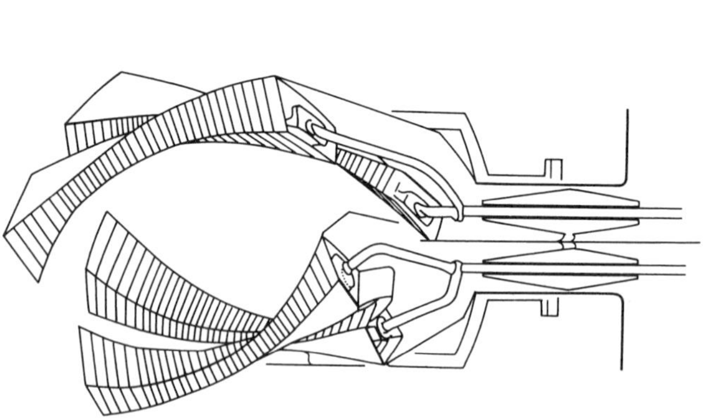

Fig. 8.10 A schematic view of an ICRF antenna with a helical structure in the Heliotron E. RF feed-throughs are shown on the right-hand side.

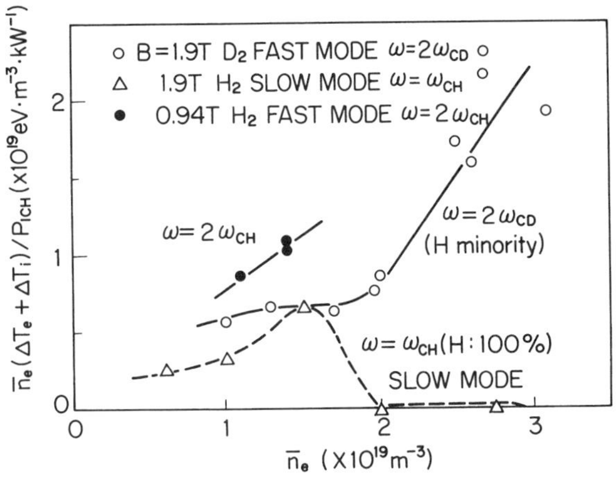

Fig. 8.11 The energy increase normalized using ICRF power versus $\bar{n}_e$ in the Heliotron E for both slow and fast modes (waves). The RF wave frequency is 26.7 MHz. Here minority heating is for $B_0 = 1.9\,\mathrm{T}$ and $\bar{n}_H/(\bar{n}_H + \bar{n}_D) = 0.02$, and second harmonic heating is for $B_0 = 1.9\,\mathrm{T}$ and $\bar{n}_D \simeq 0$.

$10^{19}\,\mathrm{m}^{-3}$). This result is qualitatively consistent with the cold plasma wave theory. The slow wave (SW) heating was used in particular for a low-density plasma to produce a high-T_i plasma that belonged to the low ion collisionality regime. The maximum ion temprature obtained was in excess of 1 keV at $\bar{n}_e = 5 \times 10^{18}\,\mathrm{m}^{-3}$.

Increases in electron and majority ion temperatures were studied by changing the minority proton ratio for fast wave (FW) heating. The characteristics of the minority heating can be qualitatively explained by mode conversion theory. With regard to the ion heating, useful information was obtained from signals from the neutral particle analyzer (NPA). The energy spectra of the minority and majority ions are shown in Fig. 8.12. In the minority proton spectrum, the tail temperature was 3.2 keV and the energy content of the tail component was substantially large. To gain an insight into the heating mechanism for ions, the ratio of the energy content of the high-energy tail ions divided by that of the bulk ions is calculated. It is defined by $R = \int f_{tail}(E)EdE / \int f_{bulk}(E)EdE$, where $f(E)$ is an ion energy distribution function. The peak value of R for protons is obtained at $\bar{n}_H/(\bar{n}_H + \bar{n}_D) \simeq 0.05$. For a low proton ratio, the fast wave (FW) excited from the high-field side can easily cross the evanescent layer to the central region and then be absorbed at the cyclotron resonance layer or in the neighborhood of the left-hand cutoff surface.

One remarkable result from the Heliotron E ICRF experiments is that the currentless toroidal plasma was sustained for $\bar{n}_H/(\bar{n}_H + \bar{n}_D) = 0.1$–$0.2$ with $P_{RF} = 700\,\mathrm{kW}$. The plasma parameters were $\bar{n}_e \simeq 2 - 3 \times 10^{19}\,\mathrm{m}^{-3}$, T_e (by ECE) $\simeq 300\,\mathrm{eV}$, and T_i (by NPA) $\simeq 350\,\mathrm{eV}$.

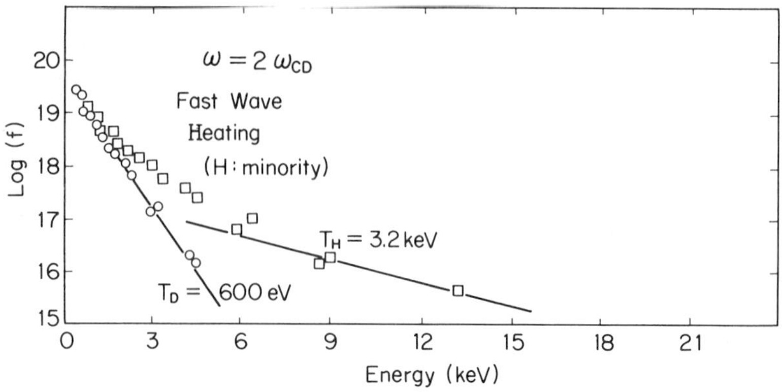

Fig. 8.12 Ion energy distributions measured by a neutral particle analyzer in case of fast wave heating for the hydrogen–deuterium plasma. Here $P_{RF} = 1.5\,\mathrm{MW}$, $B_0 = 1.9\,\mathrm{T}$, $\bar{n}_e \simeq 2.0 \times 10^{19}\,\mathrm{m}^{-3}$, and the proton is the minority ion.

In future fusion reactors, ion–ion hybrid heating may be applied to D–T plasmas or D–^{3}He plasmas. In the latter case, one significant advantage is that the neutrons produced by fusion reactions can be reduced substantially, although the ion temperature required for ignition increases several times compared to D–T plasmas. Thus reactors based on the D-^{3}He fusion reaction are called advanced reactors.

In stellarators, high-power NBI heating was first successful in both the Heliotron E and the Wenderstein VII-A (a classic stellarator device at the Max-Planck Institute, Germany, in the 1980s). When a neutral beam is injected into a toroidal plasma, neutrals become charged particles due to the ionization process and the charge exchange reaction, and the neutral particle profile is given by

$$\begin{aligned} v_B \frac{dN_B}{ds} = &- n_e \langle \sigma_{Ie} | v_B - v_{Te} | \rangle N_B \\ &- \sum_i n_i \langle \sigma_{Ii} | v_B - v_{Ti} | \rangle N_B \\ &- \sum_i n_i \langle \sigma_{cx} | v_B - v_{Ti} | \rangle NB, \end{aligned} \tag{8.119}$$

where σ_{Ie}, σ_{Ii} and σ_{cx} are cross-sections of the electron ionization, ion ionization, and charge exchange with ions, respectively, v_B is the neutral beam velocity, N_B is the number of neutral beam particles, and s is the length along the beam path. In (8.119), summations are taken for all ion species, including impurities. The solution of (8.119) is given by

$$N_B(s) = N_B(0) \exp\left[- \int_0^s \frac{ds'}{\lambda_B(s')} \right], \tag{8.120}$$

where

$$\lambda_B \equiv \frac{v_B}{n_e \langle \sigma v \rangle_T} \tag{8.121}$$

and

$$\langle \sigma v \rangle_T \equiv \langle \sigma_{Ie} v_e \rangle + \sum_i \frac{n_i}{n_e} \langle \sigma_{Ii} v_B \rangle + \sum_i \frac{n_i}{n_e} \langle \sigma_{cx} v_B \rangle . \tag{8.122}$$

Here λ_B is called the mean free path of beam neutrals and $v_{Ti} \ll v_B \ll v_{Te}$ is assumed in (8.122). The separated cross-sections for ionization of the injected hydrogen neutral beam (H^0) are shown as a function of energy in Fig. 8.13. In the case of electron ionization, $\langle \sigma_{Ie} v_e \rangle / v_B$ is shown for three Maxwellian electron velocity distributions at $T_e = 100\,\mathrm{eV}$, 1 keV, and 10 keV. It is noted that $v_{Te} \sim v_B$ for $E_B \gtrsim 400\,\mathrm{keV}$ at $T_e = 100\,\mathrm{eV}$, where E_B is a neutral beam energy. Also, since $v_B \gg v_{Ti}$, σ_{cx} and σ_{Ii} depend on v_B. Thus the cross-sections for a deuterium neutral beam (D^0) with an energy of E_B becomes equal to those for a H^0 beam with $E_B/2$. For the tritium neutral beam (T^0) case, the cross-sections correspond to a H^0 beam with $E_B/3$. Figure 8.13 suggests that the contribution of the charge exchange reaction becomes substantial for D^0 and T^0 beams with $E_B \lesssim 100$–200 keV.

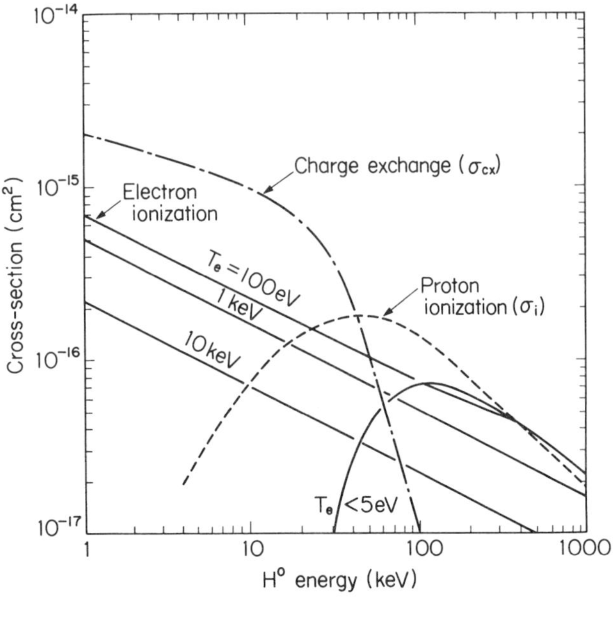

Fig. 8.13 Separated cross-sections for the ionization of an injected hydrogen neutral beam (H_0) as a function of beam energy. In the case of electron ionization, $\langle \sigma_{Ie} v_e \rangle / v_B$ is shown for a Maxwellian electron velocity distribution at temperature T_e.

In the neutral beam heating experiment, the injection direction becomes an important device parameter. There are three types of injection; co-injection, perpendicular injection, and counter-injection. "Co-" and "counter-" injection are related to the direction of the virtual plasma current producing the vacuum rotational transform in heliotrons. In tokamaks they simply depend on the direction of toroidal plasma current. As can be seen from Fig. 8.14, the maximum flight path of a neutral beam, ℓ, is $2a_p$ for perpendicular injection, while $2[(R_0 + a_p)^2 - R_c^2]^{1/2} \simeq 2a_p(2R_0/a_p)^{1/2}$ for parallel ("co-" and "counter-") injection, where R_0 is the major radius, R_c is the radius of the flight path ($\sim R_0$), and a_p is the plasma radius. In order to reduce the shine-through loss of the neutral beam power and heat the plasma efficiently, it is better to make ℓ sufficiently long, and to localize the birth profile of high energy ions inferred from (8.120) near the magnetic axis region. When the plasma radius becomes sufficiently large, perpendicular injection is allowed.

The next important problem is that of particle orbits of high-energy ions (see chapter 6). In the case of perpendicular injection, many trapped particles with small parallel velocities $v_\parallel \simeq 0$ are born inside the plasma column. Since the excursion of the trapped particle orbit is generally large and may exceed the plasma radius for particular pitch angles in the velocity space with $(v_\parallel, v_\perp)$ (the velocity space loss region), the energy deposition is degraded and its profile has a tendency to become broad, where $v_\parallel(v_\perp)$ is a parallel (perpendicular) velocity with respect to the magnetic field vector **B**. In heliotrons the excursion of the trapped particle orbit depends sensitively on the magnetic configuration—on the rotational transform, the aspect ratio, and the local magnetic field strength. Since there is a large aspect ratio $R/\bar{a} \sim 10\,(R \simeq 2.2\,\text{m}$ and $\bar{a} \simeq 0.2\,\text{m})$ and a large rotational transform $\iota(0) \sim 0.5$ and $\iota(\bar{a}) \sim 2.0$ in the Heliotron E, with

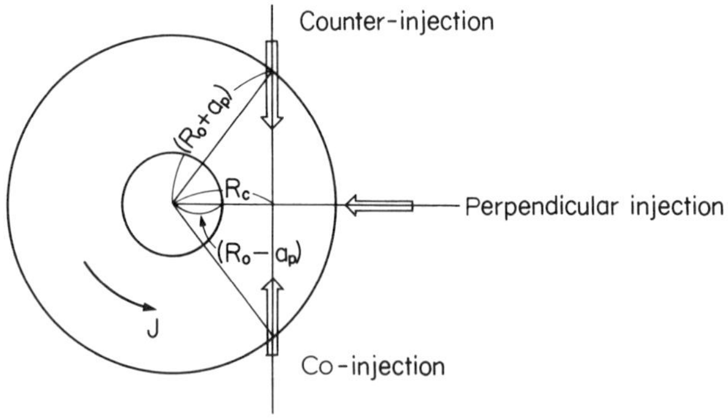

Fig. 8.14 The top view of a torus, showing the directions of co-injection, perpendicular injection and counter-injection. "Co" and "counter" depend on the direction of the plasma current in tokamaks, denoted by J. In heliotrons, they depend on the direction of the virtual plasma current producing the vacuum rotational transform.

$|\mathbf{B}| = 1.9\,\mathrm{T}$ at the magnetic axis, the width of the trapped particle excursion can be less than $\bar{a}/2$ for several 10 keV particles.

Here we consider high-energy particle orbits in the case of co-injection or counter-injection in a tokamak. For simplicity, the magnetic field is assumed to be $\mathbf{B} = (0, B_\theta, B_\phi)$, where $B_\phi = B_0(1 - r\cos\theta/R)$ in the toroidal coordinates (r, θ, ϕ). In tokamaks the poloidal field B_θ produced by the plasma current is much smaller than the toroidal field B_ϕ. Thus the drift motion due to inhomogeneity of the magnetic field is given by neglecting B_θ, and untrapped ions drift vertically or in the z direction with

$$v_d = \frac{m_i(v_\parallel^2 + v_\perp^2/2)}{eB_0R} \simeq \text{constant.} \tag{8.123}$$

The drift motion of untrapped particle in the (x, z) plane (or in the poloidal cross-section) is described by

$$\begin{cases} \dfrac{dx}{dt} = -v_\parallel \dfrac{B_\theta}{B_0}\dfrac{z}{r} \\ \dfrac{dz}{dt} = v_\parallel \dfrac{B_\theta}{B_0}\dfrac{x}{r} + v_d, \end{cases} \tag{8.124}$$

where the origin $r = 0$ is the center of the circular magnetic surface, and $x = r\cos\theta$ and $z = r\sin\theta$. From (8.124), we obtain

$$\left(x + \frac{B_0 v_d}{B_\theta v_\parallel}\right) dx + z\,dz = 0. \tag{8.125}$$

Then the untrapped particle orbit is given by

$$\left(x + \frac{rB_0 v_d}{B_\theta v_\parallel}\right)^2 + z^2 = \text{constant.} \tag{8.126}$$

Thus the particle orbit becomes a circle with a shift of $\Delta = -(rB_0v_d/B_\theta v_\parallel)$ from the origin. However, this shift is not large, since

$$|\Delta| = \frac{rB_0 v_d}{B_\theta v_\parallel} \sim \frac{rB_0}{B_\theta v_\parallel}\left(\frac{m_i v_\parallel^2}{eB_0R}\right) \sim q\rho_L < r \tag{8.127}$$

for $v_\parallel \simeq v_T$. The direction of the shift depends on the direction of the toroidal plasma current J_ϕ. It is noted that in the right-hand coordinates (r, θ, ϕ), $B_\theta < 0 (B_\theta > 0)$ for $J_\phi > 0 (J_\phi < 0)$. For $J_\phi > 0$ and $v_\parallel > 0$ (parallel to the plasma current), $\Delta > 0$, and for $J_\phi > 0$ and $v_\parallel < 0$ (anti-parallel to the plasma current), $\Delta < 0$. Also, for $J_\phi < 0$ and $v_\parallel > 0$ (anti-parallel to the plasma current), $\Delta < 0$, and for $J_\phi < 0$ and $v_\parallel < 0$ (parallel to the plasma current),

$\Delta > 0$. Thus untrapped ions produced by co- (counter-) injection move in the same (opposite) direction as the plasma current, $\Delta > 0$ ($\Delta < 0$).

The high-energy ions generated by the NBI transfer momentum and energy to background plasmas due to Coulomb collisions with electrons and ions. The transfer rate of the particle energy is given by

$$\left\langle \frac{dW}{dx} \right\rangle = -2\pi Z_B^2 e^4 \ell n\Lambda \left[\frac{A_B}{W} \sum_j \frac{n_j Z_j^2}{A_j} + \frac{4}{3\sqrt{\pi}} \frac{W^{1/2}}{T_e^{3/2}} \left(\frac{A_e}{A_B} \right)^{1/2} n_e \right], \qquad (8.128)$$

when the background ions and electrons have Maxwellian distributions, where A_B, A_j, and A_e are the atomic weights of the injected beam ions, plasma ions, and electrons, respectively. Equation (8.128) can be shown in the form

$$\left\langle \frac{dW}{dx} \right\rangle = -\frac{\lambda_1}{W} - \lambda_2 W^{1/2}, \qquad (8.129)$$

where

$$\lambda_1 = 1.30 \times 10^{-13}\, A_B Z_B^2 \ln\Lambda \sum_j \frac{n_j Z_j^2}{A_j}, \qquad (8.130)$$

and

$$\lambda_2 = 2.28 \times 10^{-15}\, \frac{Z_B^2}{A_B^{1/2}} \frac{n_e \ln\Lambda}{T_e^{3/2}}, \qquad (8.131)$$

where W and T_e are in eV and x is in cm. Here Z_B denotes the charge of beam ion. The term proportional to λ_1 is related to ion collisions, and the term proportional to λ_2 is related to electron collisions. Here we have noted that

$$\left\langle \frac{dW}{dt} \right\rangle = \left\langle v \frac{dW}{dx} \right\rangle = \sqrt{\frac{2W}{m_B}} \left\langle \frac{dW}{dx} \right\rangle = 1.39 \times 10^6 \left(\frac{W}{A_B} \right)^{1/2} \left\langle \frac{dW}{dx} \right\rangle, \qquad (8.132)$$

where $W = m_B v^2/2$. From (8.129), for $\lambda_1/W_c = \lambda_2 W_c^{1/2}$, the energy loss due to ions becomes equal to that due to electrons, which gives

$$W_c = \left(\frac{\lambda_1}{\lambda_2} \right)^{2/3} \simeq 14.8 T_e \left[\frac{A_B^{3/2}}{n_e} \sum_j \frac{n_j Z_j^2}{A_j} \right]^{2/3}. \qquad (8.133)$$

The characteristic time of the energy decay to a background thermal energy level is given by

$$\tau_{s\ell} = -\int_0^W \frac{dW}{\dfrac{dW}{dt}} = \frac{\tau_s}{3} \ln\left[1 + \left(\frac{W}{W_c} \right)^{3/2} \right], \qquad (8.134)$$

where τ_s is Spitzer's slowing down time,

$$\tau_s[\text{sec}] = 6.28 \times 10^8 \frac{A_B T_e^{3/2}}{Z_B^2 n_e \ln \Lambda}, \tag{8.135}$$

where T_e is in eV and n_e is in cm^{-3}. For $W \gg W_c$ the electron frictional force becomes dominant in the slowing down process and (8.134) reduces to

$$\tau_{s\ell} = \frac{\tau_s}{2} \ln\left(\frac{W}{W_c}\right). \tag{8.136}$$

Thus it is considered that the beam energy decays exponentially according to $dW/dt = -2W/\tau_s$ or $W = W_0 e^{-t/(\tau_s/2)}$. For $W = W_c$, (8.134) gives $\tau_{s\ell} = (\tau_s/3)\ln 2 \simeq 0.23\tau_s$.

The ratio of deposition energy to ions during the slowing down process with respect to the total transferred energy is given by

$$G_i = \frac{1}{W}\int_0^W \frac{\lambda_1/W}{\lambda_1/W + \lambda_2 W^{1/2}}\, dW = \frac{W_c}{W}\int_0^{W/W_c} \frac{dy}{1+y^{3/2}}, \tag{8.137}$$

and the ratio given by deposition energy to electrons divided with the total transferred energy is given by

$$\begin{aligned} G_e &= \frac{1}{W}\int_0^W \frac{\lambda_2 W^{1/2}}{\lambda_1/W + \lambda_2 W^{1/2}}\, dW \\ &= \frac{1}{W W_c^{1/2}}\int_0^{W/W_c} \frac{y^{3/2}}{1+y^{3/2}}\, dy. \end{aligned} \tag{8.138}$$

In Fig. 8.15, G_i and G_e are plotted as a function of W/W_c. At $W = W_c$, 74.7% of the initial neutral beam energy is deposited to background ions, while at $W = 3W_c$ only 44.1% of the initial neutral beam energy is deposited to them; or electron heating becomes possible.

The NBI heating rate in the Heliotron E was measured by varying the density. The input power was varied by changing the number of injectors and/or the accelerating voltage. The increase in the ion temprature, ΔT_i, is plotted as a function of the absorbed NBI power, P_{abs}, in Fig. 8.16. P_{abs} is normalized by the average density, in units of $10^{13}\ \text{cm}^{-3}$. Here the absorbed power corresponds to the power of the neutral beam particles ionized in the plasma column. The loss power due to unconfined ion orbits and charge exchange is not subtracted from P_{abs}. The ionization or charge exchange efficiency of beam neutrals for $\bar{n}_e = 2\text{–}6 \times 10^{13}\ \text{cm}^{-3}$ lies roughly between 20% and 60% at an acceleration voltage of 25keV, and the orbit loss is less than 10%. As shown in Fig. 8.16, for a fairly wide range of NBI power and plasma parameters, we have obtained the following scaling of the ion heating efficiency:

$$\Delta T_i = 2.0[\text{eV}] P_{abs}[\text{kW}]/\bar{n}_e[10^{13}\ \text{cm}^{-3}]. \tag{8.139}$$

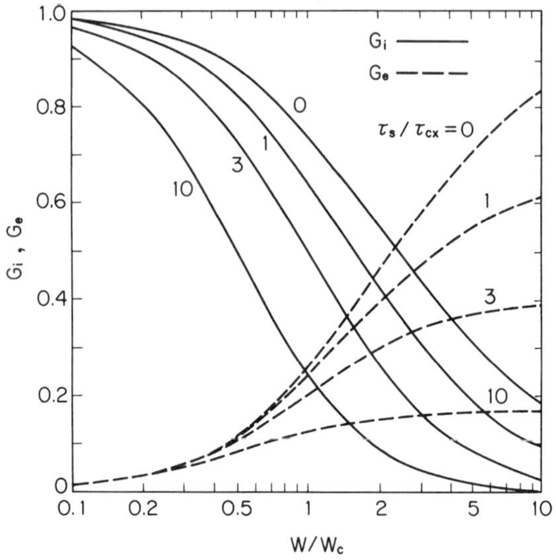

Fig. 8.15 G_i and G_e versus W/W_c. G_i is denoted by the continuous line and G_e by the dotted line.

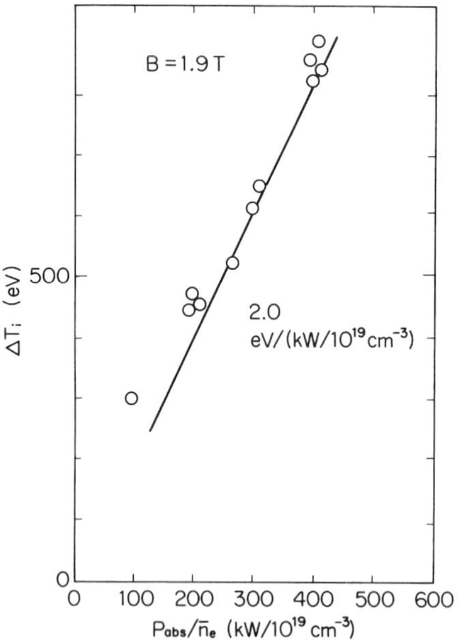

Fig. 8.16 Ion heating versus absorbed power normalized by the average density, $P_{abs}/\bar{n}_e\,[\mathrm{kW}/10^{13}\,\mathrm{cm}^{-3}]$.

The increase in the ion temperature is almost the same as that of the electrons in the high-density regime. The important feature of NBI heating in the Heliotron E is that perpendicular injection works fairly well to heat and sustain currentless plasmas without the ECRH.

8.3 THE LHD SCALING LAW OF HELIOTRON PLASMAS

Recent experiments of the Heliotron E (H-E), Heliotron DR (H-DR), ATF, and CHS, set out in Table 8.1, have shown that the energy transport of currentless plasmas confined in heliotron devices is governed by anomalous transport. Here "anomalous" means that the magnitude and parameter dependence of the energy transport coefficient are different from the neoclassical theory explained in chapter 7. This situation is quite similar to the energy transport in tokamaks. According to intensive studies of anomalous transport, the understanding of the physics has progressed significantly, but there is still no definite anomalous transport theory to explain the experimental results consistently. Theoretical approaches to anomalous transport will be explained briefly in section 8.7. Thus, the alternate approach to estimating the energy confinement time τ_E in a particular device is the empirical scaling law. During the design work for the LHD, the so-called LHD scaling law was obtained. It is also noted that the empirical scaling law provides useful information in the investigation of the physical mechanism of anomalous transport. Apparently, the LHD scaling law is consistent with the gyro-reduced Bohm-type transport coefficient, which suggests that the origin of the anomalous transport is drift wave turbulence.

The confinement time is calculated using average plasma parameters such as the electron and ion temperatures, T_e and T_i, and the electron density, n, in the following way:

$$\tau_E = \tfrac{3}{2} n(T_e + T_i)(2\pi^2 a^2 R)/P, \tag{8.140}$$

where a [m] and R [m] are the average minor radius and major radius, respectively, and P [MW] is the absorbed power for plasma heating. It is noted that when the thermal diffusivity χ is given, τ_E is also obtained as $\tau_E \propto a^2/\chi$, where the numerical coefficient depends on the profile of χ. For example, we have

$$\tau_E^{PL} = 1.4 P^{-3/5} n^{3/5} B^{4/5} a^2 R \iota^{2/5} \tag{8.141}$$

with neoclassical thermal diffusivity for the plateau regime ($\nu_{bh} < \nu < \nu_p$). The definitions of the symbols and their units are as follows: confinement time, τ_E [sec]; average electron density, n [10^{20} m^{-3}]; magnetic field, B [T]; rotational transform, ι; collision frequency, ν [sec^{-1}]; ν_p [sec^{-1}] $= \iota v_T/R$; thermal velocity, v_T [m/sec]; ν_{bh} [sec^{-1}] $= \epsilon_h^{3/2}\omega_{bh}$; helical ripple, ϵ_h; and bounce frequency of a trapped particle in the helical ripple, ω_{bh}. It is noted that the temperature dependence of χ is eliminated with the expression (8.140).

For the neoclassical ripple transport regime with $\nu < \nu_{bh}$, we obtain

$$\tau_E^R = 0.11 P^{-7/9} n B^{4/9} a^2 R^{11/9} \epsilon_h^{-1/3}, \tag{8.142}$$

where ϵ_h is taken at $r = a/2$. It is noted that the numerical coefficients 1.4 and 0.11 in (8.141) and (8.142) are just for rough estimation. It is also noted that the whole plasma region is not governed by the neoclassical plateau or ripple transport only.

In the Heliotron E, the empirical confinement time scaling laws are

$$\tau_E^{NBI} = 1.6 \times 10^{-2} P^{-0.64} n^{0.54} \tag{8.143}$$

for NBI heated plasmas at $B - 1.9$ T, and

$$\tau_E^{ECRH} = 1.7 \times 10^{-2} P^{-0.53} n^{0.66} B^{0.53} \tag{8.144}$$

for ECRH plasmas. It is interested that the P and n dependencies in (8.143) and (8.144) are similar to the plateau regime scaling law (8.141), but that the experimental values of τ_E are less than τ_E^{PL}. It is also interesting that the P dependence in (8.143) and (8.144) is similar to that in the L-mode scaling law of tokamaks.

It is expected that τ_E of heliotron plasmas depends on the rotational transform $\iota(\Psi)$, the shear proportional to $d\iota(\Psi)/d\Psi$, and the helical ripple, $\epsilon_h(\Psi)$, where Ψ is the magnetic flux function. However, the dependencies of τ_E on these quantities are fairly weak and no clear scalings have been obtained so far. In the Heliotron E and Heliotron DR experiments, τ_E was improved by a factor of about 20% by shifting the magnetic axis inside with an additional vertical magnetic field. From the scaling law point of view, these results are included in the scaling laws of (8.143) and (8.144) by increasing the coefficient by a factor of 1.2–1.3.

Recent tokamak results show that an isotope effect exists in the confinement time, or that deuterium plasmas give a better confinement time compared to hydrogen plasmas under the same experimental conditions. On the contrary, in the Heliotron E experiment, such an isotope effect is very weak. Thus the scaling laws (8.143) and (8.144) can be applicable to both hydrogen and deuterium plasmas.

In order to estimate the dependence of τ_E on a and R, other stellarator experimental results in the WVII-A Stellarator and the L-2 Stellarator are referred to. The best fitted scaling law for stellarator and heliotron devices is given by regression analysis as

$$\tau_E^{LHD} = 0.17 P^{-0.58} n^{0.69} B^{0.84} a^{2.0} R^{0.75}, \tag{8.145}$$

and the comparison between the scaling law (8.145) and the experimental results used to derive τ_E^{LHD} is shown in Fig. 8.17. Since this scaling law was

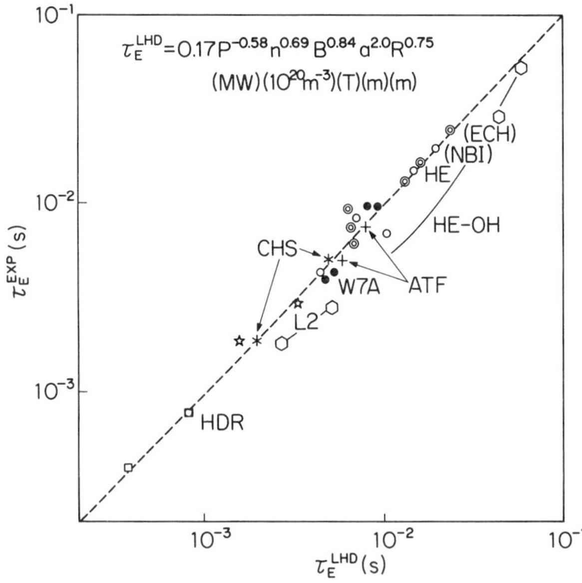

Fig. 8.17 The experimental energy confinement time versus estimation by the LHD scaling law.

used for designing the Large Helical Device (LHD), it has recently been called the LHD scaling law. From τ_E^{LHD}, the scaling of the temperature T [keV], and the figure of merit for evaluating fusion experimental devices, $n\tau_E T$, are given by

$$T^{LHD} = 0.18P^{0.42}n^{-0.31}B^{0.84}R^{-0.25}, \tag{8.146}$$

$$(n\tau_E T)^{LHD} = 3.1 \times 10^{-2}P^{-0.16}n^{1.4}B^{1.7}a^{2.0}R^{0.5}, \tag{8.147}$$

where $T = T_e = T_i$ is assumed.

The characteristic of the LHD scaling law is the fairly strong dependence on density n and magnetic field B. In order to investigate the origin of the anomalous transport corresponding to the LHD scaling law, it is useful to compare τ_E^{LHD} with τ_E given by Bohm-type transport, $\chi_B \propto (T/eB)$, and gyro-reduced Bohm-type transport, $\chi_{GB} \propto (\rho_s/a)(T/eB)$, where χ is the thermal transport coefficient and ρ_s is the effective ion Larmor radius, given by $\rho_s = (T_e/m_i)^{1.2}/(eB/m_i)$, which is equal to ion Larmor radius for $T_i = T_e$. It is noted that ρ_s usually appears in drift wave theory (see section 8.7). By using χ_B and χ_{GB} in the expression $\tau_E \propto a^2/\chi$ and eliminating the temperature dependence using (8.140), we obtain

$$\tau_E^B = 0.94P^{-0.5}n^{0.5}B^{0.5}a^{2.8}R^{-0.5} \tag{8.148}$$

for Bohm-type transport, and

$$\tau_E^{GB} = 0.31 P^{-0.6} n^{0.6} B^{0.8} a^2 R \tag{8.149}$$

for gyro-reduced Bohm-type transport. It is also noted that the numerical coefficients 0.94 and 0.31 are just for rough estimation of the absolute confinement time. It is seen that the LHD scaling law is closer to τ_E^{GB}, which suggests that a kind of drift wave turbulence may become a candidate for explaining the anomalous transport in heliotron plasmas. The spatial scale length "ρ_s" is considered to be the correlation length of the turbulence that drives the anomalous transport.

8.4 THE BOOTSTRAP CURRENT, AND PLASMA ROTATION IN HELIOTRON DEVICES

In recent heliotron and torsatron experiments, "currentless" plasma confinement has become standard. Here, "currentless" means that the net plasma current is exactly zero. However, experimental results show small net plasma currents of the order of several kA. For NBI plasmas both the bootstrap current and the beam-driven current may contribute to the observed net plasma current. For ECRH plasmas the bootstrap current becomes the probable origin of the net current. The direction and the order of magnitude of the observed net current are consistent with those expected for the bootstrap current from neoclassical theory (see chapter 7).

The existence of the bootstrap current has been confirmed in several tokamaks. In particular, in large tokamaks several hundreds of kA of bootstrap current has already been obtained, and the maximum ratio of the bootstrap current to the total plasma current was about 70% in the JT-60U Tokamak. It is important to improve our understanding of the physics of the bootstrap current and to find ways of controlling it in toroidal devices, both to reduce the current drive power in tokamaks and to realize purely currentless operation in heliotron and torsatron devices.

The toroidal currents in the ATF ECRH discharges have been compared with neoclassical bootstrap current theory (see chapter 7). In order to carry out a comparison between theory and experiment, the quadrupole field was varied while the magnetic axis position was fixed at $R_0 = 2.08$ m. The variation of the parameter corresponding to the magnitude of the quadrupole field led to a change in the toroidally averaged plasma shape from horizontally elongated ($f_m < 0$) to vertically elongated ($f_m > 0$), where f_m is a measure of the quadrupole field. Through the scan of the quadrupole field, the averaged electron density was kept constant at $\bar{n}_e = 5.5 \times 10^{18}\,\mathrm{m}^{-3}$, as shown in Fig. 8.18. The stored energy peaked at about $f_m = +0.10$. Both the electron density and the temperature profiles broadened in going from the $f_m < 0$ to the $f_m > 0$ case. The measured toroidal current (solid points) is shown as a function of the quadrupole field in Fig. 8.19. The current decreases systematically with increasing vertical elongation and becomes negative at about $f_m = +0.15$. The bootstrap

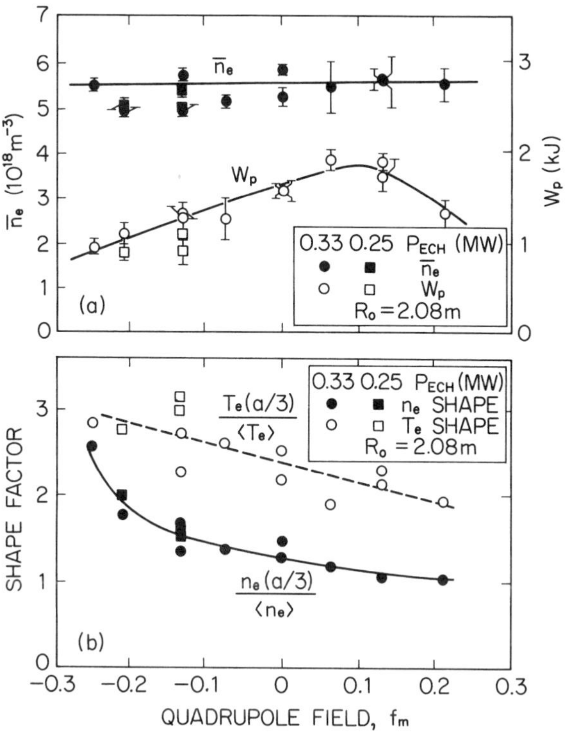

Fig. 8.18 Plasma responses when the quadrupole field is varied in ATF ECRH plasmas: (a) variation of the internal plasma energy W_p at fixed $\bar{n}_e$; (b) variation of the electron temperature and density profiles measured at $a/3$.

predictions (open points) are calculated on the basis of the experimental temperature and density profiles. The geometrical factor $\langle G_{BS} \rangle$ (see (7.161)) is calculated by using an analytic expression for the finite-collisionality correction that interpolates between different collisionality regimes. The agreement between experiment and theory is fairly good: the difference may be caused by uncertainties in the density and temperature profiles and in the impurity content.

These results are also significant in that they verify the intended ability to control the toroidal current through the quadrupole field in heliotron configurations. It is also possible to control the plasma current to a zero net value, as shown in Fig. 8.19.

Recently, the plasma rotation in the toroidal and poloidal direction and the ion temperature have been measured using charge exchange recombination spectroscopy. In those diagnostics, visible light is measured using optical fibers, and spatial resolution of the order of 5 mm can be realized in the multi-channel system. For fully ionized carbon ions, the charge exchange reaction

$$\begin{aligned} H^0 + C^{6+} &\rightarrow H^+ + C^{5+}(n = 8) \\ &\rightarrow H^+ + C^{5+}(n = 7) + h\nu \end{aligned} \tag{8.150}$$

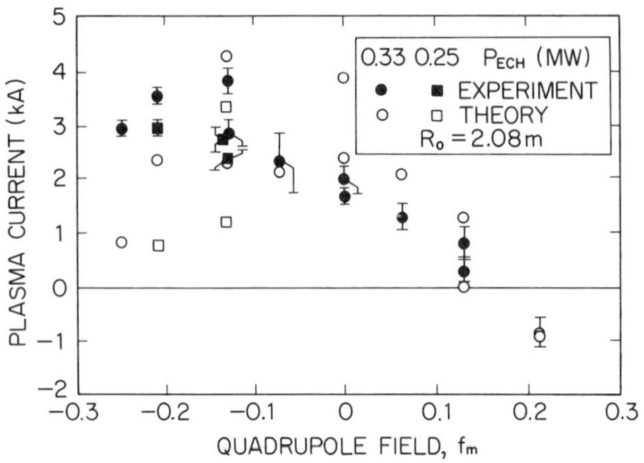

Fig. 8.19 The measured toroidal currents and neoclassical predictions for the bootstrap current when the quadrupole is varied in the ATF.

produces visible light with a wavelength of 529.05nm ($n = 8 \rightarrow n = 7$) for the hydrogen neutrals H^0, where n is a principal quantum number showing the energy state. One advantage for the light of 529.05 nm is that the decay rate in the optical fiber is weaker than the light from transition, $n = 7 \rightarrow n = 6$, which has a large radiation cross-section. Also, the Doppler broadening of 529.05 nm is larger. The hydrogen neutrals H^0 are usually injected with a neutral beam injector for measurement.

In the CHS, the rotation velocity profiles in the toroidal and poloidal directions were measured when the neutral beam was in the co-injection and the counter-injection modes (see section 8.2). Similar NBI power and plasma parameters were chosen here for both injection cases.

In heliotron configurations, the toroidal rotation is damped by the parallel viscosity due to the magnetic ripples in the toroidal direction (see chapter 7). Thus the contribution of the toroidal rotation to the radial electric field is fairly small. The radial electric field profiles induced by the poloidal and toroidal rotations are shown separately in Fig. 8.20. It is seen that effect of the toroidal rotation velocity V_ϕ on the radial electric field is small. The magnetic axis position in the vacuum magnetic configuration changes the negative radial electric field significantly. When the magnetic axis is located at 89.9 cm, the radial electric field, $|V_\theta \times B_\phi|$, becomes larger in the counter-injection case. However, when the magnetic axis is located at 97.4 cm, co-injection gives a larger value of $|V_\theta \times B_\phi|$ than counter-injection. These results are explained in the following way. When fast ions are produced by the co-injection neutral beams, the drift orbits of untrapped ions shift outside compared to vacuum magnetic surfaces. On the contrary, when fast ions are produced by the counter-injection neutral beams, they shift inside compared to vacuum magnetic surfaces. Thus, in the case of the magnetic axis at 97.4 cm, the co-injected fast ions move to the outside of the plasma column and collide with the

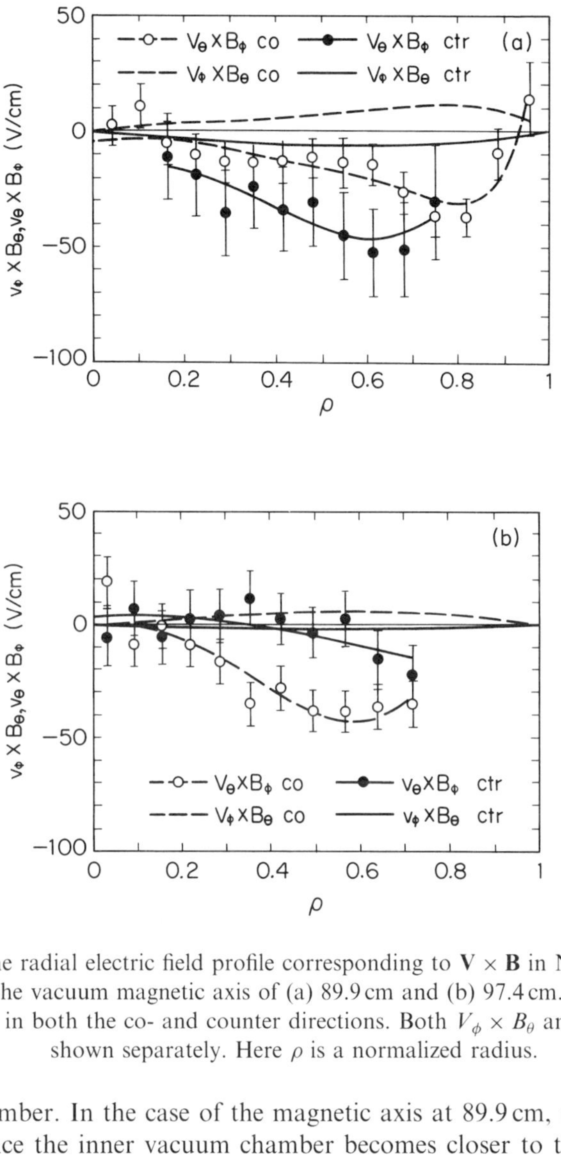

Fig. 8.20 The radial electric field profile corresponding to $\mathbf{V} \times \mathbf{B}$ in NBI plasmas of the CHS at the vacuum magnetic axis of (a) 89.9 cm and (b) 97.4 cm. Neutral beams are injected in both the co- and counter directions. Both $V_\phi \times B_\theta$ and $V_\theta \times B_\phi$ are shown separately. Here ρ is a normalized radius.

vacuum chamber. In the case of the magnetic axis at 89.9 cm, the situation is reversed, since the inner vacuum chamber becomes closer to the plasma column. Since this behavior of the fast ion drift orbits is consistent with the tendency shown in Fig. 8.20, it is considered that the fast ion loss produces the negative radial electrical field by destroying charge neutrality in the CHS.

From the equation of motion for impurity ions, we obtain

$$E_r = \frac{1}{n_I e Z_I} \frac{dP_I}{dr} - (V_\theta V_\phi - V_\phi B_\theta) \tag{8.151}$$

in cylindrical coordinates (r, θ, ϕ). It is considered that the first term on the RHS of (8.151) is small for fully ionized carbon ions with $Z_I = 6$. When it is negligible, the radial electric field E_r is obtained by measuring V_θ and V_ϕ. In the CHS, $E_r(\rho)$ was measured in the low-density NBI plasmas with the second harmonic ECRH. When the second harmonic ECRH power was 85 kW, $E_r > 0$ was obtained, as shown in Fig. 8.21. This result seems to be consistent with the electron root given by neoclassical transport theory for heliotrons. In the CHS the improvement in confinement due to this electron root or positive electric field has not been seen clearly. The reason for the appearance of the positive electric field is considered as the enhanced orbit loss of electrons caused by the second harmonic ECRH, which destroys the charge neutrality and produces a positive electric field. As a next step, it is necessary to study the parameter range of V_θ or E_r to suppress the anomalous or turbulent transport in heliotrons.

8.5 PELLET INJECTION AND THE DENSITY LIMIT IN THE HELIOTRON E

A small cylindrical or spherical frozen quantity of hydrogen or deuterium called a pellet was injected into tokamak plasmas in the late 1970s. Originally, the pellet injection was invented to supply fuel in fusion reactors. Since the edge plasma temperatures in fusion reactors are much higher than the ionization energy of deuterium or tritium, fuel particles added by gas-puffing are ionized in the surface region, and it seems difficult to obtain a peaked density profile. However, recent tokamak experiments show that broad but monotonically decreasing density profiles are realized with gas-puffing only. This result may be related to the inward particle diffusion process in tokamak plasmas. When inward diffusion exists, it is not necessary to inject the pellet

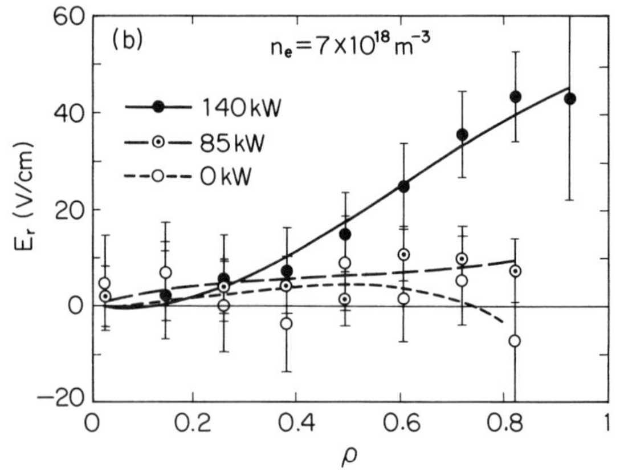

Fig. 8.21 The radial electric field (E_r) profile in the case of second harmonic electron cyclotron resonance heating (ECRH) for low-density NBI plasmas ($\bar{n}_e = 7 \times 10^{18}\,\mathrm{m}^{-3}$) in the CHS. Here ρ is a normalized radius. The additional ECRH power is changed.

into the central region of a toroidal plasma to obtain peaked density profiles. We already have a pellet injector with a maximum speed in excess of 2 km/sec. In present pellet injectors, the pellet is accelerated by the centrifugal force or the high-pressure hydrogen gas, as in an air gun. However, this may not be sufficient to inject the pellet into the central region of the fusion reactor. Typically, 10 km/sec will be required for fusion reactors, even in the presence of inward diffusion. Therefore, a pellet injector with a maximum speed of the order of 10 km/sec has become a research and development target.

One significant experimental result was that discharges with highly peaked density profiles realized by pellet injection in tokamaks showed improvement of energy confinement compared to those with broad density profiles. Although the physical mechanism of the improvement in confinement is not clear, it is usable in the ignition phase of the fusion reactor. A pellet ablation cloud measured by the emission of a D_α line in the JFT-2M Tokamak is shown in Fig. 8.22. The maximum density obtained by pellet injection in the Heliotron E, which was realized after the wall had been carbonized, is shown in Fig. 8.23. The obtained maximum average density is about $2 \times 10^{20}\,\mathrm{m}^{-3}$. Due to pellet injection, the average density increases abruplty and the high-density state is maintained during the discharge. Here the diamagnetic signal corresponding to the plasma pressure becomes stationary. The radiation loss given by the bolometric measurement increases gradually. The soft X-ray signal I_{SX} depends sensitively on the electron temperature, T_e. The decrease in I_{SX} is correlated with the decrease in T_e at the injection of the pellet. It is noted that impurity ions also affect the soft X-ray signal.

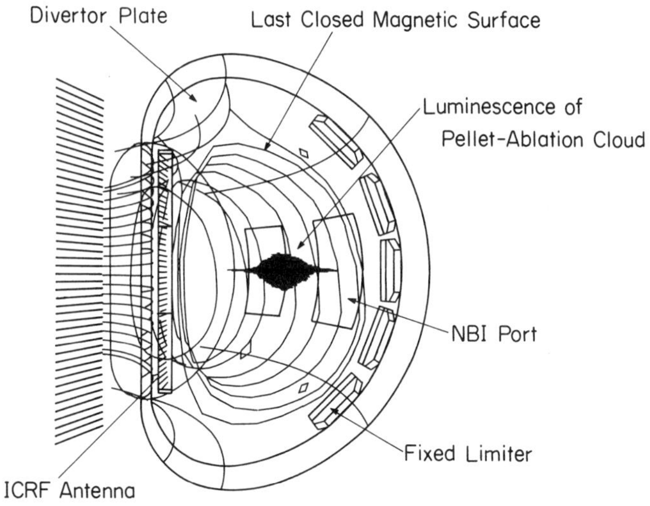

Fig. 8.22 A schematic picture of D_2 pellet ablation in the JFT-2M Tokamak. The pellet velocity is 833 m/s, the toroidal field is $B = 1.3\,\mathrm{T}$, the plasma current is $I_p = 274\,\mathrm{kA}$, and the NBI heating power is 0.88 MW. The pellet is injected from the right-hand side.

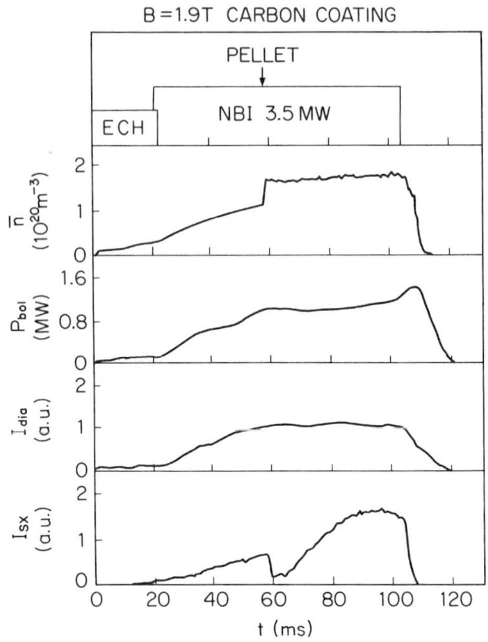

Fig. 8.23 Pellet injection into a high-density plasma with an NBI of 3.5 MW in the Heliotron E, which corresponds to the discharge of the maximum density. The time evolution of the average density ($\bar{n}_e$), the total radiation power ($P_{bo\ell}$), the diamagnetic signal (I_{dia}), and soft X-ray radiation from the central region (I_{sx}) are shown.

One of the important subjects in magnetic fusion research is the density limit. Here we discuss the density limit of heliotron plasmas. It is found that the density limit is directly related to the heating power. When high-power NBI heating was applied, high-density plasmas exceeding $10^{20}\,\mathrm{m}^{-3}$ were obtained fairly easily in the Heliotron E. By considering the power balance between the input power and the radiation loss, $P \simeq P_b V_p$, the density limit becomes proportional to $(P/a^2R)^{1/2}$, since P_b is assumed to be the bremsstrahlung loss and $P_b \propto n_e^2$ for $n_i = n_e$. Here $V_p = 2\pi^2 a^2 R$ is the plasma volume. It is also found that the density limit depends on the magnitude of the magnetic field, B, and seems proportional to $B^{1/2}$. Thus an empirical formula for the density limit of stellarator and heliotron devices, n_C^S, is given by

$$n_C^S \propto (PB/a^2R)^{1/2}. \tag{8.152}$$

It is noted that the scaling law for the density limit (8.152) tends to agree fairly well with the experimental results. Here we comment on the density limit in tokamak plasmas. The density limit is related to the Murakami and Hugill parameter,

$$n_C^T \propto B/Rq, \tag{8.153}$$

although low-q discharges are vulnerable to major disruption. In expression (8.153) there is no heating power dependence, although n_C^T depends on the plasma current through the safety factor, q. Concerning the heating power dependence of n_C^T, the scaling law of $n_C^S \propto P^{1/2}$ given in (8.152) is also approximately applicable to tokamak plasmas when the additional heating power exceeds the ohmic heating power.

We return to pellet injection and show an analysis of pellet ablation based on the Milora–Foster model. The Parks–Turnbull model, which is similar, is also frequently used in pellet ablation studies. These models are based on the shielding due to neutral gas. It has recently been pointed out that the plasma shielding due to a cold plasma along a magnetic field line is also important. Also, high-energy particle effects on the ablation rate have been studied intensively, and are particularly important in future ignited plasmas.

At the spherical pellet surface, the mass ablation rate, dm/dt, is given by

$$dm/dt = 4\pi r_p^2 \rho_s |dr_p/dt|, \tag{8.154}$$

where ρ_s is the solid density of pellet and $|dr_p/dt|$ is the retarding velocity of the pellet surface at $r = r_p$. In a plasma with density n_e and temperature T_e, the directed electron flux is given by

$$\Gamma_e = n_e(eT_e/2\pi m_e)^{1/2}, \tag{8.155}$$

and the average electron energy is

$$E_e = \frac{2\pi}{e}\int \tfrac{1}{2} m_e v^2 f_M(v) v^2 dv = \tfrac{3}{2} T_e,$$

where $f_M(v)$ is the Maxwellian distribution function. Here $m_e = 9.108 \times 10^{-31}$ kg is the electron mass and $e = 1.602 \times 10^{-19}$ J/eV, which is required to change the unit from eV to J (Joule). Also, the directional electron energy flux is given by

$$q_e = n_e(2e/\pi m_e)^{1/2} T_e^{3/2} = \tfrac{4}{3}\,\Gamma_e E_e. \tag{8.156}$$

In pellet injection experiments, electrons have an energy distribution that produces the energy influx to the pellet surface. However, mono-energetic electrons with E_e are often assumed to evaluate the total energy flux. It is then given by

$$dQ/dt = q_{ep} A_e = \tfrac{4}{3}\,\Gamma_{ep} E_{ep} A_e, \tag{8.157}$$

where A_e is the effective cross-section of the pellet for the electron influx. When the electron Larmor radius is much smaller than the pellet radius r_p, $A_e = 2\pi r_p^2$, since electrons move approximately along magnetic field lines. Thus the energy balance at the pellet surface is shown as

$$dQ/dt = \lambda\, dm/dt, \tag{8.158}$$

where λ is the energy to ablate a pellet of unit mass. In order to determine λ precisely we need to consider the details of the phase transition at the pellet surface. However, the value of λ is not sensitive for the ablation rate, since a shielding neutral cloud is formed against the incident electron energy flux, and this adjusts the ablation rate. Thus it is more important to know the relation between the thickness of the neutral cloud and the shielding effect for incident electrons.

The incident electrons usually suffer both elastic and inelastic collisions. It is considered that the decrease in the electron particle flux due to these scatterings is small, when electron energies are fairly high. Thus we assume that the electron particle flux is constant, or $\Gamma_{ep} = \Gamma_{e\infty}$, although the electron energy flux decreases where $\Gamma_{e\infty}$ is the background electron particle flux. Then, from (8.154), (8.157), and (8.158), we obtain

$$\left|\frac{dr_p}{dt}\right| = \frac{q_{ep}A_e}{4\pi r_p^2 \lambda \rho_s} = \frac{(4/3)E_{ep}\Gamma_{e\infty}A_e}{4\pi r_p^2 \lambda \rho_s}. \tag{8.159}$$

When r_p, $T_{e\infty}$, and $n_{e\infty}$ (or $\Gamma_{e\infty}$ from (8.155)) are given, E_{ep} is necessary to give the retardation velocity of the pellet surface.

Now we consider the fluid equations for the neutral cloud. The mass, momentum, and energy conservation equations are written as

$$\frac{d}{dr}(\rho v r^2) = 0, \tag{8.160}$$

$$\frac{1}{r^2}\frac{d}{dr}(\rho v^2 r^2) + \frac{dP}{dr} = 0, \tag{8.161}$$

$$\frac{1}{r^2}\frac{d}{dr}\left\{\rho v r^2\left(\frac{\gamma}{\gamma-1}\frac{P}{\rho} + \frac{1}{2}v^2\right)\right\} = \rho e W_e \tag{8.162}$$

in a stationary state of a spherically symmetric cloud, where ρ, v, P, $\gamma P/\{(\gamma-1)\rho\}$, and W_e are the mass density, the velocity, the pressure, the enthalpy, and the heat source per unit mass, respectively. In (8.162), γ is the ratio of specific heats. Equating the energy loss of the incident electrons to the pellet ablation heat source yields

$$W_e = \frac{1}{\rho}\frac{dq_e}{dr} = \frac{4}{3}\frac{\Gamma_{e\infty}}{\rho}\frac{dE_e}{dr}, \tag{8.163}$$

from (8.156). We describe the decrease in the incident electron energy in the ablation cloud as

$$\frac{dE_e}{dr} = \frac{\rho(r)}{m}L(E_e), \tag{8.164}$$

where m is the mass of the cloud and $L(E_e)$ denotes the energy loss function due to inelastic collisions, which is empirically given as

$$L(E_e) = (2.38 \times 10^{18} + 4 \times 10^{15}E_e + 2 \times 10^{21}E_e^{-2})^{-1}, \tag{8.165}$$

where the unit of $L(E_e)$ is eV m^2. Then the heat source is also shown as

$$W_e = \frac{4}{3}\frac{\Gamma_{e\infty}L(E_e)}{m}. \tag{8.166}$$

The continuity equation (8.160) can be solved easily:

$$\rho v r^2 = \rho_s r_p^2 \left|\frac{dr_p}{dt}\right| = \frac{1}{4\pi}\frac{dm}{dt} = \text{constant}, \tag{8.167}$$

where (8.154) has been used. Introducing the Mach number, $M = v/C_s$ into (8.161) yields

$$\frac{d}{dr}(\rho r^2)^{-1} + \frac{r^2}{\gamma}\frac{d}{dr}(M^2 \rho r^4)^{-1} = 0, \tag{8.168}$$

where (8.167) has been used, and where C_s is a sound velocity. Then, by substituting (8.167) and (8.168) into (8.162), we obtain

$$\begin{aligned}(1 - M^2)\frac{dM}{dr} &= \tfrac{1}{2}\rho_s^{-3}\left|\frac{dr_p}{dt}\right|^{-3} eW_e r_p^{-3}(\gamma - 1)\rho^3 r^6 M^2(\gamma M^2 + 1)\\ &\quad - \frac{2M}{r}\left(1 + \frac{\gamma - 1}{2}M^2\right).\end{aligned} \tag{8.169}$$

In order to obtain nondimensional forms of (8.168) and (8.169), $\hat{r} = r/r_p$ and $\hat{\rho} = \rho/\rho_p$ are introduced as follows:

$$\frac{d}{d\hat{r}}(\hat{\rho}\hat{r}^2)^{-1} + \frac{\hat{r}^2}{\gamma}\frac{d}{d\hat{r}}(M^2\hat{\rho}\hat{r}^4)^{-1} = 0, \tag{8.170}$$

$$(1 - M^2)\frac{dM}{d\hat{r}} = \xi\hat{\rho}^3\hat{r}^6 M^3(\gamma M^2 + 1) - \frac{2M}{\hat{r}}\left(1 + \frac{\gamma - 1}{2}M^2\right), \tag{8.171}$$

where

$$\xi \equiv \frac{\gamma - 1}{2}\left(\frac{\rho_p}{\rho_s}\right)^3 eW_e r_p^4 \left|\frac{dr_p}{dt}\right|^{-3} \tag{8.172}$$

and ρ_p is the neutral density at the pellet surface. It is noted that the first term on the RHS of (8.171) describes the heating due to electrons and the second term describes the cooling due to expansion of the neutral cloud. The parameter ξ in (8.172) is called the heating parameter. The heat source W_e in (8.166) is proportional to $L(E_e)$, which depends on spatial position. However, since the spatial dependence of $L(E_e)$ is weak, $\langle L(e_e)\rangle \simeq L(E_{e\infty}/2)$ is assumed here. It is considered that the error due to this approximation is small.

Then ξ is given as

$$\xi \simeq \langle \xi \rangle = \frac{\gamma - 1}{2} \left(\frac{\rho_p}{\rho_s} \right)^3 e \langle W_e \rangle r_p^4 \left| \frac{dr_p}{dt} \right|^{-3}$$
$$= \frac{4}{3} \frac{\gamma - 1}{2} \left(\frac{\rho_p}{\rho_s} \right)^3 \frac{eL(E_{e\infty}/2)}{m} \Gamma_{e\infty} r_p^4 \left| \frac{dr_p}{dt} \right|^{-3} = \text{constant.} \tag{8.173}$$

When ξ is given, (8.170) and (8.171) can be solved numerically for $\hat{\rho}$ and M under appropriate boundary conditions.

If there is no heating of the ablation cloud due to electrons, or $\xi = 0$, the RHS of (8.171) becomes negative and $M = 1$ at the pellet surface. The radial velocity increases continuously as a free spherical expansion. On the other hand, when strong heating of the ablation cloud exists, the cloud pressure increases near the pellet surface and the local sound velocity exceeds the flow velocity of the cloud, v, or $M < 1$ is satisfied. The asymptotic value of M at $r \to \infty$ is $\sqrt{5/\gamma} > 1$ (supersonic flow). Thus, there is a sonic radius r_s corresponding to $M = 1$. The cloud flow is subsonic for $r < r_s$ and becomes supersonic for $r > r_s$. In (8.171), the LHS becomes zero for $M = 1$ and the RHS must vanish simultaneously, since either $dM/d\hat{r}$ is finite or the Mach number changes continuously. In order to obtain a consistent transonic solution, the Mach number at the pellet surface must be adjusted to make the RHS of (8.171) zero at $M = 1$. Now we note the following relation from (8.164):

$$\int_{E_{ep}}^{E_{e\infty}} \frac{dE_e}{L(E_e)} = \frac{1}{m} \int_{r_p}^{\infty} \rho(r) dr \tag{8.174}$$

between the unknown value of E_{ep} and the cloud density $\rho(r)$. From numerical solutions of (8.170) and (8.171), the relation

$$\int_1^{\infty} \hat{\rho} d\hat{r} \simeq (1.25\xi^{1/3})^{-1} \tag{8.175}$$

has been obtained. Substituting (8.173) into (8.175) yields

$$\left| \frac{dr_p}{dt} \right| = 1.25 \left(\frac{\rho_p}{\rho_s} \right) \left(\frac{\gamma - 1}{2} e \langle W_e \rangle r_p^4 \right)^{1/3} \int_1^{\infty} \hat{\rho} d\hat{r} \tag{8.176}$$

or

$$\left| \frac{dr_p}{dt} \right| = \frac{1.25m}{r_p \rho_s} \left(\frac{\gamma - 1}{2} e \langle W_e \rangle r_p^4 \right)^{1/3} \int_{E_{ep}}^{E_{e\infty}} \frac{dE_e}{L(E_e)}. \tag{8.177}$$

We note that (8.176) determines the relation between the retardation velocity of the pellet surface and the line density of the ablation cloud. We now obtain two relations for $|dr_p/dt|$ and E_{ep}, given by (8.159) and (8.177). By solving these equations simultaneously we can determine dr_p/dt: this is called the Milora–Foster model. Usually, the solution depends weakly on E_{ep} and is determined essentialy by (8.177).

Here we show a numerical method to calculate the ablation rate profile using the Milora–Foster model. A toroidal plasma is divided into N meshes as shown in Fig. 8.24. The time required to pass the Jth mesh, Δt_J, is determined by the mesh size, Δx_J, and the pellet velocity, v_{pel}:

$$\Delta t_J = \Delta x_J / v_{pel}. \tag{8.178}$$

Since Δt_J is of the order of 10 μsec, it is assumed that the pellet moves in a plasma with uniform density and temperature.

First, $|dr_p/dt|$ is eliminated from (8.159) and (8.177). We obtain

$$\frac{E_{ep}\Gamma_{e\infty}A_e}{3\pi r_p^2 \lambda \rho_s} = \frac{1.25m}{r_p \rho_s}\left(\frac{\gamma - 1}{2} e\langle W_e \rangle r_p\right)^{1/3} \int_{E_{ep}}^{E_{e\infty}} \frac{dE_e}{L(E_e)} \tag{8.179}$$

which gives a cubic-order equation

$$c' f^2 + E_{e\infty}(f^2 - f) + 8.51 \times 10^{-4} E_{e\infty}^2 (f^3 - f) + \frac{8.51 \times 10^2}{E_{e\infty}}(f - 1) = 0 \tag{8.180}$$

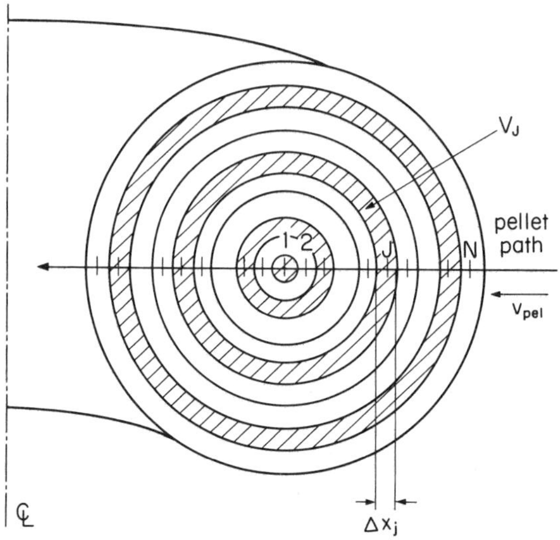

Fig. 8.24 Spatial meshes of the toroidal plasma. Here Δx_J is the path length of the pellet through the Jth mesh, and V_J is the volume of Jth mesh.

for $f = E_{ep}/E_{e\infty}$, where $L(E_e)$ given by (8.165) has been used explicitly and

$$c' = 1.42 \times 10^{-14} \frac{n_{e\infty}^{2/3} E_{e\infty}^{4/3} r_p^{2/3} M'}{\lambda' [L(E_{e\infty}/2)]^{1/3}}. \qquad (8.181)$$

In the definition of c', M' is the average atomic mass and $\lambda' \simeq 0.005\,\text{eV}$ is the sublimation energy for one atom. By obtaining a solution satisfying $0 < f = E_{ep}/E_{e\infty} < 1$ and substituting it in (8.159), an ordinary differential equation,

$$dr_p/dt = \dot{R}(r_p), \qquad (8.182)$$

is given. Integrating (8.182) over Δt_J gives the decrease of r_p within the interval of Δt_J, which gives the total number of neutral particles deposited in the Jth mesh. By dividing the total number of deposition particles by the volume of the Jth mesh, the increase in the local density can be estimated. The typical mesh size of Δx_J is taken to be about ten times the initial pellet size.

The temprature of the background plasma decreases, since a substantial electron heat flux is consumed to ablate the solid hydrogen (or deuterium) pellet. This cooling process is determined by the electron–electron collision time, τ_{ee}, since the ablation cloud generates new cold electrons around the pellet. It is considered that when Δt_J is longer than τ_{ee}, the plasma cooling affects the ablation process or limits the ablation rate due to the decrease in the electron heat flux to the pellet surface. This process is called a self-limiting ablation mechanism. By assuming that the interaction between the pellet ablation and the plasma cooling is adiabatic, the changes in the electron density and electron temperature are evaluated by the following equations:

$$\left.\frac{dn_e}{dt}\right|_J = \frac{\dot{N}_J}{V_J}, \qquad (8.183)$$

$$\left.\frac{3}{2}\frac{d(n_e T_e)}{dt}\right|_J = -\varepsilon_I \frac{\dot{N}_J}{V_J}, \qquad (8.184)$$

$$\dot{N}_J = \frac{4\pi \rho_s r_p^2 \dot{R}}{m}, \qquad (8.185)$$

where $\dot{N}_J$ is the particle deposition rate in the Jth mesh, V_J is the volume of the Jth mesh and ε_I is the average energy needed to dissociate and ionize one neutral particle. By solving (8.182), (8.183), and (8.184) simultaneously, the pellet radius, r_p, electron temperature, T_e, and electron density, n_e, of ambient plasma are obtained consistently. When the pellet passes the magnetic axis, the self-limiting ablation process significantly reduces the ablation rate, since the pellet meets ambient plasma that has already been cooled by pellet ablation. As an example, we consider a toroidal plasma with the electron density and temperature profiles

$$n_e(r)[\text{m}^{-3}] = 6 \times 10^{19}\left(1 - \left(\frac{r}{a}\right)^2\right) + 5 \times 10^{17}, \qquad (8.186)$$

$$T_e(v)[\mathrm{ev}] = 600\left(1 - \left(\frac{r}{a}\right)^2\right) + 10, \tag{8.187}$$

where the major radius R is 220 cm and the minor radius a is 20 cm. We assume that a spherical pellet with 5.86×10^{19} neutral particles is injected with a velocity of 200–800 m/sec. This pellet size corresponds to a cylindrical pellet with a diameter of 1 mm and an axial length of 1.4 mm. The ablation rate profile obtained by the Milora–Foster model associated with the self-limiting ablation process is shown in Fig. 8.25. It can be seen that the ablation rate is suppressed in the region of the magnetic axis. This is understandable from the self-limiting ablation in the smallest-volume mesh, where both the density increase and the temperature decrease are largest. In Fig. 8.26 are shown the radial profiles of n_e

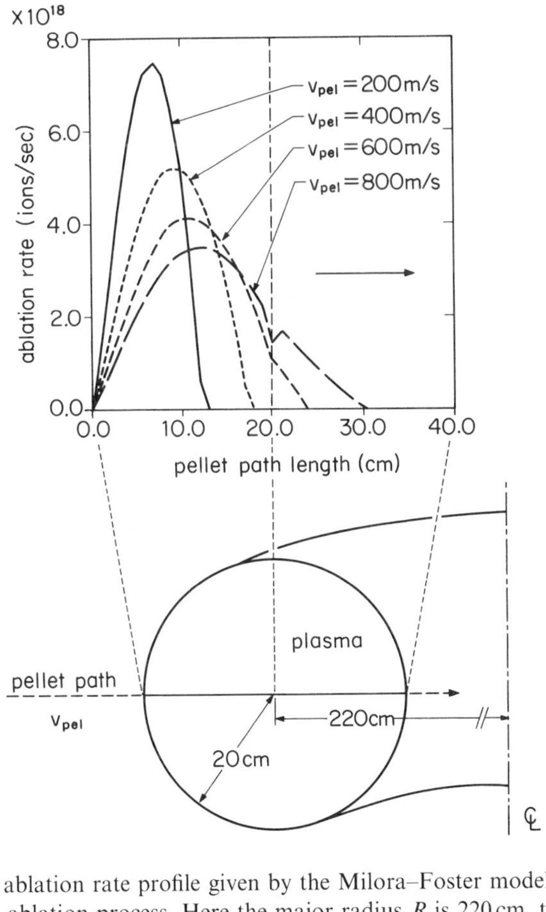

Fig. 8.25 The ablation rate profile given by the Milora–Foster model associated with a self-limiting ablation process. Here the major radius R is 220 cm, the minor radius is 20 cm, the central electron temperature $T_e(0)$ is 610 eV, and the central electron density is $n_e(0) = 6.05 \times 10^{19}\ \mathrm{m}^{-3}$. The pellet injection velocity v_{pel} is varied from 200 m/sec to 800 m/sec.

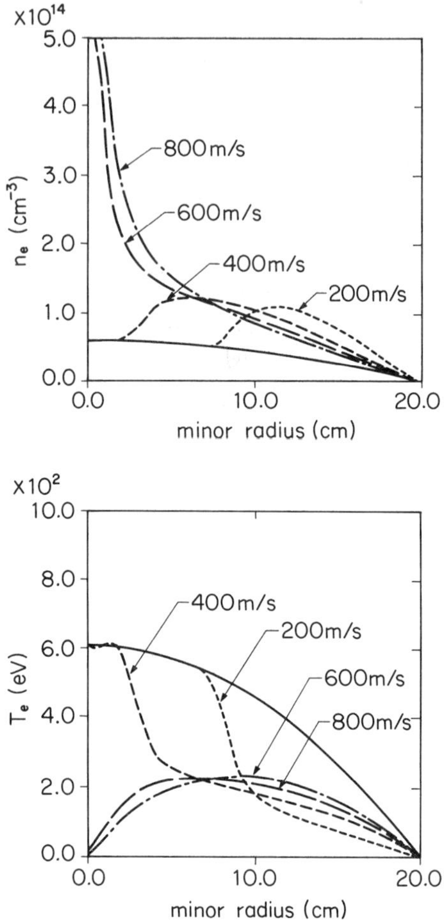

Fig. 8.26 The radial distribution of the electron density and the electron temperature after pellet injection corresponding to Fig. 8.25.

and T_e corresponding to the pellet injections shown in Fig. 8.25. When the pellet passes the magnetic axis region, a highly peaked density profile and a hollow electron temperature profile are obtained. When $v_{pel} = 800\,\text{m/sec}$, about 70% of the total neutral particles are deposited at $r/a \leq 0.5$.

8.6 PRESSURE GRADIENT DRIVEN TURBULENCE

Recent experimental results show that both electron and ion thermal transport are anomalous or deviate significantly from the predictions of neoclassical transport theory. Compared to tokamaks, since the database on ion thermal transport in heliotron plasmas is limited, we consider electron thermal transport first. In heliotron plasmas, electron thermal transport may be governed by the following mechanisms: the neoclassical transport process; drift wave turbulence; and resistive interchange turbulence. As discussed in chapter 7, helical magnetic ripples enhance the neoclassical transport in the rare-collisional $1/\nu$ regime, and the contribution of neoclassical ripple transport is considered to be substantial in the high-temperature central region of heliotron plasmas. However, since it depends inversely on the collisionality, there is a tendency for the ripple transport coefficient $\chi_e^{neo}(\bar{r})$ to decrease in the outer low-temperature region. This behavior is opposite to the experimentally obtained transport coefficient, $\chi_e^{exp}(\bar{r})$, where $\bar{r}$ is the average radius of the noncircular magnetic surface. For example, $\chi_e^{exp}(\bar{r})$ is calculated from the power balance equation as

$$\chi_e^{exp}(\bar{r}) = \frac{\int_0^{\bar{r}} (P_{ECH}(r') - P_{ei}(r'))r'dr'}{\bar{r}dT_e/d\bar{r}}, \tag{8.188}$$

for ECRH plasmas, where $P_{ECH}(\bar{r})$ is the power deposition profile of electron cyclotron waves, $P_{ei}(\bar{r})$ is the Coulomb energy transfer from electrons to ions, and $T_e(\bar{r})$ is the electron temperature profile. The other candidate to explain the transport coefficient in the central region is the anomalous transport driven by drift wave turbulence. This is related to the fact that neoclassical ripple transport cannot explain $\chi_e^{exp}(\bar{r})$ in the central region completely, and a discrepancy between $\chi_e^{neo}(\bar{r})$ and $\chi_e^{exp}(\bar{r})$ is often seen. The existence of turbulent fluctuations is confirmed by density fluctuation measurements in ECRH-produced ATF plasmas, as shown in Fig. 8.27. Here, the normalized density fluctuations

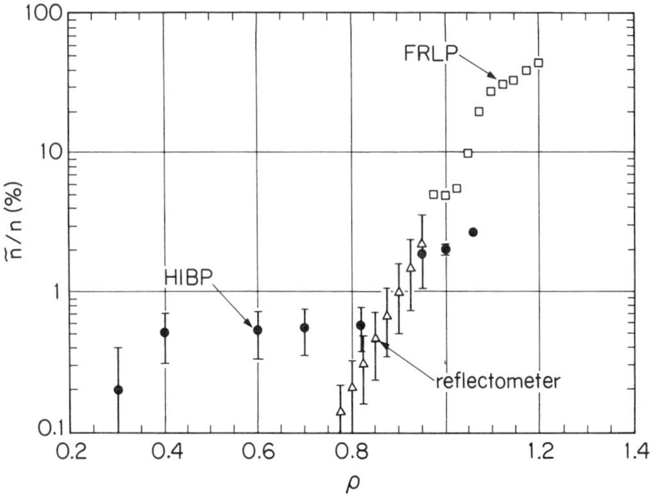

Fig. 8.27 Normalized density fluctuations as a function of the radius in ECRH-produced ATF plasmas with a heavy ion beam probe (HIBP), a reflectometer, and a fast reciprocating Langmuir probe (FRLP).

$\tilde{n}/n$ are plotted as a function of the average radius. Usually, turbulence is characterized by broad spectra in both frequency space and wave number space, and by large fluctuation amplitudes. The density fluctuations in the region $0.3 \lesssim \rho \lesssim 1.0$ are measured with the heavy ion beam probe (HIBP), where $\rho = \bar{r}/\bar{a}$ and $\bar{a}$ is the average radius of the last closed magnetic surface. HIBP exploits orbits of high-energy heavy ions such as Tl^{++} and Cs^{++}, which are obtained by ionization of Tl^{+} and Cs^{+} injected at several hundreds of keV into toroidal plasmas. HIBP is also able to measure the electrostatic potential φ, since the heavy ion orbits are affected by the electric field through the change in electric potential from $e\varphi$ to $2e\varphi$.

The density fluctuations in the region $0.8 \lesssim \rho \lesssim 0.95$ are measured using reflectometry, and those in the edge plasma at $0.95 \lesssim \rho \lesssim 1.2$ are measured using fast reciprocating Langmuir probes. The fluctuation amplitudes are less than 1% in the plasma interior, but grow to about 10% in the vicinity of the last closed magnetic surface. They continue to increase outside the last closed magnetic surface, attaining a level as large as about 50% at $\rho \simeq 1.2$. This behavior correlates well with the increase of $\chi_e^{exp}(\bar{r})$ toward the plasma surface.

The drift waves are usually destabilized by Landau damping and trapped particle effects. However, the density fluctuations in Fig. 8.27 cannot be explained by drift wave turbulence only. In particular, the characteristics of the turbulence in the edge plasma region $0.95 \lesssim \rho \lesssim 1.2$ might be different from those in the core plasma region, $0.3 \lesssim \rho \lesssim 0.5$, since the saturation levels of the density fluctuation are significantly different. One approach is to include a radiation effect due to impurities or charge exchange effects due to neutral particles in the drift wave theory. At present, there is no completely successful theory to explain the increase of $\chi_e^{exp}(\bar{r})$ along the radial direction based on drift wave turbulence.

One of the crucial instabilities in heliotron plasmas is the resistive interchange modes destabilized in the magnetic hill region by the pressure gradient. It is expected that the anomalous transport caused by resistive interchange turbulence will play a role in the region $0.7 \lesssim \rho \lesssim 1.0$. Theoretical fluctuation levels based on resistive interchange turbulence have a tendency to coincide with the measurements.

The fluctuation-driven radial particle flux is given by

$$\Gamma_r = \langle \tilde{n}_e \tilde{v}_r \rangle = \frac{\langle \tilde{n}_e \tilde{E}_r \rangle}{B}, \tag{8.189}$$

where $\langle \cdot \rangle$ denotes the average over the magnetic surface and $\tilde{n}_e$ and $\tilde{E}_r$ are the electron density fluctuation and the electric field fluctuation, respectively. It is also noted that Langmuir probes can measure both the density and electric potential (or electric field) fluctuations simultaneously. Since the resistive interchange modes are driven by the pressure gradient in the magnetic hill region, the driving force becomes weak near the last closed magnetic surface, where the pressure gradient is fairly small. Thus the anomalous particle flux (8.189) for the resistive interchange turbulence also decreases there, which is different from the behavior of experimental results.

Here we discuss the trapped ion mode and the ion temperature gradient driven mode. Both instabilities are considered to induce anomalous ion transport. Since the trapped ion mode is associated with differences in particle motion between the trapped and untrapped particles, one expects that this instability will be sensitive to Coulomb collisions. The important mechanism is that trapped particles can be changed to untrapped particles or can be expelled into the untrapped region of velocity space. Thus the perturbation depends on the effective collision frequency, ν_{eff}. When $\varepsilon = \bar{r}/R \ll 1$, the pitch angle in the velocity space, θ_c, which separates the region of trapped particles from untrapped particles, is of the order of $\sqrt{\varepsilon}$. Thus the fraction of trapped particles is correspondingly small. In the Coulomb collision term for trapped particles, we retain only the dominant term with the second derivative of the particle distribution function:

$$C(f_j) = \nu_j v_{Tj}^2 \Delta_v f_j, \tag{8.190}$$

where Δ_v is the Laplacian in the velocity space, ν_j is the collision frequency for particles of species j (for electrons $j = e$ and for ions $j = i$), and $v_{Tj}^2 = 2T_j/m_j$ is the thermal velocity. The largest term in Δ_v becomes the second derivative with respect to the pitch angle, and the collision term (8.190) is approximated as $C(f_j) \simeq -\nu_j v_{Tj}^2 f_j/(v_{Tj}\theta_c)^2 = -\nu_j f_j/\theta_c^2 \simeq -\nu_j f_j/\varepsilon$. Thus $\nu_{eff}^i = \nu_j/\varepsilon$.

The continuity equation for the trapped particle density, n_{jt}, is written as

$$\frac{\partial n_{jt}}{\partial t} + \frac{\hat{z} \times \nabla\varphi}{B} \cdot \nabla n_{jt} = -\nu_{eff}^j (n_{jt} - \sqrt{\varepsilon} n_0), \tag{8.191}$$

where the RHS describes the variation of trapped particles due to Coulomb collisions, and φ is an electrostatic potential. It is noted that the number density of trapped particles at an equilibrium is given by $n_{jt} = \sqrt{\varepsilon} n_0$. If the number of trapped particles exceeds this value, they will be ejected from the trapped region of velocity space by the Coulomb collisions. If their number becomes smaller, the opposite process takes place and the untrapped particles fill up the trapped region of velocity space. It is also noted that magnetic drift is neglected here, since it is not important for dissipative trapped particle instability.

We may simplify the problem by averaging (8.191) over the magnetic line of force: the equation then becomes two-dimensional. The electrostatic potential in (8.191) is determined from the condition of quasi-neutrality:

$$n_{it} - \frac{e\varphi}{T_i} n_0(1 - \sqrt{\varepsilon}) = n_{et} + \frac{e\varphi}{T_e} n_0(1 - \sqrt{\varepsilon}). \tag{8.192}$$

We take a slab plasma model to solve the two-dimensional problem, where z is the direction of the magnetic field line and y corresponds to the poloidal direction. For the inhomogeneous direction or x-direction, we use the local approximation, or the fact that the perturbation is localized on the surface with the largest driving force or the largest density gradient. For the perturbation in the form $\varphi \propto \exp(-i\omega t + ik_y y)$, linearizing (8.191) and (8.192) yields the dispersion equation for the trapped ion mode:

$$\frac{1}{\sqrt{\varepsilon}}\left(\frac{1}{T_i}+\frac{1}{T_e}\right)=\frac{1}{T_i}\frac{\omega-\omega_{*i}}{\omega+i\nu_i/\varepsilon}+\frac{1}{T_e}\frac{\omega+\omega_{*e}}{\omega+i\nu_e/\varepsilon}, \tag{8.193}$$

where $\omega_{*j} = (k_y T_j/eBn_0)(dn_0/dx)$. We have assumed that $\nu^i_{eff} \ll |\omega_{*i}|$ and $\nu^e_{eff} \ll \omega_{*e}$. Here we neglect ω compared to ω_{*e} and ω_{*i} in the numerator, and retain only $i\nu_e/\varepsilon$ in the denominator of the last term. This approximation means that the contribution of the trapped ions is large compared to that of the trapped electrons. We then obtain

$$\omega = \frac{\sqrt{\varepsilon}}{(1+T_e/T_i)}\omega_{*e} + i\frac{\varepsilon^2}{(1+T_e/T_i)^2}\frac{(\omega_{*e})^2}{\nu_e} - \frac{i\nu_i}{\varepsilon}, \tag{8.194}$$

where $\omega_{*e} = -(T_e/T_i)w_{*i}$ has been used. Thus the electron collisions give rise to the growth of the pertubation, while the ion collisions have a stabilizing effect. For $\nu_e\nu_i < \varepsilon^3(\omega_{*e})^2/(1+T_e/T_i)^2$, the (dissipative) trapped ion mode becomes unstable. As ν_e is reduced, the growth rate γ increases.

For the ion-temperature gradient driven (ITG) mode, we consider an inhomogeneous plasma immersed in the uniform magnetic field. The following equations describe the ITG mode:

$$\frac{\partial n_i}{\partial t} + \frac{\mathbf{b}\times\nabla\varphi}{B}\cdot\nabla n_i + \mathbf{b}\cdot\nabla(n_i v_{\|i}) = 0, \tag{8.195}$$

$$m_i n_i\left(\frac{\partial v_{\|i}}{\partial t} + v_{\|i}\mathbf{b}\cdot\nabla v_{\|i} + \frac{\mathbf{b}\times\nabla\varphi}{B}\cdot\nabla v_{\|i}\right) = -\mathbf{b}\cdot\nabla P_i - en_i\mathbf{b}\cdot\nabla\varphi, \tag{8.196}$$

$$\frac{\partial P_i}{\partial t} + \frac{\mathbf{b}\times\nabla\varphi}{B}\cdot\nabla P_i + v_{\|i}\mathbf{b}\cdot\nabla P_i + \tfrac{5}{3}P_i\mathbf{b}\cdot\nabla v_{\|i} = 0, \tag{8.197}$$

where **b** is a unit vector in the direction of the magnetic field. In the ion continuity equation (8.195) and the equation of parallel ion motion (8.196), the velocity perpendicular to the magnetic field line includes only $\mathbf{E}\times\mathbf{B}$ drift, which comes from the fact that the diamagnetic drift contribution is canceled by the gyro-viscosity term. In the ion thermal transport equation (8.197), the compressibility is included in the last term. For electrons, the inertia is negligible, and

$$-\mathbf{b}\cdot\nabla P_e + en_e\mathbf{b}\cdot\nabla\varphi = 0 \tag{8.198}$$

is employed under the condition of $T_{0e} =$ constant.

Since $n_0 = n_{e0} = n_{i0}$ and T_{0i} are inhomogeneous in the x-direction, the linearized equations obtained from (8.195)–(8.197) yield

$$\frac{\tilde{n}_i}{n_0} = -\frac{k_y}{B\omega n_0}\frac{dn_0}{dx}\tilde{\varphi} + \frac{k_\parallel \tilde{v}_{\parallel i}}{\omega}, \tag{8.199}$$

$$\tilde{v}_{\parallel i} = \frac{k_\parallel e}{m_i \omega}\left(\tilde{\varphi} + \frac{\tilde{P}_i}{en_0}\right), \tag{8.200}$$

$$\tilde{P}_i = -\frac{k_y}{B\omega}\frac{dP_{0i}}{dx}\tilde{\varphi}, \tag{8.201}$$

where the perturbed quantities are described as $\{\tilde{n}_i, \tilde{v}_{\parallel i}, \tilde{P}_i, \tilde{\varphi}\}\exp(-i\omega t + ik_y y + ik_\parallel z)$ under the local approximation. Here k_y and $k_\parallel$ are wave numbers in the y-direction and along the magnetic field line, respectively. Also, incompressibility is assumed by neglecting the last term in (8.197) for a while. Equation (8.198) gives the Boltzmann relation, $\tilde{n}_e/n_0 = e\tilde{\varphi}/T_e$. By assuming charge neutrality, $\tilde{n}_e = \tilde{n}_i$, we obtain the dispersion relation

$$1 - \frac{\omega_{*e}}{\omega} - \frac{k_\parallel^2 T_{0e}}{m_i \omega^2}\left(1 - \frac{\omega_{pi}^*}{\omega}\right) = 0, \tag{8.202}$$

where $\omega_{pi}^* = (k_y/en_0B)dP_{0i}/dx)$. For $\omega \ll \omega_{*e} = -\omega_{*i}$ and $T_{0e} = T_{0i}$, this dispersion relation is approximated as

$$\omega^2 = -\frac{k_\parallel^2 T_{0e}}{m_i}(1 + \eta_i), \tag{8.203}$$

where $\eta_i \equiv d\ln T_{0i}/d\ln n_0$. In this case, one of the roots indicates instability for $\eta_i > -1$. On the other hand, for $\eta_i \gg 1$ and $\omega_{*e} \ll \omega \ll \omega_{Ti}^*$, (8.202) becomes

$$\omega^3 = -\frac{k_\parallel^2 T_{0e}}{m_i}\omega_{T_i}^* \tag{8.204}$$

which also includes the unstable solution, where $\omega_{Ti}^* = (k_y/eB)(dT_{0i}/dx)$.

Here we include the compressibility by retaining the last term in (8.197), which gives the dispersion relation

$$1 - \frac{\omega_{*e}}{\omega} - \frac{k_\parallel^2 T_{0e}}{m_i \omega^2}\left(1 - \frac{\omega_{pi}^*}{\omega}\right) - \frac{5}{3}\frac{k_\parallel^2 T_{0i}}{m_i \omega^2}\left(1 - \frac{\omega_{*e}}{\omega}\right) = 0, \tag{8.205}$$

where $\tilde{v}_{\parallel i}$ obtained from (8.199) has been substituted into the new term that has appeared in (8.201). For $\omega \ll \omega_{*e} = -\omega_{*i}$ and $T_{0e} = T_{0i}$, (8.205) reduces to

$$\omega^2 = \frac{k_\parallel^2 T_{0i}}{m_i}\left(\tfrac{2}{3} - \eta_i\right). \tag{8.206}$$

In this case, the instability appears for $\eta_i > 2/3$. These unstable modes are called the ITG mode or the η_i-mode in the slab geometry. The η_i-mode turbulence in the realistic toroidal geometry is considered to be the origin of the anomalous ion thermal transport in tokamaks.

8.7 MIXING LENGTH THEORY AND SCALE INVARIANCE

In the theoretical model for the anomalous transport discussed in section 8.6, one of the useful ingredients is mixing length theory. For fully developed isotropic turbulence, the saturation level of the density fluctuation is expected to be

$$\frac{\tilde{n}}{n} \simeq \frac{1}{\bar{k}L_n}, \tag{8.207}$$

which is often called the mixing length theory, where $\bar{k}$ is an average wave number and L_n is a characteristic length of background density gradient. A comparison between the experimentally obtained $\tilde{n}/n$ and the RHS of (8.207) is shown in Fig. 8.28, which includes experimental results from several tokamaks. Here $r_s = 1/\bar{k}$ is introduced. The straight dot and dashed line corresponds to (8.207). It seems that the mixing length theory is valid.

A recent topic of turbulent fluctuation study concerns the shear flow effects on the turbulent behavior and the associated anomalous transport. For the L–H transition of tokamak plasmas, enhancement of the poloidal flow velocity shear seems to be correlated with reduction in the density fluctuation. In Fig. 8.29 are shown the radial profiles of the phase velocity of density fluctuation, V_{ph}, near the plasma edge region for various tokamaks (TEXT, ASDEX, TEXT-U, TFTR, and Pheadrus-T) and stellarators (ATF, and W7-AS). Here the phase velocity is approximately equal to the $\mathbf{E} \times \mathbf{B}$ drift velocity due to the radial electric field, E_r/B. The ordinate is the distance from the radius of $V_{ph} = 0$. It is noted that velocities as high as 5 km/sec are reported in W7-AS. When $\partial V_{ph}/\partial r \simeq 1/B(\partial E_r/\partial r)$ is high, any turbulent structures experience a flow velocity shear over a radial correlation length, which has a tendency to suppress the turbulence.

We briefly discuss MHD equilibrium inlcuding the plasma flow velocity $\mathbf{u}$ for an axisymmetric toroidal plasma. We assume the plasma to be confined within a perfectly conducting chamber with the boundary conditions of zero normal components of $\mathbf{u}$ and $\mathbf{B}$ at the wall (see chapter 4). First, we note that the plasma flow must be within the magnetic surfaces. This can be derived from Faraday's law, $\nabla \times (\mathbf{u} \times \mathbf{B}) = 0$, which implies that

$$\mathbf{u} \times \mathbf{B} = \nabla f, \tag{8.208}$$

where a multivalued function is allowed for f. However, since ∇f is single-valued, the jump in f is constant when it is continued once the long way around

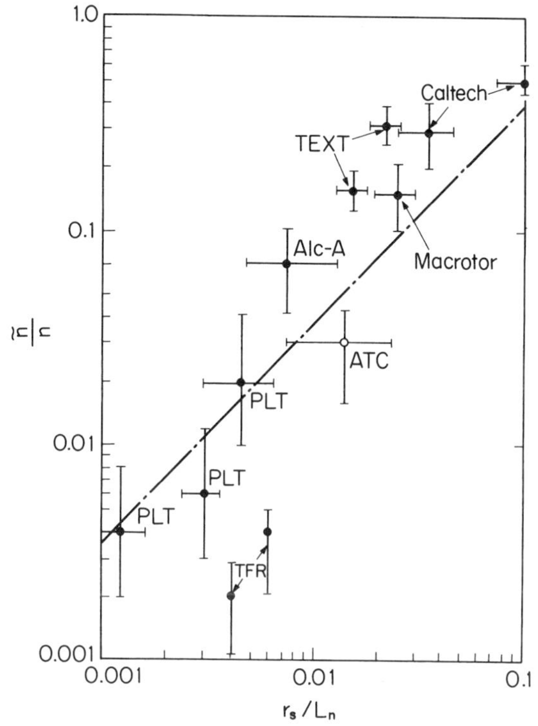

Fig. 8.28 The dependence of density fluctuation level, $\tilde{n}/n$, on r_s/L_n for various tokamak devices. The straight line denotes the relationship $\tilde{n}/n = r_s/L_n$, where r_s and L_n denote the characteristic length of the density fluctuation and that of background density gradient, respectively.

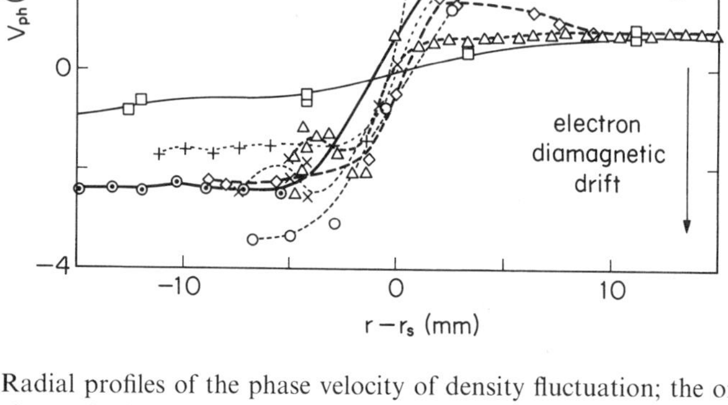

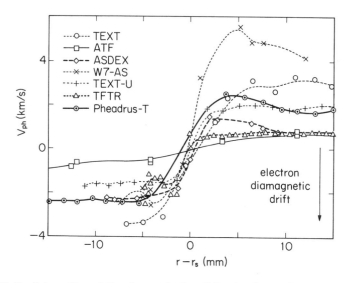

Fig. 8.29 Radial profiles of the phase velocity of density fluctuation; the ordinate denoteds the distance from the radius of zero phase velocity.

the torus, regardless of the contour of continuation. Thus $f = \chi + c\phi$, where χ is the single-valued, c is a constant, and ϕ is a toroidal angle that increases by 2π when continued once the long way around the torus. Thus $\mathbf{u} \times \mathbf{B} = \nabla\chi + c\nabla\phi$. Now we multiply this relation to $d\ell$, a length element vector, and integrate once along a closed toroidal contour that lies on the plasma boundary. Since $\mathbf{u} \times \mathbf{B}$ is normal to the boundary surface, the integral gives $2\pi c = 0$, and hence f is single-valued. We note that (8.208) now implies that $\mathbf{B} \cdot \nabla f = 0$ and $\mathbf{u} \cdot \nabla f = 0$, which means that the plasma flow exists within the magnetic surface defined by $f =$ constant. The axisymmetric magnetic field is assumed in the form $\mathbf{B} = \mathbf{B}_p + \mathbf{B}_\phi = \nabla\phi \times \nabla\Psi + B_\phi \mathbf{e}_\phi$, where $\nabla\phi = \mathbf{e}_\phi / r$ and $\Psi(r, z)$ is a poloidal flux function describing the magnetic surface in cylindrical coordinates (r, ϕ, z) (see chapter 4). By noting that $\Psi =$ constant also describes a magnetic surface, we have $\nabla f = -\Omega(\Psi)\nabla\Psi$. Then the constraint $\mathbf{u} \cdot \nabla f = 0$ implies $\mathbf{u}_p = \lambda \mathbf{B}_p$, where λ is a scalar function and the subscript p refers to the poloidal component. The $\nabla\Psi$-component of (8.208) yields $u_\phi = \lambda B_\phi + r\Omega$. From the continuity equation $\nabla \cdot (\rho \mathbf{u}) = 0$, we obtain $\mathbf{B}_p \cdot \nabla(\lambda\rho) = 0$ or $\lambda = \Phi(\Psi)/\rho$ with an arbitrary function $\Phi(\Psi)$, where ρ is a mass density. Thus the flow velocity is given by

$$\mathbf{u} = \frac{\Phi(\Psi)}{\rho} \mathbf{B} + r\Omega(\Psi)\mathbf{e}_\phi. \tag{8.209}$$

By noting that the adiabatic pressure law gives $\mathbf{u} \cdot \nabla S = 0$ in a stationary state, $\mathbf{B} \cdot \nabla S = 0$ is obtained in case of $\Phi \neq 0$ for an axisymmetric torus. Here S denotes a function of the specific entropy $S = P\rho^{-\gamma}$, where P is a pressure and γ is the ratio of specific heats. Since $\mathbf{B} \cdot \nabla S = 0$ is equivalent to $S = S(\Psi)$, we note that P is not a function of Ψ only.

The ϕ component of the momentum equation,

$$\rho(\mathbf{u} \cdot \nabla)\mathbf{u} + \nabla P = \mathbf{J} \times \mathbf{B}, \tag{8.210}$$

gives $(\nabla\phi \times \nabla\Psi) \cdot \nabla[r(B_\phi - \Phi u_\phi)] = 0$, which implies that

$$\mathbf{B} \cdot \nabla[r(B_\phi - \Phi u_\phi)] = 0 \tag{8.211}$$

for the axisymmetric torus or

$$B_\phi = \frac{1}{1 - \Phi^2/\rho}\left(\frac{I(\Psi)}{r} + r\Phi\Omega\right) \tag{8.212}$$

for an arbitrary function $I(\Psi)$. In the derivation of (8.211) we have used $\Phi(\Psi)$ and $\mathbf{B} \times (\nabla \times \mathbf{B}) \cdot \mathbf{e}_\phi = [(\mathbf{B} \cdot \nabla)(rB_\phi)]/r$. Multiplying the momentum equation (8.210) by $\mathbf{B}/\rho$ and noting that

$$\frac{1}{\rho}\,\mathbf{B}\cdot\nabla P = \gamma p^{\gamma-2} S\mathbf{B}\cdot\nabla\rho = \mathbf{B}\cdot\nabla[\gamma S\rho^{\gamma-1}/(\gamma-1)], \tag{8.213}$$

$$\begin{aligned}\mathbf{B}\cdot[(\mathbf{u}\cdot\nabla)\mathbf{u}] &= \mathbf{B}\cdot\nabla(\mathbf{u}\cdot\mathbf{u}/2) + \nabla f\cdot\nabla\times\mathbf{u}\\ &= \mathbf{B}\cdot\nabla(u^2/2) - \Omega(\Psi)\nabla\Psi\cdot\nabla\times\mathbf{u}\\ &= \mathbf{B}\cdot\nabla(u^2/2 - ru_\phi\Omega),\end{aligned} \tag{8.214}$$

we can obtain another equation of the form $\mathbf{B}\cdot\nabla f = 0$, which yields

$$\frac{\Phi^2}{2\rho^2}\,B^2 - \frac{1}{2}\,r^2\Omega^2 + \frac{\gamma}{\gamma-1}\,S(\Psi)\rho^{\gamma-1} = H(\Psi) \tag{8.215}$$

for an arbitrary function $H(\Psi)$. In the RHS of (8.214), (8.209) is substituted into $\mathbf{u}$ to obtain the last expression. This equation is the analog of Bernoulli's law in fluid dynamics. With some manipulation, the $\nabla\Psi$ component of (8.210) yields

$$\begin{aligned}\nabla\cdot\left[\left(1-\frac{\Phi^2}{\rho}\right)\frac{1}{r^2}\,\nabla\Psi\right] + \mathbf{u}\cdot\mathbf{B}\,\frac{d\Phi}{d\Psi} + r\rho u_\phi\,\frac{d\Omega}{d\Psi}&\\ + \frac{1}{r}\,B_\phi\,\frac{dI}{d\Psi} + \rho\,\frac{dH}{d\Psi} - \frac{1}{\gamma-1}\,\rho^\gamma\,\frac{dS}{d\Psi} &= 0.\end{aligned} \tag{8.216}$$

The solution of (8.215) and (8.216) depends on the five arbitrary functions of Ψ; $\Phi(\Psi)$, $\Omega(\Psi)$, $I(\Psi)$, $H(\Psi)$, and $S(\Psi)$. Also, the relations (8.209) and (8.212) must be used. This is the analog of the usual Grad–Shafranov equation for static equilibria, which is obtained by assuming $\Phi(\Psi) = 0$ and $\Omega(\Psi) = 0$ in (8.215) and (8.216). In this case two arbitrary functions of Ψ, $I(\Psi)$ and $S(\Psi)$, remain, and $S(\Psi)$ can be transformed to $P(\Psi)$. It should be noted that (8.216) is not always elliptic with the increase of flow velocity, and solving it with a boundary condition for Ψ will not be appropriate in such a case. This is similar to the situation in fluid dynamics when a supersonic flow appears. Stability of a rotating toroidal plasma is an interesting future subject.

Now we consider the two-dimensional Navier–Stokes equation

$$\frac{\partial}{\partial t}\,\nabla_\perp^2\varphi + \nabla_\perp\varphi\times\hat{\mathbf{z}}\cdot\nabla_\perp^2\varphi = \mu\nabla_\perp^4\varphi, \tag{8.217}$$

where φ is a stream function and μ is a kinematic viscosity. We assume that two-dimensional fluid turbulence is described by (8.217), and that a fluctuation energy spectrum, $E(k)$, has the form

$$E(k) = \sum C_{stu} k^s \epsilon^t \mu^u. \tag{8.218}$$

We note that $E(k)$ and k have dimensions, $[E(k)] = [L]^3[T]^{-2}$ and $[k] = [L]^{-1}$, and the local turbulent energy source ϵ and the kinematic viscosity μ corresponding to the local energy sink have dimensions, $[\epsilon] = [L]^2[T]^{-3}$ and $[\mu] = [L]^2[T]^{-1}$.

By following the dimensional analysis, $E(k)$, k, ϵ, and μ are expressed as

$$\begin{cases} E(k) \to \alpha_1^3 \alpha_2^{-2} E(k), \; k \to \alpha_1^{-1} k, \\ \epsilon \to \alpha_1^2 \alpha_2^{-3} \epsilon, \; \mu \to \alpha_1^2 \alpha_2^{-1} \mu, \end{cases} \tag{8.219}$$

where scale transformation of length is given by α_1, and that of time is given by α_2. We note that both transformations are introduced without taking into account of the Navier–Stokes equation. From (8.218) and the relations (8.219), we obtain

$$E(k) = k\mu^2 F(\epsilon/k^4\mu^3), \tag{8.220}$$

where F is a function of $\epsilon/k^4\mu^3$. By following Kolmogorov's conjecture that the energy spectrum $E(k)$ depends only on k and ϵ in the inertial range, we obtain $F(x) = x^{2/3}$. Then (8.220) becomes

$$E(k) = \epsilon^{2/3} k^{-5/3}. \tag{8.221}$$

On the other hand, the Navier–Stokes equation (8.217) has the following transformations for the isotropic turbulence in the x–y plane,

$$\begin{cases} t \to \beta_1 t, \; x \to \beta_2 x, \; y \to \beta_2 y, \\ \varphi \to \beta_1^{-1} \beta_2 \varphi, \; \mu \to \beta_1^{-1} \beta_2^2 \mu, \end{cases} \tag{8.222}$$

which do not change the form of the Navier–Stokes equation. By comparing (8.222) with (8.219), we find that β_1 corresponds to α_2 and β_2 to α_1. Thus we can obtain the well-known Kolmogorov spectrum (8.221) with the scale transformations (8.222).

The invariance of scale transformation has a foundation in the Buckingham *Pi* theorem. First, we assume that a physical law has a relation

$$f(q_1, q_2, \cdots, q_m) = 0 \tag{8.223}$$

among m quantities, $q_1, q_2, \cdots, q_m$, which are dimensional. Usually, the dimensions are expressed in terms of certain selected fundamental dimensions $L_1, L_2, \cdots, L_n$ ($n < m$), as

$$[q_i] = L_1^{a_{1i}} L_2^{a_{2i}} \cdots L_n^{a_{ni}}, \tag{8.224}$$

with exponents $a_{1i}, a_{2i}, \cdots, a_{ni}$. If $[q_i] = 1$, then q_i is a dimensionless quantity. The matrix obtained from (8.224) is called the dimension matrix,

$$\mathbf{A} = \begin{bmatrix} a_{11} & \cdots\cdots & a_{1m} \\ \vdots & & \vdots \\ a_{n1} & \cdots\cdots & a_{nm} \end{bmatrix}. \tag{8.225}$$

The elements in the jth column give the exponents for q_j in terms of the powers of $L_1, L_2, \cdots, L_n$, as shown in (8.224). Any fundamental dimension L_i can be changed by multiplication with an appropriate conversion factor $\lambda_i > 0$, which gives $\bar{L}_i$ in a new system of units:

$$\bar{L}_i = \lambda_i L_i \quad (i = 1, \cdots n). \tag{8.226}$$

If q has a dimension $[q] = L_1^{b_1} L_2^{b_2} \cdots L_n^{b_n}$, with exponents $b_1, b_2, \cdots, b_n$, then $\bar{q} = \lambda_1^{b_1} \lambda_2^{b_2} \cdots L_n^{b_n}\ q$ gives the quantity in the new system of units. Here we introduce a mathematical definition: the physical law (8.223) is unit free if we have $f(\bar{q}_1, \bar{q}_2, \cdots, \bar{q}_m) = 0$ for all choices of positive real numbers $\lambda_1, \lambda_2, \cdots, \lambda_n$. Then the *Pi* theorem can be described in the following way. Let $f(q_1, q_2, \cdots, q_m) = 0$ be a unit-free physical law amongst the *dimensional* quantities $q_1, q_2 \cdots, q_m$. Also, let $L_1, L_2, \cdots, L_n$ $(n < m)$ be the fundamental dimensions as shown in (8.224) and let $r = \text{rank}\ \mathbf{A}$ be the rank of the dimension matrix (8.225). Then there exist $m - r$ independent *dimensionless* quantities, $\Pi_1, \cdots, \Pi_{m-r}$, which can be formed from $q_1, \cdots, \mathrm{q}_m$, and the physical law becomes equivalent to

$$F(\Pi_1, \Pi_2, \cdots, \Pi_{m-r}) = 0, \tag{8.227}$$

which is expressed only in terms of the dimensionless quantities.

From knowledge of linear algebra, we can understand that, among m quantities $q_1, \cdots, q_m$, there are $m - r$ independent dimensionless variables that can be formed from $q_1, \cdots, q_m$, where r is the rank of the dimension matrix $\mathbf{A}$. Also, if $\Pi_1, \cdots, \Pi_{m-r}$ are the $m - r$ dimensionless variables, then $f(q_1, q_2, \cdots, q_m) = 0$ becomes equivalent to the relation (8.227). We will demonstrate this statement for a simple example with $m = 4$, $n = 2$, and $r = 2$.

A unit-free physical law is shown as

$$f(q_1, q_2, q_3, q_4) = 0, \tag{8.228}$$

with

$$[q_i] = L_1^{a_{1i}} L_2^{a_{2i}}, \quad i = 1, \cdots 4. \tag{8.229}$$

Then the dimension matrix is written as

$$\mathbf{A} = \begin{pmatrix} a_{11} & a_{12} & a_{13} & a_{14} \\ a_{21} & a_{22} & a_{23} & a_{24} \end{pmatrix} \tag{8.230}$$

with $r = \text{rank}\ \mathbf{A} = 2$. If Π is a dimensionless variable, then it is shown as

$$\Pi = q_1^{\alpha_1} q_2^{\alpha_2} q_3^{\alpha_3} q_4^{\alpha_4} \tag{8.231}$$

with exponents $\alpha_1, \alpha_2, \alpha_3$, and α_4. Substituting (8.229) into (8.231) yields

$$\begin{pmatrix} a_{11} \\ a_{21} \end{pmatrix}\alpha_1 + \begin{pmatrix} a_{12} \\ a_{22} \end{pmatrix}\alpha_2 + \begin{pmatrix} a_{13} \\ a_{23} \end{pmatrix}\alpha_3 + \begin{pmatrix} a_{14} \\ a_{24} \end{pmatrix}\alpha_4 = \begin{pmatrix} 0 \\ 0 \end{pmatrix}. \tag{8.232}$$

Without loss of generality, we can assume that the first two columns of **A** are linearly independent. In this case, columns three and four can be written in the form of linear combination of the first two columns,

$$\begin{pmatrix} a_{13} \\ a_{23} \end{pmatrix} = C_{31}\begin{pmatrix} a_{11} \\ a_{21} \end{pmatrix} + C_{32}\begin{pmatrix} a_{12} \\ a_{22} \end{pmatrix} \tag{8.233}$$

$$\begin{pmatrix} a_{14} \\ a_{24} \end{pmatrix} = C_{41}\begin{pmatrix} a_{11} \\ a_{21} \end{pmatrix} + C_{42}\begin{pmatrix} a_{12} \\ a_{22} \end{pmatrix}, \tag{8.234}$$

where C_{31}, C_{32}, C_{41}, and C_{42} are constants. Substituting (8.233) and (8.234) into (8.232) yields

$$\begin{cases} \alpha_1 + C_{31}\alpha_3 + C_{41}\alpha_4 = 0 \\ \alpha_2 + C_{32}\alpha_3 + C_{42}\alpha_4 = 0. \end{cases} \tag{8.235}$$

By solving α_1 and α_2 in terms of α_3 and α_4, we obtain

$$\begin{pmatrix} \alpha_1 \\ \alpha_2 \\ \alpha_3 \\ \alpha_4 \end{pmatrix} = \alpha_3 \begin{pmatrix} -C_{31} \\ -C_{32} \\ 1 \\ 0 \end{pmatrix} + \alpha_4 \begin{pmatrix} -C_{41} \\ -C_{42} \\ 0 \\ 1 \end{pmatrix}. \tag{8.236}$$

The two vectors appearing on the RHS of (8.236) represent two linearly independent solutions of (8.233) and (8.234). Thus the two dimensionless quantities are

$$\begin{cases} \Pi_1 = q_1^{-C_{31}} q_2^{-C_{32}} q_3 \\ \Pi_2 = q_1^{-C_{41}} q_2^{-C_{42}} q_4, \end{cases} \tag{8.237}$$

and the physical law is shown as

$$G(q_1, q_2, \Pi_1, \Pi_2) \equiv f(q_1, q_2, \Pi_1 q_1^{C_{31}} q_2^{C_{32}}, \Pi_2 q_1^{C_{41}} q_2^{C_{42}}) = 0 \tag{8.238}$$

by introducing a function G.

Next, we show that the physical law $G(q_1, q_2, \Pi_1, \Pi_2)$ is equivalent to $G(1, 1, \Pi_1, \Pi_2) = 0$ or $F(\Pi_1, \Pi_2) = 0$. Since $G(q_1, q_2, \Pi_1, \Pi_2) = 0$ is a unit-free law, we can write it as

$$G(\bar{q}_1, \bar{q}_2, \Pi_1, \Pi_2) = 0, \tag{8.239}$$

where $\bar{q}_1 = \lambda_1^{a_{11}}\lambda_2^{a_{21}} q_1$ and $\bar{q}_2 = \lambda_1^{a_{12}}\lambda_2^{a_{22}} q_2$ for a choice of conversion factor $\lambda_1, \lambda_2 > 0$. If $\lambda_1^{a_{11}}\lambda_2^{a_{21}} q_1 = 1$ and $\lambda_1^{a_{12}}\lambda_2^{a_{22}} q_2 = 1$ can be satisfied, we obtain $G(1, 1, \Pi_1, \Pi_2)$ from (8.239). These equations are shown as

$$\begin{pmatrix} a_{12} & a_{21} \\ a_{12} & a_{22} \end{pmatrix} \begin{pmatrix} \ln \lambda_1 \\ \ln \lambda_2 \end{pmatrix} = - \begin{pmatrix} \ln q_1 \\ \ln q_2 \end{pmatrix}, \tag{8.240}$$

which gives unique solutions for λ_1 and λ_2, since we have assumed that the two first columns of (8.230) are linearly independent.

Here we apply the scale invariance to resistive interchange turbulence in order to estimate the anomalous transport coefficient. In chapter 3 we have derived reduced MHD (RMHD) equations to describe the interchange turbulence; (3.200), (3.212), and (3.221). We introduce a slab plasma model with local coordinates (x, y, ζ) defined by $(r - r_0)dq/dr \to x$, $z - q\theta \to y$, and $\theta \to \zeta$, or the field-aligned system, where (r, θ, z) correspond to the cylindrical coordinates. Here r_0 denotes a reference radial position. Two distinct length scales exist in the system. The fluctuating quantities vary rapidly in the x- and y-directions but slow in the ζ-direction. Here the temperature is assumed to be a constant T_0. Then the fluctuations of the magnetic flux function, the stream function, and the density are given by $\Psi = \Psi_0 + \tilde{\Psi}$, $u = \tilde{u}$, and $n = n_0(1 + \tilde{n})$, where n_0 and Ψ_0 are the equilibrium density and the magnetic flux function providing the rotational transform. The normalized RMHD equations can be written as for $\tilde{\Psi}(= \tilde{\Psi}/a^2 B_0)$, $\tilde{u}(= \tilde{u}/a^2 V_A)$, and $\tilde{n}$:

$$\frac{\partial \tilde{\Psi}}{\partial t} = \frac{\bar{B}_0}{q\bar{R}} \nabla_\parallel \tilde{u} + \bar{\eta}\nabla_\perp^2 \tilde{\Psi}, \tag{8.241}$$

$$\frac{d}{dt}\nabla_\perp^2 \tilde{u} = \frac{\bar{B}_0}{q\bar{R}} \nabla_\parallel \nabla_\perp^2 \tilde{\Psi} + \frac{q^2 s}{\bar{R}} [\tilde{\Psi}, \nabla_\perp^2 \tilde{\Psi}] - \frac{\bar{\beta}\bar{B}_0^2 \bar{a}}{q\bar{R}} \frac{\partial \tilde{n}}{\partial y}, \tag{8.242}$$

$$\frac{d\tilde{n}}{dt} - \frac{\bar{\kappa} q \bar{a}}{\bar{R}} \frac{\partial \tilde{u}}{\partial y} = 0, \tag{8.243}$$

where

$$\nabla_\parallel = \frac{\partial}{\partial \zeta}, \tag{8.244}$$

$$\nabla_\perp^2 = q^2 s^2 \left(\frac{\partial}{\partial x} - \zeta \frac{\partial}{\partial y} \right)^2 + q^2 \frac{\partial^2}{\partial y^2}, \tag{8.245}$$

$$\frac{d}{dt} = \frac{\partial}{\partial t} + \frac{q^2 s}{\bar{R}} + \frac{q^2 s}{\bar{R}} [\tilde{u}, \ ] \tag{8.246}$$

and $\bar{\kappa} = (r/n_0)(dn_0/dr)$, $\bar{\beta} = \beta_p r d\Omega/dr$, $\bar{\eta} = \eta R/V_A a^2$, $q = 1/\iota$, and $s = (r/q)(dq/dr)$. Here x and y are normalized by a, and t by R/V_A, where R is a major radius. Also, ζ is non-dimensional, and β_p is a poloidal beta. In (8.242) and (8.246), the Poisson bracket has been used (see chapter 3). The quantity s

in (8.242) is called the shear parameter, and V_A is the Alfvén velocity. For normalized quantities, we simply take $\bar{B}_0 = \bar{R} = \bar{a} = 1$.

For low-β plasmas, we take the electrostatic limit and assume localized modes around $r = r_0$ with $|\partial/\partial x| \gg |\partial/\partial y|$. Then (8.241)–(8.243) are reduced to

$$\frac{1}{q}\nabla_\| \tilde{u} = \bar{\eta} q^2 s^2 \frac{\partial^2}{\partial x^2}\tilde{\Psi}, \tag{8.247}$$

$$\frac{d}{dt} q^2 s^2 \frac{\partial^2}{\partial x^2}\tilde{u} = \frac{1}{q}\nabla_\| q^2 s^2 \frac{\partial^2}{\partial x^2}\tilde{\Psi} - \frac{\bar{\beta}}{q}\frac{\partial \tilde{n}}{\partial y}, \tag{8.248}$$

$$\frac{\partial \tilde{n}}{\partial t} + q^2 s\left(\frac{\partial \tilde{u}}{\partial x}\frac{\partial \tilde{n}}{\partial y} - \frac{\partial \tilde{u}}{\partial y}\frac{\partial \tilde{n}}{\partial x}\right) - \bar{\kappa} q \frac{\partial \tilde{u}}{\partial y} = 0. \tag{8.249}$$

There are the following scale transformations that make (8.247)–(8.249) invariant:

$$\begin{array}{lll} t \to \lambda_1^{-1/2}\lambda_2^{-1/2}\lambda_3\lambda_4^{-1/2} t & x \to \lambda_3 x & y \to \lambda_3 y, \\ s \to \lambda_2 s & \tilde{u} \to \lambda_1^{-3/2}\lambda_2^{-1/2}\lambda_3\lambda_4^{1/2}\tilde{u} & q \to \lambda_1 q, \\ \tilde{\Psi} \to \lambda_4 \tilde{\Psi} & \bar{\eta} \to \lambda_1^{-9/2}\lambda_2^{-5/2}\lambda_3^3\lambda_4^{-1/2}\bar{\eta} & \tilde{n} \to \lambda_5 \tilde{n}, \\ \bar{\kappa} \to \lambda_1\lambda_2\lambda_3^{-1}\lambda_5\bar{\kappa} & \bar{\beta} \to \lambda_1^2\lambda_2^2\lambda_3^{-1}\lambda_5^{-1}\bar{\beta}, & \end{array} \tag{8.250}$$

which implies the existence of five independent scale transformations: $(\lambda_1, \lambda_2, \lambda_3, \lambda_4, \lambda_5) = (\lambda, 1, 1, 1, 1)$; $(\lambda_1, \lambda_2, \lambda_3, \lambda_4, \lambda_5) = (1, \lambda, 1, 1, 1)$; $(\lambda_1, \lambda_2, \lambda_3, \lambda_4, \lambda_5) = (1, 1, \lambda, 1, 1)$; $(\lambda_1, \lambda_2, \lambda_3, \lambda_4, \lambda_5) = (1, 1, 1, \lambda, 1)$; and $(\lambda_1, \lambda_2, \lambda_3, \lambda_4, \lambda_5) = (1, 1, 1, 1, \lambda)$.

Since the radial transport coefficient is determined by the radial correlation length Δx and the correlation time Δt, it is assumed to depend on the equilibrium quantities, s, q, $\bar{\eta}$, $\bar{\kappa}$, and $\bar{\beta}$, in the following way:

$$D \equiv (\Delta x)^2/\Delta t = s^\alpha q^\beta \bar{\eta}^\gamma \bar{\kappa}^\delta \bar{\beta}^\varepsilon, \tag{8.251}$$

where α, β, γ, δ and ε are constants. It should be noted that a numerical coefficient is assumed, with a value of unity. When Δx is related to the $\mathbf{E} \times \mathbf{B}$ drift motion, the nondimensional transport coefficient D in (8.251) describes convective cross-field transport. Since there are only five scale transformations for the five unknown constants, a functional dependence remains in (8.251) as

$$D_0^s = \bar{\eta}\bar{\kappa}\bar{\beta}/s^2. \tag{8.252}$$

This result has been obtained by requiring that (8.251) does not change after the scale transformations (8.250) are substituted. This is called scale invariance.

When magnetic field fluctuations are caused by pressure gradient driven turbulence, there is another mechanism for the anomalous transport. It occurs when the magnetic perturbations destroy the nested flux surfaces and create the stochastic magnetic field. We consider collisional as well as collisionless transport in the stochastic magnetic field. If the plasma is collisional and the mean free path is less than the correlation length of the perturbed magnetic field lines, L_c, the turbulent transport is given by

$$D_1 = \frac{v_{Te}^2}{\nu_e}\left(\frac{\delta B_r}{B_0}\right)^2, \tag{8.253}$$

where v_{Te} is the electron thermal velocity, ν_e is the electron–ion collision frequency, and δB_r is the radial component of perturbed magnetic field. This estimation is based on the assumption that $\Delta r \simeq (v_{Te}/\nu_e)(\delta B_r/B_0)$ and $\Delta t \simeq 1/\nu_e$. It is noted that v_{Te}^2/ν_e is the collisional parallel transport along the magnetic field line. On the other hand, when the mean free path is greater than L_c, the turbulent transport in the stochastic magnetic field is given by

$$D_2 = v_{Te} L_c \left(\frac{\delta B_r}{B_0}\right)^2. \tag{8.254}$$

It is noted that the correlation length is estimated as $\Delta r \simeq (\delta B_r/B_0)L_c$ and the correlation time as $\Delta t \simeq L_c/v_{Te}$.

In order to obtain the anomalous transport in the stochastic magnetic field, D_1 and D_2, we use the relation

$$\left(\frac{\delta B_r}{B_0}\right)^2 = \frac{a^2 q^2}{R^2}\left(\frac{\partial\tilde{\Psi}}{\partial y}\right)^2, \tag{8.255}$$

which is obtained from $\delta \mathbf{B}_\perp = \nabla\tilde{\Psi} \times \hat{\mathbf{z}}/R$. By using the scale transformations (8.250) for the relation $(\partial\tilde{\Psi}/\partial y)^2 = s^\alpha q^\beta \bar{\eta}^\gamma \bar{\kappa}^\delta \bar{\beta}^\varepsilon$, we find that

$$(\partial\tilde{\Psi}/\partial y)^2 = \bar{\eta} q^{-3} (\bar{\kappa}\bar{\beta}/s^2)^{5/2}, \tag{8.256}$$

by requiring scale invariance. From (8.253), (8.255), and (8.256),

$$D_1^s = \left(\frac{v_{Te}^2}{\nu_e}\right)\frac{\bar{\eta} a^2}{q R^2}\left(\frac{\bar{\kappa}\bar{\beta}}{s^2}\right)^{5/2} \tag{8.257}$$

is obtained. Since $\bar{\eta}$ is proportional to the electron–ion collision frequency ν_e, D_1^s does not depend on the collision frequency. To determine D_2, we need $\Delta r \simeq (\delta B_r/B_0)L_c$, which is equal to $a\Delta x/qs$. By assuming that $\Delta x = s^\alpha q^\beta \bar{\eta}^\gamma \bar{\kappa}^\delta \bar{\beta}^\varepsilon$, the scale transformations (8.250) give

$$\Delta x = \bar{\eta}^{1/2} q^{3/2} \bar{\kappa}^{1/4} s^{1/2} \bar{\beta}^{1/4} \tag{8.258}$$

under scale invariance. From (8.254), (8.255), (8.256), and (8.258),

$$D_2^s = (v_{Te}a^2/R)\bar{\eta}(\bar{\kappa}\bar{\beta}/s^2)^{3/2} \tag{8.259}$$

is obtained.

Finally, we discuss the interrelations between the scale invariance and mixing length theory. As already explained in section 5.5 (see (5.214) for the growth rate and (5.215) for the scale length), linear stability analysis of the resistive interchange mode with a poloidal mode number m gives the following relations:

$$\Delta x^L \simeq r_0\bar{\eta}^{1/3}G_*^{1/6}(|ms|)^{-1/3}, \tag{8.260}$$

$$\Delta t^L \simeq \tau_{hp}\bar{\eta}^{-1/3}G_*^{-2/3}(|ms|)^{-2/3}, \tag{8.261}$$

where $G_* = \bar{\kappa}\bar{\beta}/s^2$ and $\tau_{hp} = (\mu_0\rho_0)^{1/2}r_0/B_p$ is the poloidal Alfvén transit time, where B_p is the poloidal magnetic field. Here the suffix L denotes linear stability theory. We note that, by applying the scale transformations (8.250) to $\Delta x = s^\alpha q^\beta \bar{\eta}^\gamma \bar{\kappa}^\delta \bar{\beta}^\varepsilon$ and $\Delta t = s^\alpha q^\beta \bar{\eta}^\gamma \bar{\kappa}^\delta \bar{\beta}^\varepsilon$, $(\Delta x)^s = s^{1/2}q^{3/2}\bar{\eta}^{1/2}\bar{\kappa}^{1/4}\bar{\beta}^{1/4}$ given in (8.258) and $(\Delta t)^s = sq\bar{\kappa}^{-1/2}\bar{\beta}^{-1/2}$ are obtained under scale invariance. Apparently, (8.260) and (8.261) are different from $(\Delta x)^s$ and $(\Delta t)^s$. It is noted that the former expressions are obtained from linearized equations and the latter ones from nonlinear equations. However, we note that both approaches give an identical transport coefficient, as $(\Delta x^L)^2/\Delta t^L = (\Delta x^s)^2/(q^2 s^2 \Delta t^s) = \bar{\eta}\bar{\kappa}\bar{\beta}/s^2$. Thus the transport coefficient (8.252) given by scale invariance can be reproduced by mixing length theory. In studying the nondimensional parameter dependence of the transport coefficient in tokamaks and stellarators, scale invariance seems useful in comparing theoretical models with experimental results.

BIBLIOGRAPHY

Connor, J. W., and Taylor, J. B. (1984). Resistive fluid turbulence and energy confinement. *Phys. Fluids*, **27**, 2676.

Fielding, S. J., Hugill, J., McCracken, G. M., Paul, J. W. M., Prentice, R., and Stott, P. E. (1977). High-density discharges with gettered torus walls in DITE. *Nucl. Fusion*, **17**, 1382.

Hameiri, E. (1983). The equilibrium and stability of rotating plasmas. *Phys. Fluids*, **26**, 230.

Harris, J. H., Kaneko, H., Besshou, S. (1984). Magnetohydrodynamic activity in high-β currentless plasmas in Heliotron-E. *Phys. Rev. Letters*, **53**, 2242.

Houlberg, W. A., Iskra, M. A., Howe, H. C., and Attenberger, S. E. (1979). Pellet-A computer routine for modeling pellet fueling in tokamak plasmas. *Report ORNL/TM-6549*. Oak Ridge National Laboratory.

Ida, K., Miura, Y., and Itoh, S.-I. (1994). Physics mechanism to determine the radial electric field and its radial structure in a torus plasma. *J. Plasma Fusion Res.*, **70**, 514.

Idei, H., Ida, K., and Sanuki, H. (1994). Formation of positive radial electric field by electron cyclotron heating in compact helical system. *Phys. Plasmas*, **1**, 3400.

Iiyoshi, A., Motojima, O., and Sato, M. (1982). Confinement of a currentless plasma in the Heliotron-E. *Phys. Rev. Letters*, **48**, 745.

Isler, R. C., Aceto, S., Baylor, L. R. et al. (1992). Effects of magnetic geometry, fluctuations, and electric fields on confinement in the Advanced Toroidal Facility. *Phys. Fluids*, **B4**, 2104.

Kaufmann, M., Lackner, K., Lengyel, L., and Schneider, W. (1986). Plasma shielding of hydrogen pellets. *Nucl. Fusion*, **26**, 171.

Kaye, S. M., and Goldstone, R. J. (1985). Global energy confinement scaling for neutral-beam-heated tokamaks. *Nucl. Fusion*, **25**, 65.

Kikuchi, M., and Azumi, M. (1995). Experimental evidence for the bootstrap current in a tokamak. *Plasma Phys. Contr. Fusion*, **37**, 1215.

Liewer, P. C. (1985). Measurements of microturbulence in tokamaks and comparison with theories of turbulence and anomalous transport. *Nucl. Fusion*, **25**, 543.

Milora, S. L., and Foster, C. A. (1978). A revised neutral gas shielding model for pellet–plasma interactions. *IEEE Trans. Plasma Sci.*, **PS-6**, 578.

Miura, Y., Kasai, S., Sengoku, S. et al. (1986). Characteristics of improved confinement and high beta in pellet and neutral-beam injected single null divertor discharges of the JFT-2M tokamak. *Report JAERI-M-86-148*. Japan Atomic Energy Research Institute.

Murakami, M., Callen, J. D., and Berry, L. A. (1976). Some observation of maximum densities in tokamak experiments. *Nucl. Fusion*, **16**, 347.

Murakami, M., Aceto, S. C., Anabitarte, E. et al. (1990). Energy confinement and bootstrap current studies in the Advanced Toroidal Facility, in *Proceedings of the IAEA Conference on Plasma Physics and Controlled Nuclear Fusion Research*, Washington, vol. 2, p. 455.

Mutoh, T., Okada, H., Motojima, O. et al. (1984). ICRF heating of currentless plasmas in Heliotron E. *Nucl. Fusion*, **24**, 1003.

Nakamura, Y., Nishihara, H., and Wakatani, M. (1986). An analysis of the ablation rate for solid pellet injected into neutral beam heated toroidal plasmas. *Nucl. Fusion*, **26**, 907.

Obiki, T., Mizuuchi, T., Zushi, H. et al. (1988). Recent experiments on Heliotron E, in *Proceedings of the IAEA Conference on Plasma Physics and Controlled Nuclear Fusion Research, Nice*, vol. 2, p. 337.

Parks, P. B., and Turnbull, R. J. (1978). Effect of transonic flow in the ablation cloud on the lifetime of a solid hydrogen pellet in a plasma. *Phys. Fluids*, **21**, 1735.

Stix, T. H. (1962). *The theory of plasma waves*. McGraw-Hill, New York.

Stix, T. H. (1972). Heating of toroidal plasmas by netural injection. *Plasma Phys.*, **14**, 367.

Sudo, S., Motojima, O., Sato, M. et al. (1985). Pellet injection experiment on NBI current-free plasmas in Heliotron E. *Nucl. Fusion*, **25**, 94.

Sudo, S., Takeiri, Y., Zushi, H. et al. (1990) Scalings of energy confinement and density limit in stellarator/heliotron devices. *Nucl. Fusion*, **30**, 11.

Sweetman, D. R. (1973). Ignition condition in Tokamak experiments and role of neutral beam injection heating. *Nucl. Fusion*, **13**, 157.

Takeuchi, K., Wakatani, M., and Hashimoto, H. (1985). Ray tracing analysis of current-free ECRH plasmas in Heliotron E. *J. Phys. Soc. Japan*, **54**, 2915.

Tsui, H. Y. W., Wootton, A. J., Bell, J. D. et al. (1993). A comparison of edge turbulence in tokamaks, stellarators, and reversed-field pinches. *Phys. Fluids*, **B5**, 2491.

Uo, K., Iiyoshi, A., Obiki, T. et al. (1986). Studies of currentless Heliotron plasma, in *Proceedings of the IAEA Conference on Plasma Physics and Controlled Nuclear Fusion Research, Kyoto*, vol. 2, p. 355.

Wakatani, M., and Sudo, S. (1996). Overview of Heliotron E results. *Plasma Phys. Contr. Fusion*, **38**, 937.

Yagi, M., Wakatani, M., and Hasegawa, A. (1987). Transport driven by *g* modes and resistive drift waves based on scale invariance. *J. Phys. Soc. Japan*, **56**, 973.

Yagi, M., Wakatani, M., and Shaing, K. C. (1988). Inter-relation between scale invariance and mixing length theory based on *g* and rippling mode turbulent transport. *J. Phys. Soc. Japan*, **57**, 117.

Young, K. M. (1974). The C-stellarator—a review of confinement. *Plasma Phys.*, **16**, 119.

9

THE STEADY-STATE FUSION REACTOR

9.1 INTRODUCTION

First, we discuss the condition for ignition based on the global power balance in section 9.2. In magnetically confined ignited plasmas, heating by neutral beam injection (NBI) or by a radio frequency (RF) electromagnetic wave is replaced by heating due to high-energy charged particles produced by fusion reactions. In deuterium (D) and tritium (T) plasmas, the behavior of alpha particles is the most important issue, as shown in section 9.3. When the heating due to alpha particles, or alpha heating, exceeds the overall energy loss, net energy production is possible by the D–T fusion reaction in the ignited plasma. A total fusion energy, including neutrons, of the order of 3 GW is typically expected in the fusion reactor of the near future.

The most significant advantage of the stellarator or heliotron type of fusion reactor is the potential to realize steady-state operation economically. When the magnetic coils generating a three-dimensional magnetic configuration optimized to confine a burning plasma are made of superconducting materials, continuous operation will be realized under a continuous supply of fuel or a deuterium and tritium mixture. The fuel is ionized, and then heated by alpha particles directly or indirectly, through electrons and ions. In order to maintain continuous operation, the thermalized alpha particles known as alpha ash must be exhausted. However, since the alpha ash consists of charged particles, it is confined by the magnetic field. Some specific alpha particles can be lost due to the presence of the velocity space loss region (see Fig. 6.8) or the direct orbit loss from the inner region. If this process is substantial, the confinement of bulk fuel plasma might be degraded. Therefore, the direct orbit loss is considered to be negligible and particle transport across magnetic surfaces becomes the dominant process by which to carry the alpha ash from the central region to the peripheral region. After crossing the last closed magnetic surface, the alpha ash is carried to the divertor along the magnetic field line. The divertor has the function of neutralizing charged particles and pumping them out, as will be explained in section 9.4. The divertor is also designed to control the heat flux transported from the main plasma. When a high heat flux of the order of more than 10 MW/m^2 is deposited on the divertor plate, its material will be damaged substantially. In order to maintain the divertor for a sufficiently long period, high-density ($\sim 10^{20}\,m^{-3}$) and low-temperature plasmas ($\sim 10\,eV$) must be kept in front of the plate. The heat flux can be controlled by the radiation loss in such plasmas. This type of divertor is called a gas divertor. If the divertor works ideally, the high heat flux, the alpha ash, the impurity, and the fuel ions are controlled or pumped out to realize steady-state operation.

It is considered that the first-generation fusion reactor will use the D–T mixture as a fuel. In this case 14 MeV neutrons occupy a dominant part of the fusion energy. In order to thermalize these nuetrons, a radiation shield and blanket with a thickness of the order of 1 m is necessary. On the other hand, 14 MeV neutrons produce enormous radiated materials if there is no careful choice of shielding materials. In order to reduce the radiation level, low-activation material is required for the components that face the plasma. In section 9.5, a conceptual design of a heliotron reactor is discussed briefly. Finally, a comparison is made between the heliotron reactor and the tokamak reactor.

9.2 FUSION REACTIONS AND POWER BALANCE

The fusion of two light nuclei, A and B, forms a compound nucleus in an excited state that then decays into reaction products C and D; $A + B \rightarrow C + D$. The mass difference $\Delta m = (m_A + m_B) - (m_C + m_D) > 0$ is converted into kinetic energy according to Einstein's formula $\Delta E = (\Delta m)c^2$, where c is the speed of light.

In order to realize the fusion reaction, the two nuclei must overcome the long-range Coulomb repulsion force and approach sufficiently close to each other. For fusion to occur as a result of random collisions between two nuclei, they must be made sufficiently energetic. Usually, energies of the order of 10–100 keV or temperatures of 10^8–10^9 K are required. At these high temperatures, light atoms, such as hydrogen, deuterium, helium, and so on, are completely stripped of their orbital electrons. This globally neutral state of positively charged light nuclei and electrons is a thermonuclear plasma.

The rate at which fusion reactions take place between light nuclei of species a and b in the thermonuclear plasma is

$$n_a n_b \langle \sigma v \rangle_f = n_a n_b \int f_a(\mathbf{v}_a) f_b(\mathbf{v}_b) |\mathbf{v}_a - \mathbf{v}_b| \sigma_f(|\mathbf{v}_a - \mathbf{v}_b|) d^3 v_a d^3 v_b, \tag{9.1}$$

where n_i is the density, $\mathbf{v}_i$ is the velocity, and f_i is the velocity distribution function of species i, and σ_f is the fusion cross-section. The velocity distribution function of fuel ions in a magnetically confined thermonuclear plasma may be represented by a Maxwellian distribution

$$f_i(\mathbf{v}) = \left(\frac{m_i}{2\pi k_B T_i} \right)^{3/2} e^{-m_i v^2 / 2 k_B T_i}, \tag{9.2}$$

where T_i and m_i are the temperature and mass, respectively, and k_B is the Boltzmann constant. It is expected that Coulomb collisions will cause fuel light ion species in the thermonuclear plasma to have about the same temperature, so that the reaction rate $\langle\sigma v\rangle_f$ in (9.1) can be evaluated as a function of a single temperature, $T = T_a = T_b$.

Fusion reaction rates $\langle\sigma v\rangle_f$ for D–T, D–D, and D–^{3}He reactions for thermonuclear plasmas are shown in Fig. 9.1:

$$\mathrm{D} + \mathrm{T} \rightarrow {}^4\mathrm{He} + \mathrm{n}(14.1\,\mathrm{MeV}) + 17.6\,\mathrm{MeV}, \tag{9.3}$$

$$\begin{aligned} \mathrm{D} + \mathrm{D} &\rightarrow \mathrm{T} + \mathrm{p} + 4.03\,\mathrm{MeV}, \\ &\rightarrow {}^3\mathrm{He} + \mathrm{n}(2.45\,\mathrm{MeV}) + 3.27\,\mathrm{MeV}, \end{aligned} \tag{9.4}$$

$$\mathrm{D} + {}^3\mathrm{He} \rightarrow {}^4\mathrm{He} + \mathrm{p} + 18.3\,\mathrm{MeV}, \tag{9.5}$$

where D, T, and He denote deuterium, tritium, and helium, respectively. Also n denotes a neutron and p denotes a proton. At temperatures below the threshold values, 4 keV for the D–T reaction, 35 keV for the D–D reaction, and 30 keV for the D–^{3}He reaction, the reaction rates are negligible. In the D–D reaction in (9.4), the two branches occur with almost equal probability. It is known that there are many other possible fusion reactions; however, they are not suitable for economical energy production from the point of view of threshold energies higher than those given in (9.3)–(9.5), or from the point of view of natural resources.

We may identify the principal challenges of fusion research from these data. The plasma must be heated to thermonuclear temperatures (10^8–10^9 K) and confined for a sufficiently long time for the thermonuclear energy production

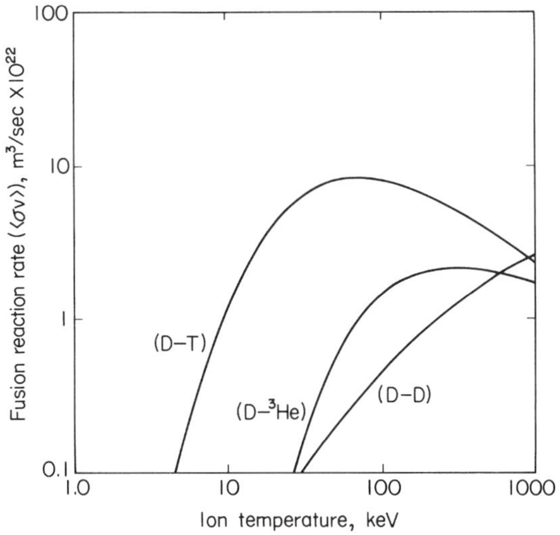

Fig. 9.1 Fusion reaction rates versus ion temperature.

to exceed the energy required to heat the plasma. From a simple energy balance, it can be stated that

$$\tfrac{1}{4}\, n^2 \langle \sigma v \rangle_f U_f > 3nk_B T/\tau_E, \tag{9.6}$$

where n is the ion density, T is the temperature, τ_E is the energy confinement time, and U_f is the fusion reaction energy release. The inequality (9.6) can be used as a breakeven criterion for the scientific feasibility of fusion power. Using the physical constants of a D–T thermonuclear plasma, (9.6) can be written as

$$n\tau_E > 3 - 4 \times 10^{20}\ \mathrm{s/m^3}. \tag{9.7}$$

Since no conceivable material could confine a thermonuclear plasma, confinement other than by walls is necessary. Two basically different approaches to achieving this energy breakeven are being investigated. In the first approach, magnetic fields are designed to confine plasmas stably in a magnetic trap. The goals of magnetic confinement research are to achieve plasma densities of about $10^{20}\ \mathrm{m^{-3}}$ and an energy confinement time of about a few seconds. Magnetically confined plasmas can be heated to thermonuclear temperatures by several different means. In stellarator and heliotron reactors, heating by the absorption of microwave or radio frequency waves, and heating due to the slowing down of injected energetic particles, are being investigated intensively, since there is no large net current for ohmic heating.

The second approach to achieving energy breakdown in a thermonuclear plasma is to make the plasma sufficiently dense that enough energy is released in the 10–100 nsec that corresponds to the time scale of plasma disintegration with inertia (inertial confinement). The basic idea is to compress small pellets of the order of several 10^{-1} cm of D–T fuels to densities of the order of 10^{27}–$10^{28}\ \mathrm{m^{-3}}$ and to heat them to thermonuclear temperatures. The compression and heating is to be accomplished by focusing multiple beams upon the fuel pellets. Several types of beams are currently under development for this purpose: lasers, relativistic electrons, light ions, and heavy ions. At the present time, lasers are the most highly developed method, and the progress in obtaining extremely high-density plasmas is significant.

The most promising reaction for near-term fusion reactors is D–T fusion. This reaction produces a 3.52 MeV alpha particle which slows down in the fuel plasma, and thus heats it. A neutron of 14.1 MeV escapes from the plasma and ultimately deposits its energy in the surrounding material in the form of heat. Neutrons at this energy and at the fluence level characteristic of a fusion reactor will cause substantial material damage. Thus the development of materials is also a challenge for fusion research.

As the fuel resource, deteurium exists at 0.0153 atom percent in seawater and can be extracted. On the other hand, tritium must be produced, since it β-decays with a half-life of 12.5 years. The neutron produced in the D–T reaction can be used to produce tritium, with capture by a lithium nucleus. ^{6}Li, which exists at about 7.5 atom percent in natural lithium, has an (n, α) reaction cross-section for thermal neutrons of about 950 barns, or $950 \times 10^{-24}\,\mathrm{cm}^2$. The more plentiful ^{7}Li has an $(n, \alpha n')$ reaction, although the cross-section is much smaller. Thus the fuel cycle is closed by adding the reactions

$$\mathrm{n} + {}^6\mathrm{Li} \rightarrow \mathrm{T} + {}^4\mathrm{He} \tag{9.8}$$

and

$$\mathrm{n} + {}^7\mathrm{Li} \rightarrow \mathrm{T} + {}^4\mathrm{He} + \mathrm{n}' \tag{9.9}$$

to the D–T reaction (9.3). Since one D–T reaction yields 17.6 MeV or 7.83×10^{-19} kWh, the estimated lithium resources contain of the order of 6×10^{16} kWh of thermal energy. It is considered that this number is comparable to the accessible energy content of the estimated world-wide uranium resurces, based on the fission reaction.

Here we again consider the global power balance in the steady state,

$$P_\alpha + P_{add} = P_R + P_{TR}, \tag{9.10}$$

where the alpha heating power is

$$P_\alpha = \tfrac{1}{4} n_i^2 \langle \sigma v \rangle_f U_\alpha, \tag{9.11}$$

the energy loss due to transport is

$$P_{TR} = \frac{3 n_i k_B T}{\tau_E}, \tag{9.12}$$

and the radiation rate is

$$P_R = n_i^2 W_{rad}, \tag{9.13}$$

where n_i is an ion density equal to an electron density. In (9.10), P_{add} is the additional heating power, $U_\alpha = 3.52$ MeV, and W_{rad} corresponds to the bremsstrahlung radiation due to the electron interaction with ions and the cyclotron radiation.

Defining the ratio of the additional to the alpha heating as $\xi = P_{add}/P_\alpha$, (9.10) can be written as

$$n_i \tau_E = \frac{3 k_B T}{\langle \sigma v \rangle_f U_\alpha (1 + \xi)/4 - W_{rad}}. \tag{9.14}$$

This quantity depends on the plasma temperature, both explicitly and implicitly through $\langle \sigma v \rangle_f$ and W_{rad}, and on the additional heating. Equation (9.14) is plotted in Fig. 9.2 for a 50%D–50%T plasma in which the impurity content is negligible.

When a plasma can achieve a power balance without additional heating (or $\xi = 0$), this is termed ignition. An ignited plasma is one of the goals of the present fusion research into stellarator and heliotron devices. When $\xi \neq 0$,

$$Q = \frac{P_\alpha + P_n \varepsilon_B}{P_{add}} = \frac{\left(1 + f\dfrac{U_\alpha}{U_B}\right)(1 + 4\varepsilon_B)}{\xi} \tag{9.15}$$

describes a power amplification factor in reacting D–T plasmas. The quantity P_{add} is the power injected into the plasma. When P_{add} is in the form of neutral beams at energy U_B, a fraction f of the injected particles will undergo beam-thermal particle fusion, thus enhancing the additional heating of the plasma by $(1 + fU_\alpha/U_B)P_{add}$. The fusion neutral power, P_n, is enhanced by a factor of ε_B due to exothermic nuclear reactions in the blanket.

It is clear from Fig. 9.2 that the energy confinement time required for a power balance can be reduced by increasing the ratio of the additional to the

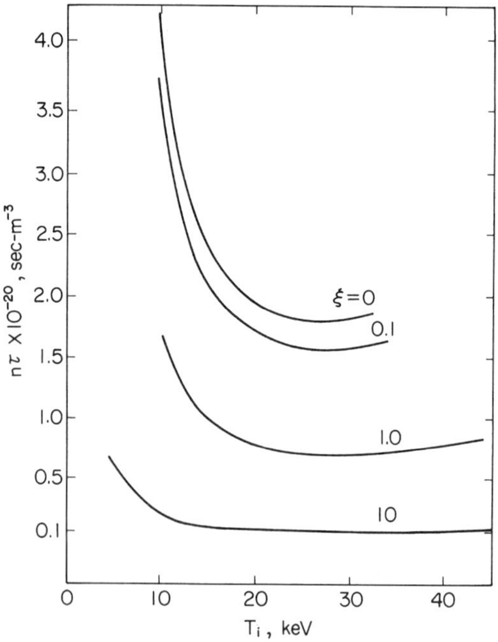

Fig. 9.2 $n_i\tau_E$ required for a power balance in a 50%D–50%T plasma. Here, $\xi = P_{add}/P_\alpha$.

alpha heating power. A minimum criterion for a power-producing fusion reactor is that the entire fusion plant produces more electrical power than it consumes. A power balance in the fusion plant usually requires $Q \gtrsim 15$–10 for net power production. It is noted that the value of Q also depends on the ratio of the D–T mixture. The presence of impurities significantly alters the possibility of achieving ignition. We can estimate an upper bound on the impurity concentration that is tolerable in an ignited plasma at $\xi = 0$ by neglecting the transport loss term in (9.10) and solving

$$\tfrac{1}{4}\langle\sigma v\rangle_f U_\alpha = W_{rad} \tag{9.16}$$

for a given n_z/n_i, where n_z is an impurity density. A simple fit that approximates the features of the detailed radiative power loss calculations reasonably well is

$$P_R(\mathrm{MW/m^3}) = 4.8 \times 10^{-43} n_e n_z \times \times \left(Z^2 T_e^{1/2} + \frac{3.792 \times 10^{-2} Z^4}{T_e^{1/2}} + \frac{8.604 \times 10^{-4} Z^6}{T_e^{3/2}} \right), \tag{9.17}$$

where Z is the atomic number of the impurity, n_e and n_z have units of $\mathrm{m^{-3}}$, and T_e has units of keV. The three terms in (9.17) represent the bremsstrahlung, the line radiation, and the recombination radiation processes, respectively. From (9.16) and (9.17), the maximum impurity concentration n_z/n_i for ignition is shown in Fig. 9.3 as a function of the atomic number of the impurity species. It is apparent from Fig. 9.3 that contamination by intermediate- to high-Z impurities has severe effects on the ability to achieve an ignited plasma.

The economic attractiveness of fusion reactors will depend upon the achievable power density. It is estimated using

$$P_T(\mathrm{MW/m^3}) = 4 \times 10^{-20}(14.1\varepsilon_B + 3.52(1+\xi)) n_i^2 \langle\sigma v\rangle_f, \tag{9.18}$$

where n_i has units of $\mathrm{m^{-3}}$ and $\langle\sigma v\rangle_f$ has units of $\mathrm{m^3/sec}$. This fusion power density includes the alpha energy enhanced by the additional heating and the neutron energy enhanced by a factor of ε_B, due to exothermic reactions in the blanket surrounding the toroidal plasma. Roughly, $P_T \sim 1\,\mathrm{MW/m^3}$ is typical for magnetic fusion in the tokamak, stellarator, and heliotron.

We now examine a problem that will be encountered in maintaining a steady-state power balance in a thermonuclear plasma. We can use the overall power balance given by (9.10) as

$$dU/dt = (P_\alpha + P_{add}) - (P_R + P_{TR}), \tag{9.19}$$

where $U = 3n_i k_B T$ is the plasma internal energy. Let us assume that a power balance has been achieved at the beginning of a burn cycle, and consider the

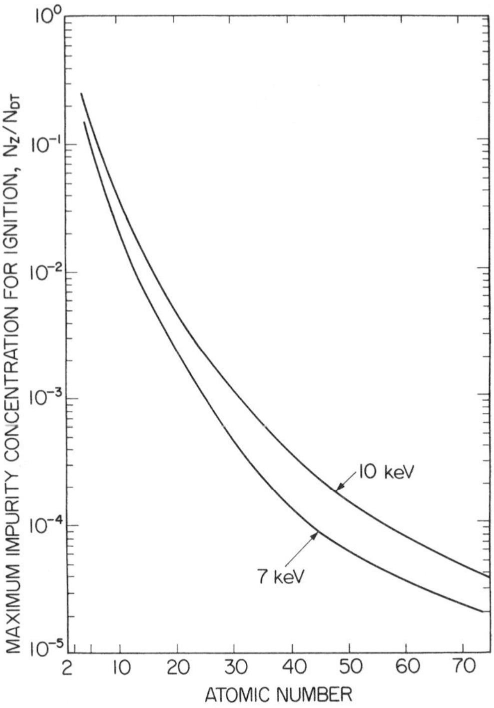

Fig. 9.3 The maximum tolerable impurity concentration in an ignited plasma.

effect of a small temperature perturbation. Linearizing (9.19) in the neighborhood of the power balance condition (9.10) yields

$$\frac{d\tilde{U}}{dt} = \left(\frac{\partial P_\alpha}{\partial T}\,\Delta T + \frac{\partial P_{add}}{\partial t}\,\Delta T\right) - \left(\frac{\partial P_R}{\partial T}\,\Delta T + \frac{\partial P_{TR}}{\partial T}\,\Delta T\right), \tag{9.20}$$

where $\tilde{U} = 3n_i k_B \Delta T$. If the RHS of (9.20) is positive, the response to the perturbation is such as to increase the temperature perturbation. Thus the power balance is unstable for an increase in temperature ΔT, if

$$\left[\left(\frac{\partial P_\alpha}{\partial T} + \frac{\partial P_{add}}{\partial T}\right) - \left(\frac{\partial P_R}{\partial T} + \frac{\partial P_{TR}}{\partial T}\right)\right] > 0. \tag{9.21}$$

This is called thermal instability. Because the fusion cross-section increases sharply with temperature for D–T fusion (see Fig. 9.1), the alpha heating term is destabilizing. The additional heating term in (9.21) is zero in the case of neutral beam injection. The radiation loss term is generally positive, and thus stabilizing, for low-Z impurities that are fully stripped, but can be negative, and thus destabilizing, for high-Z impurities that are only partially ionized. The transport loss term can be stabilizing or destabilizing depending on whether τ_E varies inversely ($\tau_E \propto T^{-\alpha}, \alpha > 0$) or directly (e.g., neoclassical banana diffusion with $\tau_E \propto T^{1/2}$) with temperature. When thermal instability occurs, there are several ways to control it: one is to change the ratio of the D–T density by fueling, and the other is to enhance the direct alpha particle loss by shifting the plasma position outward with the vertical field (see chapter 7). Fusion reactors must be operated in a stable state, in the absence of thermal instability.

9.3 THE ALPHA PARTICLE DISTRIBUTION FUNCTION AND ALPHA DENSITY

Alpha particles are born with an energy of 3.52 MeV due to the D–T reactions. The birth profile $S = n_T n_D \langle \sigma v \rangle_f$ is estimated by using the D–T reaction rate shown in Fig. 9.1 for a given ion temperature and density profile. Here an approximation to $\langle \sigma v \rangle_f$ is

$$\langle \sigma v \rangle_f = \exp(a_1/T_i^r + a_2 + a_3 T_i + a_4 T_i^2 + a_5 T_i^3 + a_6 T_i^4), \tag{9.22}$$

where $r = 0.2935$, $a_1 = -21.378$, $a_2 = -25.204$, $a_3 = -7.101 \times 10^{-2}$, $a_4 = 1.938 \times 10^{-4}$, $a_5 = 4.925 \times 10^{-6}$, $a_6 = -3.984 \times 10^{-8}$, and T_i is in keV. This birth profile may be measured by neutron tomography for 14 MeV neutrons, since neutrons produced by D–T reactions are exactly equal to alpha particles only at the birth point. Other diagnostic systems to measure alpha particles directly are also being developed, on the basis of charge exchange reactions or collective Thomson scattering.

In order to heat a D–T plasma efficiently, we must confine almost all of the alpha particles. When they are confined, they interact with D and T ions and electrons and slow down by transferring a part of their energy to the background plasma. The slowing down distribution of alpha particles is estimated by assuming that the alpha particle density is much lower than the electron density. At the thermonuclear temperatures of the D–T plasma, $T \simeq 10$–20 keV, the velocity of an alpha particle is in the range $v_{T_i} < v_\alpha < v_{T_e}$, where v_{T_i} and v_{T_e} are the thermal velocities of ions and electrons, respectively. The alpha particles are decelerated due to Coulomb collisions as

$$\frac{dU_\alpha}{dt} = -\frac{2U_*}{\tau_s}\left(\frac{U_\alpha}{U_*} + \left(\frac{U_*}{U_\alpha}\right)^{1/2}\right), \tag{9.23}$$

where $U_\alpha = m_\alpha v_\alpha^2/2$, and the slowing down time is

$$\begin{aligned} \tau_s &= 3 m_e m_\alpha v_{T_e}^3 / (16\sqrt{\pi} Z_\alpha^2 e^4 n_e \ln \Lambda_e) \\ &\simeq 3.7 \times 10^{16} (T_e[\mathrm{eV}])^{3/2} / n_e[\mathrm{cm}^{-3}]. \end{aligned} \tag{9.24}$$

$U_* = m_\alpha v_c^2/2$ is the particle energy at which the rates of slowing down by ions and electrons become equal, where

$$v_c = [(3\sqrt{\pi}m_e/(4m_p \ln \Lambda_e)) \sum_j Z_j^2 \ln \Lambda_j n_j/A_j n_e]^{1/3} v_{T_e}, \tag{9.25}$$

$$\ln \Lambda_e = 23.9 + \ln(T_e[\text{eV}]/n_e[\text{cm}^{-3}]^{1/2}), \tag{9.26}$$

$$\ln \Lambda_j = 14.2 + \ln((T_e[\text{eV}]/n_e[\text{cm}^{-3}])^{1/2} A_\alpha A_j v_0/(A_\alpha + A_j)). \tag{9.27}$$

Here $m_\alpha v_0^2/2 = 3.52\,\text{MeV}$, m_p is the proton mass, and A_j is the atomic mass number. Then we can use the simple linearized Fokker–Planck equation without pitch angle scattering,

$$\frac{\partial f_\alpha}{\partial t} - \left(\frac{v^3 + v_c^3}{v^2 \tau_s}\right)\frac{\partial f_\alpha}{\partial v} - \frac{3 f_\alpha}{\tau_s} = S\delta(v - v_0)/4\pi v_0^2 - f_\alpha/\tau_L, \tag{9.28}$$

where the second term on the LHS corresponds to (9.23) or

$$dv/dt = -(v^3 + v_c^3)/v^2\tau_s. \tag{9.29}$$

In (9.28), S is the total alpha particle source rate (in cm^{-3}/sec). The last term is a loss term that can be used to represent, for example, ripple losses of alpha particles or losses due to the ergodic magnetic field. Here, for simplicity, we assume τ_L to be independent of the pitch angle and the energy. If $\tau_L \lesssim \tau_s$, substantial alpha-particle energy is lost from the reacting plasma. To solve (9.28) we first replace f_α by

$$F = f_\alpha \exp\left[\int_{t_0}^{t} (1/\tau_L - 3/\tau_s)dt'\right], \tag{9.30}$$

where t_0 is an arbitrary constant, and (9.28) becomes

$$\frac{\partial F}{\partial v} - \frac{v^2\tau_s}{v^3 + v_c^3}\frac{\partial F}{\partial t}$$
$$= -S\tau_s V^2 \delta(v - v_0)\exp\left[\int_{t_0}^{t} (1/\tau_L - 3/\tau_s)dt'\right]/[4\pi v_0^2(v^3 + v_c^3)], \tag{9.31}$$

where v_c, τ_s, and S are functions of time through their dependence on temperature and density. Equation (9.31) is solved as

$$F(v, t) = \int_{v}^{v_0} dv' S(t')\delta(v' - v_0)v'^2 \tau_s(t') \times$$
$$\times \exp\left[\int_{t_0}^{t'} (1/\tau_L - 3/\tau_s(t''))dt''\right]/[4\pi v_0^2(v_c^3(t') + v'^3)], \tag{9.32}$$

where the path of integration is along the characteristic given by (9.29) and $F(v_0, t^*) = 0$ is assumed. Completing the integration over v' in (9.32) gives the expression for the alpha particle distribution:

$$f_\alpha(v, t) = S(t^*)\tau_s(t^*)\exp\left[-\int_{t^*}^{t}(1/\tau_L - 3/\tau_s(t'))dt'\right]\frac{1}{[4\pi(v_0^3 + v_c^3(t*))]}, \tag{9.33}$$

where $t' = t^*$ is the birth time of an alpha particle that has velocity v at time t. An expression for t^* is obtained from solution of the characteristic equation (9.29), but an analytic solution is not easily obtained when τ_s and v_c depend on time. When τ_s and v_c are constant in time, the time integration in (9.33) can be carried out analytically. With the use of (9.29), (9.33) becomes

$$\begin{aligned} f_\alpha(v, t) &= \frac{S\tau_s}{[4\pi(v_0^3 + v_c^3)]}\exp\left[(1/\tau_L - 3/\tau_s)\int_{v_0}^{v}\frac{v'^2\tau_s}{v'^3 + v_c^3}dv'\right] \\ &= \frac{S\tau_s}{[4\pi(v_0^3 + v_c^3)]}\left(\frac{v^3 + v_c^3}{v_0^3 + v_c^3}\right)^{(\tau_s/3\tau_L - 1)} \\ &= \frac{S\tau_s}{[4\pi(v^3 + v_c^3)]}\left(\frac{v^3 + v_c^3}{v_0^3 + v_c^3}\right)^{\tau_s/3\tau_L}. \end{aligned} \tag{9.34}$$

This is called the classical steady-state slowing down distribution. From the point of view of alpha-particle driven microinstabilities, the alpha particle distribution is stable if $\partial f_\alpha/\partial v < 0$ for all v. It is noted that if $\tau_L < \tau_s/3$, the steady-state distribution (9.34) can be inverted in velocity space. When alpha particles slow down to an energy comparable with the background plasma temperature, the Maxwellian part appears in the alpha particle distribution. However, this is not included in (9.34).

The important question is the alpha particle density corresponding to the Maxwellian part, or alpha ash in steady-state operation. To evaluate the alpha ash density or the helium density, we introduce a parameter $\hat{\rho} = \tau_\alpha^*/\tau_E$, where τ_α^* is global helium confinement time including the recycling process and τ_E is the energy confinement time. For a total number of alpha particles N_α in a plasma volume V, the outward alpha flux from the plasma surface, Γ_α^0, and the recycled alpha flux from the wall, Γ_α^R, are related as follows:

$$\tau_\alpha^* = \frac{N_\alpha}{\Gamma_\alpha^0 - \Gamma_\alpha^R}. \tag{9.35}$$

In order to simplify the analysis, we assume:

(i) a uniform plasma temperature T;
(ii) one species of impurity ion;
(iii) that alpha particles are lost from the volume V after 3.5 MeV energy is deposited to the background plasma,

(iv) that for the energy balance, the contribution of newly born alpha particles is negligible.

Since plasma is quasi-neutral, or $n_e = n_i + Zn_z + 2n_{He}$, the total charged particle numbers are $n_{tot} = n_e + n_i + n_z + n_{He} = n_e g_{tot}$. Here, $g_i = n_i/n_e = 1 - Zg_Z - 2g_{He}$. The power balance becomes

$$\tfrac{1}{4}n_e^2 g_i^2 U_\alpha \langle\sigma v\rangle_f = \tfrac{3}{2}\frac{n_e g_{tot} k_B T}{\tau_E} + W_{rad}(T, g_Z, g_{He})n_e^2. \tag{9.36}$$

Here, the second term in the RHS of (9.36) denotes the radiation loss. Equation (9.36) can be written as

$$n_e \tau_E = \tfrac{3}{2} g_{tot} T \bigg/ \left(\frac{\langle\sigma v\rangle_f}{4} g_i^2 U_\alpha - W_{rad}\right). \tag{9.37}$$

We consider the particle balance of alpha particles, $n_{He}/\tau_\alpha^* = n_i^2\langle\sigma v\rangle_f/4$, which gives

$$n_e \tau_E = \frac{4 g_{He}}{\hat{\rho} g_i^2 \langle\sigma v\rangle_f}. \tag{9.38}$$

Equating (9.37) to (9.38) yields

$$4\left(\frac{U_\alpha}{\hat{\rho}} + \frac{3T}{2}\right) g_{He}^3 - 4\left(\frac{U_\alpha}{\hat{\rho}} + \frac{9T}{2}\right) g_{He}^2 + \left(\frac{U_\alpha}{\hat{\rho}} + \frac{27}{2}T\right) g_{He} - 3T = 0, \tag{9.39}$$

for the case of $g_Z = 0$ and $W_{rad} \simeq 0$. For (9.39), we find two solutions; $g_{He} = 0.5$ and $g_{He} = 3T/(U_\alpha/\hat{\rho} + 3T/2)$. The former solution, $g_{He} = 0.5$, denotes a pure helium plasma, while the latter gives

$$T < \frac{U_\alpha}{28.5\hat{\rho}} \simeq \frac{123[\text{keV}]}{\hat{\rho}} \tag{9.40}$$

to realize $g_{He} < 0.1$. For a finite radiation loss solutions of (9.37) and (9.38) for g_{He} (or f_{He} in the figure) as a function of T, where $\hat{\rho}$ is a parameter, are shown in Fig. 9.4. It can be seen that for finite $\hat{\rho}$ there are two solutions at constant T for ignition. One is the case of high helium or alpha ash density: the other corresponds to the case of low alpha ash density.

Here we divide the alpha particles into two types; one is the newly born alpha due to the D–T reaction and the other is the recycled alpha from the wall and the divertor. The particle balances are

$$\frac{dN_{\alpha 1}}{dt} = -\frac{N_{\alpha 1}}{\tau_{\alpha 1}} + \frac{N_D N_T \langle\sigma v\rangle_f}{V}, \tag{9.41}$$

$$\frac{dN_{\alpha 2}}{dt} = -\frac{N_{\alpha 2}}{\tau_{\alpha 2}} + R_{eff}\left(\frac{N_{\alpha 1}}{\tau_{\alpha 1}} + \frac{N_{\alpha 2}}{\tau_{\alpha 2}}\right), \tag{9.42}$$

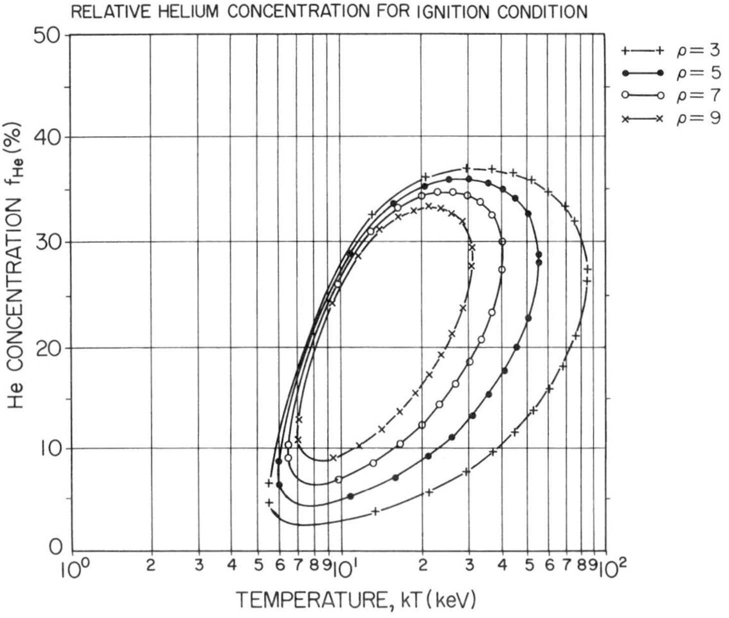

Fig. 9.4 The helium concentration $f_{\rm He}$ versus the plasma temperature T at ignition for a constant $\hat{\rho} = \tau_\alpha^*/\tau_E$.

where the $N_{\alpha 1}$ are the first-generation alpha particles, the $N_{\alpha 2}$ are the recycled alpha particles, V is the plasma volume and R_{eff} is the effective recycling coefficient, given by $R_{eff} = \Gamma_\alpha^R/\Gamma_\alpha^0$. Both $\tau_{\alpha 1}$ and $\tau_{\alpha 2}$ are alpha particle confinement times, and N_D and N_T are the total deuterium and tritium numbers, respectively. In the stationary state, the solutions of (9.41) and (9.42) are

$$\begin{cases} N_{\alpha 1} = \dfrac{\tau_{\alpha 1} N_D N_T \langle \sigma v \rangle_f}{V} \\ N_{\alpha 2} = \dfrac{R_{eff}}{1 - R_{eff}} \dfrac{\tau_{\alpha 2}}{\tau_{\alpha 1}} N_{\alpha 1}. \end{cases} \tag{9.43}$$

Then the total number of alpha particles is

$$N_\alpha = N_{\alpha 1} + N_{\alpha 2} = \frac{N_D N_T \langle \sigma v \rangle_f}{V} \left[\tau_{\alpha 1} + \frac{R_{eff}}{1 - R_{eff}} \tau_{\alpha 2} \right] \tag{9.44}$$

and the global alpha particle confinement time is

$$\tau_\alpha^* = \tau_{\alpha 1} + \frac{R_{eff}}{1 - R_{eff}} \tau_{\alpha 2}. \tag{9.45}$$

The effective pumping efficiency is given by $\epsilon_{eff} = 1 - R_{eff}$, and is related to the recycling rate on the first wall.

9.4 THE ALPHA-DRIVEN TOROIDAL ALFVÉN EIGENMODE

In this section we discuss alpha-driven MHD fluctuations; particularly, the Toroidal Alfvén Eigenmode (TAE mode) in a tokamak with a circular cross-section. If we apply the stellarator expansion and averaging in the toroidal direction, we may obtain a similar TAE mode in heliotron devices. The energetic component is included in our calculation by a gyrofluid model with Landau closure. Here we write the linearized Ohm's law in a toroidal geometry as follows (see (3.200)),

$$\frac{\partial \tilde{\Psi}}{\partial t} = \frac{R}{B_\phi} \mathbf{B}_0 \cdot \nabla \tilde{\varphi}, \tag{9.46}$$

where $\tilde{\Psi}$ and $\tilde{\varphi}$ are the fluctuations of the magnetic flux and the electrostatic potential, respectively, $\mathbf{B}_0$ is the equilibrium magnetic field and B_ϕ is the toroidal component. We will neglect the toroidal component of the velocity compared to the perpendicular component, as in the usual reduced MHD model. The energetic alpha component is assumed to enter only through the pressure gradient term in the equation of motion. Here, for simplicity, we retain only the alpha contribution to the pressure. The linearized vorticity equation is written as (see (3.212))

$$\frac{\partial \tilde{U}}{\partial t} = -\left[R\mathbf{B}_0 \cdot \nabla\left(\frac{\tilde{J}_\phi}{B_\phi}\right) + \nabla\tilde{\Psi} \times \mathbf{e}_\parallel \cdot \nabla\left(\frac{J_{\phi 0}}{B_\phi}\right)\right] + \left[\mathbf{e}_\parallel \times \nabla\left(\frac{R}{B_\phi}\right)\right] \cdot \nabla \tilde{P}_\alpha, \tag{9.47}$$

where

$$\tilde{U} = \mathbf{e}_\parallel \cdot \left[\nabla \times \left(\frac{\rho_0 R}{B_\varphi^2} \nabla_\perp \tilde{\varphi} \times \mathbf{e}_\parallel\right)\right] \tag{9.48}$$

and the unit vector along the magnetic field line is $\mathbf{e}_\parallel$. The density of the background plasma is ρ_0. In the present approximation, $\mathbf{e}_\parallel \simeq \mathbf{e}_\phi$, where $\mathbf{e}_\phi$ is a unit vector in the toroidal direction. The fluctuation of the toroidal current is $\tilde{J}_\phi$ and that of the alpha pressure is $\tilde{P}_\alpha$. The toroidal current is related to the magnetic flux by

$$\mu_0 J_{\phi 0} = R\nabla_\perp \cdot \left(\frac{1}{R^2} \nabla_\perp \Psi_0\right) \tag{9.49}$$

and

$$\mu_0 \tilde{J}_\phi = R\nabla_\perp \cdot \left(\frac{1}{R^2} \nabla_\perp \tilde{\Psi}\right), \tag{9.50}$$

where $\mathbf{B}_0 = \nabla\Psi_0 \times \mathbf{e}_\phi / R + B_\phi \mathbf{e}_\phi$ is assumed. In order to complete the fluid moment model for the TAE mode, it is essential to adapt the energetic species equations to toroidal geometry, because the existence of the TAE mode depends on toroidal couplings to produce gaps in the shear Alfvén spectrum.

For simplicity, we assume concentric magnetic flux surfaces. Expanding the toroidal effect to first order in the inverse aspect ratio, $r/R < 1$, and retaining only the two dominant poloidal modes for the TAE mode, we then arrive at the following two coupled second-order eigenmode equations for the Fourier component proportional to $\exp(-i\omega t + im\theta - in\phi)$ of the quantity $E = \tilde{\varphi}/r$:

$$\left[\frac{d}{dr} r^3 \left(\frac{\omega^2}{V_A^2} - k_{\|m}^2\right)\frac{d}{dr} - (m^2 - 1) r \left(\frac{\omega^2}{V_A^2}\, k_{\|m}^2\right) + \left(\frac{\omega^2}{V_A^2}\right)' r^2\right] E_m + \left(\epsilon \frac{d}{dr}\frac{\omega^2}{V_A^2}\frac{r^4}{a}\frac{d}{dr}\right) E_{m+1} = 0, \tag{9.51}$$

$$\left[\frac{d}{dr} r^3 \left(\frac{\omega^2}{V_A^2} - k_{\|m+1}^2\right)\frac{d}{dr} - [(m+1)^2 - 1] r \left(\frac{\omega^2}{V_A^2}\, k_{\|m+1}^2\right) + \left(\frac{\omega^2}{V_A^2}\right)' r^2\right] E_{m+1} + \left(\epsilon \frac{d}{dr}\frac{\omega^2}{V_A^2}\frac{r^4}{a}\frac{d}{dr}\right) E_m = 0, \tag{9.52}$$

where V_A is the Alfvén velocity, $\epsilon = 3a/2R$, the prime denotes radial differentiation, and the subscripts m and $m+1$ denote the two dominant poloidal mode numbers. Also, $k_{\|m} = (n - m/q)/R$ is the parallel wave number, with R the major radius and q the safety factor. In deriving (9.51) and (9.52) the second and third terms in the RHS of (9.47) are neglected, and (9.46) and (9.50) are substituted by writing $\tilde{J}_\phi$ for $\tilde{\phi}$. Equations (9.51) and (9.52) describe the ideal MHD limit. In cylindrical geometry ($\epsilon = 0$), the two poloidal modes E_m and E_{m+1} are decoupled. Then (9.51) and (9.52) are singular at $\omega_1^2 = k_{\|m}^2 V_A^2$ and $\omega_2^2 = k_{\|m+1}^2 V_A^2$, respectively, which gives the two cylindrical shear Alfvén continua. In toroidal tokamak geometry, (9.51) and (9.52) are coupled due to the finite toroidicity, and the poloidal mode numbers are no longer good "quantum" numbers. The toroidal shear Alfvén continuum can be obtained by setting the determinant of the coefficients of the second-order derivative terms equal to zero, which yields the following two branches:

$$\omega_\pm^2 = \frac{k_{\|m}^2 V_A^2 + k_{\|m+1}^2 V_A^2 \pm \sqrt{(k_{\|m}^2 V_A^2 - k_{\|m+1}^2 V_A^2)^2 + 4\epsilon^2 x^2 k_{\|m}^2 V_A^2 k_{\|m+1}^2 V_A^2}}{2(1 - \epsilon^2 x^2)}, \tag{9.53}$$

where $x = r/a$ is the normalized radius. At the crossing point of the two cylindrical continua, where $k_{\|m} = k_{\|m+1}$ or $q = (2m+1)/2n$, a gap appears, the width of which is $\Delta\omega = \omega_+ - \omega_- \simeq 2\epsilon x |k_{\|m} V_A|$, evaluated at the crossing point. In Fig. 9.5 is shown the toroidal continuum (solid curves) given by (9.53) for $m = -1$ and $m = -2$ with $n = -1$ and $\epsilon = 0.375$, with a constant density profile and $q(r) = 1 + (r/a)^2$; the cylindrical continua (dashed curves) are also shown. The corresponding TAE mode structure is shown in Fig. 9.6; its eigenfrequency, $\omega = 0.93(|k_{\|m} V_A|)_{q=3/2}$, lies inside the continuum gap. The eigenmode has been obtained by numerically solving (9.51) and (9.52) with the shooting method by imposing boundary conditions at $r = 0$ and $r = a$. At $r = 0$, $E_m \propto r^{m-1}$, and $E_m = 0$ at $r = a$. Note that the mode peaks at the location of the gap; i.e., near the crossover point of the cylindrical continua.

Two linear fluid equations for the alpha density and the alpha parallel flow velocity are shown as

$$\frac{\partial \tilde{n}_\alpha}{\partial t} = \frac{T_{0\alpha}}{M_\alpha \omega_{c\alpha}} \Omega_d(\tilde{n}_\alpha) - n_{0\alpha} \nabla_\|^{(0)} \tilde{V}_{\|\alpha} + \frac{q_\alpha n_{0\alpha}}{M_\alpha \omega_{c\alpha}} \Omega_d(\tilde{\varphi}) - \frac{q_\alpha}{M_\alpha \omega_{c\alpha}} \frac{dn_{0\alpha}}{dr} \frac{1}{r} \frac{\partial \tilde{\varphi}}{\partial \theta} + D_{n\alpha} \nabla_{\perp 0}^2 \tilde{n}_\alpha, \tag{9.54}$$

$$\frac{\partial \tilde{V}_{\|\alpha}}{\partial t} = \frac{T_{0\alpha}}{M_\alpha \omega_{c\alpha}} \Omega_d(\tilde{V}_{\|\alpha}) - \left(\frac{\pi}{2}\right)^{1/2} V_{T\alpha} |\nabla_\|^{(0)}| \tilde{V}_{\|\alpha} - \frac{V_{T\alpha}^2}{n_{0\alpha}} \nabla_\|^{(0)} \tilde{n}_\alpha - \frac{q_\alpha v_{T\alpha}^2}{M_\alpha \omega_{c\alpha} R n_{0\alpha}} \frac{dn_{0\alpha}}{dr} \frac{1}{r} \frac{\partial \tilde{\Psi}}{\partial \theta} + D_{V\|\alpha} \nabla_{\perp 0}^2 \tilde{V}_{\|\alpha}, \tag{9.55}$$

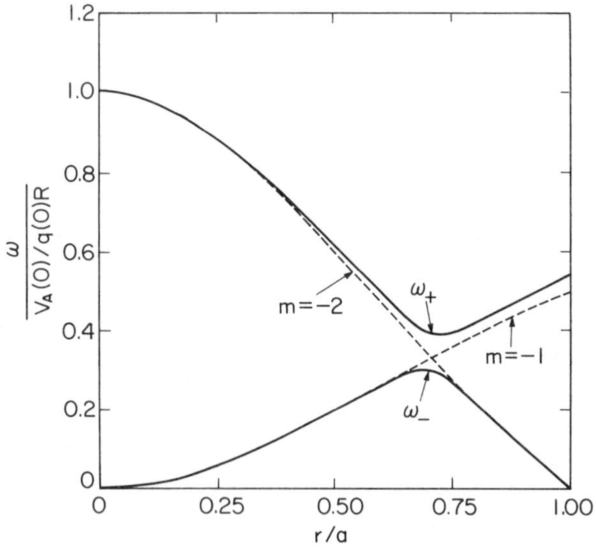

Fig. 9.5 The toroidal shear Alfvén continuous spectrum with a gap for the safety factor profile $q = 1 + (r/a)^2$ and a constant density profile. The cylindrical limit spectra (dashed curves) for $m = -1$ and $m = -2$ with $n = -1$ cross at the magnetic surface of $q = 3/2$. Here, $\omega_\pm$ are given by (9.53).

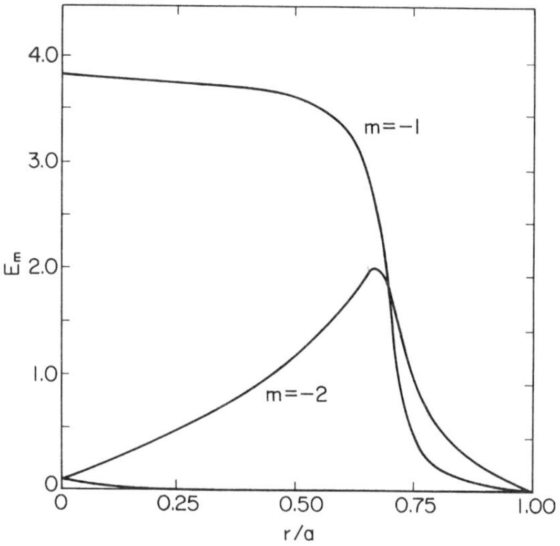

Fig. 9.6 Radial profiles of the dominant poloidal modes ($m = -1$ and $m = -2$) of the $n = -1$ TAE mode for the same equilibrium as in Fig. 9.4.

where Ω_d is the operator

$$\Omega_d \equiv \frac{R^2}{2}\frac{\partial}{\partial r}\left(\frac{1}{R^2}\right)\frac{1}{r}\frac{\partial}{\partial\theta} - \frac{R^2}{2r}\frac{\partial}{\partial\theta}\left(\frac{1}{R^2}\right)\frac{\partial}{\partial r}. \tag{9.56}$$

Here $\omega_{c\alpha} = q_\alpha B_0/M_\alpha$ is the alpha cyclotron frequency, $n_{0\alpha}$ is the equilibrium alpha density, $T_{0\alpha}$ is the equilibrium alpha temperature, q_α is the charge of the alpha particle, and $M_\alpha V_{T\alpha}^2/2 = T_{0\alpha}$. The $|\mathbf{V}_\parallel^{(0)}|$ operator has a simple interpretation only for the case in which θ and ϕ are Fourier analyzed, in which case it is the absolute value of the parallel wave number $|k_\parallel|$ (the gyro-Landau fluid model). Diffusion terms with diffusion coefficients $D_{n\alpha}$ and $D_{V\parallel\alpha}$ are included in the alpha density and parallel flow evolution equations for numerical reasons. These are kept at a level at which they do not have a significant effect on the answers obtained. Equations (9.46), (9.47), (9.48), (9.50), (9.54), and (9.55), with the relation $\tilde{P}_\alpha = T_{0\alpha}\tilde{n}_\alpha$, complete our system of linear time evolution equations. These equations are solved in toroidal geometry with a three-dimensional initial value code, which evolves the four scalar fields, $\tilde{\Psi}$, $\tilde{U}$, $\tilde{n}_\alpha$, and $\tilde{V}_{\parallel\alpha}$. All quantities are expanded in Fourier harmonics in θ and ϕ, retaining both sine and cosine components, and discretized on a radial grid. The time evolution is followed until an asymptotic, linear, exponentially growing state is achieved. Because the initial value code can only find unstable cases and the system of equations self-consistently includes both the ideal MHD component and the energetic species response, the solutions remain well defined even at locations at which the ideal MHD continua are encountered.

The growth rate of a TAE mode can be derived from the two-pole model, where a Maxwellian distribution is assumed for the energetic species. We first simplify (9.54) and (9.55) by evaluating that all quantities including radial derivatives are uniform in the radial direction, in addition to setting $E_\parallel = 0$, $k_y \rho_\alpha \simeq 0$, and $\partial/\partial t = -i\omega$, where ρ_α is the Larmor radius of the alpha particle. The alpha particle perturbed density and parallel velocity moment equations may then be written as

$$(\omega - \omega_{D\alpha}) \frac{\tilde{n}_\alpha}{n_{0\alpha}} = k_\parallel \tilde{V}_{\parallel\alpha} + \frac{q_a}{T_{0\alpha}} (\omega_{D\alpha} - \omega_{*\alpha})\tilde{\varphi}, \tag{9.57}$$

$$\left(\omega - \omega_{D\alpha} + i\sqrt{\frac{\pi}{2}} |k_\parallel| V_{T\alpha}\right) M_\alpha n_{0\alpha} \tilde{V}_\parallel$$
$$= k_\parallel T_{0\alpha}\left(\tilde{n}_\alpha + n_{0\alpha} \frac{q_\alpha \tilde{\varphi}}{T_{0\alpha}}\right) + n_{0\alpha}(\omega_{*\alpha} - \omega) q_\alpha A_\parallel, \tag{9.58}$$

where $\omega_{D\alpha} = k_y T_{0\alpha}/q_\alpha B L_c$, $\omega_{*\alpha} = k_y T_{0\alpha}/q_\alpha B L_n$, and k_y is a wave number in the θ-direction, or $k_y = m/r$. Here, L_c and L_n are the characteristic length of curvature and density gradient, respectively. It is noted that $\tilde{\Psi} = -RA_\parallel$ and $E_\parallel = -\nabla_\parallel \tilde{\varphi} - \partial A_\parallel/\partial t$ are used in (9.58). With Ohm's law, $A_\parallel = k_\parallel \tilde{\varphi}/\omega$ or $E_\parallel = 0$, and the definitions $\xi = (\omega - \omega_{D\alpha})/(\sqrt{2} k_\parallel V_{T\alpha})$, $V = \tilde{V}_{\parallel\alpha}/\sqrt{2} V_{T\alpha}$, and $\Phi = q_\alpha \tilde{\varphi}/T_{0\alpha}$, (9.57) and (9.58) become

$$\xi \frac{\tilde{n}_\alpha}{n_{0\alpha}} = V + \left(\frac{\omega_{D\alpha} - \omega_{*\alpha}}{\sqrt{2} k_\parallel V_{T\alpha}}\right)\Phi, \tag{9.59}$$

$$\left(\xi + \frac{i\sqrt{\pi}}{2}\right) V = \tfrac{1}{2}\left[\left(\frac{\tilde{n}_\alpha}{n_{0\alpha}} + \Phi\right) - \left(1 - \frac{\omega_{*\alpha}}{\omega}\right)\Phi\right]. \tag{9.60}$$

Solving (9.59) for V and substituting it into (9.60) leads to

$$\frac{\tilde{n}_\alpha}{n_{0\alpha}} = \left(\xi^2 + \frac{i\sqrt{\pi}}{2}\xi - \tfrac{1}{2}\right)^{-1}\left[\tfrac{1}{2}\frac{\omega_{*\alpha}}{\omega} + \left(\xi + \frac{i\sqrt{\pi}}{2}\right)\left(\frac{\omega_{D\alpha} - \omega_{*\alpha}}{\sqrt{2} k_\parallel V_{T\alpha}}\right)\right]\Phi, \tag{9.61}$$

which may be written as

$$\frac{\tilde{n}_\alpha}{n_{0\alpha}} = -\left[\frac{\omega_{*\alpha}}{\omega} + \frac{\omega_{D\alpha}}{\sqrt{2} k_\parallel V_{T\alpha}}\left(1 - \frac{\omega_{*\alpha}}{\omega}\right) Z^{(2)}(\xi)\right]\Phi \tag{9.62}$$

by generating the term $(\omega_{*\alpha}/\omega)\Phi$, where

$$Z^{(2)}(\xi) = \left(\xi + \frac{i\sqrt{\pi}}{2}\right) \Big/ \left(\tfrac{1}{2} - \frac{i\sqrt{\pi}}{2}\xi - \xi^2\right). \tag{9.63}$$

The dispersion relation is then obtained by coupling the perturbed density from (9.62) to the vorticity equation (9.47). Since we are primarily interested in the energetic particle instability drive, we retain only the contributions of the alpha pressure on the RHS in (9.47);

$$b_s\left(1-\frac{k_\parallel^2 V_A^2}{\omega^2}\right)\frac{|e|\tilde{\varphi}}{T_{0e}}=\frac{\omega_{De}}{\omega}\left(\frac{n_{0\alpha}}{n_{0e}}\frac{T_{0\alpha}}{T_{0e}}\right)\left(\frac{\tilde{n}_\alpha}{n_{0\alpha}}\right), \tag{9.64}$$

where $b_s = k_\perp^2\rho_s^2 \simeq k_y^2\rho_s^2$, $\rho_s^2 = T_{0e}/(m_i\omega_{ci}^2)$, $\omega_{De} = k_y T_{0e}/(|e|BL_c)$, and n_{0e} and T_{0e} are the electron density and temperature, respectively. Equation (9.64) has been obtained by assuming that $\nabla(J_{\phi 0}/B_\phi) \simeq 0$ and using (9.46) and (9.50). Substituting (9.62) into (9.64) then leads to the following dispersion relation:

$$b_s\left(1-\frac{k_\parallel^2 V_A^2}{\omega^2}\right)=-\frac{q_\alpha}{|e|}\frac{\omega_{*\alpha}\omega_{De}}{\omega^2}\frac{n_{0\alpha}}{n_{0e}}+\left(\frac{n_{0\alpha}T_{0\alpha}}{n_{0e}T_{0e}}\right)\frac{\omega_{De}^2}{\omega^2}\left(\frac{\omega_{*\alpha}}{\omega}-1\right)\xi Z^{(2)}(\xi), \tag{9.65}$$

where $\xi = \omega/(\sqrt{2}k_\parallel V_{T\alpha})$. Here, $\omega_{D\alpha}$ has been neglected in comparison to ω in the argument of the $Z^{(2)}$ function, since this is generally small for the parameters of interest for the TAE mode.

Next, we note that the first term on the RHS in (9.65) is small compared to $b_s k_\parallel^2 V_A^2/\omega^2$. In addition, since $\omega_{De} \propto (1/B)dB/dr \propto \cos\theta$, a poloidal average of this term will vanish. However, the second term on the RHS is proportional to $\omega_{De}^2 \propto \cos^2\theta$ and has a poloidal average of 1/2. Multiplying through by ω and taking the imaginary part then yields

$$\gamma=\frac{\langle\omega_{De}^2\rangle}{\omega_r b_s}\frac{n_{0\alpha}T_{0\alpha}}{n_{0e}T_{0e}}\,\mathrm{Im}(Q_\alpha) \tag{9.66}$$

with

$$Q_\alpha=\left(\frac{\omega_{*\alpha}}{\omega_r}-1\right)\xi Z^{(2)}(\xi),$$

and $\langle\omega_{De}^2\rangle^{1/2} = (k_y T_{0e})/(2|e|BR)$ for $L_c \simeq R$. Here we have used the fact that $\gamma \ll \omega_r$ for the TAE mode, and we have retained only terms to first order in γ/ω_r, where ω_r denotes a real frequency. Taking the imaginary part of (9.63), writing $n_{0\alpha}T_{0\alpha}\langle\omega_{De}^2\rangle/(b_s n_{0e}T_{0e}) = \beta_\alpha V_A^2/(4R^2)$, and using $\omega_r^{TAE} = k_\parallel V_A \simeq V_A/(3R)$ for the $(m, n) = (1, 1)$ mode, we then obtain a final expression for the growth rate:

$$\gamma=\frac{9\sqrt{\pi}}{4}\beta_\alpha(\omega_{*\alpha}-\omega_r^{TAE})\frac{\xi}{(1-2\xi^2)^2+\pi\xi^2}. \tag{9.67}$$

The TAE modes can be unstable for $\omega_{*\alpha} > k_\parallel V_A$ with a sufficiently large β_α.

First TAE modes were studied in the high-energy NBI experiment in low magnetic field tokamaks. When the density is high and the magnetic field is low, the Alfvén velocity can be decreased to less than the beam injection velocity. In the TFTR and DIII-D Tokamaks, such experiments demonstrated the existence of unstable TAE modes and showed that the neutron production rate decreases and the high energy ion loss increases due to the onset of the unstable TAE modes.

The alpha particles may effect sawtooth oscillations, as discussed in chapter 5. Since the MHD instability with $(m, n) = (1, 1)$ is the most probable candidate to induce the sawtooth, theories for high-energy particle effects on the $m = 1$ internal kink or tearing mode have been developed. Here we briefly discuss how the high-energy ions modify the $m = 1$ internal kink mode.

In tokamaks, the $m = 1$ internal kink mode becomes unstable in the cylindrical model for the radial displacement $\xi(r)$ shown in Fig. 9.7, when the $q = 1$ surface appears inside the plasma column. The displacement $\xi(r)$ is shown as

$$\xi(r) = \begin{cases} \xi_\infty = \text{constant} & (r < r_s) \\ 0 & (r > r_s), \end{cases} \tag{9.68}$$

where $r = r_s$ denotes the $q = 1$ surface. Since ξ is rapidly changing but continuous at $r \simeq r_s$, boundary layer analysis is usually applied to obtain the eigenvalue or growth rate and the eigenfunction.

The inertia term is given by

$$I = \frac{\pi}{2} \int_0^a \rho\gamma^2 |\xi|^2 r dr, \tag{9.69}$$

where ρ is the mass density and γ is the growth rate. The total perturbed energy is $\delta W_F + I = 0$, where δW_F is the same as that given in (5.69). In the cylindrical plasma, the Euler–Lagrange equation for $\xi(r)$ that minimizes the total perturbed energy is

$$\frac{d}{dr}\left(r^3[(\mathbf{k} \cdot \mathbf{B})^2 + \rho\gamma^2]\frac{d\xi}{dr}\right) - g\xi = 0, \tag{9.70}$$

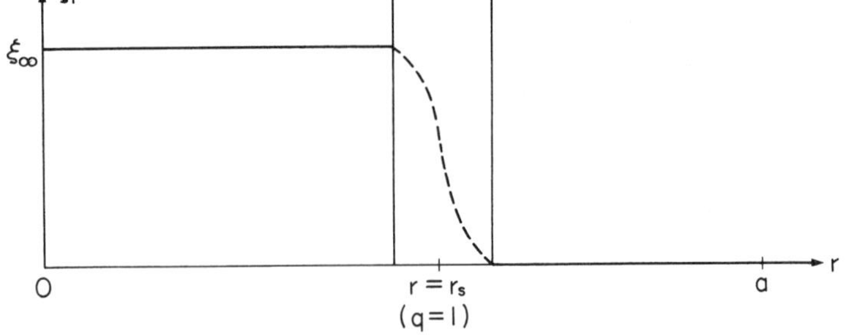

Fig. 9.7 The mode structure of the radial displacement $\xi_r = \xi(r)$ of the $m = 1$ internal kink mode.

where g is given explicitly in (5.84). We may solve this equation by introducing a small parameter $g \sim r^2/R^2 \sim \delta$ and considering a "singular layer" where $r - r_s = x \sim \delta$. The inertia term will be significant only in the singular layer, and thus $\gamma \sim \delta$. Outside the singular layer, the solutions correct to the order of δ are

$$\frac{d\xi}{dr} = \begin{cases} \dfrac{\xi_\infty}{r^3(\mathbf{k}\cdot\mathbf{B})^2} \displaystyle\int_0^r g dr & r < r_s \\ \dfrac{\text{constant}}{r^3(\mathbf{k}\cdot\mathbf{B})^2}, & r > r_s. \end{cases} \tag{9.71}$$

Within the singular layer, we may neglect the contribution from g. Writing $\mathbf{k}\cdot\mathbf{B} = (\mathbf{k}\cdot\mathbf{B})'_{r_s} x$ and choosing the solution that has $\xi \to \xi_\infty$ as $x \to -\infty$ and $\xi \to 0$ as $x \to \infty$, we obtain

$$\xi = \tfrac{1}{2}\xi_\infty \left\{ 1 - \frac{2}{\pi} \tan^{-1}(|(\mathbf{k}\cdot\mathbf{B})'_{r_s} x| / \gamma \rho_s^{1/2}) \right\}, \tag{9.72}$$

where $\rho_s = \rho(r_s)$. We equate $d\xi/dr$ obtained from (9.72) for $x \to -\infty$ with $d\xi/dr$ obtained from the upper expression in (9.71) for $r \to r_s$. This yields a growth rate

$$\gamma = -\frac{\pi}{\rho_s^{1/2} |(\mathbf{k}\cdot\mathbf{B})'_{r_s}| r_s^3} \int_0^{r_s} g dr. \tag{9.73}$$

This expression can be extended to include the high-energy ion contribution by introducing a high-energy perturbed potential energy, δW_K, in $\int_0^{r_s} g dr$.

Here we again consider the linearized momentum balance equation

$$\rho\omega^2 \xi = \nabla \cdot \overset{\leftrightarrow}{\tilde{P}} + \mathbf{B} \times (\nabla \times \tilde{\mathbf{B}})/\mu_0 + \tilde{\mathbf{B}} \times (\nabla \times \mathbf{B})/\mu_0. \tag{9.74}$$

In order to calculate the kinetic effects perturbatively, we construct a quadratic form from (9.74). First, we separate the perturbed pressure tensor into two parts: $\overset{\leftrightarrow}{\tilde{P}} = \tilde{P}_F \overset{\leftrightarrow}{I} + \overset{\leftrightarrow}{\tilde{P}}_K$, where $\tilde{P}_F \overset{\leftrightarrow}{I}$ is the isotropic fluid part that comes from the adiabatic response of the perturbed distribution function $\tilde{f}$, and $\overset{\leftrightarrow}{\tilde{P}}_K$ is the kinetic part that comes from the nonadiabatic response of $\tilde{f}$. We take an inner product of (9.74) with ξ^* and integrate over the whole plasma volume to obtain the energy integral,

$$\omega^2 \delta K = \delta W_F + \delta W_K, \tag{9.75}$$

where the superscript asterisk denotes a complex conjugate and

$$\delta K = \int \rho|\xi|^2 dV, \tag{9.76}$$

$$\delta W_F = \int \xi^* \cdot (\nabla \tilde{P}_F + \tilde{\mathbf{B}}/\mu_0 \times (\nabla \times \mathbf{B}) + \mathbf{B} \times (\nabla \times \tilde{\mathbf{B}})/\mu_0) dV, \tag{9.77}$$

$$\delta W_K = \int \xi^* \cdot \nabla \cdot \overset{\leftrightarrow}{\tilde{P}}_K \, dV. \tag{9.78}$$

It is noted that δK and δW_f comes from the ideal MHD equation, whereas δW_K represents the correction due to kinetic effects.

The perturbation of the energetic particle pressure tensor can be written as

$$\overset{\leftrightarrow}{\tilde{P}}_h = \tilde{P}_{\perp h} \overset{\leftrightarrow}{I} + (\tilde{P}_{\| h} - \tilde{P}_{\perp h})\mathbf{e}_\| \mathbf{e}_\| + (P_{\| h} - P_{\perp h})(\mathbf{e}_\| \tilde{\mathbf{e}}_\| + \tilde{\mathbf{e}}_\| \mathbf{e}_\|). \tag{9.79}$$

Since we are dealing with low-frequency modes, or $|\omega| \ll \Omega_i$, the pressure tensor is still diagonal and has only two perturbed components:

$$\begin{cases} \tilde{P}_{\| h} = m_h \int v_\|^2 \tilde{f} d^3 v \\ \tilde{P}_{\perp h} = m_h \int \dfrac{v_\perp^2}{2} \tilde{f} d^3 v, \end{cases} \tag{9.80}$$

where $\tilde{f}$ is the gyrophase-averaged perturbed distribution function. It is permissible to substitute $\overset{\leftrightarrow}{\tilde{P}}_h$ without the third term into $\overset{\leftrightarrow}{\tilde{P}}_k$ in (9.78).

In the PDX Tokamak experiment the MHD fluctuations destabilized by energetic ions were first observed, when NBI was injected in the direction perpendicular to the toroidal plasma (see Fig. 9.8). This instability is called the fishbone instability. Theoretically, the contribution from δW_k shown in (9.78) destabilizies the $(m, n) = (1, 1)$ internal mode and the envelope of magnetic fluctuations are seen to take the form of a "fishbone." In D–T burning tokamak plasmas, alpha particles may excite the fishbone instability for some parameter ranges, which may induce rapid loss of alpha particles and degrade the alpha heating. When this effect is not serious, the fishbone instability can be used to pump out the alpha ash.

9.5 THE DIVERTOR PHYSICS OF THE STELLARATOR AND HELIOTRON

Since stellarator and heliotron devices aim to realize a steadily operating fusion reactor, the particle and energy balances must be satisfied under continuous fueling. Fuel particles injected by gas-puffing or pellet injection are heated to a thermonuclear temperature by alpha heating, and then alpha particles and

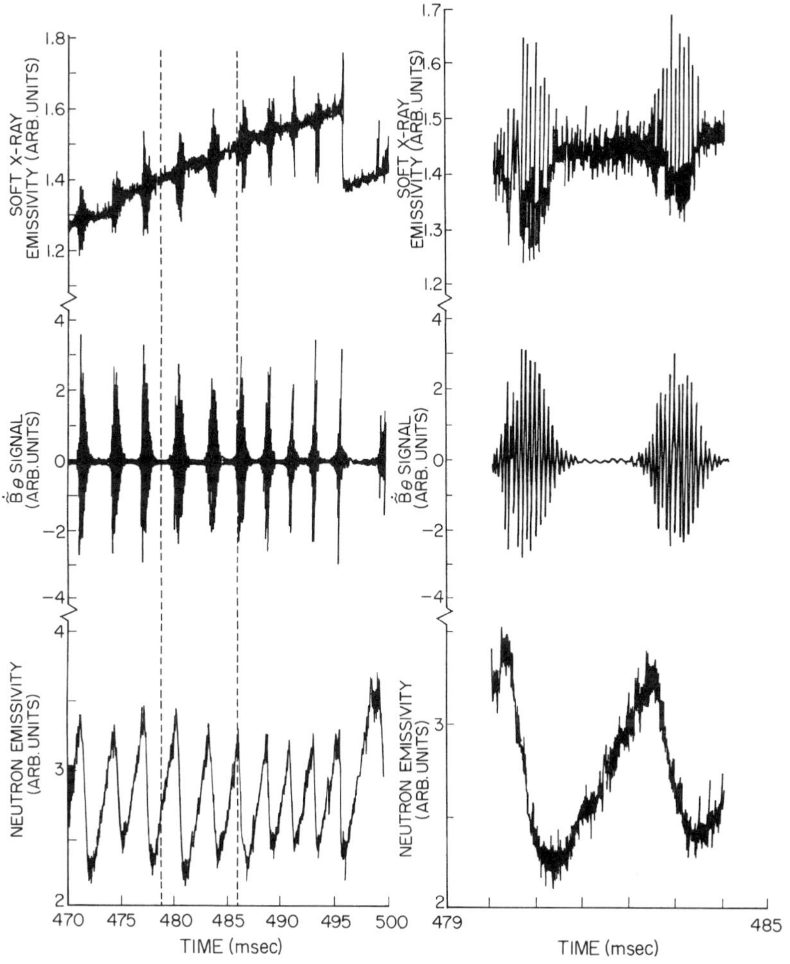

Fig. 9.8 Soft X-ray emissivity, fluctuation of the poloidal magnetic field, $d\tilde{B}_\theta/dt$, and fluctuation of the neutron emissivity of the fishbone instability observed in the PDX Tokamak.

neutrons are generated by the D–T fusion reaction. It is expected that almost all alpha particles will be confined and that they will give up almost all of their kinetic energy to the background fuel plasmas during the slowing down process, up to the background plasma temperature. The slowing-down alphas or alpha ash must be exhausted to maintain a steady-state particle balance. The energy deposited to the background fuel plasma is finally lost by crossing the last closed magnetic surface through transport or radiation. In order to manipulate the particle and energy transport after crossing the last closed magnetic surface, a divertor is necessary. The divertor is composed of magnetic field lines connecting between the last closed magnetic surface and the divertor plate.

In the straight stellarator approximation, a clear separatrix surface is usually generated, as shown in section 2.5. The behavior of the magnetic field lines outside the separatrix surface is an important subject in the divertor physics of the stellarator and heliotron. In the toroidal device, the separatrix surface is destroyed by the toroidal effects and the volume inside the last closed magnetic surface decreases compared to the volume inside the separatrix surface in the straight stellarator limit.

In Fig. 9.9 is shown the length of the magnetic field line from the point outside the last closed magnetic surface to the vacuum chamber in the Heliotron E device, as a function of the starting position for following the magnetic field line. For horizontally elongated magnetic surfaces, the starting position is taken along the major radius. The magnetic surfaces are controlled with an additional toroidal field characterized by α^* (see (2.156)). The magnetic field is assumed to be curl-free here. The decay of the length of the magnetic field line is not monotonic. This behavior is derived from magnetic

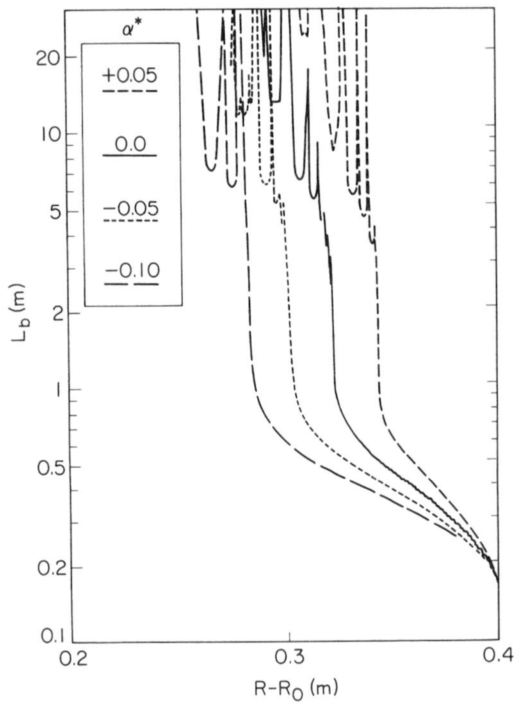

Fig. 9.9 The connection length L_b of the magnetic field line between the starting position outside the last closed magnetic surface $R - R_0$, and the wall of the vacuum chamber. $R - R_0$ is measured along the major radius from the magnetic axis $R = R_0$. Here, the magnetic surfaces have a horizontally elongated elliptic shape. The parameter α^* corresponds to the additional toroidal field that changes the position of the last closed magnetic surface.

island structures that survive in the stochastic magnetic field region. It should be noted that the stochastic behavior of the magnetic field line is seen only when the vacuum chamber is removed. It is impossible to see the stochasticity in the projection points of a magnetic field line (see Fig. 2.1) with a finite length of magnetic field line, as shown in Fig. 9.9. According to characteristics of the divertor plasmas discussed later in this section, it is essential to generate a temperature gradient along the magnetic field line with a fairly short length of magnetic field line. From divertor experiments in tokamaks, about 10 m of magnetic field line is necessary to obtain a high-density and low-temperature divertor plasma in front of the divertor plate. In the Heliotron E case, magnetic field lines of the order of 10 m exist just outside the last closed magnetic surface. It is expected to be possible to obtain a divertor plasma similar to that of a tokamak with the divertor in the Large Helical Device (LHD), since the divertor chamber and the baffle will be suitably equipped (see Fig. 9.10).

From here on, we will analyze the divertor plasma using a simple fluid model. Since plasma flow along magnetic field lines is essential, we will apply a one-dimensional fluid model to the divertor plasma. By imposing an ambipolar condition and charge neutrality on the divertor plasma, $n_i = n_e = n$ and $u_i = u_e = u$ are assumed, where $n_i(n_e)$ and $u_i(u_e)$ are the ion (electron) density and the ion (electron) flow velocity along the magnetic field line. By denoting the coordinate along magnetic field line by ℓ, the continuity equation of one-dimensional plasma becomes

$$\frac{\partial n}{\partial t} + \frac{\partial}{\partial \ell}(nu) = S_n, \tag{9.81}$$

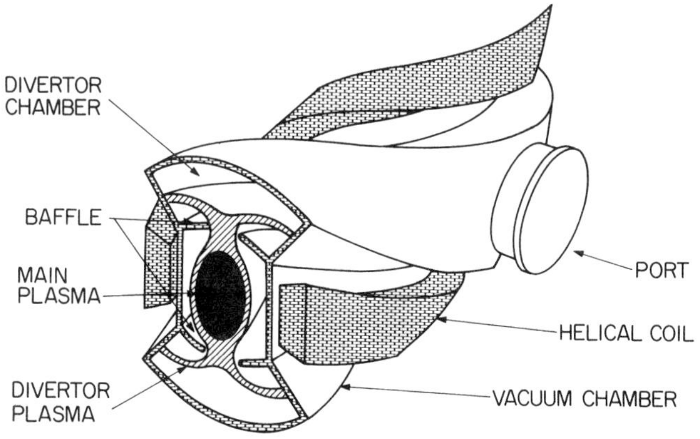

Fig. 9.10 A schematic picture of the Large Helical Device (LHD). In the poloidal cross-section, the divertor chamber and baffle are shown.

where $S_n = n_e n_0 \langle \sigma v \rangle_I$, n_0 denotes neutral density, and $\langle \sigma v \rangle_I$ denotes the ionization rate, which is approximately given by

$$\log_{10}\langle \sigma v \rangle_I = \begin{cases} -3.054x - 15.72e^{-x} + 1.603e^{-x^2} & (T_e \le 20\,\mathrm{eV}) \\ -0.5151x - \dfrac{2.563}{x} - 5.231 & (T_e > 20\,\mathrm{eV}) \end{cases} \tag{9.82}$$

for hydrogens, where $x = \log_{10} T_e$.

Neutral particles are approximately described by a fluid equation along the magnetic field line projected on the poloidal plane. This approximation is crude for stellarator and heliotron devices, although it is fairly good for tokamaks. The best way to treat neutral particles is by the Monte Carlo method, including ionization and charge exchange reactions, since it is not difficult to take the three-dimensional structure of the stellarator or heliotron plasma into account. It is also noted the fluid model is valid for neutrals when the mean free path is shorter than the characteristic length of the divertor plasma. When a high-density and low-temperature plasma is formed in front of the divertor plate, the fluid description becomes appropriate. The continuity equation of neutrals becomes

$$\frac{\partial n_0}{\partial t} + \frac{\partial}{\partial X}(n_0 u_0) = -S_n, \tag{9.83}$$

where the length on the poloidal plane is given by $X = \ell(B_p/B_T) = \ell b_p$, and where B_p and B_T are the poloidal and toroidal magnetic fields of the tokamak. In the stellarator or heliotron, B_p is estimated from the rotational transform at the last closed magnetic surface, $\iota(a) = (R/a)(B_p/B_T)$, where a is an average radius.

The stationary particle flux along a magnetic field line going to a divertor plate is given by (9.81):

$$\Gamma(d) = \Gamma(0) + \int_0^d S_n d\ell, \tag{9.84}$$

where $\ell = 0$ denotes the divertor throat position and $\ell = d$ denotes the divertor plate. In (9.84), $\Gamma(d)$ is the particle flux at the divertor plate and $\Gamma(0)$ is the flux at the divertor throat position. By introducing the amplification rate, $R_m = \Gamma(d)/\Gamma(0)$, (9.84) gives

$$R_m = 1 + \frac{1}{\Gamma(0)} \int_0^d S_n d\ell > 1. \tag{9.85}$$

From the boundary conditions at the divertor plate, the heat flux of the electrons is given by $Q_e(d) = 2\alpha T_e(d)\Gamma(d)$, where α denotes the contribution of secondary electrons at $\ell = d$. The electron temperature at the divertor plate is estimated as $T_e(d) = Q_e(d)/(2\alpha\Gamma(d)) \simeq Q_e(0)/(2\alpha\Gamma(0)R_m) \propto 1/R_m$. By noting that $Q_e(d) \leq Q_e(0)$ and $R_m \geq 1$, the electron temperature at $\ell = d$ decreases with an increase in R_m. In the fluid model the boundary condition for the flow velocity at $\ell = d$ is given by Bohm condition $u = C_s$, where C_s is the ion sound velocity. This relation comes from the fact that the fluid description is valid for $u \leq C_s$, and that the shear condition corresponding to a monotonic electric potential variation near the divertor plate is $u \geq C_s$. Thus $\Gamma(d) = n_e(d)u_e(d) \propto n_e(d)(T_e(d))^{1/2}$ gives $n_e(d) \propto R_m^{3/2}$. In case of $R_m \gg 1$, electrons coming from the main plasma through the divertor throat hit the divertor plate more than once. This recycling process is essential to obtain low-temperature and high-density divertor plasmas.

The steady-state particle flux of the divertor plasma can be obtained from (9.81) and (9.83) by assuming that $\Gamma_i = \Gamma_e = \Gamma$. This means that the effect of the parallel electric field is negligible, which is a good approximation for high-density plasmas. From (9.81),

$$\frac{\partial(nu)}{\partial\ell} = S_n = \frac{n_0 u_0}{\lambda_0}, \tag{9.86}$$

where u_0 is the flow velocity of neutrals and λ_0 is a mean free path of ionization. From (9.83),

$$\frac{\partial}{\partial X}(n_0 u_0) = -\frac{n_0 u_0}{\lambda_0 b_p} \tag{9.87}$$

is obtained for neutrals, since $X = \ell b_p$ is the length on the poloidal plane.

Here we impose boundary conditions at the divertor plate,

$$Q_i(d) = 2T_i(d)\Gamma(d), \tag{9.88}$$

$$Q_e(d) = 2\alpha T_e(d)\Gamma(d), \tag{9.89}$$

$$u(d) = \{(T_e(d) + T_i(d))/m_i\}^{1/2}, \tag{9.90}$$

where T_i and m_i are the ion temperature and ion mass, respectively. It is noted that condition (9.90) corresponds to the Bohm condition. At the divertor plate,

$$\Gamma_0(d) = -\delta\Gamma(d) \tag{9.91}$$

is imposed, where Γ_0 is the neutral flux and δ is the recycling rate.

In order to solve (9.86) and (9.87) analytically, we assume that:

(i) $\langle\sigma v\rangle_I$ is spatially constant;
(ii) u_0 is also spatially constant and a neutral temperature T_0 is assumed equal to $3T_e(a)/2$;
(iii) $n_e \simeq$ constant except in the region in front of the divertor plate.

With (i) and (ii) we obtain

$$\Gamma(x) = \Gamma(0)\left[1 + \frac{\delta \exp\left(-\frac{d}{\lambda_0}\right)\left\{\exp\left(\frac{x}{\lambda_0}\right) - 1\right\}}{1 - \delta\left\{1 - \exp\left(-\frac{d}{\lambda_0}\right)\right\}}\right], \tag{9.92}$$

from (9.86) and (9.87). By substituting $x = d$, the solution (9.92) becomes

$$\Gamma(d) = \frac{\Gamma(0)}{1 - \delta\left\{1 - \exp\left(-\frac{d}{\lambda_0}\right)\right\}}. \tag{9.93}$$

Introducing $g = Q_i(d)/Q_e(d)$ and $\alpha' = 1 + \alpha g$ yields

$$u(d) = \left(\frac{\alpha' T_e(d)}{m_i}\right)^{1/2}. \tag{9.94}$$

Here we note that the mean free path of neutrals, λ_0, is given as

$$\frac{d}{\lambda_0} = \frac{dn_e\langle\sigma v\rangle_I}{u_0} = \left(\frac{T_2}{T_e(d)}\right)^2, \tag{9.95}$$

where

$$T_2^2 \equiv \frac{dQ_e(d)\langle\sigma v\rangle_I m_i}{2\alpha\sqrt{3\alpha'}}. \tag{9.96}$$

Substituting (9.95) into (9.93) yields

$$\frac{T_1}{T_2} = f(y) = \frac{y}{1 - \delta\left\{1 - \exp\left(-\frac{1}{y^2}\right)\right\}}, \tag{9.97}$$

where $T_1 \equiv Q_e(d)/(2\alpha\Gamma(0))$ and $y = T_e(d)/T_2$. T_1 is the temperature at the divertor plate assuming no recycling and T_2 is the temperature corresponding to $\lambda_0 = d$. $f(y)$ is shown as a function of y for various values of δ in Fig. 9.11. According to the value of T_1/T_2, there are two cases for the solution $T_1/T_2 = f(y)$.

(I) When $T_1/T_2 > 1$ and $\delta \lesssim 1$, there are three types of solution.

(i) For $y > 1$, $T_e(d) > T_2$ and $T_2 \lesssim T_1$. This solution corresponds to high-temperature and low-density divertor plasmas when the neutral particle flux is small and almost all neutrals are pumped out from the divertor.

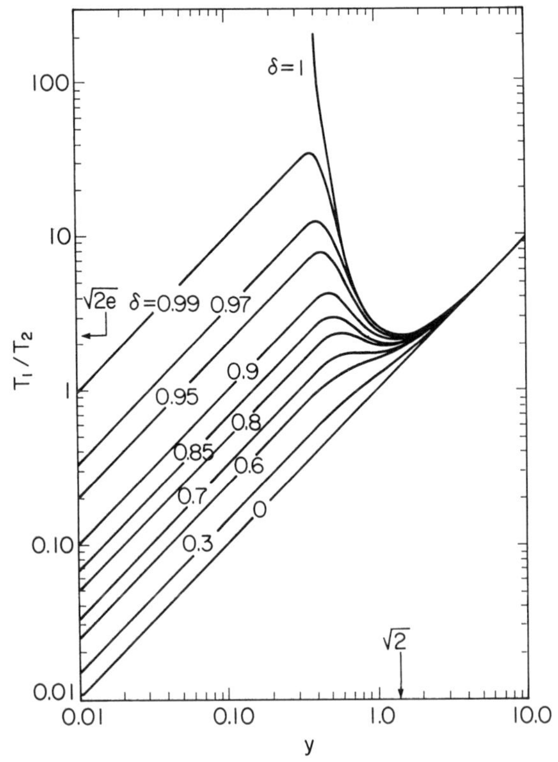

Fig. 9.11 T_1/T_2 versus $y = T_e(d)/T_2$ for various recycling rates δ.

(ii) In the region of $y \lesssim 1$ or $T_e(d) \lesssim T_2$, the plasma temperature in front of the divertor plate decreases and the plasma density is increased due to the ionization of neutrals and the recycling.

(iii) For $y \simeq 0.01 \sim 0.1$ and $\delta \simeq 1$, $T_e(d) \sim (1-\delta)T_2 \lesssim (1-\delta)T_1$, and the plasma temperature in front of the divertor plate decreases signficantly due to the high recycling rate.

(II) When $T_1/T_2 < (2e)^{1/2}$, there is only one solution that corresponds to low-temperature and high-density plasmas in front of the divertor plate. So far, we have assumed $\delta =$ constant. However, the change of plasma density in front of the divertor plate may affect the ionization rate of neutrals, which changes the recycling rate δ. In order to study this problem, we replace the pumping rate, $1-\delta$, with $\exp(-b/\lambda_0)$, where b denotes the characteristic width of the divertor plasma at the divertor plate. With this change, (9.97) becomes

$$\frac{T_1}{T_2} = f(y) = \frac{y}{1 - \left\{1 - \exp\left(-\frac{\Delta}{y^2}\right)\right\}\left\{1 - \exp\left(-\frac{1}{y^2}\right)\right\}}, \tag{9.98}$$

where $\Delta \equiv b/d$. In Fig. 9.12, dependence of $f(y)$ on Δ is shown. According to this figure, there are two solutions of y for a given T_1/T_2; one is $T_e(d) > T_2$ and $T_2 \lesssim T_1$, which is the same as I(i) above. The recycling effect is negligible here. The other corresponds to low-temperature and high-density plasmas in front of the divertor plate.

It is possible to write (9.98) with a variable corresponding to the density, $z = (n(d)/n_2)^{2/3}$;

$$\left(\frac{n_2}{n_1}\right)^{2/3} = g(z) = \frac{1}{z} \frac{1}{1 - \{1 - \exp(-\Delta z^2)\}\{1 - \exp(-z^2)\}}, \tag{9.99}$$

where n_1 and n_2 denote densities corresponding to T_1 and T_2, respectively. The dependency of $g(z)$ on z is shown in Fig. 9.13 for various values of Δ, which also demonstrates the existence of two solutions of z for a given $(n_2/n_1)^{3/2}$. It is noted that, for a fixed $\Delta(\neq 0)$, the higher-density solution is limited by an asymptotic maximum value.

We will show an example of three stationary solutions for the divertor plasma, when the influx $\Gamma(0)$ to the divertor is fixed. In Fig. 9.14 is shown the electron temperature and electron density at the divertor plate as a function of incident ion flux, which was first shown by Saito *et al.* (1985). This figure is a different expression of Fig. 9.11, showing three solutions in (9.97).

In many tokamak experiments with the divertor, high-density ($n_{ed} \lesssim 10^{20}\,\mathrm{m}^{-3}$) and low-temperature ($T_{ed} \lesssim 10\,\mathrm{eV}$) divertor plasmas have been obtained, and it is clearly shown that the heat flux from the last closed

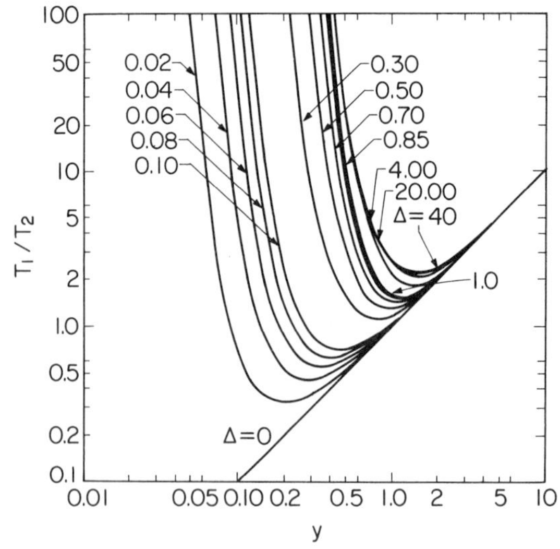

Fig. 9.12 T_1/T_2 versus $y = T_e(d)/T_2$ for various values of Δ. Δ is related to the pumping rate, $1 - \delta$.

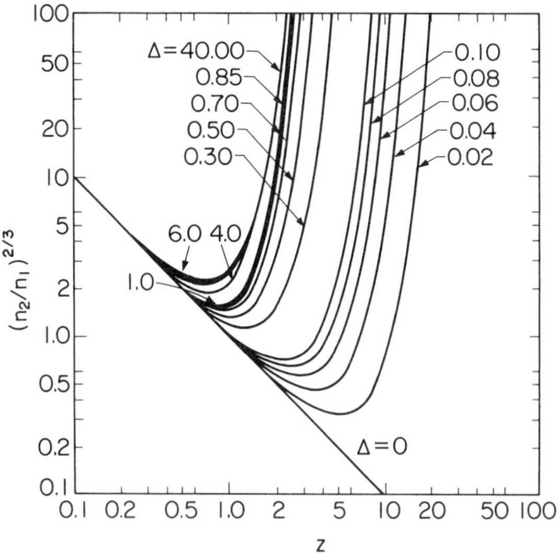

Fig. 9.13 $(n_2/n_1)^{2/3}$ versus $z = (n(d)/n_2)^{2/3}$ for various values of Δ.

magnetic surface can be controlled without increasing the impurity ions in the main plasma. However, no such an experiment exists for heliotron devices so far. It is expected that high-density and low-temperature divertor plasmas will be produced in the Large Helical Device with the helical divertor and the buffle as shown in Fig. 9.10.

9.6 THE CHARACTERISTICS OF THE STEADY FUSION REACTOR

The most significant advantage of the heliotron reactor is continuous operation, which means that the reactor works for several months or a year without intermission. The steady fusion reactor is characterized in the following way:

(i) superconducting coils,
(ii) ignition of the D and T mixture,
(iii) impurity and helium ash control,
(iv) cooling of the divertor plate,
(v) erosion of components that face the plasma.

We shall discuss these points in turn below.

(i) So far, all heliotron devices have the $\ell = 2$ continuous helical coil. The largest $\ell = 2$ superconducting helical coil with $R = 3.9\,\text{m}$ and $a_c = 0.94\,\text{m}$ will be built for the LHD, where R and a_c are the major and minor radii of the helical coil, respectively. A heliotron fusion reactor will be equipped similarly, with an $\ell = 2$ continuous superconducting helical coil. However, there is the criticism that the continuous helical coil will be difficult to maintain and repair.

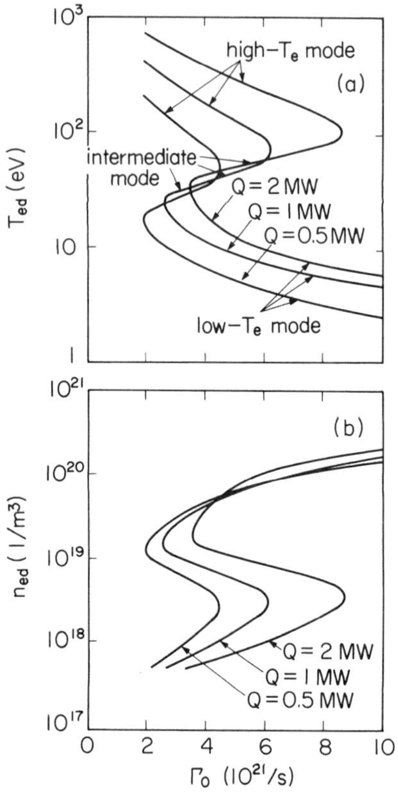

Fig. 9.14 (a) The electron temperature at the divertor plate T_{ed} versus the incident ion flux $\Gamma(0)$. (b) The electron density at the divertor plate n_{ed} versus the incident ion flux $\Gamma(0)$ for various heating powers Q. Three types of solution, the high-T_e mode, the intermediate mode, and the low-T_e mode, are seen.

In the case of a continuous coil, we need the technology to build superconducting helical coils that will work for about 30 years without serious difficulties. Another possibility is to develop a superconducting modular coil system to generate magnetic surfaces that are almost the same as those in a heliotron reactor with an $\ell = 2$ continuous helical coil. There are several studies of vacuum magnetic surfaces for modular heliotron devices; however, experimental results are limited. Modular coil systems are being investigated for the advanced stellarators W7-AS and W7-X. W7-AS with the modular coil system has a similar size to the Heliotron E. A small stellarator, IMS, with modular coils has been studied at the University of Wisconsin, although both the density and the temperature of the IMS plasmas are low. It is noted that the magnetic energy of the LHD is about 0.9 GJ and that of the W7-X is about 0.6 GJ, although the plasma volume is different—about 23 m^3 for the LHD and about 32 m^3 for the W7-X.

(ii) In Fig. 9.15 is shown the schematic evolution of burning plasma in a heliotron reactor. After the superconducting helical coil generates a magnetic field of 5–6 T, fuel particles of a D and T mixture are injected using gas-puffing. The breakdown of neutral particles is induced by electron cyclotron heating, with a frequency of 140–170 GHz and a power level of 5–10 MW. After a build-up of plasma density suffcient for neutral beam injection, NBI heating starts, with an injection energy of about 500 keV and a total power of 50–100 MW. With an increase of the alpha heating power, the average plasma temperature increases up to 10 keV and the average plasma density is controlled in the range $1\text{–}2 \times 10^{20}\ \mathrm{m}^{-3}$. The action of the divertor also starts. When the alpha heating power reaches of the order of 100 MW, ignition begins and the NBI heating is shut down. This time scale is about several hundreds of seconds. The average plasma temperature is fixed to obtain a total fusion power of about 3 GW. In this case, the alpha heating power is about 600 MW. In the burning phase, the fuel plasma density, the plasma temperature and the alpha ash density will be monitored and controlled to keep the burning state steady. The expected electric output power is about 1 GW.

(iii) One serious problem is the impurity level in the steady burning state. The species and content of the impurity depend on materials of the plasma-facing components, such as the divertor plate and the first wall. Here erosion of the divertor plate and the first wall due to ion sputtering is inevitable. In order to minimize the erosion of the plasma-facing components and keep the impurity

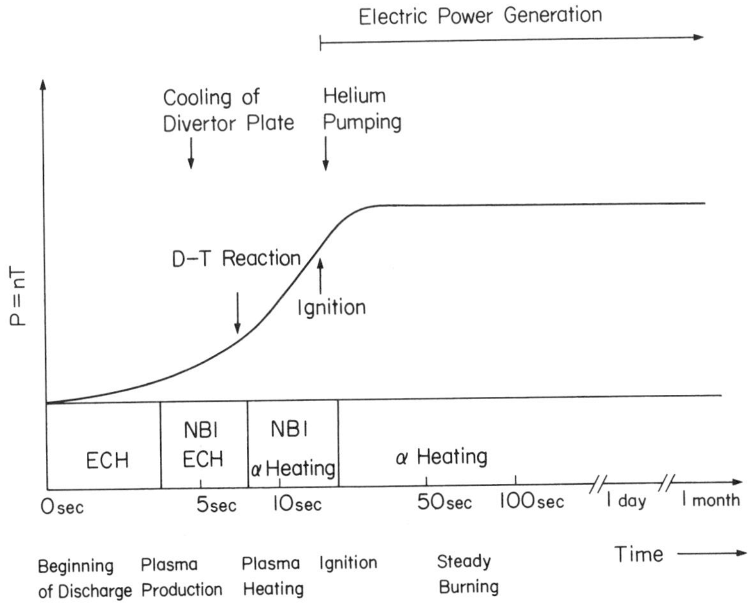

Fig. 9.15 The expected time evolution of a steady fusion reactor.

content in the burning plasma at a low level, careful selection of materials is important, and this will be one of the main subjects in the thermonculear experimental test reactor after ignition is realized in the near future. Control of the helium density or alpha ash depends on the scaling of the particle confinement time τ_p, and on the ratio between τ_p and the energy confinement time τ_E. When $\tau_p/\tau_E \gg 1$, it is not easy to keep the helium density below 10% of the electron density, as shown in Fig. 9.4. Here, we have assumed that the alpha ash confinement time is equal to the fuel particle confinement time. Also, the ignition parameter $n\tau_E$ becomes large with an increase in the helium density. This is related to dilution of the fuel particle density, which decreases the fusion reaction rate. In order to compensate for the reduction of fusion power or alpha heating, a longer confinement time is required. Another disadvantage of the accumulation of alpha ash or an increase in β_α is that pressure-driven MHD instabilities such as interchange or ballooning modes become more dangerous than in the case with negligible alpha ash density. In other words, the critical beta of the fuel plasma pressure decreases with an increase in β_α. It is also noted that, when improved energy confinement is obtained, the particle confinement is usually improved. Such behavior is clearly seen in the H- (high confinement) mode in both tokamaks and stellarators. In order to make a compact fusion reactor, the H-mode is more desirable than the L- (low confinement) mode, if the plasma density is controlled to be constant with the aid of an ELMs (Edge Localized Modes). The ELMs degrade both the particle and energy confinement somewhat. It is not clear whether or not ELMs are controllable, since they may be triggered by resistive or ideal MHD instabilities. As we discussed in chapter 6, heliotron devices usually have a fairly large loss cone of velocity space in the edge region of the plasma column. This may help to reduce the alpha ash density, although there is a possibility of alpha heating reduction.

(iv) In the first D–T experiment in the JET, the temperature of the divertor plate increased to 1300–1800°C after 1–2 sec of high-power NBI heating, and the carbon density increased, which stops the H-mode or improved confinement state. In the design of the ITER (International Thermonuclear Experimental Reactor), the divertor must manipulate a high heat flux of the order of 10–20 $\mathrm{MW/m^2}$ for a long pulse of 500–1000 sec. The divertor plate must be cooled to keep its temperature sufficiently low. In order to convert the electron heat flux to radiation, gas-puffing of deuterium neutrals and light impurity injection have been considered in the divertor chamber. However, since the main plasma must be kept sufficiently clean, conversion efficiency of the heat flux to radiation by injecting impurity ions must be tested in the present large tokamaks and the LHD.

(v) Even if the electron heat flux can be converted into radiation, there remains ion sputtering, which erodes the plasma-facing components significantly. Thus they will have to be exchanged periodically in both the toakamak and the heliotron reactor. The plasma-facing components, including the divertor plate, also suffer damage from neutron fluxes. There are environmental problems to be solved: the materials of the plasma-facing components become radioactive and the decay time of the radioactivity is extremely long. In order to reduce radioactivity, vanadium and aluminum are considered to be attractive materials.

Finally, we briefly estimate the size of a heliotron reactor. From the LHD scaling law discussed in chapter 8, a major radius of about 20 m and a minor radius of about 2 m will be necessary. With improvements in confinement of a factor of 1.5–2.0, compared to the LHD scaling law (see section 8.3), the size of a heliotron reactor will become compact. Since the aspect ratio of the heliotron device is not small compared to tokamaks, the major radius of the heliotron reactor should become larger than that of the tokamak reactor. In the design of the ITER, it is about 8 m. In order to obtain a confinement time of several seconds, the minor radius of the heliotron reactor will have to be comparable to, or a little smaller than, that of the ITER.

BIBLIOGRAPHY

Cheng, C. Z. (1990). Alpha-particle effects on low-*n* magnetohydrodynamic modes. *Fusion Technol.*, **18**, 443.

Cordey, J. G., Goldstone, R. J., and Mikkelsen, D. R. (1981). A generalized sufficient condition for velocity-space instability of fusion product distributions and application to heating of D–T tokamak reactors. *Nucl. Fusion*, **21**, 581.

Heidbrink, W. W., Strait, E. J., Doyle, E., Sager, G., and Snider, R. T. (1991). An investigation of beam driven Alfvén instabilities in the DIII-D tokamak. *Nucl. Fusion*, **31**, 1635.

JET Team (1992). Fusion energy production from a deuterium–tritium plasma in the JET tokamak. *Nucl. Fusion*, **32**, 187.

McGuire, K. M., Goldstone, R. J., Bell, M. et al. (1983). Study of high-beta magnetohydrodynamic modes and fast-ion losses in PDX. *Phys. Rev. Lett.*, **50**, 891.

McGuire, K. M., Adler, H., Alling, P. et al. (1995). Review of deuterium–tritium results from the Tokamak Fusion Test Reactor. *Phys. Plasmas*, **2**, 2176.

Reiter, D., Wolf, G. H., and Kever, H. (1990). Burn condition, helium particle confinement and exhaust efficiency. *Nucl. Fusion*, **30**, 2141.

Saito, S., Kobayashi, T., Sugihara, M., Hirayama, T., and Fujisawa, N. (1985). Bifurcated confinement solutions for diverted tokamaks. *Nucl. Fusion*, **25**, 828.

Spong, P. A., Carreras, B. A., and Hedrick, C. L. (1992). Linearized gyrofluid model of the alpha-destabilized toroidal Alfvén eigenmode with continuum damping effects. *Phys. Fluids*, **B4**, 3316.

Stacey, W. M., Jr. (1981). *Fusion plasma analysis*. Wiley-Interscience, New York.

Stangeby, P. C., and McCracken, G. M. (1990). Plasma boundary phenomena in tokamaks. *Nucl. Fusion*, **30**, 1225.

Van Dam, J. W., Fu, G.-Y., and Cheng, C. Z. (1990). Thermonuclear instability of global-type shear Alfvén modes. *Fusion Technol.*, **18**, 461.

Wong, K. L., Fonck, R. J., Paul, S. F., et al. (1991). Excitation of toroidal Alfvén eigenmodes in TFTR. *Phys. Rev. Letters*, **66**, 1874.

INDEX

PHYSICS LIBRARY

1-MONTH